WHERE DID THE TOWERS GO

Cover image credit: Detective Greg Semendinger, New York City Police Aviation Unit.

This and other photos by Greg Semendinger were included in my submissions to NIST and in my federal qui tam case in early 2007, but I did not know who the photographer was until February 10, 2010, when ABC news publicized these as never-before seen photos[1], and later changed the title[2]. Curiously, this release was just days after the United States Supreme Court denied my case, a case in which I included this photo with the caption, "This is not a collapse and it is fraudulent to have so stated. This may be evidence of criminal intent to deceive the public. It is also evidence of the use of DEW."[3,4]

---

[1] Dramatic images of World Trade Centre collapse on 9/11 released for first time, By Philip Delves Broughton, Last updated at 12:01 AM on 12th February 2010, *http://www.dailymail.co.uk/news/worldnews/article-1249885/New-World-Trade-Center-9-11-aerial-images-ABC-News.html*

[2] Chilling Aerial Photos of 9/11 Attack Released, By Ula Ilnytzky and Colleen Long Associated Press Writers, New York, February 10, 2010 (AP), *http://abcnews.go.com/US/wireStory?id=9796098*

[3] This whistleblower case filed April 25, 2007, was based on my Request for Corrections (RFC) to the National Institute of Standards and Technology (NIST), (March 16, 2007), *http://www.ocio.os.doc.gov/ITPolicyandPrograms/Information_Quality/PROD01_002619*

[4] (March 29, 2007), *http://www.ocio.os.doc.gov/ITPolicyandPrograms/Information_Quality/PROD01_002619*

WHERE DID THE TOWERS GO

?

EVIDENCE OF DIRECTED FREE-ENERGY TECHNOLOGY ON 9/11

by JUDY WOOD, B.S., M.S., Ph.D.

Published by The New Investigation, *www.WhereDidtheTowersGo.com*
 Published 2010
Printed in China

This book contains evidence submitted to NIST in a Request for Corrections (RFC) *(March 16, 2007), www.drjudywood.com/articles/NIST/NIST_RFC.html, www.ocio.os.doc.gov/ITPolicyandPrograms/Information_Quality/PROD01_002619*

This evidence was also submitted in a federal qui tam case against the contractors for NIST's report (NCSTAR 1) on WTC1 and WTC2 for science fraud. (April 25, 2007)
United States District Court, Southern District of New York, Docket Number: (07-cv-3314),
United States Court of Appeals for the Second Circuit, Docket Number: (08-3799-cv),
United States Supreme Court Docket Number: (09-548),
www.supremecourt.gov/Search.aspx?FileName=/docketfiles/09-548.htm,
www.drjudywood.com/articles/NIST/Qui_Tam_Wood.shtml

This evidence was also submitted to NIST in reference to WTC7: (Feb., Sept., 2008)
http://wtc.nist.gov/comments08/,
www.drjudywood.com/pdf/080915_NIST_comments.pdf,
www.drjudywood.com/pdf/080229_FletcherMcAllister.pdf,
www.drjudywood.com/pdf/080229_AFFIDAVITtight.pdf,

Information about the images is on page xxxv, and identified with each image.

Library of Congress Control Number: 2010916516

14 13 12 11 10 9 8 7 6 18 19 20 21 22

ISBN-13: 978-0-615-41256-6
ISBN-10: 0615412564

This book may be ordered from the website at
www.WhereDidtheTowersGo.com or through *www.drjudywood.com*

The webpages associated with this book are: *www.drjudywood.com/towers*

To the memory of those who lost their lives on 9/11, and those who lost their lives as a result of their selfless help in all of the aftermaths of this event,

And to those who are currently struggling with the ongoing effects of this event,

And to the first responders, photographers, and all who lived through it,

And to the world.

Table of Contents

List of Figures

8. Dustification **131**

9. Where Did the Buildings Go? 171

10. Holes **197**

11. Toasted Cars **213**

12. Tissue Beams and Tortilla Chips 247

13. Weird Fires 257

19. Earth's Magnetic Field on 9/11 — 413

20. Tesla-Hurricane-Magnetometer Correlation — 431

List of Tables

List of Map Legends

FOREWORD

The book you hold in your hands is the most important book of the twenty-first century. Let me explain why I say such a thing. *Where Did the Towers Go?* is a work, assuming that its content and message are properly and fairly heeded, that offers a starting point from which those who genuinely want to do it can begin, first, to rein in and then, perhaps, even end the wanton criminality and destructiveness of a set of American policies that took as their justification and starting point the horrific events of September 11, 2001.

It is now almost a decade since 9/11 took place, and in all that time no *unassailable*, *permanent*, or, in pragmatic terms, *politically influential* progress has been made in determining exactly and irrefutably *what took place* on that day—or what *did not* take place.

But now Dr. Judy Wood, in this unique, powerful, landmark work of forensic scientific investigation, provides us at last with that determination: She shows us *what did happen* on 9/11. Although Dr. Wood's scientific training and understanding are deep and complex, she has the gift of being able, without compromise, to express ideas of the greatest complexity in terms readily understandable to any interested and attentive lay person.

More must be said about these subjects in a minute, but this all-important fact remains: Those who read Dr. Wood's book fairly, openly, and thoroughly will take away with them the gift of *knowing* once and for all what happened on 9/11. They will take away the gift of knowing that they *have at last been shown the truth clearly and plainly*, no matter how different this truth may be from what they have been told for many years by supposedly higher authorities, from the government itself on through newspapers, journalists, progressive radio programs and commentators, even figures from the so-called "9/11 truth movement." Dr. Wood's book will give all those who read it carefully a solid foundation for the courage to believe not what they may have been told by one authority or another on any level and for many years, but to believe instead what their own minds, their own eyes, and their own reason tell them: That is, scientific truth as revealed through close forensic study of *all* of the evidence that has been left behind. As Dr. Wood says again and again, she arrives at truth through the study of *evidence*. The truth is not what anyone, no matter who they are, might say it is. To the place where the evidence leads, and to that place alone—*that* is where the truth is.

Where Did the Towers Go? is not the work of a day. In her first chapter, Dr. Wood tells us that her study of 9/11 really began on that calamitous September day itself, when she "realized that what was being seen and heard on television was contradictory and appeared to violate the laws of physics." This means, as I write these words, that Dr. Wood has been a student of 9/11 for eight-and-a-half years. Yet the *preparation* for that study took even longer. Dr. Wood, after all, holds a B.S. in

Civil Engineering, an M.S. in Engineering Mechanics (Applied Physics), and a Ph.D. in Materials Engineering Science—degrees that speak to nothing less than an adult lifetime dedicated to scientific analysis and observation.

Dr. Wood's areas of special focus within physics and engineering will strike readers also for their obvious suitability to study of 9/11. Dr. Wood's M.S. thesis involved the development of a Fizeau interferometer to study the effects of material defects on the thermal expansion behavior of composite materials. Her Ph.D. dissertation (in words from her web site) "involved the development of an experimental method to measure thermal stresses in bimaterial joints using moiré interferometry." Careful readers of *Where Did the Towers Go?* will quickly understand the remarkable compatibility between the subject of Dr. Wood's dissertation and its applicability to her analyses of 9/11. The same is true of certain of the courses she taught when she was a member of the faculty at Clemson University. These included Experimental Stress Analysis, Engineering Mechanics, Mechanics of Materials (the Strength of Materials), and (though not at Clemson) Strength of Materials Testing.

It's difficult to imagine an academic preparation *more* logically relevant to a study of 9/11 than Dr. Wood's—to a study, that is, not of the history of 9/11, not of the origins of it, not of the motives for it, but, simply, solely, and only to a study of what *happened*, literally, in and to the World Trade Center buildings on 9/11.

There is another element of Dr. Wood's research that qualifies her even more exactly for work of the kind described in this book. Here is a passage from Dr. Wood's web site:

> One of Dr. Wood's research interests is biomimicry, or applying the mechanical structures of biological materials to engineering design using engineering materials. Other recent research has investigated the deformation behavior of materials and structures with complex geometries and complex material properties, such as fiber-reinforced composite materials and biological materials. Dr. Wood is an expert in the use of moiré interferometry, a full-field optical method that is used in stress analysis, as well as materials characterization and other types of interference. In recent years, Dr. Wood and her students have developed optical systems with various wavelengths and waveguides. Dr. Wood has over 60 technical publications in refereed journals, conference proceedings, and edited monographs and special technical reports.

A word used here—"interferometry"—will become familiar to readers as they move into Dr. Wood's book. When preceded by "moiré," the word refers to "a full-field optical method that is used in stress analysis." The web site adds that Dr. Wood is also an expert in the use of "other types of interference." Their applicability to the study of 9/11 is made clear, again, in this description, from Dr. Wood herself, of her special areas of research:

> The main focus of my research has been in the area of experimental mechanics and optical interferometry, which is referred to as photomechanics. That is, all of my graduate work and research has been in the area of interferometry to study material behavior. Photomechanics, an area of experimental mechanics, is the use

> of optical images and optical interferometry to determine material characteristics. So, it is second nature for me to see anomalies in material behavior when looking at photographic images. Also, being an experimentalist using interferometry, I have occasionally encountered unexpected phenomena that presented themselves as puzzles. Solving these puzzles has provided me with a wide range of experience with anomalous material characteristics and the interference of electromagnetic energy.

It's safe to say that less than a majority of Americans know very much about Nikola Tesla (1856-1943), the historic figure whose story must be introduced at this point. Tesla is under-recognized in the United States partly because of his victimization by profit-driven interests opposed to his work—and opposed especially to his development of a way to harness free energy.[1] Though little known in the United States, Tesla was the world's greatest pioneering genius in the early harnessing of electricity, the development of alternating current, the study of field effects—interferometry—and, as mentioned, the development of access to free energy—that is, access to and the harnessing of energy drawn from force fields or even from the plasma present everywhere in the cosmos.

Mentioning Tesla at this point is necessary for the very good reason that Dr. Judy Wood, in *Where Did the Towers Go?*, shows that the power used to destroy the WTC buildings on 9/11—a power sufficient to turn more than 1,000,000 tons of building material into dust—is power derived from force fields, or directed energy, power of the kind that was pioneeringly studied by Nikola Tesla and that now, obviously, has been advanced by others for the most destructive of purposes rather than for the benevolent, socially meliorative uses for which it is equally well suited.

In short, Tesla's energy, imagined by him as something useful for the nurturing or even the saving of human society, has instead, since his death, been weaponized. The simple fact is that 9/11 was planned and staged as a demonstration to the world of the enormity of that power in its weaponized form.

Over the past six years, as she revealed to the public the details of her research piece by piece,[2] Dr. Wood often found herself the subject of extreme abuse from every quarter of the so-called "9/11 truth community." I have followed Dr. Wood's work over those six years, and I would like to say a few words about what she has been doing and, implicitly, about the way her work has been received.

Dr. Wood is not, in actuality, herself a part of the "9/11 truth community." Even if at one time she may have naturally considered herself to be so, this is no longer the case. The "movement"—something I have been a student of since mid-2003—has itself grown so politicized, so thoroughly infiltrated by figures and forces whose aim is to generate internal division in order to generate not progress but paralysis and stasis; that, as I said earlier, this "movement" has been made incapable, over almost a decade, of producing any *unassailable*, any *permanent*, or any *politically influential* evidence of *what really happened* on September 11, 2001.

Dr. Wood herself has been regularly and sometimes spectacularly victimized, smeared, attacked, marginalized, and misrepresented by figures and groups putatively "inside" the 9/11 truth movement. It is even the case that a student of Dr. Wood's, a gifted young man dedicated to the purpose and progress of her work, was murdered in cold blood, as also was another similar person before him. In spite of these crimes, violations, and attacks, however, Dr. Wood remained devoted unflinchingly to her research, and here, now, with its completion and with the publication of *Where Did the Towers Go?*, she brings the paralysis and bloody in-fighting of the truth movement to an end.

She has been able to bring about this enormous achievement—for which the entire world must certainly be grateful—by refusing to speculate in "opinion" or "belief" and by refusing to argue about (or even to *raise*) subjects or questions of the sorts that for years have led to paralysis and logjam, questions such as *who* planned and executed the attacks of 9/11, or *why* they did so, or who *knew* about this or that aspect of the operation, or *when* they knew, or *where* someone was and *when* they were there, and on and on.

On the contrary, Dr. Wood has worked and works now solely and only as an observing scientist. She comes to no conclusions whatsoever other than those that emerge logically, in accordance with the scientific method in which she is trained, conclusions that cannot be logically escaped or avoided after close and objective study of *all* available evidence. At the same time, such conclusions are *never* allowed by Dr. Wood, again in accordance with scientific method, to be in excess of what is supported by the evidence.

Let us make a list of the things that Dr. Wood proves in *Where Did the Towers Go?*—proves not just beyond reasonable doubt, but beyond *any doubt whatsoever*:

> 1) That the "official" or "government" explanation for the destruction of the World Trade Center on 9/11 is, scientifically, false through and through.
>
> 2) That the WTC buildings were not destroyed by heat generated from burning jet fuel or from the conventional "burning" of any other substance or substances.
>
> 3) That the WTC buildings were not destroyed by mini-nuclear weaponry.
>
> 4) That the WTC buildings were not destroyed by conventional explosives of any kind, be they TNT, C4 or RDX, nor were they destroyed by welding materials such as thermite, thermate, or "nano-thermite."
>
> 5) That there was in fact no high heat at all involved either in bringing about the destruction of the buildings or generated by the destruction of them.

And now let us turn to what Dr. Wood proves *beyond any reasonable doubt.*

She proves that the *kinds* of evidence left behind after the destruction—including "fires" that emit no high heat and have no apparent source ("Weird Fires"); glowing steel beams and molten metal, *neither* of them emitting high heat; the levitation and flipping of extremely heavy objects, including automobiles and other vehicles; patterns of scorching that *cannot* have been caused by conventional "fire" ("Toasted Cars"); the sudden exploding of objects, people, vehicles, and steel tanks; the near-complete absence of *rubble* after the towers' destruction, but instead the presence of entire buildings'-worth of *dust*, both airborne and heavier-than-air ("Dustification")—

Dr. Wood proves that these and other kinds of evidence *cannot* have been created by conventional oxygen-fed fire, by conventional explosives, or by nuclear fission. At the same time, however, she shows *that all of them are in keeping with the patterns and traits of directed-energy power*, of force-fields directed into interference with one another in ways following the scientific logic of Nikola Tesla's thought and experimentation—*and* in ways also paralleling the work of contemporary Canadian scientist and experimenter John Hutchison, who, following Tesla's lead, has for many years produced again and again and again "the Hutchison Effect," creating results that include weird fires (having no apparent fuel); the bending, splintering, or fissuring of bars and rods of heavy metal; the coring-out, from *inside*, of thick metal rods; and the repeated *levitation* of objects.[3]

These same effects, similar to the Hutchison effect[4] but on an exponentially massive scale, are what occurred at the World Trade Center on September 11, 2001. The implications of this fact, however unbelievable they may seem initially, are of a powerful and obvious importance to every living being in the world. That a power of this magnitude and intensity, a power drawn from other energy already *existing*—that a power of this enormity has been demonstrated to the world for the first time and on this scale *not* as a force potentially advantageous to human life, planetary health, and social well-being but, instead, as a *weaponized* force capable of unprecedented and incalculable destruction and ruin—this is a fact undeniably sobering to every thinking and feeling human being.

Thanks to the painstaking and unflagging work of Dr. Judy Wood—and thanks to her book, this book that you are about to read—the long debate about *what* happened on 9/11 will now end. The next step is to decide how to respond to the truth that, here, we have once and for all been shown. The implications of Dr. Wood's work are every bit as world-embracing and absolute in their importance as was the introduction of weaponized nuclear fission over half a century ago, and in fact even more so. Dr. Wood herself has referred to 9/11 as *The New Hiroshima.* To follow the now-known implications of directed energy weaponry with the greatest of care, to do so with expedience, clarity, justice, and, above all, with the aim of doing only the highest service to the well-being of mankind, the earth, and the future of both—these are the tasks laid out for us by Dr. Wood's magisterial, humane, paradigm-changing work. It is up to us—who else, after all, is there?—to take these matters up now that Dr. Wood has shown us the immensity of their importance.

She herself, near the end of her book, says something of a similar nature. It's appropriate that I close not with my words, but with hers:

> He who controls the energy, controls the people. Control of energy, depending on what that energy is, can either destroy or sustain the planet.
>
> We have a choice. And the choice is real. We can live happily and fruitfully and productively, or we can destroy the planet and die, every last one of us, along with every living being on this planet.

—Eric Larsen
—March 2010

---

[1] For an excellent introduction to the story of the maligning of Tesla and the suppression of his work, see Rand Clifford's excellent "From Reptiles to Humans: A Three-Brain Odyssey" *http://www.starchiefpress.com/articles/article42.html*

[2] *http://drjudywood.co.uk/*

[3] *http://www.thehutchisoneffect.com*

[4] Although the two, the Hutchison Effect and the phenomena seen on 9/11, share parallel origins in physics and produce results that are similar in some observable ways, there is no question of their being accurately or fairly called the same thing. Just as Tesla never developed *his* ideas with the thought of weaponization, neither has John Hutchison worked with such a thought in mind.

Author's Preface

Faced with intolerable ideas, or with intolerable acts, people in very large numbers have begun simply denying them, declaring them "unreal" and thus with a word striking them out of existence. ...But the pattern itself of not seeing is inescapable, evident to anyone who looks. —Eric Larsen, *A Nation Gone Blind*[1]

For the record, I do not believe that our government is responsible for executing the events of 9/11/01 – nor do I believe that our government is not responsible for executing the events of 9/11/01. This is not a case of *belief.* This is a crime that should be solved by a forensic study of the evidence. Before it can be determined *who* did it, it must first be determined *what* was done and *how* it was done.

The order of crime solving is to determine

1) *WHAT* happened, then
2) *HOW* it happened (e.g., by what weapon), then
3) *WHO* did it. And only then can we address
4) *WHY* they did it (i.e. motive).

Let us remember what is required to convict someone of a crime. You cannot convict someone of a crime based on *belief.* You cannot convict someone of a crime if you don't even know what crime to charge them with. If you accuse someone of murder using a gun, you'd better be sure the body has a bullet hole in it.

And yet before noon on 9/11/01, we were told *who* had done it and *how* it had been done, this before any investigation had even been conducted to determine *what* had been done. As of this publication only one person[2]—myself, Dr. Judy Wood[3]—has conducted a comprehensive investigation to determine *what* happened to the World Trade Center (WTC) complex, a question that is part of a federal case I filed[4]. It might be surprising for readers to learn that The National Institute of Standards and Technology (NIST) did not analyze *what* happened to the WTC, the very first step in any scientific forensics investigation. That is, NIST did not analyze the *collapse* of the World Trade Center towers, *despite* the fact their report is entitled, *NIST NCSTAR 1—Final Report on the Collapse of the World Trade Center Towers.* NIST's mandate from Congress was to

> 1. Determine why and how WTC 1 and WTC 2 collapsed following the initial impacts of the aircraft and why and how WTC 7 collapsed.[5]

Yet two pages later, in a footnote, the NIST report says that

> The focus of the investigation was on the sequence of events from the instance of aircraft impact to the initiation of collapse for each tower. For brevity in this report, this sequence is referred to as the "probable collapse sequence," although it does not actually include the structural behavior of the tower after the conditions for collapse initiation were reached and collapse became inevitable.[6]

The NIST report,[7] that is, merely offered a probable [hypothetical] 'collapse sequence' purporting to explain the sequence of events leading up to the 'collapse' of the WTC towers. Yet NIST did not "determine *why and how* WTC 1 and WTC 2 'collapsed' following the initial impacts of the aircraft," which was their mandate. Had NIST determined *"why and how"* the towers were destroyed, they would have first determined *what* happened by dealing with phenomena that are empirically confirmed to have occurred. As is glaringly evident, they did not do this.

I challenged NIST[8] on their scientifically-flawed report,[9] noting that the images presented in their report, as well as their "probable [hypothetical] 'collapse' sequence" violated the laws of physics. In their written reply to me they openly acknowledged that they had not analyzed the collapse.[10]

> As stated in NCSTAR 1, NIST only investigated the factors leading to the initiation of the collapses of the WTC towers, not the collapses themselves.[10]

That is, the NIST personnel admitted their report to be a fraud. Their position is that if they did not analyze the "collapse," they need not address why their "probable [hypothetical] 'collapse' sequence" in fact violates the laws of physics. They are willing to accept responsibility only for saying that the building obeyed the laws of physics *before* it was destroyed. This document, in which NIST states that it did not analyze the "collapse," is part of my legal case and is available in documents posted on my website.[11]

A large portion of the sub-report, NCSTAR1-6, contains information that *appears* to be the product of a detailed analysis of what happened after the building's destruction was initiated. But in response to my informing them that their *apparent analysis* violated the laws of physics, NIST, as said, stated that they had not analyzed the collapse, despite thousands of pages giving the *appearance of an analysis.* It is incongruent for NIST to report on something that they acknowledge they did not analyze. The entire NIST report, including its title (*NIST NCSTAR 1—Final Report on the Collapse of the World Trade Center Towers*), is a deception.

Dr. Morgan Reynolds, in the case he filed,[11] addressed how this crime was *not* committed with airplanes. Remember, to convict someone of a crime, you need to prove *how* the crime was committed. It may surprise you to learn that there is no actual, verifiable evidence confirming that airplanes crashed at any of the four locations on 9/11/01. However, as Dr. Reynolds shows, there is an abundance of evidence to the contrary.[11] That does not mean there were no airplanes. It only means that no evidence of the alleged airplanes was found at the crime scenes. It also does not mean that eyewitnesses were dishonest or did not see what they believed were airplanes. But what this does mean is that there is a significant contradiction between the physical evidence and the story we were given. You cannot legally convict someone of murder using a gun if the body has no bullet holes in it, no matter how many people thought they saw the accused shoot the gun. Once again, you cannot convict someone of a crime based on *belief.* Otherwise magic tricks could be used to convict anyone of a crime, and we end up in a similar situation to the original Salem witch hunts, where people were tried and executed without there being any evidence of the accusations made against them.

Many people have speculated as to *who* committed the crimes of 9/11 and/or *how* they did so. But without addressing *what* happened, speculation of this kind is nothing more than *conspiracy theory*, a phrase that also describes the box-cutter story we were given before noon on 9/11/01. My own research, *not* speculation, is a forensics investigation of *what* happened to the WTC complex on 9/11/01. I don't address *who* did it, nor am I concerned with that question. Before issues of that kind can be addressed, we must first determine *what* happened, and that is the objective of my research. By definition, research that is purely empirical cannot be about and has nothing to do with *conspiracy theory* of any kind. The fact that others (in the mainstream media, the alternative media, and the so-called "9/11 truth movement") promote various theories about 9/11 is irrelevant to my research. On the other hand, to determine *what* happened, we must address *all* of the available evidence. Anyone declaring *who* did what or *how* they did it before they have determined *what* was done is merely promoting either speculation or propaganda. The popular chant, "9/11 was an inside job," is, scientifically speaking, no different from the chant that "19 bad guys with box cutters did it." Neither one is the result of a scientific investigation supported by evidence that would be admissible in court. Neither identifies *what* crime was committed or *how* it was committed.

So let us consider the body of empirical evidence that must be explained in order to determine *what* happened.[12] What is presented here is not a theory and it is not speculation. It is evidence. Here, then, is the evidence of *what* happened on 9/11/01.

---

[1] Eric Larsen, *A Nation Gone Blind: America in an Age of Simplification and Deceit*, *http://www.ericlarsen.net/nation.excerpt.html*

[2] Only non-classified documents in the public domain are considered.

[3] B.S. (Civil Engineering, 1981) (Structural Engineering), M.S. (Engineering Mechanics (Applied Physics, 1983), and Ph.D. (Materials Engineering Science, 1992) from the Department of Engineering Science and Mechanics at Virginia Polytechnic Institute and State University in Blacksburg, Virginia, *http://drjudywood.com/articles/a/bio/Wood_Bio.html*

[4] United States District Court, Southern District of New York, Docket Number: (07-cv-3314), United States Court of Appeals for the Second Circuit, Docket Number: (08-3799-cv), Supreme Court Docket Number: (09-548), *http://www.supremecourt.gov/Search.aspx?FileName=/docketfiles/09-548.htm*. But this case presents a dilemma for the courts as it involves *someone's* classified technology, no matter whose classified technology it was. A civil case involving classified technology cannot be held behind closed doors without publicly acknowledging this fact. Perhaps this is why the United States Court of Appeals, in their written decision, respectfully acknowledged that the law (FERA) applied to this case, but "for the ease of" dismissing the case, they were ignoring this law. See: *http://www.drjudywood.com/articles/NIST/Qui_Tam_Wood.shtml*

[5] *NIST NCSTAR 1 – Final Report on the Collapse of the World Trade Center Towers*, September 2005, E.1 Genesis of this investigation, p. xxxv (p. 37), *http://wtc.nist.gov/reports_october05.htm*

[6] *NIST NCSTAR 1 – Final Report on the Collapse of the World Trade Center Towers*, September 2005, E.2 Approach, p. xxxvii (p. 39) footnote[!], *http://wtc.nist.gov/reports_october05.htm*

[7] *NIST NCSTAR 1 – Final Report on the Collapse of the World Trade Center Towers*, September 2005, http://wtc.nist.gov/reports_october05.htm

[8] *http://ocio.os.doc.gov/ITPolicyandPrograms/Information_Quality/PROD01_002619*

[9] To my amazement, I was the first person to challenge NIST on their report's absence of an analysis to "determine *why and how* the WTC 'collapsed,'" which qualified me to file a qui tam case for science fraud. *http://ocio.os.doc.gov/ITPolicyandPrograms/Information_Quality/PROD01_002619*

[10] Response to Request for Correction from Dr. Judy Wood, dated March 16, 2007, *http://ocio.os.doc.gov/ITPolicyandPrograms/Information_Quality/ssLINK/PROD01_004161*, *http://drjudywood.com/articles/NIST/*

Qui_Tam_Wood.html

[11] *http://drjudywood.com/articles/NIST/Qui_Tam_Wood.html*

[12] *http://drjudywood.com/wtc/index.html#index*

Images

History, Documentation, Research, Education, Criticism, and Fair Use.

September 11th was among the greatest, if not the greatest historical event in all of known history. Rarely does a day go by that national headlines do not include a story with implications based upon this single day. We were told, *"everything changed on 9/11"* [1,2,3,4,5,6] but few even know what physically happened on that day in history that *changed everything* and what phenomena we witnessed.

My academic background in Civil Engineering, Engineering Mechanics (Applied Physics), and Materials Engineering Science, with experimental mechanics and interferometry being central to my research, has provided me with very unique qualifications and expertise. My unique education and training put me in a position where I could not ignore what I saw on 9/11. I waited for others to come forward, including those in official positions, photographers, media, journalists, and those with similar expertise as mine, but no one came forward. My conscience left me no other choice than to come forward myself and conduct forensic research into 9/11, based on the empirical evidence of this phenomena.

The empirical evidence of this event was captured in the many photographs taken by courageous individuals. These photographs establish a very important evidentiary basis that, using scientific methods, allow a scientist to arrive at supportable scientific conclusions. My unique background allowed me to interpret the many available photographs, including the selected photographs used in this book, and provide constructive criticism of relevant evidence not previously disclosed. I have commented on each of the photographs regarding what material evidence they disclose to us to assist in our understanding of 9/11, structuring this book as a textbook for teaching and presentation. This book expands upon the scholarship and knowledge of the empirical evidence to support each student's understanding of the phenomena that we witnessed on 9/11.

My intellectual integrity prevents me from calling this a collapse. This is why I have chosen to stand up. My conscience leaves me no other choice.[7]
Photo Credit: Greg Semendinger, NYPD

The photos contained in this book fall under the protections of the "fair use" doctrine, as they are being used "[F]or purposes such as criticism, comment, news reporting, teaching (including multiple copies for classroom use), scholarship, or research, is not an infringement of copyright." [8]

By applying my research expertise to the analysis of these photographs, I have been able to compile empirical evidence and conduct an independent investigation to explain these never-before-seen phenomena of monumental proportions. This provides us all with a greater knowledge and understanding of that horrific day, September 11th, 2001. Nikola Tesla considered this technology to be a double-edged sword that could destroy our planet or be a boom for humanity.[9] Let us use it for good and provide free energy for all.

---

[1] Jordan Debree and Lee Wang, "Defending the Home Front: The Military's New Role," October 10, 2006, *PBS Frontline,* WGBH educational foundation, *http://www.pbs.org/wgbh/pages/frontline/enemywithin/reality/military.html*

[2] Jay Rosen, "What if Everything Changed for American Journalists on September 11th? My Speculations," AUGUST 13, 2004, *http://archive.pressthink.org/2004/08/13/after_911.html*

[3] Everything changed on 9/11, October 13, 2004, *http://rhetorica.net/archives/2711.html*

[4] Brad Carlton, "How Bush Hit the 'Trifecta' on 9/11--and the Public Lost Big-Time," The Baltimore Chronicle and The Sentinel, June 12, 2002, *http://baltimorechronicle.com/trifecta_jun02.shtml*

[5] Eyal Press, The Wisdom of the Public, *The Nation*, September 25, 2009, *http://www.thenation.com/blog/wisdom-public*

[6] Jeff Eason, "Diving Board Disappearances: Americans Forget FDR's Admonition About Fear," June 22, 2006, *http://www.mountaintimes.com/columns/0622_sweet_tea.php3*

[7] *http://911wtc.freehostia.com/gallery/originalimages/GJS-WTC28.jpg*

[8] 17 U.S.C. § 107.

[9] Phil, We owe what we do today to Michael Faraday and Nikola Tesla (1857-1943) *http://www.phils.com.au/tesla.htm*

If you listen to the evidence carefully enough, it will speak to you and tell you exactly what happened. If you don't know what happened, keep listening to the evidence until you do. The evidence always tells the truth. The key is not to allow yourself to be distracted away from seeing what the evidence is telling you.[1]

Empirical evidence is the truth that theory must mimic.[2]

---

[1] My own motto.

[2] A powerful statement by someone who has taught me well.

1.
Introduction

I found out something I never knew:
I found out that my world was not the real world.
—Robert F. Kennedy, 1968

On 9/11, I realized that what was being seen and heard on television was contradictory and appeared to violate the laws of physics. I remember watching the TV in the faculty conference room. The TV kept playing the same film over and over, showing what appeared to be a building unraveling like a sweater.

Figure 1. (9/11/01) A Tower goes down. But does this look like a collapse? It doesn't even look like a typical controlled demolition.
Source: http://911wtc.freehostia.com/gallery/originalimages/GJS-WTC28.jpg

I had never seen a building unravel like a sweater, and I tried to imagine what was going on that might make it look that way. Certainly the time it took the building to go away did not make sense.

My introduction to the day's events was from a radio playing in the background while I was working at home. Before heading in to campus, I turned on the television and was amazed to find the same view of the event on every channel. My first thought was of another Orson Welles "War of the Worlds" type of scenario, except with TV this time instead of radio. In 2001, my parents lived near the Pentagon, which we were told had also been attacked. So I thought I would test my "Orson-Welles scenario" theory by calling my mother to ask if there were fighter jets overhead. I called her and found she was unaware that anything was going on. When she looked outside, she saw no fighter jets. So the two of us tended to conclude it might be an "Orson-Welles event" and I headed in to campus as a skeptic.

After spending weeks and months researching 9/11, I found a common pattern in people's responses. It seemed that the strongest influence on whether or not someone questioned the events came from how they first learned of them. When I phoned my mother, she was introduced to the event through questioning it. Meanwhile, my father was outside in his garden. A neighbor who worked at the Pentagon came home early, traumatized over what was going on. She ran over to my father to tell him how horrific these events were. He believed his neighbor, his neighbor was traumatized, and that traumatized my father as well. I feel fortunate to have seen these two very different responses.

When I was growing up, my family did not have a TV. I think this made an enormous difference in the reason I was able to see what I saw. Especially since that day, I have come to understand that when exposed to new information, I depend on my eyes and what I observe. It is very likely that when I looked at the TV that morning, I did not hear it at all. For me, it seems that if what I see and hear do not agree, my mind just shuts out the sound. That may be what allowed me to see what was actually going on.

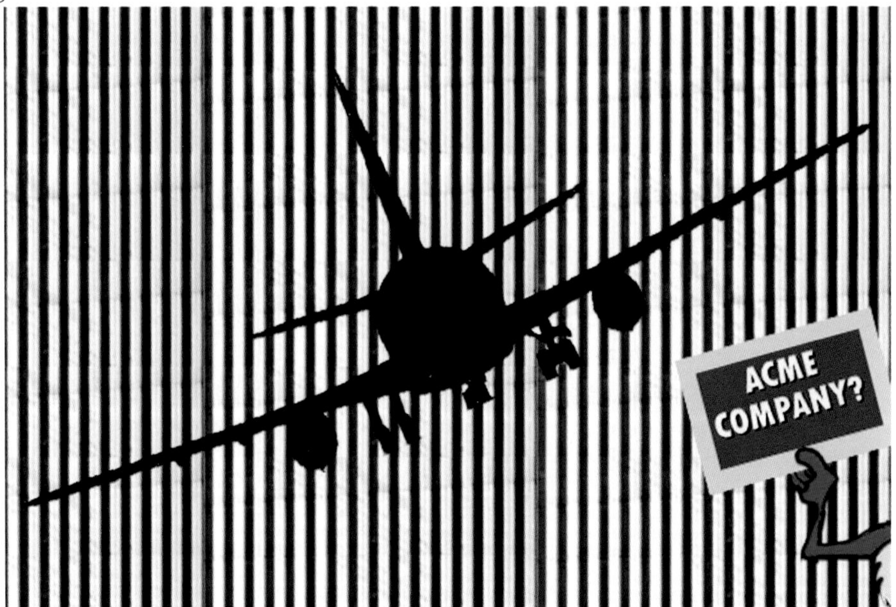

Figure 2. Reminiscent of Wile E. Coyote in a Warner Brothers cartoon.[1]
http://drjudywood.com/articles/a/GH/GH_carcrash.html

When I saw the airplane-shaped hole in one of the buildings, I began to laugh, in spite of the tragedy of the event, because I was reminded of watching a Roadrunner cartoon. I occasionally use Roadrunner cartoons to teach concepts in class. In my Mechanics of Materials class, when I introduce shear stress, I show a Roadrunner cartoon where Wile E. Coyote holds up an "ACME steel plate" to stop the Roadrunner. You see a big "whoosh" fly by. Wile E. Coyote checks his steel plate and finds a perfect hole cut through it in exactly the shape of the Roadrunner. Wile E. Coyote then gives a look as if to say, "Ahhh… so it must have been the Roadrunner that just flew through here."

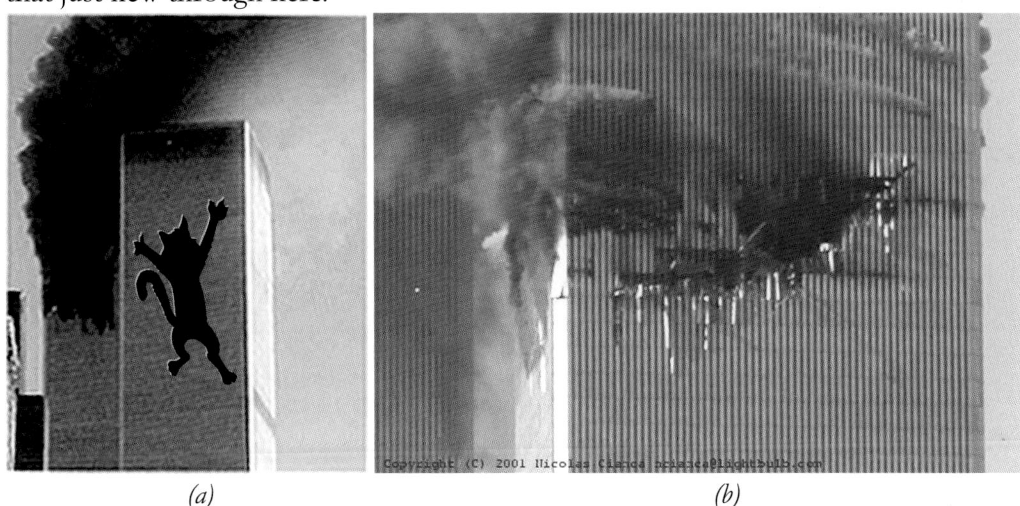

(a) *(b)*

Figure 3. The plane-shaped hole reminded me of the Roadrunner hole.
(a)http://drjudywood.com/articles/a/GH/GH_carcrash.html, (b)http://911research.com/wtc/evidence/photos/docs/wtc_fires_dsnc1775A.jpg

Because I suspected some sort of twisted joke was being played on us, I was tempted to laugh at that Roadrunner hole in the side of the tower. My colleagues looked at me like I was crazy.

A. The Pressure to Conform

Was I wrong? Buildings don't just "unravel" like sweaters, and what we saw did not look like any building failure or demolition that I'd ever seen. But why did no one else see it the way I did? I felt alone and ostracized for questioning it, so I went back to my office.

There, I wondered if I was crazy for seeing the situation differently from others. I struggled with how to resolve this. The building did look like a sweater unraveling, or, better, it looked as though it had turned into a volcano of dust. Something did not look right no matter how I tried to analyze it. I asked myself, "Am I in *The Twilight Zone?*"[2]

How would I know if reality had been "altered," or, better put, if an explanation were being offered that had little basis in physics? I decided to test my sanity by trying to calculate what I thought I had seen. I decided to calculate approximately how much time it might take the building to fall down. I did not know how the towers were actually built, so I did not try to model that. I just wanted to make a rough estimate of how long it might take under the fastest conditions. They said the buildings were about a quarter-mile tall. Thus, with a known distance, I could calculate how long it would take the roof to hit the ground, neglecting the factor of air resistance. After plugging in the approximate numbers, I arrived at 9.055 seconds (Figure 4).

$$\sqrt{\frac{(height)}{\left(\frac{gravity}{2}\right)}} = \sqrt{time^2} = time$$

$$\sqrt{\frac{\left(\frac{mile}{4}\right)\left(\frac{5{,}280\,ft}{mile}\right)}{\left(\frac{32.2\frac{ft}{s^2}}{2}\right)}} = \sqrt{81.988} = 9.055\ s$$

Figure 4. Note the actual height is 1,368-ft and 1,362-feet, for WTC1 and WTC2, respectively. But this was my first approximation.

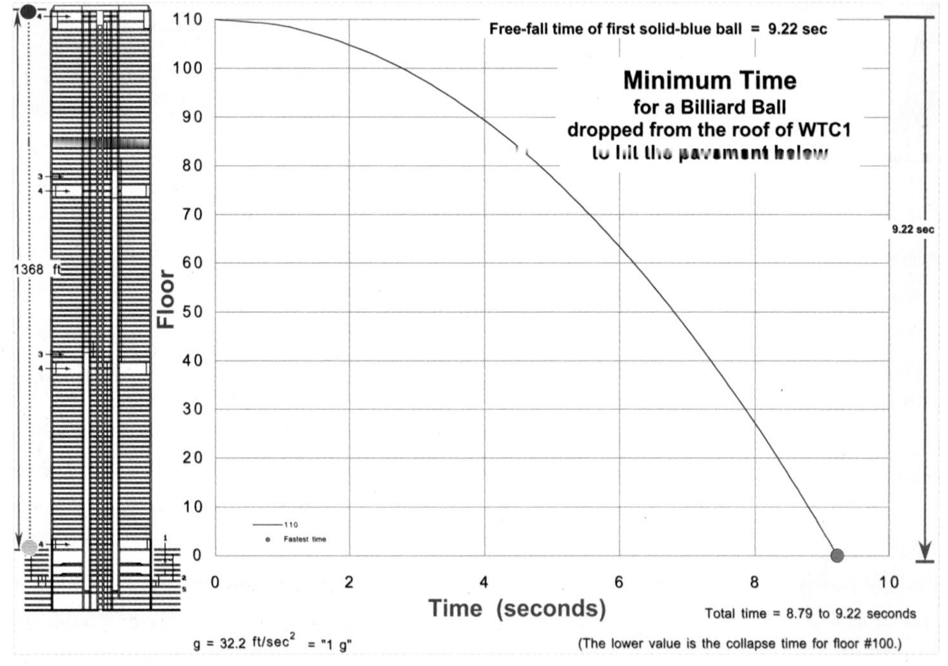

Figure 5. t = 9.22 seconds, the time it takes a billiard ball to fall 1,368 feet in a vacuum. This is the fastest time something on the roof can hit the ground (without being propelled by something like rocket motors).[3]

This result appeared to be about the same length of time the collapse had taken on the TV footage. "OK," I thought, "So maybe I'm sane after all." But then I thought, "No, wait, the roof *can't* fall straight down, at *free-fall* speed!" After all, what would it be falling *through?* It had to fall through all those other floors. So how long would the fall take in that case? I thought of an avalanche. It would also be like a game of tag-team hand-offs in a relay race. This is when what I call the "Billiard Ball

Example" (BBE) was born. I wanted to calculate a range, in order to know that it can neither take less time than t_1 nor more time then t_2, giving every advantage to the event I had seen. Or ($t_1 < t < t_2$).[4]

B. What about Momentum?

We saw a lot of dust billowing up as the towers just seemed to go "poof." Where did that dust come from? Perhaps it came from each floor shattering against the next. But if it did, each time a floor shattered into powder, it would pretty much "float away." So, once a "collapse" was initiated (assuming, for argument's sake, that that's possible), there would not have been an accumulating and increasing mass "riding the pile down," like a pile driver, through the entire height. So, if there were not an increasing mass pushing the building down, how *could* it have come down? It could not have pushed itself down with its own weight. After all, the weight is removed as the building turns to dust. That is, the force on the supports near the base of the building diminishes as the upper floors turn to dust and go away.

C. The Resistance Paradox

If a floor is pulverized by slamming into the floor below it, that floor below must be rigid in order to resist the force acting upon it. Compare a car slamming into the base of a granite cliff versus slamming into another car left out of gear in the middle of the highway. The first will meet *far* greater resistance than the second will. Consequently, if there *is* such resistance, the building *cannot* "fall" *at anything even remotely approaching free-fall speed.*

D. Timing

Consider how fast a chain-reaction *collapse* would have to progress if all 110 stories were to be destroyed in about <u>10 seconds</u>. How many times can you clap your hands in 10 seconds? Well, many more than ten, but nowhere even near one *hundred* ten. Then consider that the floors, unlike your hands, are about 12 feet apart and have a lot of material in between them that would need to be destroyed in a fraction of a second.[5]

But let's assume that something unexplainable happened and each floor did go "splat" onto the next floor, did turn to powder, and then left just barely enough energy to break the next floor loose, and so on. So, each time one floor pulverizes itself, the next floor is on its own to start moving all over again. This one-by-one baton-passing would take <u>97 seconds</u> (not accounting for air resistance) for the roof to hit the ground, if we assume there was, somehow, enough energy. (See Figure 6.)

But the building was damaged, and so, that being the case, there will be resistance in some places while in other places there will not. So let's assume that nine out of every 10 floors are missing or damaged. That means that the top floor can freely accelerate for 10 floors until it goes "splat" and pulverizes itself. We'll assume that with its last bit of energy, it knocks the next floor loose, which then picks up speed for 10 floors, until it also goes splat and pulverizes itself. In this case, it would take about <u>31 seconds</u> for the roof to hit the ground, again, not considering for air resistance. (See Figure 7.) In any case, we know that 9 out of every 10 floors were *not* missing. Bottom line: *The idea of a chain reaction of gravity collapses was physically impossible.*

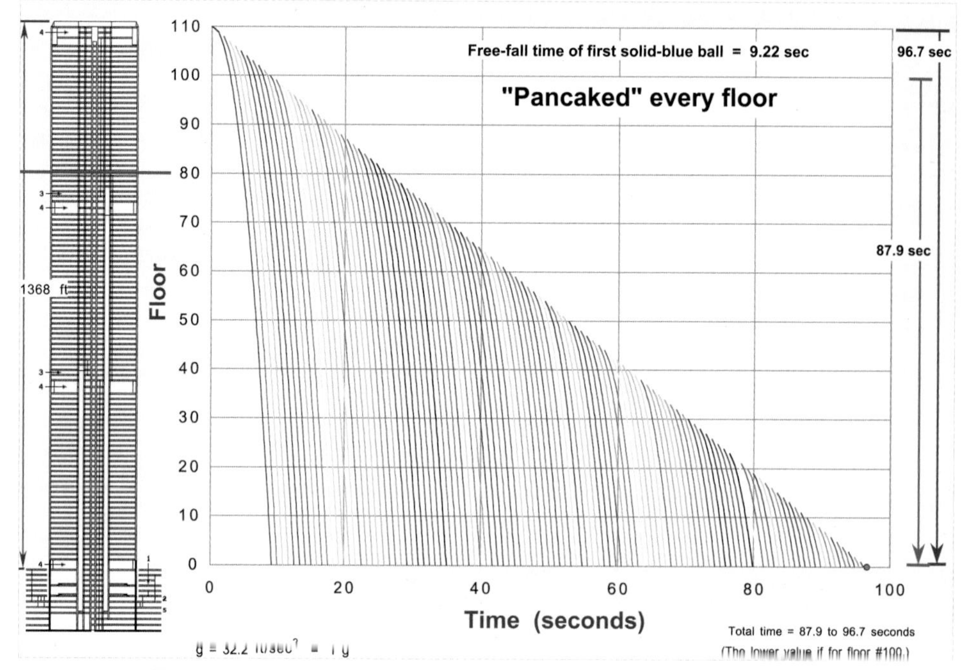

Figure 6. t = 96.7 seconds, if there was a "baton pass" at every floor.[3]

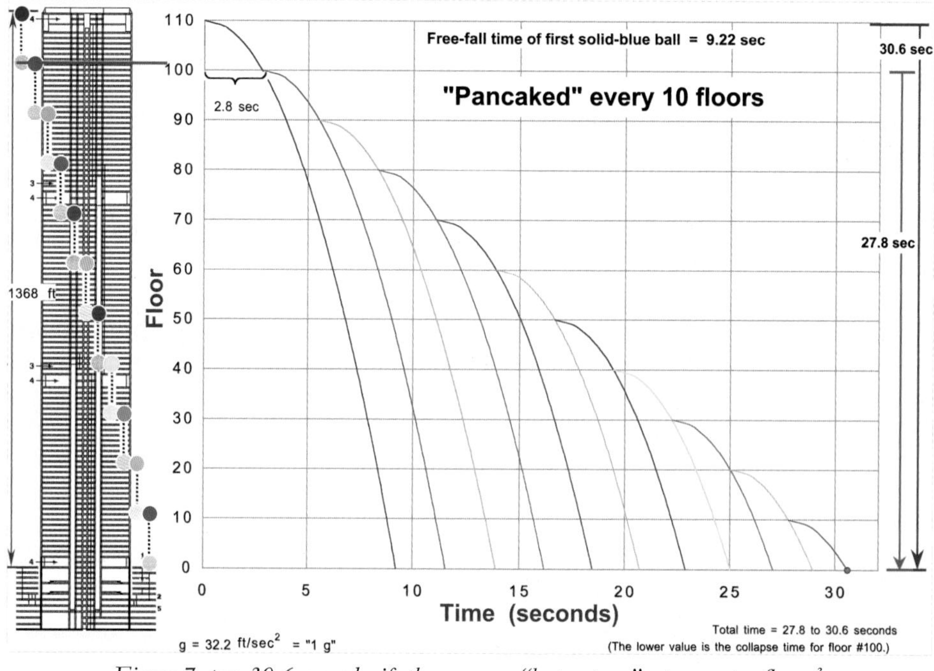

Figure 7. t = 30.6 seconds, if there was a "baton pass" at every ten floors.[3]

But the building went away in much *less* than 30 seconds. I could see this from the TV. So how could it do that? To answer this question, we begin by considering

Figure 8 below. The red ball (100th floor) cannot start moving until the blue ball (110th floor) gets there and knocks it loose. Then the orange ball cannot start moving until the red ball arrives to set it going, and so on. But, what if the red ball was *pushed* to start dropping *before* the blue ball got there? And what if the orange ball received a similar head start? And then the purple ball, then the green ball, and so on? That would be the only way to get the entire building on the ground in less than 10 seconds. The resistance of each floor that is in the way slows down the "collapse wave." So, what if we *remove* the resistance *just ahead* of the collapse wave?

To put it succinctly, *something must destroy the building, somehow,* ahead *of the "collapse wave."*

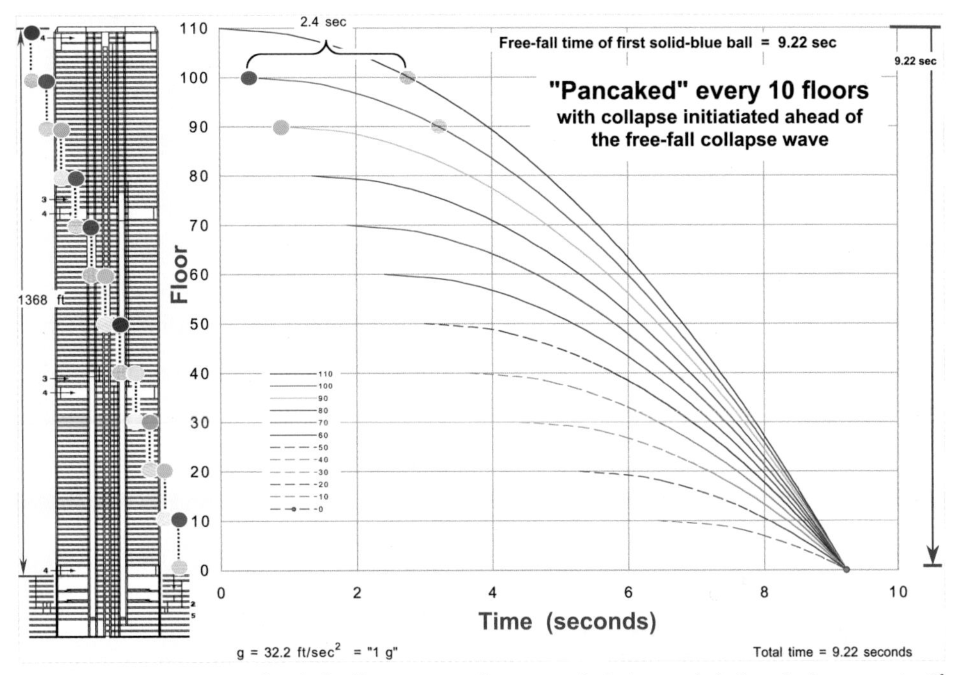

Figure 8. t = 9.22 seconds, if the "baton passes" were made 2.4 seconds before the baton arrived.[3]

It is tempting to jump to the question of *how* this was done, based on our knowledge of possibilities. But, since our knowledge base does not include *all* possibilities, doing that would be like taking a multiple-choice test without being given a correct choice. For example, if we are trying to conclude how the Towers were destroyed and are given answers having to do only with firecrackers, slingshots, and bubble gum, we're not likely to pick the correct answer. You can never be sure you have the correct choice until you know the entire what of *what* happened. All we know at this point is that the towers went away faster than they would have in a gravity-driven free-fall collapse.

To know more, we need more evidence. After all, that—evidence—is the difference between "science" and "conspiracy theory." In this book, I will present evidence of *what* happened in order to find out *how* it happened. I will not address *who did it* or *may have* done it. By definition, this book will not be about "conspiracy theories."

We cannot determine *who* did it until we determine *what* was even done and *how* it was done. Yet before noon on 9/11/01, many of us were told *who* had done it. That is, we were given a conspiracy theory, not the result of a scientific investigation. So, what's wrong with that? Nothing, except that it qualifies as a falsehood, being unverifiable by *evidence* as to *what* exactly happened, or *how*, let alone by *whom*. Doing such a thing, going after *who* did it before establishing *what* happened and *how* it happened, is a way of ensuring that the truth will be hidden for a very long time, if not forever. It is called a cover-up. I will not be part of a cover-up.

As Martin Luther King said, "A time comes when silence is betrayal."[6] That time came for me at 1:32 A.M. on November 3, 2004.[7] The next day, I told my mother that I had to look into what happened on 9/11/01 because my conscience left me no other choice. She replied, "If you do that, you won't have a career." I answered, "If I don't, no one will have a career." Something extraordinary had taken place, yet no one else was asking, *"Where did the towers go?"*

With a basic knowledge of engineering mechanics and physics, we can prove that the collapse of the World Trade Center twin towers could not have happened for the reasons given in official accounts or in most alternative "9/11 truth" explanations. The physics of the controlled demolition models, the airplane fuel models—they *do not work*. That's what this book is about: Looking at the evidence to discover what did happen.

Let us now turn to focus on the true nature of the destruction of the World Trade Center towers.

---

[1] Illustration by Nick Buchannon. This image is not from a Warner Brothers cartoon, but is reminiscent of the Roadrunner and Coyote cartoon, "*Gee Whiz-z-z-z-z-z-z*," which was released May 5, 1956. Wile E. Coyote (also known as "The Coyote") was created by animation director Chuck Jones in 1948 for Warner Brothers. Jones based the Coyote on Mark Twain's book *Roughing It* in which Twain described the coyote as "a long, slim, sick and sorry-looking skeleton" that is "a living, breathing allegory of Want. He is *always* hungry." *http://en.wikipedia.org/wiki/Wile_E._Coyote_and_Road_Runner*

[2] *The Twilight Zone* was a television series created by Rod Serling. The original series (1959-1964) had 156 episodes presented serious science fiction and abstract ideas through fantasy, science fiction, suspense, or horror. *http://en.wikipedia.org/wiki/The_Twilight_Zone*

[3] *http://drjudywood.com/articles/BBE/BilliardBalls.html*

[4] See my website at *http://drjudywood.com/articles/BBE/BilliardBalls.html.* See also the next chapter where the BBE (*Billiard Ball Example*) is outlined.

[5] Material such as desks, chairs, file cabinets, piping, computers, bookcases, sinks, toilets, lighting fixtures, water tanks, steel columns, steel and concrete floors, etc., would need to be destroyed on each floor.

[6] Martin Luther King, Jr., *Beyond Vietnam -- A Time to Break Silence,* April 4,1967, at a meeting of Clergy at Riverside Church in New York City, *http://www.americanrhetoric.com/speeches/mlkatimetobreaksilence.htm,* audio: *http://drjudywood.com/media/mlkagainstvietnam.mp3,*

[7] We were told that the voting irregularities of the 2000 election would be corrected before the 2004 election. It became apparent they had not been.

2.

The Billiard Ball Example
A Refutation of the Official Collapse Theory

We've arranged a civilization in which most crucial elements profoundly depend on science and technology. We have also arranged things so that almost no one understands science and technology. This is a prescription for disaster. We might get away with it for a while, but sooner or later this combustible mixture of ignorance and power is going to blow up in our faces. —Carl Sagan

When people think you make them think, they will like you, but when you really make them think, they will hate you. —Don Marquis

A. Introduction

Very shortly after the events of September 11, 2001, the U.S. government proclaimed with certitude that the attackers were 19 Arab suicide bombers under the guidance of one Osama bin Laden. What quickly followed were "authoritative" pronouncements, through NOVA and a few academicians, about what had brought the WTC towers down. This early public consensus of "the authorities" was that the buildings had been unable to withstand the horrific onslaught of the plane crashes and the heat of subsequent fires.

Since that time, questions have arisen about the veracity of the *Official Theory* of the events of 9/11. One area of particular interest has been the issue of the WTC tower "collapses." Could the towers have indeed been brought down as a consequence of the apparent air strikes against them?

B. The Value of Simplicity

What can you prove through simple models of complex situations? Let's say I tell you that I ran to a store (10 miles away), then
to the bank (5 more miles), then
to the dog track (7 more miles), then
to my friend's house (21 more miles), then
home ...all in 2 minutes.

To disprove my story, you could present a simple case. You could posit that the world's record for running just one mile is 3:43.13, or just under four minutes. So it does not seem possible that I could have run over 40 miles in 2 minutes. That is, it does not seem possible for me to have run 43 miles in half the time it would take the holder of the world record to run just one mile. Even if you gave me the benefit of having run all 43 miles at world-record pace, it still would not have been possible for me to have covered that distance in two minutes.

Remember, the proof need not be complicated. You don't need to prove exactly how long it *should have* taken me to run that distance. Nor do you need to prove how much longer it would have taken if I had stopped to place a bet at the dog track. To disprove my story, you need only show that the story I gave you is not physically possible.

Now, let us consider whether any of the collapse times provided to us are possible within the confines of the story we were given.

C. Elapsed Time: How long did it take the towers to disappear?

Three sources have provided data and/or opinions about how long it took for the WTC towers to collapse. (1) The Official Story, as expressed in the 9/11 Commission Report[1] (2) Columbia University's Seismology Group record of the earth-shaking associated with the *collapse*[2] (which was used by NIST in their report[3]), and (3) independent 9/11 researchers who have attempted to discern the collapse time through various methods using video analysis.

The 9/11 Commission Report[4] states, "At 9:58:59, the South Tower collapsed in ten seconds. The building collapsed into itself, causing a ferocious windstorm and creating a massive debris cloud."[5]

The August Fact Sheet (Answers to Frequently Asked Questions) by NIST states, "NIST estimated the elapsed times for the first exterior panels to strike the ground after the collapse initiated in each of the towers to be approximately 11 seconds for WTC 1 and approximately 9 seconds for WTC 2."[6] [emphasis added] The height of the South Tower (WTC2) is 1362 feet, and the height of the North Tower (WTC1) is 1368 feet, nearly the same.[7]

We will therefore assign the value of 10 seconds to the Official Story.

Columbia University's Seismology Group recorded seismic events of 10 seconds and 8 seconds, corresponding to the collapses of WTC2 and WTC1, respectively.

Information Based on Seismic Waves recorded at Palisades New York			
Seismology Group, Lamont-Doherty Earth Observatory, Columbia University			
Event	**Origin time (EDT)** (hours:minutes:seconds)	**Magnitude** (equivalent seismic)	**Duration**
"Impact 1" at WTC1	08:46:26±1	0.9	12 seconds
"Impact 2" at WTC2	09:02:54±2	0.7	6 seconds
"Collapse 1," WTC2	09:59:04±1	2.1	**10 seconds**
"Collapse 2," WTC1	10:28:31±1	2.3	**8 seconds**

Table 1. Information Based on Seismic Waves at Palisades New York[8]

Because the exact nature of what caused the towers to collapse has not been determined, it is difficult to assign a clear meaning to the geological evidence. However, for our purposes we interpret the evidence here as suggesting a fall-time in the vicinity of free-fall.

The third source of data/opinion about fall times comes from independent researchers and is based on their examination of video footage. Here we have a range of suggested fall times, from approximately 9 seconds to perhaps as long as 15 or even 18 seconds (though the longer times seem to involve questionable reasoning).[9] One problem with attempting a video analysis is that the later part of the collapse is hidden in the immense dust clouds the event produced. Additionally, there is the issue of whether or not the videos reflect the absolute time involved or are affected by subtle time distortions such as a tape's running-speed.

From the evidence available, we will assume that the fall time must have been within the range of 9 to 15 seconds.

However, our purpose in the exercise below is to identify whether the Official Story is worthy of our belief. For that reason, we will hypothetically accept the value published in *The 9/11 Commission Report*[10], which is 10 seconds, for our fall time.

Before we begin our analysis, we might ask ourselves, "Do any of these values seem reasonable as a length of time that would allow for complete collapse, from gravity alone, of buildings approximately a quarter mile high ?"

Let's calculate a few values we can use as reference.[11]

For the following, we will use the height of WTC1 as 1368 feet and consider each floor to be 12.44 feet high. (1368/110 =12.44 ft/floor).

We will assume gravity = 32.2 ft/sec^2 or 9.81 m/sec^2.

D. Case 1: Free-fall from the WTC1 Roof

Using a billiard ball as a timing device, let's consider the minimum time it would take the blue billiard ball to hit the pavement, a fall of more than 1/4 mile (see Figure 9). We'll start the timer when the ball is dropped from the roof of WTC1, and we'll assume the fall to be taking place in a vacuum, with no air resistance. (Note, large chunks of the building will have a very low surface-area-to-mass ratio, so in their case air resistance also can be neglected.)

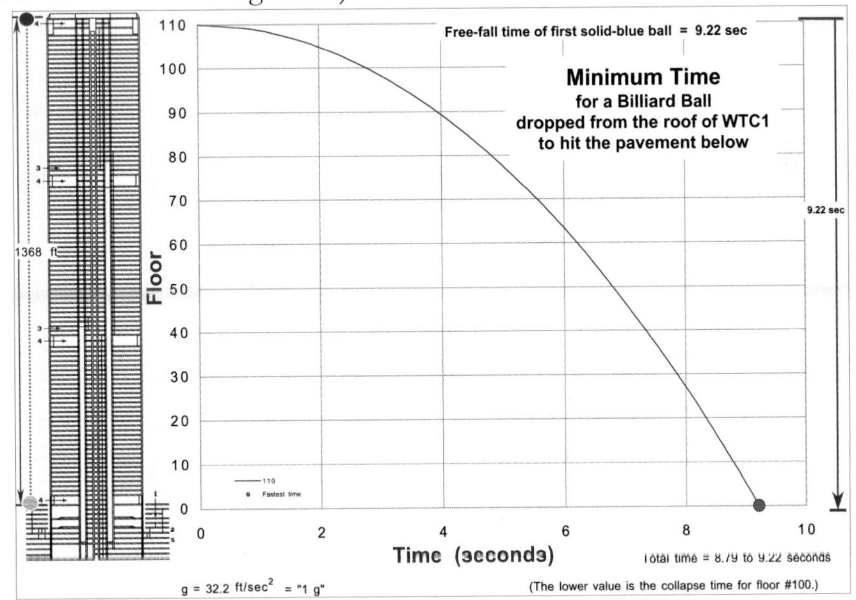

Figure 9. Minimum Time for a Billiard Ball dropped from the roof of WTC1 to hit the pavement below, assuming no air resistance. (The point of contact with the pavement is shown in red.)[12]

The (dark blue) billiard ball will accelerate from the moment it drops over the edge of the WTC1 roof. If in a vacuum, it would hit the pavement, 1368 feet below, in 9.22 seconds, shown by the blue curve in the figure, below. It will take longer if air resistance is considered, but for simplicity, we are neglecting air resistance.

You'll notice that the billiard ball begins to drop very slowly, then accelerates under gravity. In a vacuum, it will hit the pavement, 1368 ft. below, in 9.22 seconds. That is to say, unless it is propelled by an additional force, it will take at least 9.22 seconds to reach the ground.

E. The "Pancake Theory"

We need to consider the "pancake theory" because it is an integral part of the Official Story. According to the U.S. government (through FEMA and vocal supporters of the official story, such as Thomas Eagar),[13] the WTC towers fell by "pancaking," propelled from gravity alone, in 10 seconds. Our purpose here is to examine the likely veracity of this claim.

According to the pancake theory, one floor fails and falls onto the floor below, causing the second one to fail and fall on the floor below *it*, and so forth. The "pancake theory" implies that this pattern is repeated all the way to the ground floor. Even if the initial "collapse" were to begin at the 80th floor, where the building was damaged, the 30 floors above the 80th floor would still need to "pancake" in order to get the entire structure to the ground. There is no evidence of 30 floors remaining intact, so we must assume, if we are to accept the Official Story, that there was "pancaking" also in the topmost floors, although we do not know what mechanism may have caused such a phenomenon.

F. The Visual Evidence

In the case of both WTC1 and WTC2, not only did we see no block of floors remaining intact at the completion of the event; but we saw virtually no accumulation of floors at all. Rather, the video and photographic record clearly shows a pulverization of the floors throughout the event (see Figure 10).

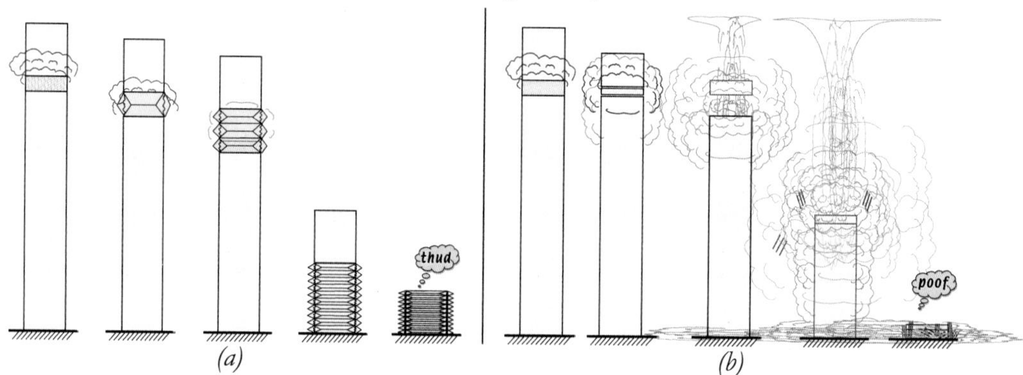

Figure 10. (a) Model A: The Floors Remain Intact and Pile Up like a Stack of Pancakes. (b) Model B: The Floors Blow Up Like an Erupting Volcano.

Thus, we cannot assume that the floors stacked up like pancakes.

Therefore, in our attempt to give credence to the Official Story, we will take the conservative approach. We will assume that a falling floor initiates the fall of the one below, while itself becoming pulverized. In other words, when one floor impacts another, the small amount of kinetic energy from the falling floor is consumed by (a) pulverizing the falling floor and (b) breaking free the next floor. In reality, there is not enough kinetic energy to do either.[14],[15],[16] But for the sake of calculating a "collapse" timing, let's assume that there were. After all, millions of people believe they saw the buildings "collapse."

In Figure 10 are two schematics. Model A represents a true "pancaking" of floors—that is, the floors remain intact and pile up like a stack of pancakes. Model B represents a disintegration of floors during the "collapse"—that is, the floors actually disintegrate from the top down. The question that now confronts us is: Which of these two models best matches the images below? Again, the question is, "Where did the towers go?"

Figure 11. There is little to no free falling debris ahead of the WTC2 "collapse wave."[17]

Figure 13a shows the rubble was not deep enough to reach the undercarriage of the black Cushman scooter, and, furthermore, the flagpoles in the background look full height. These facts show evidence of pulverization, but not of pancaking. All reports from "The Pile" confirm that, apart from the steel, nothing but fine powder remained.[18] The earthquake-induced collapse in Pakistan suggests how much more rubble and how much less dust should have been at Ground Zero if the "official" gravitational collapse story were true.

Figure 12. A Layer of uniform dust left by the "collapse" on Fulton Street.
Photo by Terry Schmitt, http://ken.ipl31.net/gallery/albums/wtc/img_1479_001.jpg

The point of these examples is that kinetic energy cannot be spent in diametrically opposite tasks; that is, it cannot be spent in "pulverization" *and* in "pancaking." But let us look closer at what the rubble-remains of a "pancaked" building failure actually look like. Figure 13b, shows the collapse of a building in Pakistan after an earthquake. The building has "pancaked," and there is still a distinctive appearance of floors in the rubble.

(a) WTC

(b) Pakistan earthquake

Figure 13. Comparison between WTC rubble and Pakistan earthquake rubble.
(a)http://reddit.com/info/iq0i/comments/ciqdw (b)Photo: Rolling Stone[19]

So, if there were enough kinetic energy for pulverization, there would be pancaking *or* pulverization, but *not both*. Energy can be spent only once. If the potential energy is spent to pulverize a floor upward and outward, it cannot be spent again to accelerate the building downward. In order to have pancaking, a sufficient downward force is required to trigger the failure of the next floor.

(a) My intellectual integrity prevents me from calling this a "collapse."

(b) The building turned into powder encountering nothing other than air.

Figure 14. Images Illustrating What Really Happened that Day.

(a)cropped:.http://911wtc.freehostia.com/gallery/originalimages/GJS-WTC28.jpg, (b)http://hyouhei03.blogzine.jp/tumuzikaze/images/2008/04/10/3.jpg,

But *if the building above that floor has been pulverized, then there can be no downward force.* As observed in the pictures below, much of the material has been ejected *upward and outward.* Any pulverized material remaining over the footprint of the building will be suspended in the air and therefore cannot contribute to a downward force slamming onto the next floor. With pulverization, furthermore, the small particles have a much larger surface-area-to-mass ratio, and thus air resistance becomes significant. As everyone can recall, the dust took many *days* to settle out of the air—not hours or minutes.

So, even though the mechanism to trigger the "pancaking" of each floor seems to be absent, let's nevertheless consider the length of time we would need to expect for such a collapse. Consider the following images (Figure 14):

To illustrate the timing for this domino effect, we will use a sequence of falling billiard balls, where each billiard ball triggers the release of the next billiard ball in the sequence. This is analogous to assuming that pulverization is instantaneous and does not slow down the process. (Note that the billiard balls are used as timing devices that are identical except for color. They are not intended to represent a kind of collision.)

G. Case 2: Progressive Collapse in Ten-Floor Intervals

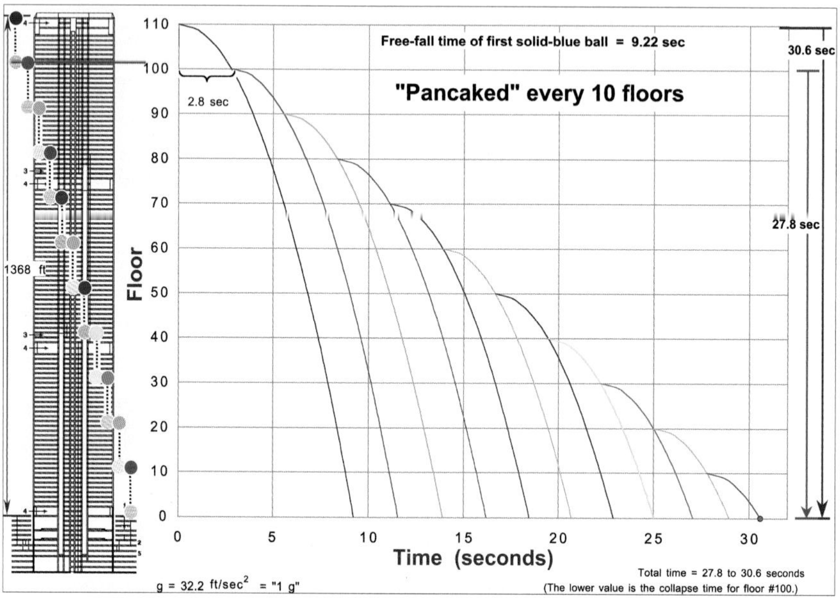

Figure 15. Minimum time for the collapse, if nine of every ten floors have been demolished prior to the "collapse."[12]

To account for the building's damaged zone, let's simulate the floor beams collapsing every 10th floor, as if something has destroyed 9 out of every 10 floors for the height of the building. This assumes no resistance within each 10-floor interval. That is, we are using the conservative approach that there is no resistance between floor impacts, and also that there was damage only to each 10-floor "package," not damage throughout the entire height of the building. Refer to Figure 15.

The clock starts when the blue ball is dropped from the roof (110th floor). Just as the blue ball passes the 100th floor, the red ball drops from the 100th floor. When the red ball passes the 90th floor, the orange ball drops from the 90th floor, etc. Notice that the red ball (at floor 100) cannot begin moving until the blue ball reaches that level, which is 2.8 seconds after the blue ball begins to drop.

This approximates the "pancaking" theory, assuming that each floor that's a part of the "pancaking" (collapsing) process provides no resistance at all. With this theory, no floor below the "pancake" can begin to move until the progressive collapse has reached that level. For example, there is no reason for the 20th floor to suddenly collapse before it is damaged.

With this model, a minimum of 30.6 seconds is required for the roof to hit the ground. Of course it would take longer if we accounted for air resistance. And it would take longer if we accounted for the structure's resistance to pulverization, since in actuality the columns at each level would absorb a great deal of the energy of the falling floors. Thus, if anything, our calculated collapse times are more generous toward the official story than they need to be.

H. Case 3: Progressive Collapse in One-Floor Intervals

Similar to Case 2, above, let us consider a floor-by-floor progressive collapse. Refer to Figure 16:

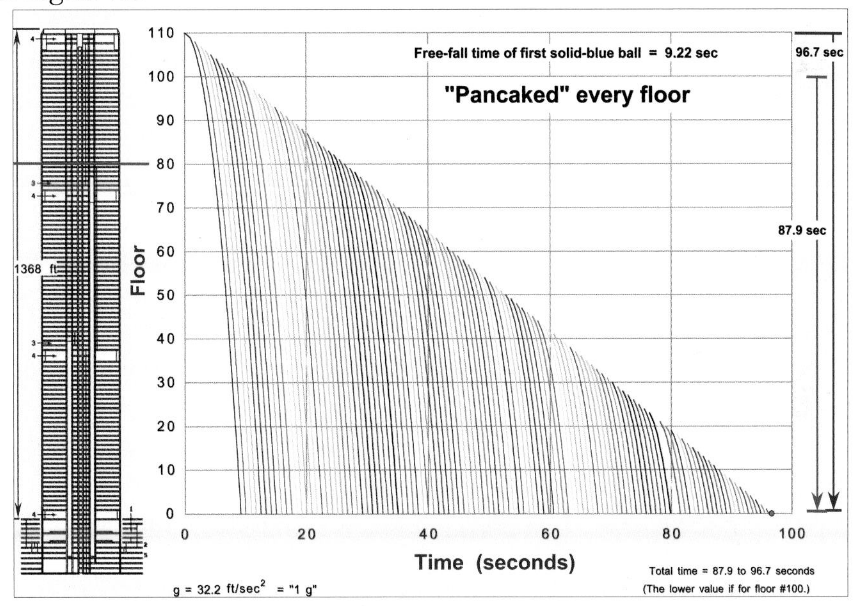

Figure 16. Minimum time for the collapse, if every floor collapsed like dominos.[12]

This figure shows that, if every floor were to pancake—in line with the official model of the "collapse"—then the time required would be approximately 100 seconds, one minute and forty seconds, almost two minutes, *not* the almost free-fall collapse rate of around ten seconds that all of us saw with our own eyes.

I. Case 4: A Progressive Collapse at Near Free-fall Speed

Now let's go on to consider a model where the same phenomenon happens, again once every 10 floors, but this time "with a little help." Consider the following chart.

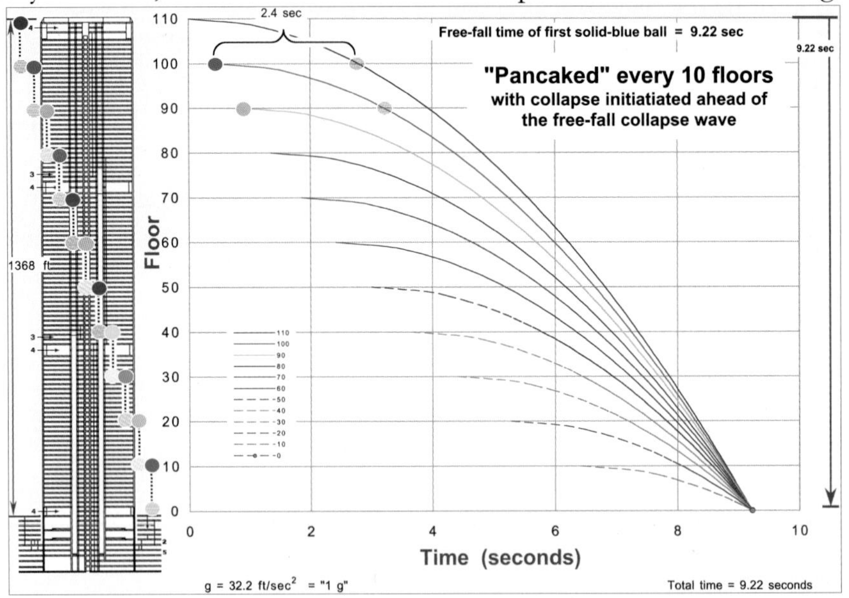

Figure 17. Minimum Time for a Billiard Ball dropped from the roof of WTC1 to hit the pavement below, assuming no air resistance.[12]

Let's say that we want to bring down the entire building in the same time a billiard ball would take in free-fall, in a vacuum, from the roof of WTC1 to the street (that is, 9.22 seconds). Now, if the entire building is to be on the ground in 9.22 seconds, each floor immediately below the "pancaking" floors must start moving *before* the downward-pressing energy in the "progressive collapse" reaches it. To illustrate this, again use the concept of the billiard balls. If the red ball (dropped from the 100th floor) is to reach the ground at the same time as the blue ball (dropped from the 110th floor), the red ball must be dropped 0.429 seconds after the blue ball is dropped. But, the blue ball will take 2.8 seconds after it is dropped just to reach the 100th floor in free fall. Therefore, the red ball will need to begin moving 2.4 seconds *before* the blue ball arrives to "trigger" the red ball's motion. That is, each of these floors will need a 2.4 second head start. But how can an upper floor be destroyed by slamming into a lower floor if the lower floor has already moved out of the way?

Nevertheless, for the building to be collapsed in about 10 seconds, each lower floor would *have* to start moving *before* a higher floor could reach it—and reach it *by the force of gravity alone.*

Is this what we all saw on the day when the buildings suddenly weren't any longer there? I believe it's very clear in the videos: The "wave" of collapse, progressing down the building, is moving faster than free-fall speed. It's unquestionable that bringing about a result like this would require something like a controlled, sequenced detonation. As we watched, the entire building was essentially turning to dust. Consider, almost at random, any middle-to-lower floor—the 40th, for example. The

40th floor needs to start moving before any of the higher floors have "free-fallen" to its point. But why would it start moving? There was no fire there to weaken it. And, if anything, there was an ever-*de*creasing load on it as the floors above it were turning to dust.

Figure 18. Horizontal plumes below the "collapse wave" in the North Tower during top-down collapse.
Photo by Richard Lethin: http://www.reservoir.com/extra/wtc/wtc-small.1056.jpg

In Figure 11, notice that WTC2 is less than half of its original height, and yet *no significant debris* has fallen ahead of the demolition wave. Now if we consider the seismic data that will be presented in Chapter 6, a related and significant question emerges:

Why does the ground rumble for *only 8 seconds* while WTC1 "disappeared?"

Answer: The part of the building shown in Figure 12 (dust and paper) does not make a thud when it hits.

This upper part of the building surely took a lot longer to hit the ground in the form of dust than it would have taken as much larger chunks of material. We know that sheets of paper have a very high surface-area-to-mass ratio and will stay aloft for long periods of time, which is why paper is an excellent material for making toy airplanes. The alert observer will notice that much of the paper here is covered with dust, indicating that this dust reached the ground *after* the paper did. In the photo shown in Figure 12, there are a few tire tracks through the dust, but not many, so the image was probably taken shortly after one (or both) of the towers had been destroyed. Also, the people in the picture look as though they've just come out of hiding, curious to see what had happened and to take photos. If there had been a strong wind blowing, it would have blown the paper away before the dust had had a chance to settle on it. As it was, paper fell more quickly, and the fact that much of the paper is covered with dust indicates the relative aerodynamic properties of this dust. That is, the high surface drag of the dust keeps it aloft longer.

Also, notice the darkness of the sky as well as the haze in the distance. The day was clear in the immediate NYC area, with no clouds in the sky—except for the dust clouds. This overcast appearance as well as the distant haze can be explained only by dust from the "collapse" still suspended in the air.

In a conventional controlled-demolition, a building's supports are knocked out and the building is broken up as it slams to the ground. In a conventional controlled-demolition, gravity breaks up the building. Here, it seems the only functional role gravity played was in gradually drawing the dust down out of the air.

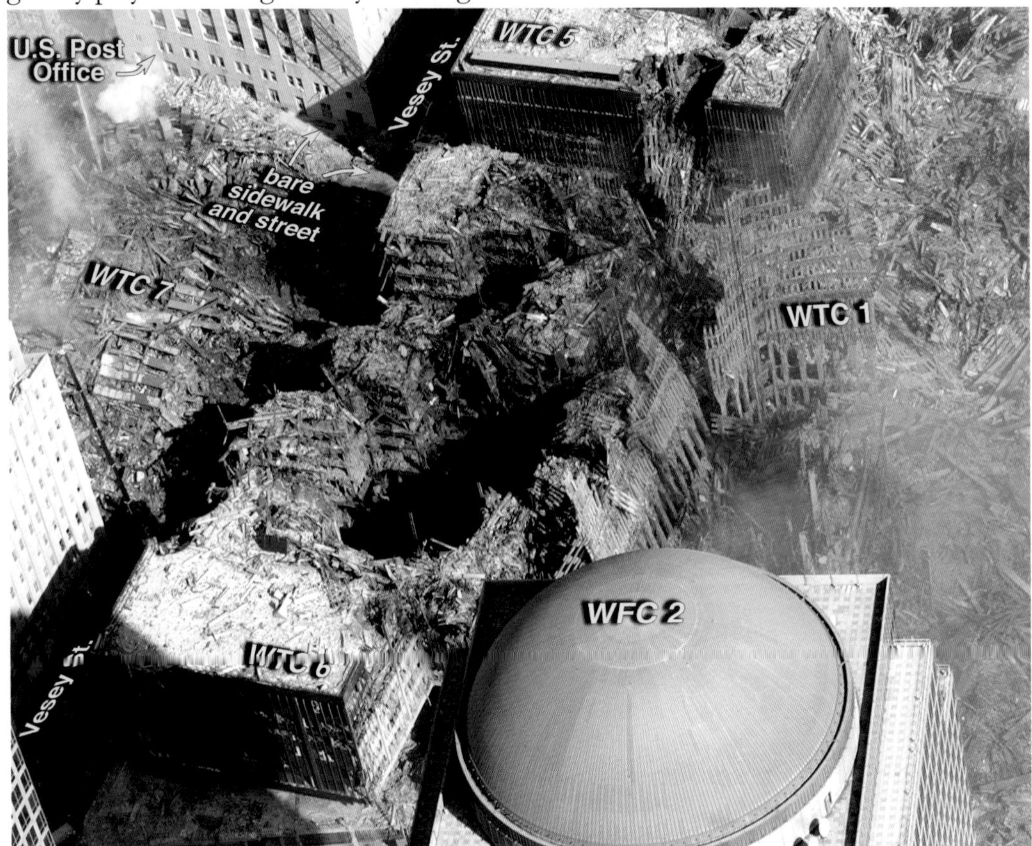

Figure 19. A view over the dome of WFC2 shows the damage to WTC6 in the center of the photo. To the left is the collapsed WTC7. Its debris stack is about five stories high.
To the right of WTC6 is the remaining north wall of WTC1, which leans toward WTC6, an eight-story building. Where did the wall go? Where did the top 102 floors of the north wall go? They did not fall on WTC6 or WTC7 because there are no steel "wheatchex" outer cladding there. Some of the core of WTC1 remains, but where is the rest of the core? The amount of steel on the ground barely covers the ground. If there is no wheatchex debris from WTC1 piled on top of WTC6, how did WTC6 get cored out from the top down?!
http://www.studyof911.com/gallery/albums/userpics/10002/132105581_a75a50d39a_o%7E0.jpg

Now, let's consider reality by way of some questions:

(1) How likely is it that all supporting structures on a given floor will fail at exactly the same time?

(2) If all supporting structures on a given floor did *not* fail at the same time, would that portion of the building tip over, or would it fall straight down into its own footprint?

(3) What is the likelihood that supporting structures on every floor would fail at exactly the same time *and* that these failures would progress through every

floor *with perfect symmetry*?

J. Conclusions

We have seen that if motion must be *restarted* at *every* floor, the total collapse time *must* be more than 10 seconds. Given that the building disintegrated from the top down, it is difficult to believe there could be much momentum to transfer, if any. We must also consider the energy required to pulverize the floor between each "pancake." After being pulverized, the surface-area/mass of each floor's composition-material is greatly increased and the air resistance becomes significant. How can this pulverized material contribute any momentum as it "hangs" in the air and floats down at a much, much slower rate than the "collapsing" floors?

In conclusion, the explanations of the collapse given by the 9/11 Commission Report and The National Institutes of Standards and Technology (NIST) are not physically possible[20]

Appendix A: Governing Equations

$$x_2 = x_1 + v_1 t + (1/2)\, g t^2$$

$$t = \sqrt{\frac{2h}{g}}$$

$$v_2 = v_1 + g t$$

where,

x_1 is the initial position
x_2 is the final position
v_1 is the initial speed
v_2 is the final speed
t is the time for this interval
g is the acceleration, the coefficient of gravity.

Appendix B: Conservation of Momentum and Energy

For those concerned about Conservation of Momentum and Conservation of Energy, consider

1. Conservation of Momentum.

The amount of momentum (p) that an object has depends on two physical quantities: the mass and the velocity of the moving object.

$$p_1 = m_1 * \mathbf{v}_1$$

Where p is the momentum, m is the mass, and v the velocity.

If momentum is conserved it can be used to calculate unknown velocities following a collision.

$$(m_1 * \mathbf{v}_1)_i + (m_2 * \mathbf{v}_2)_i = (m_1 * \mathbf{v}_1)_f + (m_2 * \mathbf{v}_2)_f$$

where the subscript i signifies initial, before the collision, and f signifies final, after the collision.

If $(m_1)_i = 0$, and $(\mathbf{v}_2)_i = 0$, then $(\mathbf{v}_2)_f$ must $=0$.
So, for conservation of momentum, there cannot be pulverization.

If we assume the second mass is initially at rest $[(\mathbf{v}_2)_i = 0]$, the equation reduces to

$$(m_1 * \mathbf{v}_1)_i = (m_1 * \mathbf{v}_1)_f + (m_2 * \mathbf{v}_2)_f$$

As you can see, if mass $m_1 = m_2$ and they "stick" together after impact, the equation reduces to,

$$(m_1 * \mathbf{v}_1)_i = (2m_1 * \mathbf{v}_{new})_f$$

or $\mathbf{v}_{new} = (1/2) * \mathbf{v}_1$

If two identical masses are colliding and sticking together, they will travel at half the speed as the original single mass. But in the case of the WTC, mass m_1 and m_2 did not stick together because there was pulverization.

2. Conservation of Energy:

In elastic collisions, the sum of kinetic energy before a collision must equal the sum of kinetic energy after the collision. This is not possible if some of the energy is used to pulverize the next floor and to fail the floor supports. So the collision is inelastic. Conservation of kinetic energy is given by the following formula:

$$(1/2)(m_1 * \mathbf{v}^2_1)_i + (1/2)(m_2 * \mathbf{v}^2_2)_i =$$
$$= (1/2)(m_1 * \mathbf{v}^2_1)_f + (1/2)(m_2 * \mathbf{v}^2_2)_f + (\textbf{Pulverize}) + (\textbf{Fail Floor Supports})$$

where (**Pulverize**) is the energy required to pulverize a floor and (**Fail Floor Supports**) is the energy required to fail the next floor.

If $(1/2)(m_1 * \mathbf{v}^2_1)_i + (1/2)(m_2 * \mathbf{v}^2_2)_i = (\textbf{Pulverize}) + (\textbf{Fail Floor Supports})$,

then $(1/2)(m_1 * \mathbf{v}^2_1)_f + (1/2)(m_2 * \mathbf{v}^2_2)_f = 0$,

so there will be no kinetic energy to transfer. But in reality,

$(1/2)(m_1 * \mathbf{v}^2_1)_i + (1/2)(m_2 * \mathbf{v}^2_2)_i < (\textbf{Pulverize}) + (\textbf{Fail Floor Supports})$, meaning the energy from the momentum is less than what is required to pulverize the concrete and fail the floor supports, so there certainly can be no excess energy to transfer.

So, for conservation of energy, we must assume there is some additional energy such that,

$$(1/2)(m_1{*}\mathbf{v}^2{}_1)_i + (1/2)(m_2{*}\mathbf{v}^2{}_2)_i + (\textbf{Additional Energy}) = (\textbf{Pulverize}) + (\textbf{Fail Floor Supports}),$$

where (**Additional Energy**) is the additional amount of energy needed to have the outcome we observed on 9/11/01.

Appendix C: Why not elastic collisions?

Assume that the top floor stays intact as a solid block weight, Block-A. Start the collapse timer when the 109th floor fails. At that instant, assume floor 108 miraculously turns to dust and disappears. Block-A can then drop at free-fall speed until it reaches the 108th floor. After Block-A travels one floor, it now has momentum. If all of the momentum is transferred from Block-A to Block-B, which is the next floor, Block-A will stop moving momentarily, even if there is no resistance for the next block to start moving.

$$(m_1 * \mathbf{v}_1)_i = (m_2 * \mathbf{v}_2)_f$$

If Block-A becomes dust and hangs in the air or is bounced upward after triggering the motion of Block-B, the mass of Block-A will not arrive in time to transfer momentum to the *next* "pancaking" between Block-B and Block-C. That is, there will be no mass to transfer the momentum. In other words, the momentum will not be increased as the "collapse" progresses.

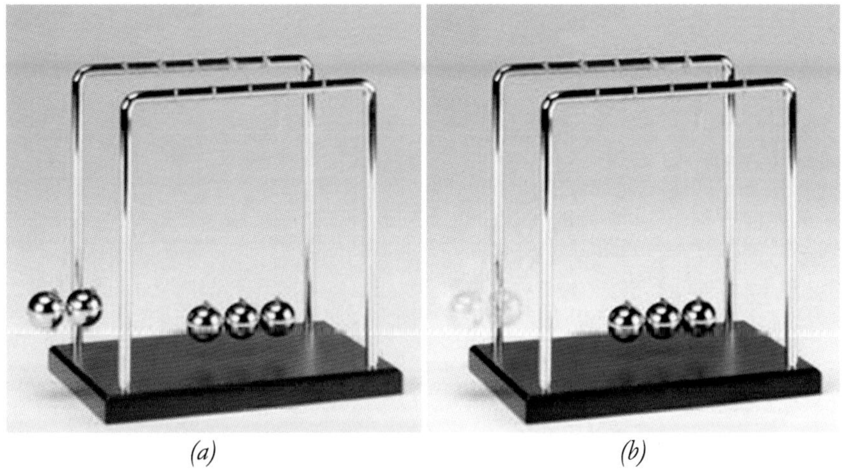

(a) (b)

Figure 20. (a) Conservation of Momentum if *there is* no *Pulverization and* no *Structural Resistance.*
(b) With Pulverization, there is no mass left to impact, so there is no momentum to transfer.

However, as we can observe, the building disintegrated from the top down and there was no block of material riding the pile down, much less an accumulation of blocks riding the pile down. Recall the physics demonstration shown in Figure 20. (I believe everyone who has finished high school has seen one of these momentum demonstrations at some point in their life.) The floors of the WTC had spaces between them. So visualize a separation between the steel balls in Figure 20a. If when the first ball hits the second, they turn to dust, there is no mass to strike the next, as represented by the ghost images in Figure 20b. If they are not there, they cannot transfer momentum.

---

[1] 8:46:40 by *The National Commission on Terrorist Attacks Upon the United States* (The 9/11 Commission), *http://www.911commission.gov/report /911Report_Ch1.htm*

[2] 8:46:26±1, 09:02:54 AM±2, From seismic data, *Lamont-Doherty Earth Observatory: http://www.ldeo.columbia.edu/LCSN/Eq/20010911_WTC/fact_sheet.htm*

[3] 08:46:30 by NIST, Page 37 (pdf page 87 of 298), *http://wtc.nist.gov/NISTNCSTAR1CollapseofTowers.pdf*

[4] *9/11 Commission Report, http://www.9-11commission..gov/report/index.htm,*

[5] *9/11 Commission Report*, Chapter 9, p. 305 of pdf. *http://www.9-11commission..gov/report/index http://www.9-11commission..gov/report /911Report_Ch9.htm, http://www.9-11commission..gov/report/911Report_Ch9.pdf.*

[6] "Fact Sheet," *Answers to Frequently Asked Questions*, The National Institutes of Standards and Technology (NIST), *http://wtc.nist.gov/pubs/ factsheets/faqs_8_2006.htm.*

[7] The height of the South Tower (WTC2) is 1362 feet, and the height of the North Tower (WTC1) is 1368 feet. *http://www.infoplease.com/spot/wtc1.html.*

[8] Seismology Group, Lamont-Doherty Earth Observatory, Columbia University. *http://www.Ideo/columbia.edu/LCSN/Eq/20010911_WTC/fact _sheet.htm*

[9] Other values: *http://911research.wtc7.net/essays/demolition/seismic.html; http://911research.wtc7.net/wtc/analysis/collapses/freefall.html#timeline.*

[10] *9/11 Commission Report, http://www.9-11commission..gov/report/index.htm,*

[11] The height of the South Tower (WTC2) is 1362 feet, and the height of the North Tower (WTC1) is 1368 feet. *http://www.infoplease.com/spot/wtc1.html.*

[12] *http://drjudywood.com/articles/BBE/BilliardBalls.html*

[13] Thomas Edgar, "The Collapse: An Engineer's Perspective," http://www.*pbs.org/wgbh/nova/wtc/collapse.html.*

[14] Wayne Trumpman (September 2005), *http://911research.wtc7.net/papers/trumpman/CoreAnalysisFinal.htm.*

[15] Jim Hoffman, *http://911research.wtc7.net/papers/dustvolume/index/index.html.*

[16] D.P. Grimmer, June 20, 2004, *http://www.physics911.net/thermite.htm.*

[17] *http://hereisnewyork.org//jpegs/photos/5245.jpg, http://www.studyof911.com/gallery/albums/userpics/10002/site1106.jpg, http://911research.wtc7.net/wtc/evidence/photos/docs/site1106.jpg,*

[18] *http://reddit.com/info/iq0i/comments/ciqdw.*

[19] *Rolling Stone*, Issue 988, December 1, 2005, p. 80.

[20] I gratefully acknowledge the comments and contributions to this chapter made by Morgan Reynolds, Jeff Strahl, and especially Alex Dent, for providing the initial motivation and encouragement to share this.

The chapter was originally orally presented in slightly different form as a paper, "The World Trade Center Towers as Bio-Inspired Structures: Characteristics of Their Design and Demise," for the 2006 Society for Experimental Mechanics Annual Conference at the Adam's Mark Hotel in St. Louis, Missouri, USA, June 7, 2006. It appeared subsequently as "A Refutation of the Official Collapse Theory" in James H. Fetzer, *The 9/11 Conspiracy: The Scamming of America* (2007), pp. 83- 100.

3.
The "Jumpers"
'It Was Like Raining People'

It's a rare person who wants to hear what he doesn't want to hear. —Dick Cavett

It was like raining people. —Michael Ober and Decosta Wright, Emergency Medical Technicians, and John Malley and Arthur Myers, Firefighters

Among the most horrific images from 9/11 is that of "The Falling Man,"[1] who came to represent the many people who fell to their death that day. These people are often referred to as "jumpers," but did they all, in fact, *jump*? And if they did, why did they do it? Once again, the question requires a closer look and examination.

Figure 21. The Falling Man, a photograph by Richard Drew, Associated Press.[2]

This falling man, thought to be Jonathan Briley, appears to be peaceful and relaxed. He does not look like someone in distress. But let us observe further.

Looking at these images can be difficult. It was too difficult for me until I realized that these people are communicating to us. They want us to hear them and they want their stories told. Once I realized this, I could not look away, for I had made them a promise to look at what they were trying to tell us. In this chapter I attempt to fulfill my promise to them.

A. Energized Launch

Figure 22. Leaving the building.
Cropped from original photograph by Jose Jimenez/Primera Hora via Getty Images, http://www.twin-towers.net/images/jumper11.jpg

This fellow in the orange shirt, shown in Figure 22, did not expect to be where he is and looks somewhat surprised to find himself there. He is energetically flapping his arms as if to regain his balance, like a gymnast on a balance beam or a last-resort attempt to fly. He did not expect to be here. Where did he come from and why is he here? From the information in Figure 23, he appears to be falling in front of the hole in WTC1. If he were jumping because of heat or smoke, he would not be flapping his arms so energetically. He doesn't look like someone who chose to jump. If he accidentally fell, why was he where he might fall? In short, he looks to me like someone who energetically launched away from something horrific without a choice.

Figure 23. The WTC1 hole.
by Nicolas Ciacca, http://www.911research.com/wtc/evidence/photos/docs/wtc_fires_dsnc1775A.jpg

It is not clear that the fellow in the orange shirt in Figure 24 is thc same fellow in the orange shirt in Figure 22, but it could be.

Figure 24. Relief.
http://www.twin-towers.net/images/wtc_jump4.jpg

Someone choosing to jump to their death would likely choose to carry something special with them, perhaps a photo of their loved one or even a briefcase or backpack. Yet it is striking that none of the people in these photos appear to be carrying anything, as if they were caught completely by surprise, perhaps on their way to or from the restroom. Consider this montage:

Figure 25. And there were many. All appear empty handed.
http://www.twin-towers.net/images/comp.jpg

And now consider *this* picture:

Figure 26. A "jumper" at some distance from the building.
http://911wte.freehostia.com/gallery/originalimages/005.JPG

The person shown falling here in Figure 26 is at a fairly substantial distance from the building. (A "mechanical floor" can be seen in the lower left corner of the photograph. A mechanical floor simply housed various mechanical plants. For each tower, floors 41-42 and 75-76 were mechanical floors.) So, he is at least 45 stories from the ground. The question is this: How did he get *this far* from the building?

Let us consider the world record for the standing broad jump, which was 11 feet 6 inches, set in 1906 by Ray C. Ewry of the New York Athletic Club.[3] For an estimate of Ewry's average speed, let's assume that he rose to a height of 2.5 feet from the ground and fell back to the ground at free-fall speed. A standing broad jump might take place at an average speed of 10 mph. Now, if the fellow in Figure 26 had jumped from the 105th floor with a launch speed of about 10 mph in the horizontal direction, it might have been possible for him to have moved approximately 100 feet from the building by the time he passed the 45th floor. That is, *a very energized standing broad jump could carry someone 100 feet from the building while dropping 60 stories.*

But if he jumped because he was too hot, would he not have been too lethargic to launch that far out from the building? If he were suffering from smoke inhalation, would he not just *stumble* out the window, providing he did not lose consciousness first? He also appears to have a piece of clothing part way off that is dangling behind him. Why would someone begin to disrobe before jumping? We don't see photographs of *any* who tumbled *against* the building. *This implies an energized launch by all the jumpers,*

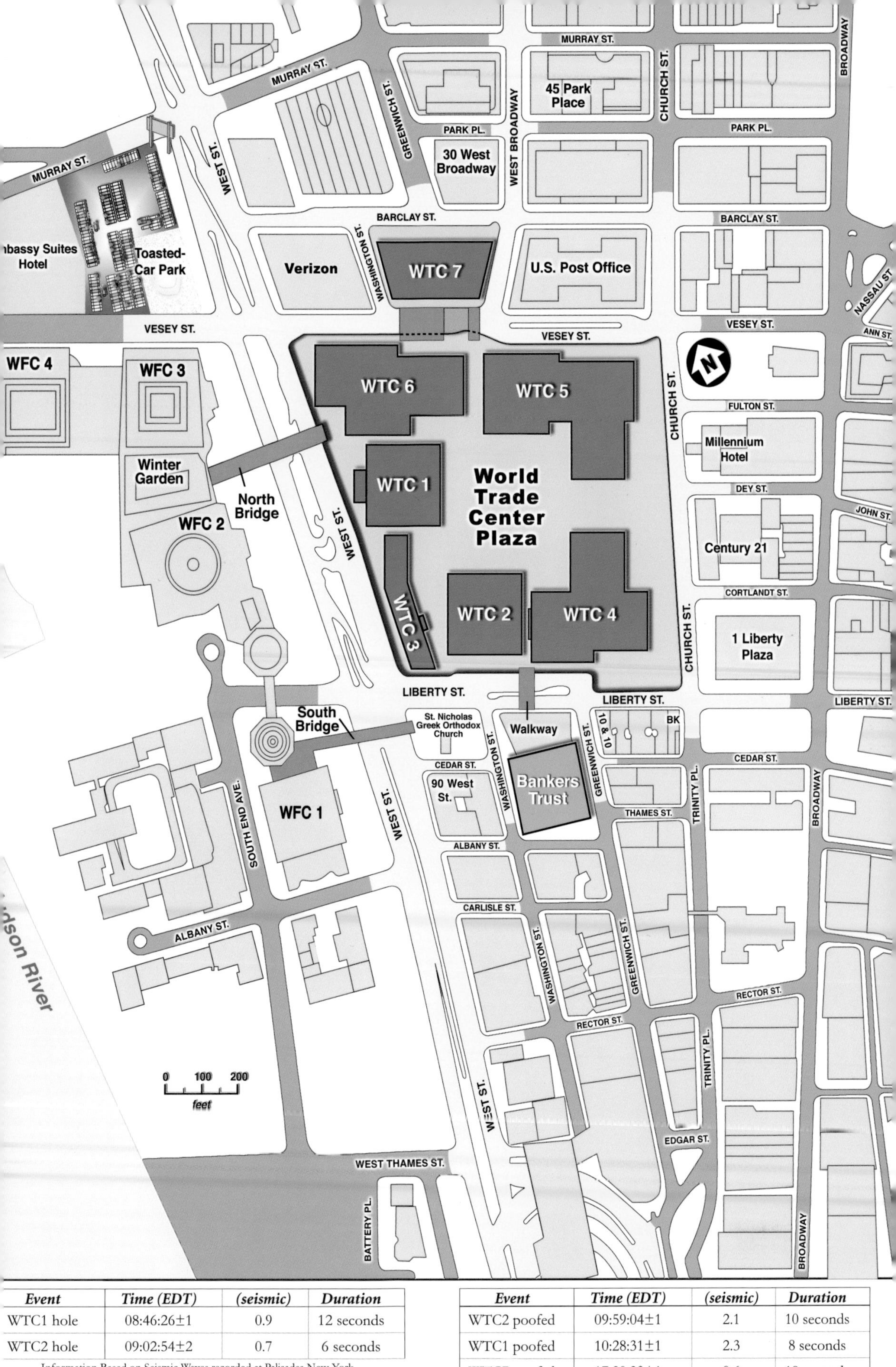

Event	*Time (EDT)*	*(seismic)*	*Duration*
WTC1 hole	08:46:26±1	0.9	12 seconds
WTC2 hole	09:02:54±2	0.7	6 seconds

Information Based on Seismic Waves recorded at Palisades New York *Seismology Group*, Lamont-Doherty Earth Observatory, Columbia University.

Event	*Time (EDT)*	*(seismic)*	*Duration*
WTC2 poofed	09:59:04±1	2.1	10 seconds
WTC1 poofed	10:28:31±1	2.3	8 seconds
WTC7 poofed	17:20:33±1	0.6	18 seconds

Building	*Stories*
WTC1	110
WTC2	110
WTC3	22
WTC4	9
WTC5	9
WTC6	8
WTC7	47
WFC1	40
WFC2	44
WFC3	54
WFC4	34
Bankers Trust	41
One Liberty Plaza	54
Millennium Hotel	59 (47)

unless wind currents immediately pulled them away from the building. Before we can address these matters, we need to examine something else.

Figure 27 shows a close-up view of the northwest corner of WTC1 some time before its total destruction. You can see people hanging on the outside of the building. The fumes emerge from the building and appear to blow away from this corner in a horizontal direction, like hair being parted in a stiff breeze. Figure 28 shows the western face of WTC1 shortly before its total destruction. Several people can be seen hanging onto the outside of the building near the northwest corner, near the upper left of the left face, indicated by the arrow. They appear to be above the 100th floor and upwind of where the apparent fires are, so they should have had fresh air. The fumes to the right of the arrowhead in Figure 28 appear to be streaming upward at about a 45-degree angle from the horizontal, not horizontally, as they do on the right side of Figure 27. It is curious why the fumes appear to be emerging differently in these two photographs.

Figure 27. A close-up view of the northwest corner.
http://www.911truth.ch/img/911%20wtc%20people%20.jpg

When Jack Gentul was interviewed in the documentary, *The Falling Man.*[4], he told of the last conversation he had with his wife, Elaine Gentul. Elaine worked on the 97th floor of WTC2 and apparently had decided, like her colleagues, to put wet clothes over her head and try to get out of the building. Jack Gentul said that her breathing sounded very labored and she was concerned about what she thought was heavy smoke. So he said to her, "Call me when you get down." Soon after that, however, Elaine ended up on the street. Jack Gentul shares his thoughts.

> I know that Elaine was found on the street in front of the building across from hers. So, whether she jumped or fell, I don't know. I believe she was alive when it happened because of that phone call. I hoped that she had succumbed to the smoke, but it doesn't seem likely. It's something I can't know.[5]

Elaine Gentul had no intention of dying and every intention of finding some way to survive the unfathomable and appalling position she found herself in. My own instinct, too, would be to wrap myself in wet clothes and then make a run for it. What went wrong?

Figure 28. The Western Face of WTC1.
http://911wtc.freehostia.com/gallery/originalimages/GJS-WTC27.jpg

B. Necessity

There were many people hanging out of the building, but one in particular, on the 105th floor of WTC1, caught my attention. This person appears to be hanging outside of the building directly above the middle of the airplane-shaped hole. The upper-right silhouette-cut-out of the "wingtip" is visible in the Figure 29.

A closer view of the top area of Figure 29 is shown in Figure 30. Upon closer examination, it can be seen that many of the people in this photo have disrobed. In particular, notice the fellow in the upper-left region of this photo. He appears to be hanging by one hand and one foot while attempting to remove his pants. Why is it so important for him to hang so many stories above the ground and remove his pants? He has not yet jumped, so it appears that he wants to live. Hanging by one hand and foot outside a window at approximately the 105th story while removing his pants indicates just how important he must have felt it was to take them off. But *why*? And why in so *precarious* a place? If he wanted merely to avoid the smoke, why didn't he simply hold his breath, duck inside the window, take his pants off, and pop back out (assuming he has some peculiar desire to dangle 100 stories above the ground to begin with)? If, on the other hand, he were trying to avoid the heat, why would he take off his pants, since they would serve as a protective layer against it, like a mitt someone uses to flip burgers over a grill?

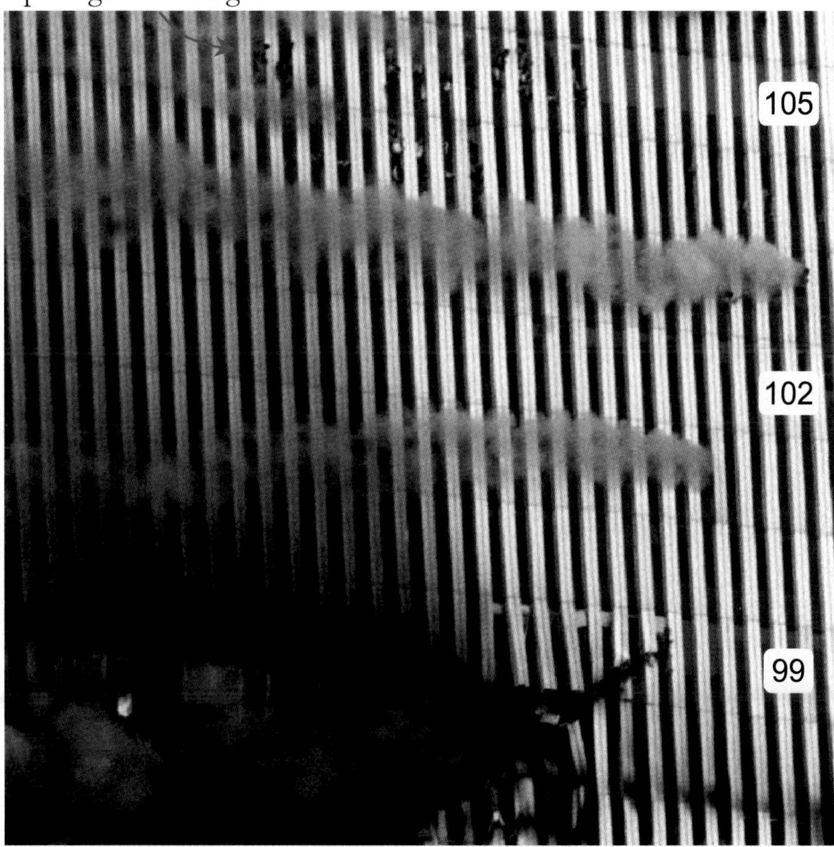

Figure 29. The people can be seen on the north face of WTC1.
Photo by Jeff Christensen/REUTERS, Figure 8-4 p.112, file page 150 of 298, http://wtc.nist.gov/NCSTAR1/PDF/NCSTAR%201-7.pdf

We were told that people were trapped on these floors due to fire and smoke. If I were on the 105th floor and there was fire and/or smoke, I might, as already said, go to the restroom or water fountain to wet my clothing and then make a run for it. And if there were a fire, the sprinklers would probably already have been activated, causing anyone in the room to become wet. But if there were no water, it is likely to have become quite warm. Thus it is a reasonable assumption that the clothing of the people on those floors would either have been made wet from the fire sprinkler system or at least damp because of profuse sweating. Either way, when looking at the picture of the people dangling outside the building in Figure 30, we may reasonably conclude that they were doing so because of some necessity, and also that they were removing their clothes because of some necessity.

At this juncture we must introduce a hypothesis in an attempt to explain this strange behavior. Consider what might be expected if some sort of energy field, such as a microwave field, had been affecting that area just inside the building. Such a field might be part of what comprises the Active Denial Systems (ADS) that are now being used for crowd control. It is equally possible that such a field was part of whatever was destroying the building. In either case, *wet clothing intensifies the pain caused by such microwaves,* as is acknowledged in an article about the ADS: "Wet clothing might sound like a good defense, but tests showed that contact with damp cloth actually intensified the effects of the beam."[6]

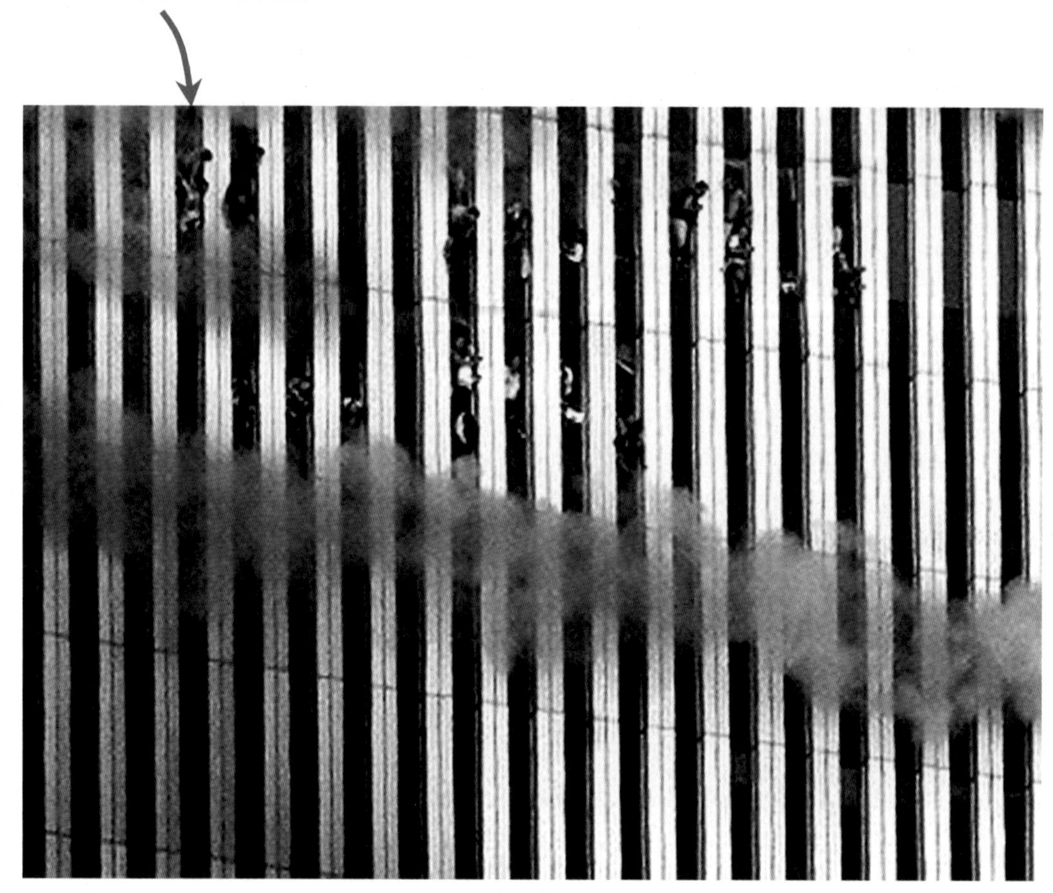

Figure 30. (9/11/01) The People.
Photo by Jeff Christensen/REUTERS
http://911wtc.freehostia.com/gallery/originalimages.0006.JPG, (Original at *http://digitaljournalist.org/issue0110/images/m05.jpg)*

Thus, the actions of people appearing to disrobe while hanging outside of the building are consistent with there being an energy field contained within the walls of the building. If you place your hand on a red-hot burner of a kitchen stove, your hand flies off of it. You do not think about it; it is an instinctive response. Similarly, perhaps few if any of those people chose to jump. They just flew out of the windows. In Figure 21, Jonathan

Briley looks relaxed and at peace, as though he just solved the worst problem of his life—and just before he realized the next one, one that he would not be able to solve. The person in Figure 26 seems to have been highly motivated to distance himself from the building, much like the hand that flies off a hot burner. But it is quite clear that *with the people hanging from the building and taking off their probably wet clothes, we have an indicator of some other factor in play besides just extreme heat and/or smoke.* In the area where these people are hanging outside of the building, there is no apparent fire or smoke emerging from the windows. There is, however, a stream of hazy fumes flowing from we know not where.

C. Stepping Around the Bodies

As we evaluate this possibility of an energy field, another important question occurs: What did eyewitnesses observe that day? Did they see an airplane hit the tower? Did they hear an airplane? Did they see airplane parts? After 9/11, along with a number of my colleagues, I began to look at some of the first-responder transcripts (discussed elsewhere in this book). Reading through them gave me a sense of seeing through the eyes of those first responders. I began to feel that I was able to see what they had seen and to go where they had gone. Throughout my reading, I kept looking for evidence that an airplane had crashed into the building. Many first respondents did not even realize that anything had happened to WTC2 until they were *told* an airplane had hit it, or until they saw an explosion. Several assumed that clothing on the street or sidewalk must have come from luggage on the plane; perhaps they were in search of some kind of evidence that would verify that a plane really had crashed into the building.

While reading through the first responder transcripts, hoping to find evidence of plane parts on the ground, I found descriptions of bodies that had fallen, body parts, and descriptions of people continuing to rain down from above. It was a horrible scene, and I did not want to see it again through reading. So I tuned it out and "stepped around" these descriptions, much in the way the first responders themselves had seemingly stepped around the bodies. If I looked at them and dwelt on them, I would not be able to do my job. Then I asked my colleagues what they thought about these hideous scenes. Two of them said that the transcripts they had read just did not happen to contain these descriptions, while another found himself stepping around the body parts as I had.

At this point, I decided to look at exactly what I had been stepping around and blinding myself to. It became clear to me what the effect of that scene must have been on the people who were there, or on any other caring person. Fire fighters and emergency medical responders desperately care about people. It is why they do what they do. From my own life experiences, I believe that feeling helpless has got to be one of the most painful human responses one can experience.

> …We walked into there, on the way into the building, there was people [sic, et passim] running around, people that was jumping, landing on the grounds. Whatever building we walked into there was a glass canopy, foyer type thing

where from the street you can see into the building. As we were walking into the building, we just heard a huge bang. As we looked up, it was someone who had jumped, God knows how far up. We got into the lobby of the building and it was like chaos, there were people everywhere. Once we got in there, I don't remember exactly who it was, I think it was one of the fire Chiefs who said "You know it's not safe, we gotta get outside." So we left to go outside, and as we stepped out of the building, *it was like, raining people.* People were just jumping from everywhere. Just all over it was bodies and parts just scattered. We walked across, I believe it was West Street, and we set up the command post over there next to the fire command post, just trying to get everything in order. At the same time we're trying to watch to see what's going on. The only thing that was going on was you could see the buildings burning and people just jumping. You could watch them fall from like the 90th floor all the way down. *It's like you go to school for so long to be able to take (care) of people and treat them and be able to fix them when there's something wrong with them, and there's nothing, they hit the ground, and that's it.*[7]

Or consider how Captain Karin Deshore "tuned out" the raining people:

...I spoke with paramedic Charl from Flushing and I said to him did you see that, he said what. I said just look up to your left. Here the bodies kept coming out of both buildings. Some of them were on fire. Some of them were moving, others were not moving and the worst part was as they hit the ground they would go like a *splush* sound. You could just see the whole body would just disintegrate into pieces and splatter all over and the sound and I saw a couple of them do that, *was just enough to make me tune out to that.*

I became more concerned with everybody in my contingency not getting hit by falling debris.[8]

Or consider Fire Marshal Steven Mosiello's statement:

What was I thinking (about) all the jumpers and everything else.

I believe at that point when I went over to get the Commissioner, *that's when the fireman of 216 was killed by a jumper when a jumper landed on him.*

I went to the building, looked for the Commissioner, and they said he had just left. So now we've crossed each other's paths. I came back to the command post across the street at Two World Financial, and the Commissioner was there talking to the Mayor.

Q. Can you describe what the plaza and the area outside looked like at that point in time?

A. Outside number One World Trade, there were jumpers. *There was a tremendous amount of bodies on the ground.* They were hitting at a rate of probably *one every 30 or 40 seconds* onto the glass atrium that was there, which is a distinct explosion-type sound when they hit, as was—I think you can close your eyes and you knew when a jumper hit the ground over there. It was very distinct. *After you saw enough of them, you just stopped looking.* [9]

Moseillo, a little later in his interview, corroborates the idea of a "rain of people":

Q: Got it.

That firefighter, by the way, just for the record, is firefighter Suhr, S-U-H-R, who was hit by a jumper.

> A. Right.
>
> I was on my way out of that building when he was hit by a jumper. There was a company coming. I don't know the company. They were walking next to the building, and I yelled at the lieutenant to get in the street with his men because look what just happened. There were jumpers coming down. He probably wasn't very aware of it at the time. Probably like everybody else he was scared and he wasn't thinking either.[10]

We have, in other words, a "rain of people," some of whom may be trying to escape smoke and heat, though not *all* of them are. *Some* of them are trying to escape something else.

D. The "Jumpers" Body Count

But *how many* people "jumped" that day? As noted above, Steven Mosiello estimated the rate of "jumper" impacts on the ground as occurring every 30-40 seconds. But firefighter John Malley had a different estimate:

> ...So the men went to the garage, and the officer went to the command post. We stood there and watched everybody jumping and waiting for our assignments, for our officer to come back.
>
> People started to jump with sucha [sic]—it was maybe one jumper every five second [sic] at one point, every ten seconds. *Then they just started jumping like one every second, two seconds. There were people just coming down like* it was raining people.[11]

There are thus various estimates for the "rain of people" that was seen. Let's say there were three people per minute "falling" from each face of WTC1. That would be 12 people per minute, or 1200 people who fell during the 102-minute event. Add to that 421 other victims (including 343 firefighters), and the total comes to 1,621. In some places it was reported that 1,594 victims had been identified by DNA and 124 by other means. It appears that the "jumpers" and firefighters could account for all of those thus identified.

The horror of it all was captured by firefighter Arthur Myers in his interview:

> Right after that then you see live people jumping. This is the first time I've ever seen people jump like this in my whole career.
>
> Q. 20 years.
>
> A. In 20 years, this is the first time I've ever witnessed this, and it was just blowing my mind.
>
> The chauffeur from 3 Engine, he was telling me, listen, don't look, just don't—I said, "How can I not look? I've never seen this before." *Just any time you thought that would be it, then you'd see more waves of people coming.* It was like raining people. *You could hear when they hit the ground, bang, bang, and the body parts just dismantling all over the place.*
>
> *At that time it just got to me. I turned around to look away from it, and I'm saying to myself these are people. Man, there are people dying here. I couldn't believe what I was seeing.*
>
> When I turned around, someone—they said the chief ordered them to move 39 Engine, my rig wasn't there. I said, "Where the hell is 39?" It's like somebody stealing your car. What they did was they took our rig and put it right in front of

> the north tower.
>
> I said, "Wait a minute. They're full of shit. They're not going to leave the rig there." [12]

Paramedic Gary Smiley captures the stunned reaction of one of his co-workers:

> ...(Emergency Medical Technician) Felton, Sean Cunliffe. I remember Felton because we started to notice people coming off the building, coming off the north tower and at first we didn't know what was going on, and then you could see that they were jumping. In fact I had to actually hold Felton because he wanted to run across the street and catch them. I told him that he couldn't do that. They were hitting the ground, of course, exploding. It was horrible. We must have—we probably saw about a dozen people jump.[13]

The view Firefighter Bertram Springstead had overlooking the Vista Hotel for five seconds was more than he needed to see.

> I remember somebody said, "You think you're having a bad day? Take a look out this window." We looked out the Trade Center window, and there was the Vista Hotel, I guess it was there. I'm not really sure what building I was looking at, but I'm pretty sure it was the roof of the Vista. There had to be 30, 40 jumpers sprayed out all over the roof. I went, "Oh, Jesus, what the hell is going on here?"
>
> As I was looking out the window, which is a total of five seconds, another jumper comes by, *kind of like clipped the edge of the roof and just vaporized. The guy just disappeared. There was no longer a body, just a big cloud of red.*
>
> Q. Wow.
>
> A. I was like, "I didn't need to see that." A total of five seconds I was looking out that window, total.[14]

In her interview, emergency medical technician Decosta Wright also corroborates the "rain of people":

> ...So after that, after the second plane hit, *it was just raining people, people were just jumping.* When the first one hit, they were just jumping periodically. You would see one man jump out. I was in—in our position, we could actually see when they hit the ground *and I seen body parts just going everywhere when they hit the ground, so when one guy hit, all you hear was boom, then you see his arms and legs, just flew right off.*
>
> Then right after that I see a woman, she was just spinning in the air. After that, when the second plane hit, that's when my partner was like—he couldn't believe my reaction, because all I was saying, all these people that was [sic] jumping out. I was just saying oh, my God, oh my God. I couldn't believe—I was like, oh, my God. I couldn't believe. It was so unreal.[15]

But the grizzliest image conjured in these testimonies is that of firefighter Kevin Martin, which is cited here to show that this "rain of people" was literally covering the ground:

> Going up West Street there were other companies, but we were distracted, there were just bodies all over the place and parts and it was just clothes and flesh. I remember a guy getting off the rig and stepping, like he was stepping—I thought he stepped in dog manure. It was bodies. Yes, I'm done.[16]

What are we to make of all this? Clearly, the behavior of the jumpers is an indicator that something else besides heat and smoke were involved in their "decisions" to jump from the buildings, if in fact they even jumped, for they were being exposed to something truly horrific. Their behavior is consistent with people exposed to an energy field, such as the Active Denial System, which uses a microwave-energy field for crowd control.

---

[1] "The Falling Man" is a story about a photograph taken by Richard Drew at 9:41:15 a.m. on September 11, 2001. The Falling Man in the photograph was unofficially identified as Jonathan Briley, a 43-year-old who worked as a sound engineer for the "Windows on the World" restaurant. Source: http://en.wikipedia.ord/wiki/The_Falling_Man.

[2] *http://www.esquire.com/cm/esquire/images/fallingman-lg.jpg, http://www.esquire.com/features/ESQ0903-SEP_FALLINGMAN,*

[3] August 5, 1906, *http://query.nytimes.com/gst/abstract.html?res=9402E0DD1F3EE733A25756C0A96E9C946797D6CF, EWRY'S WORLD'S RECORD IN STANDING LONG JUMP.*

[4] 9/11 The Falling Man, Narrator - Tim Hopper. 75 mins. 3/16/06, 0:10:30 - 0:15:00, *http://www.youtube.com/watch?v=BXnA9FjvLSU*

[5] 9/11 The Falling Man, Narrator - Tim Hopper. 75 mins. 3/16/06, 0:10:30 - 0:15:00, *http://www.youtube.com/watch?v=BXnA9FjvLSU*

[6] "Say Hello to the Goodbye Weapon," *Nonlethal Weaponry – Sandia Researchers Willy Morse and James Pacheco fine-tune the small-sized Active Denial System. http://www.wired.com/science/discoveries/multimedia/2006 /12/72134.*

[7] File No. 9110093, *World Trade Center Task Force Interview*, "EMT Michael Ober," October 16, 2001, pp. 3-4, *http://graphics8,nytimes.com/packages/html/nyregion/20050812_WTC_GRAPHIC/9110093.PDF*

[8] File No. 9110192, "Captain Karin Deshore," November 7, 2001, pp. 6-7. *http://graphics8.nytimes.com/packages/html/nyregion/20050812 _WTIC_GRAPHIC/9110192.PDF*

[9] File No. 9110141, "Fire Marshal Steven Mosiello," October 23, 2001, pp. 6-7, *http://graphics8.nytimes/com/packages/html/nyregion20050812_WTC_GRAPHIC/9110141.PDF*

[10] Ibid., pp. 12-13.

[11] File No. 9110319, *World Trade Center Task Force Interview*, "Firefighter John Malley," December 12, 2001, p. 3. *http://graphics8.nytimes.com/packages/html/nyregion/20050812_WTC_HRAPHIC/9110319.PDF*

[12] File No 9110052, *World Trace Center Task Force Interview*, "Firefighter Arthur Myers," October 11, 2001, pp. 5-6, *http://graphics8.nytimes.com/packaes/html/nyregion/20050812_WTC_GRAPHICS/9110052.PDF*

[13] File No. 911039, *World Trade Center Task Force Interview*, "Paramedic Gary Smiley," October 10, 2001, p. 10. *http://graphic8.nytimes.com/packages/html/nyregion/20050812_WTC_GRAPHIC/9110039.PDF*

[14] File No. 9110225, "Firefighter Bertram Springstead," December 4, 2001, pp. 7-8, *http://graphics8.nytimes.com/packages/pdf/nyregion/20050812_WTC_GRAPHIC/9110225.PDF*

[15] File No. 9110054, *World Trade Center Task Force Interview*, "EMT Decosta Wright," October 11, 2001, p. 3, *http://graphics8.nytimes.com/packages/html.nyregion/20050812_WTC_GRAPHIC/9110054.PDF*

[16] File No. 9110232, *World Trade Center Task Force Interview*, "Firefighter Kevin Martin," December 5, 2001, p. 10. *http://graphics8.nytimes.com/packages/html/nyregion/20050812_WTC_GRAPHIC/9110232.PDF*

4.
Magic Shows and The Power of Suggestion

We are never deceived; we deceive ourselves. —Johann Wolfgang von Goethe

You didn't hear it. You didn't see it. You won't say nothing to no one ever in your life. You never heard it. Oh how absurd it all seems without any proof. You didn't hear it. You didn't see it. You never heard it, not a word of it. You won't say nothing to no one. Never tell a soul what you know is the Truth.
—The Who, 1921-Tommy[1]

A. Unbelievable

Figure 31. Once upon a time there were two towers.[2]

Figure 32. (9/17/01) And then they went away.[3]

On 9/11, we heard it said over and over again, "It's unbelievable!" Yes, what we were shown and what we were told were unbelievable. So, how is it that the unbelievable becomes believable?

On the morning of 9/11, there were two towers, each over a quarter-mile tall and made of 500,000 tons of material. And then they went away. They were gone in a mere 8 to 10 seconds each.

B. We Know What We Saw—Or Do We?

It has been said, "The American people know what they saw with their own eyes on September 11, 2001."[4] But do they? Most of us were shown images on TV and were told what we saw.

When we go to see a magic show, we're "told"—whether literally or by some process of implication—what we are to see, and so we interpret what we see as unbelievable. But we know we've gone to see a magic show. David Copperfield has performed illusions on TV that appear to cause the Statue of Liberty *and* a freight train to disappear. But we know these are magic shows.

When *War of the Worlds* aired in 1938, some people missed the opening disclaimer letting them know that the show was fictitious. Those who missed the opening thought it was real. Why?

> *The War of the Worlds* was an episode of the American radio drama anthology series *Mercury Theatre on the Air*. It was performed as a Halloween episode of the series on October 30, 1938 and aired over the Columbia Broadcasting System radio network. Directed and narrated by Orson Welles, the episode was an adaptation of H. G. Wells' novel *The War of the Worlds*.
>
> The first two thirds of the 60-minute broadcast was presented as a series of simulated news bulletins, which suggested to many listeners that an actual Martian invasion was in progress. Compounding the issue was the fact that the Mercury Theatre on the Air was a 'sustaining show' (i.e., it ran without commercial breaks), thus adding to the dramatic effect. Although there were sensationalist accounts in the press about a supposed panic, careful research has shown that while thousands were frightened, there is no evidence that people fled their homes or otherwise took action. The news-bulletin format was decried as cruelly deceptive by some newspapers and public figures, leading to an outcry against the perpetrators of the broadcast, but the episode launched Welles to fame.
>
> Welles's adaptation was one of the Radio Project's first studies.[5]

> *The War of the Worlds* (1898), by H. G. Wells, is an early science fiction novel which describes an invasion of Earth by aliens from Mars. It is one of the earliest and best-known depictions of an alien invasion of Earth, and has influenced many others, as well as spawning several films, radio dramas, comic book adaptations, and a television series based on the story. The 1938 radio broadcast caused public outcry against the episode, as many listeners believed that an actual Martian invasion was in progress, a notable example of mass hysteria.[6]

But what has *The War of the Worlds* got to do with 9/11?

My teaching schedule for Fall 2001 began late in the day and included a long evening class on Tuesdays. Being somewhat of a night owl, I set my alarm for sometime after 9:00 AM. On 9/11, I happened to wake up before my clock-radio alarm came on and was working in the next room on my computer when the clock radio finally did come on. I vaguely heard the radio in the background and kept on working, though I wasn't paying much attention to what was being said. Then I began to notice that whatever story they were telling seemed to go on and on. It sounded as though they were telling some sort of sick joke about a drunk driver who made *a wrong turn* on the way home from a party and ran his plane into a building. But then the same thing seemed to have happened *a second time*. It sounded like a joke, except that they never got to the punch line. Eventually, I began to wonder if what they were saying might have something to do with reality. So I went downstairs and turned on the TV.

The same story was being played on every station I could pick up. There had indeed been a horrible event. But why were we getting the same story on *every* station? Then I realized also that there were no commercial breaks. That meant that this must be serious. But whatever happened to the idea of getting another perspective? When there's a plane crash, one network might be replaying the video of it while another is playing an "up close and personal story," or a "background story" about someone's loved one who left home that morning to go on a trip, or about some parts manufacturer of the aircraft, or what detailed experience some innocent bystander went through. But here we had every station playing *nearly the same video and in nearly the same way, saying essentially the same thing, over and over again without a commercial break*. To me, it did not feel right. (This is when I called my mother in the Washington, D.C., area to ask if she saw fighter jets overhead.) The mere fact that there were no commercial breaks commanded my attention as if to say, "This is more important than anything you've ever known." That is, like *War of the Worlds*, it was a *sustaining show*.[7]

How many stories have aired without commercial breaks, and what effect does that have on us?[8]

I gathered my things together and headed into campus. It must have been around 10 AM. When I got to campus and was walking up to our building, one of my graduate students met me in the street. That conversation is one I will never forget. He asked, "Dr. Wood, who is Bin Laden and what's Al-Qaeda?"

I entered our building and went up the corner stairs. My colleagues were talking about what they thought had taken place. One said, "After the USS *Cole*[9], we should have just taken them all out. These guys need to be taken out." I asked, "For what?" My colleague responded, "The towers. They're down, they're both down." Then we all went to the faculty conference room to watch the replays on TV.

What I saw was surreal. These buildings did not just "collapse," they *unraveled*—as I've said before, like sweaters. Something did not smell right, but here were my colleagues (full professors who *should* have recognized the apparent contradiction of physical principles already being put out as "the story"), with *pitchforks* in hand, rallying the troops to "go get the bad guys." It was like a done deal, open-and-shut. Meanwhile, there I was, looking at the TV monitor and thinking there was a kind of *War of the Worlds* sick joke being played on us.

If there had been commercial breaks during *War of the Worlds*, would there have been such a panic?

If there had been commercial breaks during news coverage on 9/11 would people have questioned it more?

On September 11, 2001, we were told that an airplane hit a building and caused the building to self-destruct an hour later, taking just 8 to 10 seconds to "collapse." But, as we have seen, a gravity-driven collapse in 8 to 10 seconds is physically impossible, no matter what might have initiated the gravity collapse (bombs, natural causes, and so on). In addition, we were told that two airplanes, each hitting one of the Twin Towers, had caused the total destruction of the entire complex of seven buildings—while not significantly, fatally, or totally damaging any *other* buildings! The explanation we were given was that jet fuel had ignited office material and that this fire, fed by burning office material, significantly weakened the steel-frame buildings. But steel fireplace grills don't collapse from fire.

Nearly everyone has heard of kerosene heaters. Do they *melt*?

Figure 33. Portable Kerosene heater.
http://www.amazon.com/Portable-Kerosene-Heater-Model-CV-2230/dp/B000A6D1IC,
http://ecx.images-amazon.com/images/I/41S-aBrDn6L._SS500_.jpg

The people "jumping" from the towers was a phenomenon contrary to anything firefighters had ever before observed. Even to these New York City firefighters, it was unbelievable. So, when did this phenomenon become believable?

Consider confusing and dangerous events where someone doesn't have time to evaluate the situation and yet must react. For example, if a team of police officers with rifles were to yell, "get down on the ground!" most people would immediately follow those orders. When someone yells, "Fire!" in a movie theater, one can expect a stampede to the exit even without evidence of a fire. Shouting and confusion will get most people to do what is asked of them. Most of us will acquiesce so as not to add to the confusion. People watched the jumpers. They saw it, so it must be real.

Figure 34. Kerosene tractors.
http://farm4.static.flickr.com/3141/2873386173_a9a65d4fc1.jpg?v=0

Yet we also know that gasoline burns hotter than kerosene and we know that cars do not collapse from overheated or melted engines.

(a) *(b)*

Figure 35. Wood paneling and paper don't burn hot enough to melt or significantly weaken steel.
(a)Source: northlineexpress.com, (b)Source: Century wood stoves

And wood stoves made of steel do not self-destruct from fire.

C. What Did We Really See?

So what did we really see on 9/11/01? It may or may not be what we were told we saw. What if the first reported observation was wrong, and then everyone followed it? The human mind does not operate like a tape recorder. The human mind puts together the best story it can based on what it has been presented with, and it adapts. In a confusing and unbelievable situation, people tend to look for something that makes sense.

We look for what we're accustomed to seeing or expect to see, and we use this to make what sense we can of the confusion. My heart goes out to the doctors who, amongst the confusion, tried to make sense of what they saw and what they were told (along with the rest of us).[10] In an interview, two doctors recall that patients quit coming in to the emergency rooms by noon on 9/11. Dr. Tony Dajer was asked, "Where are all the people?" He groped for an explanation. Another doctor latched on to the hope that most everyone made it out alive. These two doctors are a wonderful example of how much people care for their fellow man. And, like the rest of us, they desperately wanted—and needed—it all to make sense.

But what happens when it *doesn't* make sense?

D. Approach

We must observe the actual evidence, carefully. But how do we do that through a preconceived, conditioned, and biased perspective? One way might be to wipe the slate clean and begin with a new vocabulary, a new language, including a visual language. This may require that we ignore images that we have long been conditioned to react to in very predictable ways.

In the chapters that follow, I have purposely given each of the observed phenomena a new and unique name, using a new vocabulary that does not come preloaded with familiar connotations and therefore biased interpretations. Referring to particulate-filled air as "smoke" causes the biased or pre-conditioned observer to think of "fire" as the cause of what is seen. Instead, then, let us use a very generic term (perhaps even an elementary-school-level term) that we would not normally use in this situation, but one that describes solely and only what we do see, with no other weighted or pre-supposed meaning. Let's use not the word "smoke" but the word "fumes." A glossary of terms is provided at the end of this book for convenience.

---

[1] *http://www.lyricsfreak.com/w/who/you+didnt+hear+it_10187724.html, http://www.youtube.com/watch?v=G5GBroq6dsM*

[2] *http://img83.imageshack.us/img83/1059/1di7.jpg*

[3] *http://bocadigital.smugmug.com/photos/10695097-D.jpg*

[4] Dr. Robert M. Gates, President of Texas A&M University and former Director of the Central Intelligence Agency (CIA) and current Secretary of Defense. *http://www.infowars.com/articles/sept11/reynolds_on_jones_show.htm*

[5] *http://en.wikipedia.org/wiki/The_War_of_the_Worlds_(radio)*

[6] *http://en.wikipedia.org/wiki/War_of_the_Worlds*

[7] *http://en.wikipedia.org/wiki/The_War_of_the_Worlds_(radio), http://en.wikipedia.org/wiki/Sustaining_program*

[8] *http://www.guardian.co.uk/media/2001/sep/17/mondaymediasection.september112001*

[9] The USS *Cole* bombing was a suicide bombing attack against the U.S. Navy destroyer USS Cole (DDG 67) on October 12, 2000 while it was harbored in the Yemeni port of Aden. Seventeen American Sailors were killed. *http://en.wikipedia.org/wiki/USS_Cole_bombing*

[10] *http://www.youtube.com/watch?v=yUPJTe1KVyg*

5.
The Bathtub

We're talking about the fact that most people see what they expect to see, what they want to see, what they've been told to see, what conventional wisdom tells them to see, not what is right in front of them in its pristine condition. —Vincent Bugliosi

The World Trade Center (WTC) towers did not "collapse" on 9/11/01. They didn't have *sufficient time* to collapse because they were destroyed faster than is physically possible for a gravity-driven collapse. As we saw in Chapter 2, the evidence indicates that they were reduced to particles of dust in midair. This in itself rules out Conventional Controlled Demolition (CCD), where a building is knocked off its supports and thereafter slams to the ground.

My own journey through the actual physical evidence began when I considered the so-called WTC *"bathtub,"* the name given to the retaining wall that protected the foundations of the WTC from water, even though these retaining walls extended as far as 70 feet below the water table. If the towers had indeed *collapsed*, or if conventional bombs had blown up the building, there would have been an enormous amount of material slamming down onto the WTC bathtub and foundation. If the building had been destroyed by a gravity collapse, certainly the foundation would have been destroyed or at least severely damaged by the entire 110-story building crashing down onto it.

A. What the WTC Bathtub Is

Figure 36. There was no significant damage to the bathtub on 9/11. This picture looks west from the center of the WTC1 footprint.
http://911research.wtc7.net/wtc/arch/docs/bathtub_wall1.jpg

The World Trade Center was built on bedrock protected by this underground "bathtub" or foundation-ring extending down to bedrock, seven stories below the surface of lower Manhattan. This enclosure, which some call the "slurry wall," held the Hudson River's water away from the foundations of the Twin Towers as well as of WTC buildings 3 and 6. According to *Wall Street Journal* architecture critic Ada Louise Huxtable, this structure "...saved [all of] lower Manhattan from the waters of the Hudson River."[1] Many observers worried about whether the wall, after 9/11, would continue to do its job to prevent flooding, but, "To the relief of the engineers, there is no evidence that the 70-foot-deep retaining wall around the basements has been damaged or breached, although the collapse of the towers left one section perilously unsupported."[2] In the SPIKE TV documentary about the ironworkers at Ground Zero, one remarked, "You know, it was amazing, it didn't really damage [that much]... if they had fallen over sideways, could you imagine the damage to Lower Manhattan?"[3]

And so, on September 11 the bathtub mysteriously remained *without significant damage despite two quarter-mile-tall towers allegedly collapsing onto it.* How did the bathtub avoid significant damage despite a million tons of WTC material supposedly slamming down on it? Even if no material had directly hit the bathtub, serious seismic impacts on bedrock would have damaged walls, wall corners, and tunnels under the WTC that lead beneath the Hudson River—damage that would have been caused by seismic motion similar to that caused by an earthquake. The bathtub was not built to withstand such colossal impact, we may be assured, if only because New York is not an active seismic zone (see next chapter). Although the exact number is disputed, each tower weighed an estimated 500,000 tons. No bathtub structure could remain unscathed after a mountain of material from a quarter-mile-high building was dropped on it not just once, but *twice.* The intact structure appears to contradict the official theory of a gravity-driven collapse in which virtually the entire weight of the Twin Towers would crash into the bathtub.

B. Design and Purpose

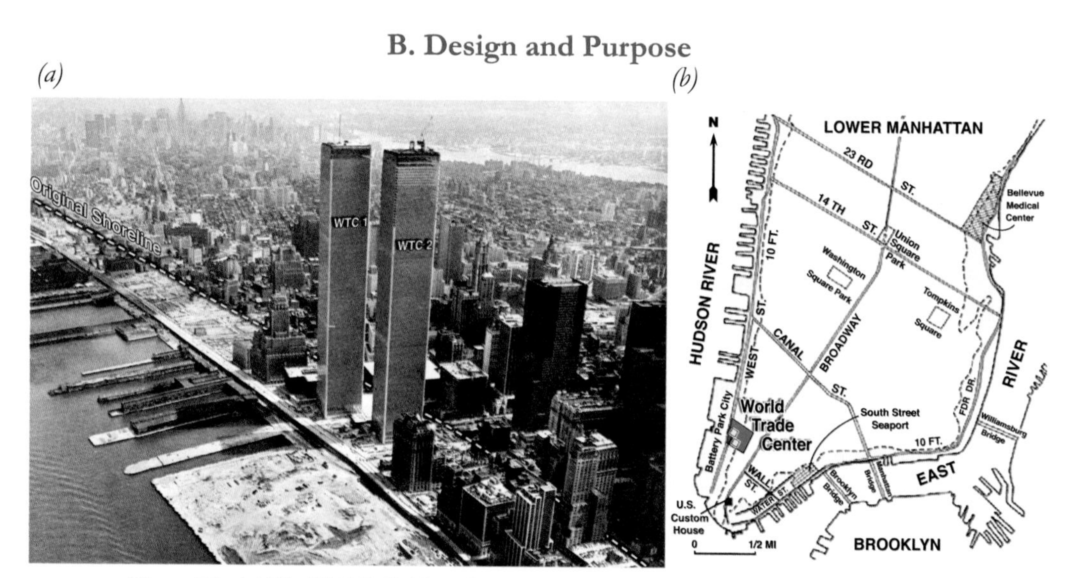

Figure 37. (a) The WTC (b) Foundations in the shaded zone are below the water table.
(a)http://farm4.static.flickr.com/3306/3409907039_2ca324a4f3_b.jpg[4],
(b)http://www.downtownexpress.com/de_150/columbia.gif

As we've seen, The World Trade Center (WTC) towers were built on bedrock below the water table. Landfill expanded the width of Manhattan as shown in below Figure 37b.

Figure 38. World Trade Center construction, towers rising.
http://data.GreatBuildings.com/gbc/images/cid_wtc_mya_WTC_const.4.jpg [5]

Figure 38 provides a view into the area, 60 to 70 feet below the water table, that was protected by the bathtub, which served as a dike, allowing the WTC towers essentially to be built in the Hudson River.

Figure 39. Design features of the "bathtub."
(1) Slurry walls form a water-tight bathtub. (2) PATH rail lines pass under WTC2. (3) Cracks in the bathtub would allow water inside.
(a)http://data.greatbuildings.com/gbc/images/cid_wtc_mya_WTC_finished2.jpg,
(b)http://news.bbc.co.uk/olmedia/1550000/images/_1553074_flooding_300inf.gif [6]

Figures 39 through 43 show diagrams of the PATH (Port Authority Trans Hudson) rail lines connecting New York and New Jersey, traveling under the Hudson and up into the bottom of the WTC bathtub. The base of the bathtub is bedrock, and the Twin Towers, rail lines, and tunnels were all anchored into that bedrock. If it were to be dramatically shaken, fissures in the tunnels would allow water to back up into the bathtub.

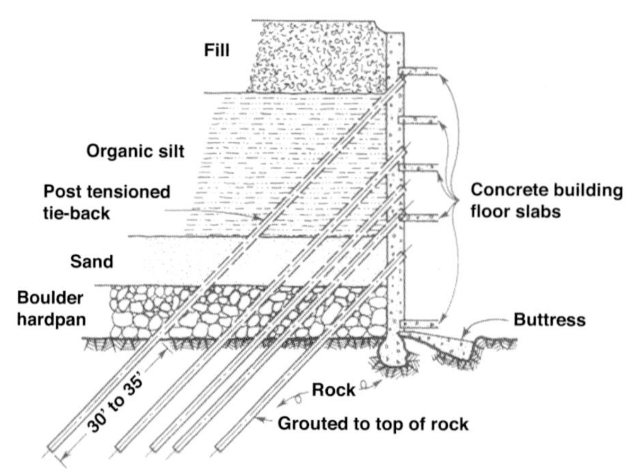

Figure 40. Typical section shows placement of tie-backs during construction. After concrete floor slabs were cured, the tension on tie-backs was released.
http://www.civil.columbia.edu/%7Eling/wtc/ce4a.jpg, http://www.time.com/time/magazine/article/0,9171,1101020527-238628,00.html

Figure 41. (1970) The below-sea-level bathtub during construction of the WTC.
http://graphics8.nytimes.com/packages/images/photo/2008/10/24/26tunnel/25540099.JPG

In Figure 40, the stoutness of the bathtub design suggests how important its great strength was to the WTC. For example, the tension tie-backs were embedded in

30-35 feet of bedrock. You can see many of the ends of tie-backs sticking out on the west wall of the bathtub in Figure 36. Figure 42 shows a map of the WTC showing where the PATH commuter railroad lines under the Hudson enter and exit. The walls of the deep bathtub are in red, and the subway line adjacent to the main WTC bathtub is indicated by the dashed blue line.

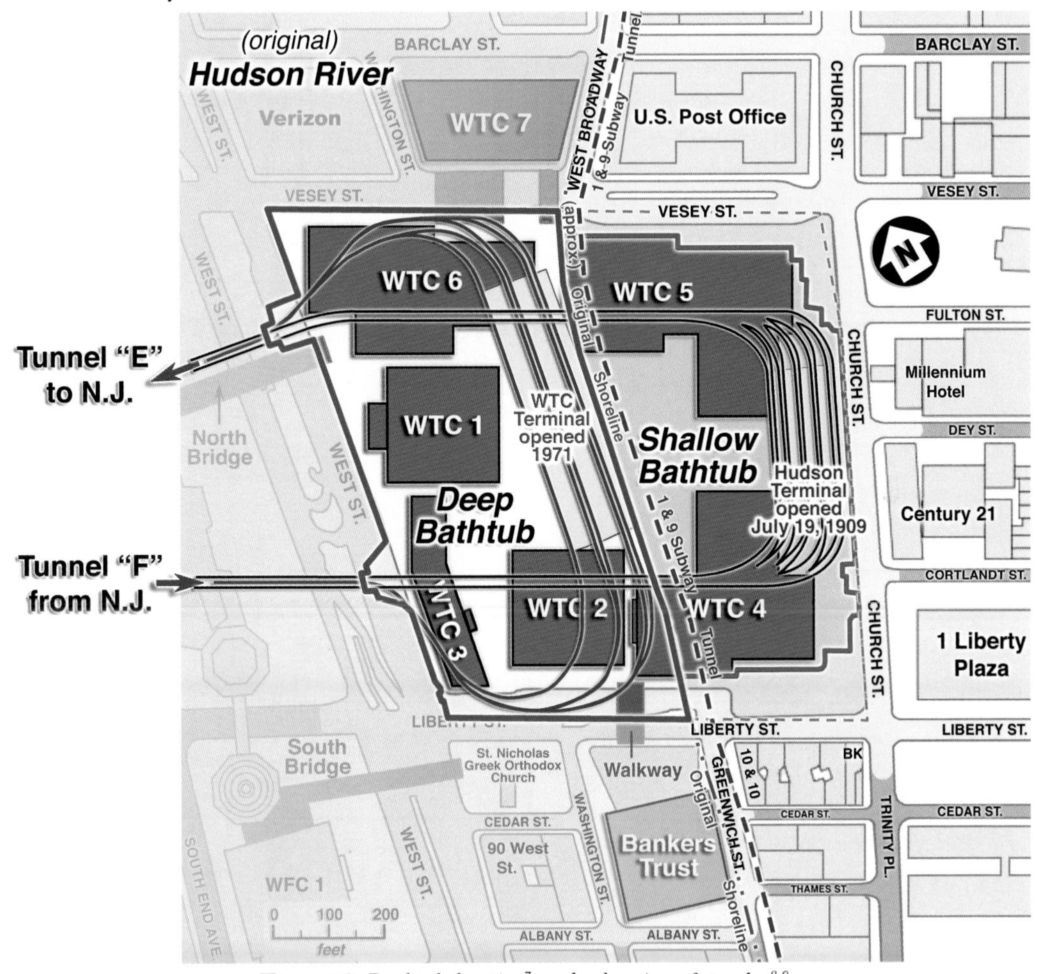

Figure 42. Bathtub location[7] *and relocation of tracks.*[8,9]

Figure 42 shows the two bathtubs under the WTC complex, with the Twin Towers standing in the deep bathtub and buildings 4 and 5 standing mostly within the shallow foundation. The subway traveled north-south through the shallow bathtub. The figure also identifies neighboring buildings, such as the WFC (World Financial Center) complex, the Verizon building, the US Post Office, Bankers Trust and others. Note that WTC7 did not stand in either bathtub. Figure 42 shows the loop taken by PATH trains within the big bathtub, and Figure 43 shows a cross section of the WTC lower levels. The PATH trains turn beneath WTC2, the station platforms parallel the bathtub east wall, and the subway, on a higher level, traveled along the east side of the mid-bathtub wall.

Figure 44 shows the new PATH complex as of August 2006, with the west wall at the top of the photo. The image gives a graphic look at how large the PATH layout is within the big bathtub, while the fact of rail lines and platforms remaining in their original locations suggests that the underground damage to PATH was not devastating (see Figure 44).

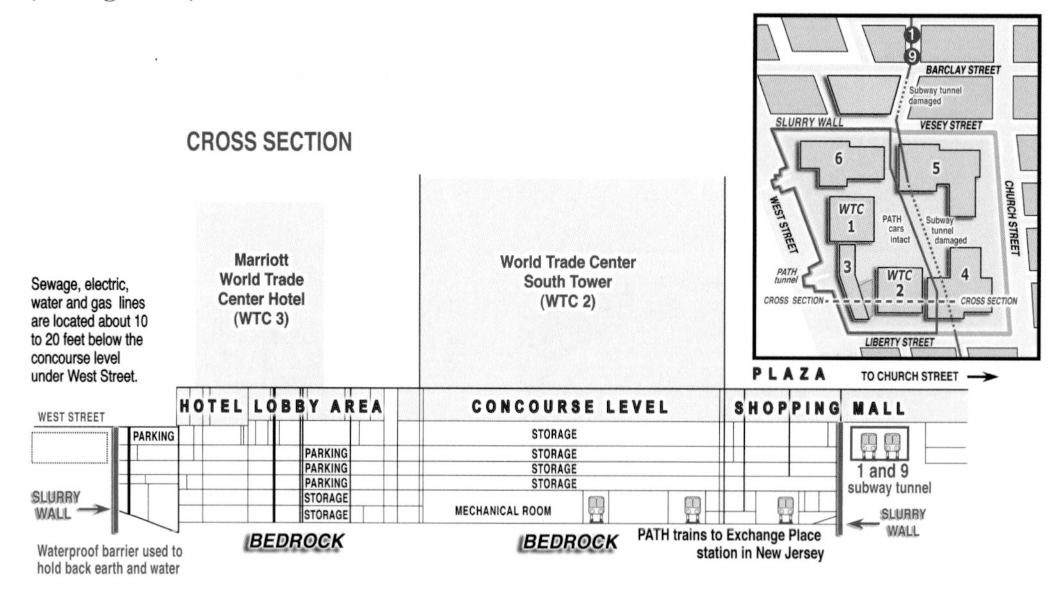

Figure 43. Cross-section of the WTC complex, highlighting buildings 2 and 3 and the seven subbasements. Note the shopping mall at the ground level, on the right, below WTC4 and above the PATH and subway rail lines.

redrawn from: http://www.terrorize.dk/911/maps/below.ground.level.gif

Figure 44. The Ground Zero site as of August 2006, showing the location of the buildings relative to the bathtub walls.

The subway line is in the small bathtub and the PATH station platform is on the other side of the big bathtub wall; no tracks were relocated. The blue dashed line locates the Figure 43 cross section.

Adjusted from: http://www.world-heritage-tour.org/america/us/newYork/WTC/map.jpg

Figure 45. The nearly cleaned-out bathtub. March 15, 2002.
The big bathtub is in the foreground and the shallow bathtub is in the background. Some superficial damage to the top of the bathtub is visible in the foreground (along the eastern wall), adjacent to where WTC 4, a 9-story building, once stood.
http://www.photolibrary.fema.gov/photodata/original/6027.jpg

C. Concern for the Bathtub

Shortly after 9/11, the BBC ran an article that expressed New Yorkers' concerns over the integrity of the bathtub in light of the official story of the twin towers collapsing on it.

1. New York at risk of flooding, Wednesday, 19 September, 2001, 19:43 GMT

> New Yorkers trying to come to terms with both massive loss of life and the devastation of Manhattan may now be facing yet another calamity—flooding.
>
> Engineers are warning that the walls of a giant concrete box, 20 metres deep, which acted as the foundations of the World Trade Center, are in danger of collapse.
>
> This would allow the Hudson River, which runs on one side of the island, to flow into the site, and from there into the entire New York subway system.[10] [emphasis added]

Yet according to *Time Magazine* nearly nine months later, when the bathtub was all cleaned out, it had not failed:

2. The Battle For Ground Zero, By Richard Lacayo, Sunday, May. 19, 2002

> Though it may never feel right to describe the place as clean, the cleanup of the World Trade Center site is done. What was "the Pile," a jagged mountain of knotted steel and concrete, is now a hole, a neatly squared-off, rectangular cavity of 16 gray-brown acres.[11]

This is amply demonstrated by the "intact" nature of the bathtub following the clean-up. See Figure 46.

Figure 46. (3/1/06) A composite photo of the bathtub, showing the area below sea level.[12] *(combined images) http://i.pbase.com/o6/51/57151/1/80214223.ZDLVztNA.IMG_1519.jpg, http://i.pbase.com/o6/51/57151/1/80214222.E69PEfED.IMG_1518.jpg,*

D. Evidence of Little Damage

The big bathtub thus suffered only minimal damage. There was no functional damage; the only damage the destruction of the WTC caused to the bathtub was superficial, as the following *New York Times* article attests,

1. Under the Towers, Ruin and Resilience

> After almost three weeks of exploration, engineers have completed the first survey of the seven-story, 16-acre basement under the ruined trade center complex and have found a varied pattern of destruction. Some areas are nothing but rubble; others seem almost undamaged. To the relief of the engineers, there is no evidence that the 70-foot-deep retaining wall around the basements has been damaged or breached, although the collapse of the towers left one section perilously unsupported.
>
> [...]
>
> The underground work, he said, has entered a "quiet period" of thinking and planning as engineers ponder how to excavate debris from the basement without damaging the retaining wall, known as the "bathtub" that keeps the nearby Hudson River out of the site. [emphasis added]
>
> [...]
>
> To keep the wall from being damaged or moved, excavation of the basement will have to proceed in stages, story by story, Mr. Tamaro said, with such tiebacks being installed along the western and southern sides of the bathtub where the basement floors are no longer providing enough support. Time estimates range from four months to a year, depending on who in the room is talking—a measure of how much uncertainty surrounds the process.[13]

Shortly after the clean-up began, it was discovered that earth-moving equipment was damaging the area around the WTC bathtub, and new rules were established. So an important question occurs: How is it possible that earth-moving equipment can do more damage to the bathtubs than the destruction and alleged collapse of two 500,000-ton buildings that stood directly over them?

Yet another article highlights this question in even greater relief. I cite the article in full in order to give the problem its full impact:

2. World Trade Center Bathtub: From Genesis to Armageddon

Prelude

In 1993, terrorists detonated a bomb in the WTC basement adjacent to a column of the north tower (WTC 1) causing damage to the floors that were supporting the slurry walls. Fortunately, the walls themselves were not damaged, did not leak, and were able to span across the damaged areas. Visual inspection of the walls in spring 2001 revealed that the walls were in good condition.

Armageddon

On September 11, 2001, terrorists again struck the WTC complex, this time causing the collapse and destruction of the majority of above-grade structures and the partial collapse of the below-grade structures. *The limits of the bathtub and the condition of the below-grade structures were not immediately evident in the aftermath of the attack.*

Initial Response

Immediately after the collapse, the New York City Department of Design and Construction established a team of engineers and contractors to assist the NYC Fire Department in its search and rescue efforts. One group of engineers, under the direction of Thornton-Tomasetti Engineers (TTE), focused on the inspection of adjacent buildings while another provided advice on below-grade structures in the WTC complex, the World Financial Center complex located to the west in the Battery Park City landfill, the PATH tubes, and the New York City subway tunnels.

As heavy equipment (e.g., 1,000-ton cranes) began to arrive at the site, it became apparent that ground rules had to be established for the safe use of the equipment outside the confines of the basement, over major utilities, over access stairs to the PATH tubes and ramps, in the streets, and over structural platforms spanning open water. The use of this heavy equipment adjacent to the slurry walls or over the basement structure itself could cause the collapse of the slurry walls or any remaining basement structures. A collapse of the slurry wall would mean inundation from the nearby Hudson River. [emphasis added]

As a first step, Mueser Rutledge Consulting Engineers (MRCE) prepared cartoon-like sketches showing the location of below-grade structures outside the slurry wall that could not be traversed by heavy equipment. The locations of four 6-foot diameter water lines were also identified. The Port Authority closed valves for two water intake lines shortly after the incident. The other two discharge water lines could backfeed river water into the basement during periods of high tide and had to be sealed as soon as possible. The sketches were provided to the Fire Department and the contractors for use in placing rescue, construction, and demolition equipment. Weidlinger Associates subsequently prepared more detailed utility drawings for the contractors.

PATH Tunnels

> Concurrent with rescue work in New York, Port Authority engineers were investigating the condition of the PATH tunnels in Jersey City, New Jersey, where the Exchange Place Station, which was at an elevation 5 feet lower than the WTC PATH Station, had served as a sump for fire water, river water, and broken water mains discharging into the bathtub. Inspection indicated that water in the tunnels between New York and New Jersey had completely filled the north tunnel at the midriver low point. *Pumps were immediately put into action to keep Exchange Place Station from flooding. As much as 3,000 gallons per minute were pumped from the north tunnel for a 12-hour period each day. Tests of the water were inconclusive as to the source; however, most was believed to come from the vast amounts of water that were poured onto the debris to extinguish continuing fires.* Within days, a 16-foot long low-strength concrete plug was placed in each tube as a seal in the event that the bathtub walls were breached and the tunnels fully flooded. The plugs were designed to withstand an 80-foot head of water pressure and will be removed once the slurry walls are fully secured (Figure 6). The Port Authority is currently preparing to remove the plugs in preparation for rehabilitation of the tunnels. [emphasis added] [14]

Figure 47. (10/10/01) Damage from earth-moving equipment along Liberty Street.
http://www.photolibrary.fema.gov/photodata/original/4262.jpg

During the cleanup, dirt and debris were piled up to form a ramp between the east bathtub and the west bathtub, allowing heavy earth-moving equipment to drive between the two sides. More than six months after the clean-up began, the ramp was removed and it was again discovered that this area of the WTC bathtub had been damaged. So, once again: how is it possible that earth-moving equipment can do more damage to this structure than the *collapse* of two 500,000-ton buildings that had stood directly over them?

This entire question was highlighted by yet another article:

3. Workers Rush To Repair Huge Hole In WTC 'Bathtub'

03/21/2002 (archives accessed Nov. 18, 2006)

> When a huge hole was dug out of Lower Manhattan to build the foundation for the World Trade Center more than 30 years ago, workers constructed the wall to keep the nearby waters of the Hudson River from seeping in through the earth. The seven-story pit is now almost all that remains of the World Trade Center after months of removing debris.
>
> Engineers say there is no imminent danger of the wall collapsing or of the pit filling with water.[15]

PATH trains resumed operation November 2003, only two years after 9/11. Water is visible in Figure 48b, for example, but *there was no flooding from the Hudson River.* The water came from fire hoses and rainwater, and it had to go somewhere.

(a) pre-9/11/01 *(b) shortly after 9/11/01* *(c) 11/23/03*

Figure 48. (a) WTC Station Platform before the 9/11. (b) WTC Station Platform after 9/11; PATH train was not crushed, (c) WTC Station Platform November 23, 2003.

(a)http://hudsoncity.net/tubes/pathtrainatwtcofficialpa250.jpg, (b)http://hudsoncity.net/tubes/pathcarwtc-250.jpg, (c)http://hudsoncity.net/tubes/temporarywtcplatformfirstlasttubetrain-250.jpg

Figure 48a shows a PATH train in the big bathtub before 9/11; Figure 48b after 9/11 shows minor non-structural platform damage, probably water damage; and Figure 48c shows the updated platform and cars, which look rather similar to their predecessors.

An exposed PATH train tunnel outside the bathtub shows no structural damage (Figure 49a) and an intact PATH train car is lifted from the bathtub (Figure 49b), showing no indication of having been crushed.

(a) dry PATH train tunnel *(b) (2/22/02)*

Figure 49. (a) No significant structural damage in this PATH tunnel. (b) PATH Train cars were not crushed and were lifted from the bathtub February 22, 2002.

(a)http://www.nj.com/cgi-bin/nph-cachecam.cgi?url=photo.live.advance.net/njo/images/1321/ZPTUN9-01.jpg&ct=10800, (b)http://img.timeinc.net/time/photoessays/groundzero/zero09.jpg

Figure 50 shows a *New York Times* sketch of alleged damage to the underground portion of the WTC within the bathtub. It seems odd that the condition at the center of the PATH platforms was "not inspected or undetermined." Why? Figure 50, for example, shows no structural damage at that section of the platform, but only water damage. It is difficult to be entirely confident that the *New York Times* sketch is an accurate picture of the damage pattern in the bathtub. Interestingly, the slurry wall on the west, or Hudson, side of each tower is damage-free, according to the *New York Times*. Also, the PATH tunnel entrances, rigidly connected to the bathtub and bedrock, are "intact or mostly intact." Only three of seven PATH cars were damaged. While the *New York Times* uses the term "crushed," it seems unlikely that three cars could be totally crushed while four train cars remained intact (see Figure 50)![16]

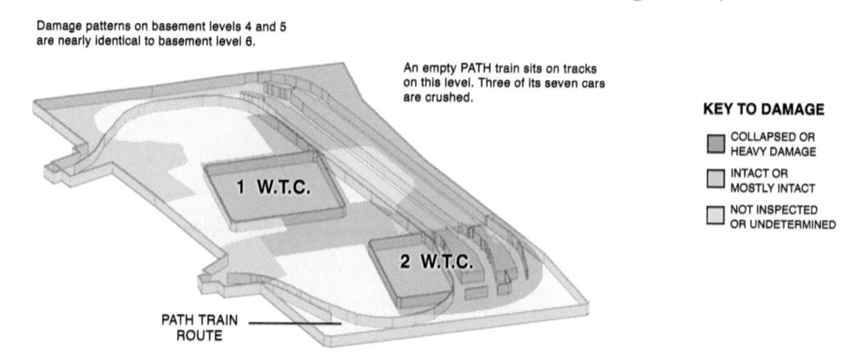

Figure 50. Four of the seven PATH train cars under WTC were not damaged
Adjusted from http://www.nytimes.com/library/national/index_WALL.html

Outside the bathtub east wall and also in the shallow bathtub, even the subway suffered surprisingly little: "Considering the devastation near the trade center, and the fact that the tunnels were only five feet below the road surface in some places, complete tunnel collapses were not as extensive as some engineers had feared."[17]

Figure 51. Warner Brothers Store in the WTC Mall, viewed from the Strawberry store.
(Viewed from Location B in Figure 183, page 190.)
http://911stories.googlepages.com/487354604jtxfjiphjx4.jpg/487354604jtxfjiphjx4-full.jpg;
http://911stories.googlepages.com/insidethenorthtower:witnessaccounts,lobb

Figure 51 shows the Warner Brothers store in the WTC shopping mall, as viewed from Strawberry, before 9/11. A photo taken the other direction (a photo of Strawberry viewed from Warner Brothers) is shown in Figure 182 (page 189), taken just after 9/11. Figure 52a shows store contents from this Warner Brothers store that was in the WTC shopping mall on the concourse level. The concourse level is in the first subbasement. The Warner Brothers characters recovered from this store are shown in Figure 52b. Note that Roadrunner does not have a scratch on him despite surviving the destruction of WTC2 above him. Yet, as shown in the cross section in Figure 43 above, the shopping mall is the first floor to be impacted. Figure 179 (page 188) shows the situation above this location (WTC4) and Figure 54 shows the situation in the Mall just under WTC4.

Figure 52. (after 9/11/01) (a) Store contents. (b) Foghorn Leghorn, Bugs Bunny, and Roadrunner.
(a)http://www.amny.com/media/photo/2006-08/24928918.jpg,
(b)http://www.amny.com/media/photo/2006-08/24929073.jpg

Figure 53. (9/19/01) Rescue workers in the WTC mall, under WTC4, near the Strawberry store.
(image lightened) http://www.photolibrary.fema.gov/photodata/original/5347.jpg

Figure 54. (9/19/01) Innovations Luggage (left) in the WTC Mall under WTC4. This is a view from Location C in Figure 183, page 190. (Also see Figure 181, page 189.)
(image lightened) http://www.photolibrary.fema.gov/photodata/original/5345.jpg

4. Pulling Building 6

During the cleanup of the WTC site, the remaining portions of WTC6 needed to be demolished. Engineers did not use explosives for fear of damaging the bathtub wall (or slurry wall), as discussed in the PBS special, *America Rebuilds.*[18] Instead, workers attached cables to the remaining structure and rocked it back and forth until it toppled over.

> **Demolition worker:** Oh, we're getting ready to pull building 6.
>
> Luis Mendex, Department of Design and Construction.
>
> We have to be very careful how we demolish building 6. We were worried about building 6 coming down and damaging the slurry wall. So we wanted that particular building to fall within a certain area.[19]

Video shows cables attached to the remains of WTC6 being pulled by grapplers in order to rock it.

> **Demolition worker:** We've got the cables attached in four different locations. They'll be pulling the building to the north. It's not every day you try to pull down an eight-story building with cables.[20]

And a construction worker or narrator, watching the event:

> There's a certain excitement in the air, bringing the last structure down of the World Trade Center.[21]

They were *all* worried about damaging the bathtub by bringing down the fragments of an eight-story building with a few sticks of dynamite, yet the alleged gravity-driven collapse of two 110-story buildings were nothing to worry about?

5. NIST: No Significant Damage done

The National Construction Safety Team (NCST) Advisory Committee met via teleconference on Thursday, December 14, 2006, from 9:00 a.m. to 11:00 a.m., to discuss their progress on investigating the "collapse" of WTC7. They invited the public to watch the live webcast of their meeting. They also allowed the public to call in with questions, an invitation that required prior arrangements since the schedule allowed time for only six calls of five minutes each. Fortunately, Jerry Leaphart, my lawyer, learned about this meeting early enough to get on the schedule. As it happened, there were only two callers, and he was one of them.

During this teleconference, the NCST Advisory Committee discussed possible causes of the fires in WTC7. The Committee was asked if the "collapse" of the towers could have ruptured a pipeline that carried fuel through the WTC complex over to WTC7.

Figure 55. There was no significant damage to the bathtub on 9/11. This picture looks west from the center of the WTC1 footprint.
http://911research.wtc7.net/wtc/arch/docs/bathtub_wall1.jpg

Dr. Shyam Sunder made the following statement about the WTC1 and WTC2 seismic signals:

> The signals' strength due to the collapse of the towers were not of any magnitude that was seismically significant from an earthquake design standpoint or from the design or a failure of a structural component or of I would say of a piping system that might be used in a structure, so ah there wasn't anything that gave us pause in terms of that being a significant seismic event to have ruptured the pipeline.[22]

After all is said and done, Figure 55 says everything most eloquently, and in a manner utterly contradictory to the official versions of the "collapse" of the WTC towers, for the bathtub here shows no signs of significant damage at all.
In short, the "Bathtub" evidence itself is powerful testimony that the official story of the twin towers' "pancaking" down and slamming into the bathtub is precisely that—a *story* that fails to fit the evidence. This means, logically, that the controlled demolition model cannot be true either.

---

[1] *The Wall Street Journal,* September 28, 2006, p. D8.

[2] Dennis Overbye, "Under the Towers, Ruin and Resilience," Science Times, The New York Times, October 9, 2001. archived: *http://www.archinode.com/WTCnytimes2.html*

[3] *http://www.total411.info/2006/10/video-metal-of-honor-ironworkers-on.html; http://www.youtube.com/watch?v=X-99CLdHWC.*

[4] *http://forum.skyscraperpage.com/showthread.php?p=4276293*

[5] *http://www.GreatBuildings.com/cgi-bin/gbi.cgi/World_Trade_Center_Images.html/cid_wtc_mya_WTC_const.4.gbi*

[6] altered from source: *http://news.bbc.co.uk/2/hi/americas/1553074.stm#graphic*

[7] *http://www.serendipity.li/wot/wtc_ch2b/fig-2-11.jpe*

[8] *http://images.nycsubway.org/i21000/img_21823.jpg*

[9] Map redrawn from p. 3 (pdf p. 53 of 298), *http://wtc.nist.gov/NISTNCSTAR1CollapseofTowers.pdf*

[10] *http://news.bbc.co.uk/2/hi/americas/1553074.stm*

[11] *http://www.time.com/time/magazine/printout/0,8816,238628,00.html*

[12] Two images were combined to provide this panoramic view.

[13] Dennis Overbye, "Under the Towers, Ruin and Resilience," Science Times, The New York Times, October 9, 2001, *http://www.nytimes.com/2001/10/09/science/physical/09WALL.html?ex=1227934800&en=ebd94d7c8b998058&ei=5070*

[14] World Trade Center "Bathtub": From Genesis to Armageddon, George J. Tamaro, *http://www.nae.edu/nae/bridgecom.nsf/weblinks/NAEW-63AS9S/$FILE/Bridge-32n1.pdf?OpenElement., http://www.nae.edu/nae/bridgecom.nsf/weblinks/CGOZ-8NLJ9?OpenDocument*

[15] Workers Rush To Repair Huge Hole In WTC 'Bathtub', By: NY1 News, 03/21/2002 (archives accessed Nov. 18, 2006), *www.ny1.com/ny1/content/index.jsp?stid=1&aid=20029*

[16] *http://www.nytimes.com/library/national/index_WALL.html. http://www.time.com/time/magazine/article/0,9171,1101020527-38628,00.html*

[17] *http://www.nytimes.com/2001/09/28/nyregion/28SUBW.html?ex=1160193600&en=8fcbf2fe869ef9a7&ei=5070*

[18] *America Rebuilds: A year at Ground Zero,* PBS video, shown on PBS Monday, September 11, 2002, *http://www.pbs.org/americarebuilds/*

[19] Segment A from the PBS video, America Rebuilds, *http://www.youtube.com/watch?v=FNEoiOP76QQ*

[20] Segment B from the PBS video, America Rebuilds, *http://www.youtube.com/watch?v=KHtcI9ge6bE*

[21] Segment A from the PBS video, America Rebuilds, *http://www.youtube.com/watch?v=FNEoiOP76QQ*

[22] The National Construction Safety Team (NCST) Advisory Committee met via teleconference, Thursday, December 14, 2006, 9:00 a.m. - 11:00 a.m., *http://wtc.nist.gov/media/NCSTACmeetingDec06.htm.* See also (Audio Segment: *WTC SeismicSignature__NCST___comment* (mp3) (132 kB)) *http://drjudywood.com/media/WTCSeismicSignatureNCSTAd.mp3*, An Audio of the entire meeting is provided here: *http://drjudywood.com/articles/NCST/NCST.html*

6.
Seismic Impact

When the only tool you own is a hammer, every problem begins to resemble a nail. —Abraham Maslow

A. Introduction

We were told that the buildings *pancaked* down in a progressive collapse, one floor slamming onto the next and gaining momentum all the way to the ground. If the towers had indeed "collapsed" in this manner, the foundation bedrock would have experienced a tremendous force hammering on it throughout the entire "collapse." If a downward force is large enough to destroy columns on a given floor, it can only do so if the other end of the column is rigidly supported. And, according to Newton's first law, the forces supporting the column will be equal to the impacting forces, up until the moment the column fails. It would be like hammering a nail into bedrock.

Figure 56. (1971) Looking north at the WTC Towers as they near completion. LIFE Magazine.[1]

The official story says that the *collapse* was caused by gravity and that the momentum of the *collapsing* floors gained speed and energy that wouldn't have dissipated until hitting bedrock. If that had been the case, the WTC bathtub, which holds back the Hudson River, would have been completely crushed. The NYC underground subway system and a very large part of Manhattan would have been devastated by flooding from the water rushing in.

Figure 57. (1971) The towers were massive. LIFE[2]

Figure 58. (9/22/01) But what remained was not massive. Time, Inc.[3]

B. Slamming to the Ground, without Disturbing Other Buildings

When the National Construction Safety Team (NCST) Advisory Committee met via teleconference on Thursday, December 14, 2006, from 9:00 a.m. to 11:00 a.m., the following statement, as we have seen, was made by Dr. Shyam Sunder about the WTC1 and WTC2 seismic signals:

> The signals' strength due to the collapse of the towers were not of any magnitude that was seismically significant from an earthquake design standpoint or from the design or a failure of a structural component or of I would say of a piping system that might be used in a structure, so ah there wasn't anything that gave us pause in terms of that being a significant seismic event to have ruptured the pipeline.[4]

That is, as part of the official investigation of WTC7, it was determined that pipelines passing under the WTC complex and entering WTC7 would not have been ruptured because the "collapse" of WTC1 and WTC2 were not significant seismic events. That is, two 500,000-ton buildings, each over a quarter-mile tall, slammed to the ground and yet did not cause a significant seismic event that would damage the pipelines under the WTC complex. But then an overwhelmingly important question occurs: Why would the destruction stop at ground level?

Windows of buildings that are damaged in earthquakes typically have long corner-to-corner cracks or edge-to-edge cracks. But the windows of the Bankers Trust (Deutsche Bank Building), located at 130 Liberty Street, directly across the street from WTC2, had round and oval holes through them. If the ground had shaken as in an earthquake, we would expect typical long cracks across the windows. Instead, the destruction of the WTC towers left small rounded holes that cannot be explained by projectile damage. See arrows in Figure 59.

Figure 59. Round holes in windows above the 10 & 10 Firehouse.
Kurt Sonnenfeld -FEMA photo

In Figure 60, the vertical axis is the east-west ground speed in nm/s. The horizontal scale is time, showing a thirty-minute interval. The top line starts at 8:40 AM EDT and records a seismic disturbance at 8:46:26 AM EDT. At 9:02:54 AM EDT, there is a smaller and shorter disturbance. The second line is a continuation of the recorded signal that begins at 9:10 AM EDT and shows no disturbance. The third line begins at 9:40 AM EDT and shows a major seismic event at 9:59:04 AM EDT. The fourth line is a continuation that begins at 10:10 AM EDT and shows a major seismic event at 10:28:31 AM EDT.[5]

The "impact" of WTC1 causes vibration that dampens out after about *18 seconds* (see the expanded segment around 8:46:26 in Figure 60). The destruction of WTC2 causes a vibration that ceases after about *8 seconds* (see the expanded segment around 9:59:04 in Figure 60). That is, when WTC1 got its hole, the building vibrated like a tuning fork. When WTC2 was demolished, the vibrations lasted only about 8 seconds. I would expect a progressive collapse to be like a relay race of failures, one shaking the next. How can there be a progressive collapse without shaking? How can there be a progressive collapse, with one floor falling down onto the next, without shaking?

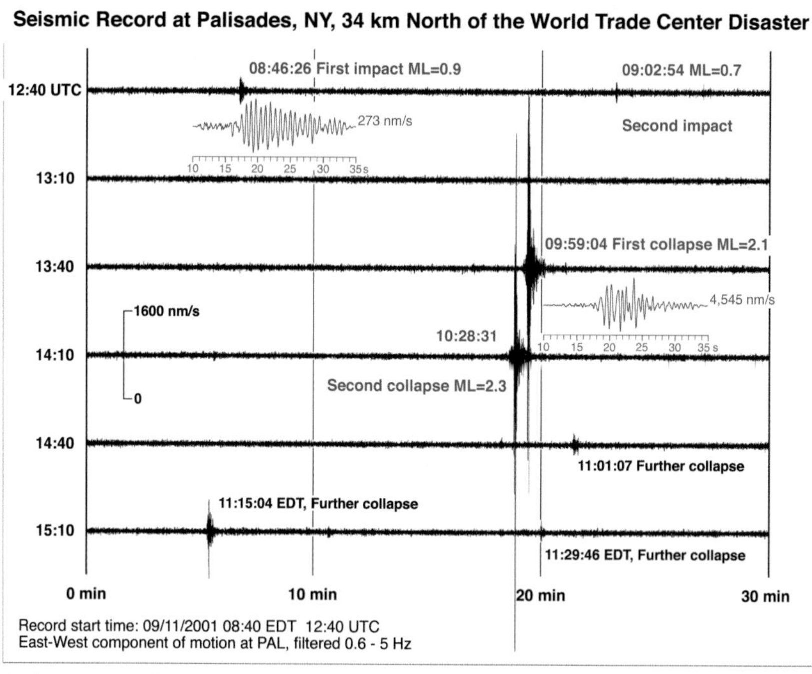

Figure 60. Seismic recordings on E-W component at Palisades for events at World Trade Center (WTC) on September 11, distance 34 km. Three hours of continuous data shown starting at 08:40 EDT (12:40 UTC).

Data were sampled at 40 times/s and passband filtered from 0.6 to 5 Hz. The two largest signals were generated by collapses of Towers 1 and 2. Eastern Daylight Time (EDT) is UTC minus 4 hours. Expanded views of first impact and first collapse shown in red. Displacement amplitude spectra in nm/s from main impacts and collapses shown at right. Sampling is done for 14-second time windows starting about 17 s after origin time. Note broadband nature of spectra for collapses 1 and 2. Their signals are similar with a correlation coefficient of about 0.9 as are those for two impacts.[6]

C. It Didn't Last Long Enough

We can see again, in Figure 61 below, seismic data showing that in the case of WTC1 the ground shook for less than 8 seconds.

But as we have previously seen, it would take over 9.22 seconds for a billiard ball dropped from the roof to hit the ground, not adjusting for air resistance. The

August Fact Sheet ("Answers to Frequently Asked Questions") put out by the National Institute of Standards and Technology (NIST) states that "NIST estimated the elapsed times for the first exterior panels to strike the ground after the collapse initiated in each of the towers to be approximately 11 seconds for WTC1 and approximately 9 seconds for WTC2."[7] The height of the South Tower (WTC2) is 1362 feet, and the height of the North Tower (WTC1) is 1368 feet, nearly the same.[8] We will therefore assign the value of 10 seconds to the Official story.

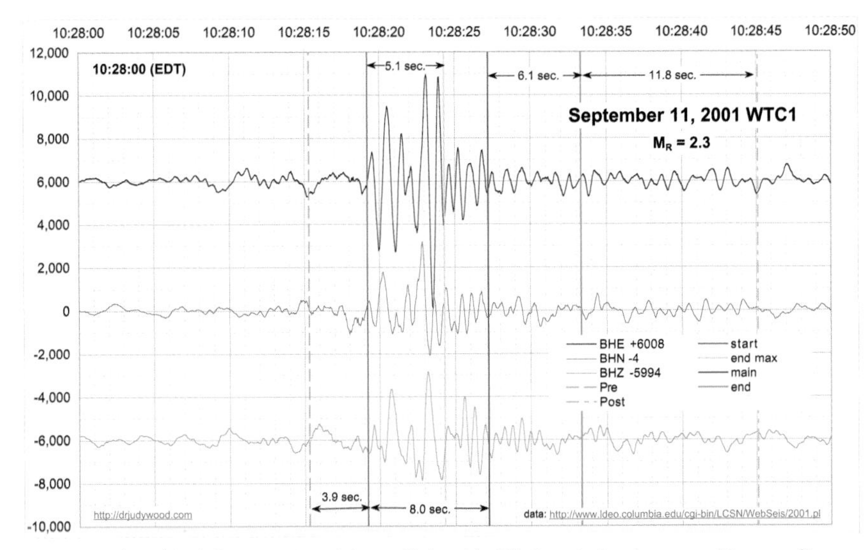

Figure 61. (9/11/01) Seismic signal from Columbia University's seismographic recording station. The time reported for the "collapse" of WTC1 was 10:28:31±1.
Data source: http://www.ldeo.columbia.edu/cgi-bin/LCSN/WebSeis/2001.pl

Columbia University's Seismology Group recorded seismic events of 10 seconds and 8 seconds in duration, corresponding to the collapses of WTC2 and WTC1 respectively.

Information Based on Seismic Waves recorded at Palisades New York			
Seismology Group, Lamont-Doherty Earth Observatory, Columbia University			
Event	**Origin time (EDT)** (hours:minutes:seconds)	**Magnitude** (equivalent seismic)	**Duration**
"Impact 1" at WTC1	08:46:26±1	0.9	12 seconds
"Impact 2" at WTC2	09:02:54±2	0.7	6 seconds
"Collapse 1," WTC2	09:59:04±1	2.1	**10 seconds**
"Collapse 2," WTC1	10:28:31±1	2.3	**8 seconds**
"Collapse 3," WTC7	17:20:33±1	0.6	**18 seconds**

Table 2. Information Based on Seismic Waves at Palisades New York.[9]

$t = \sqrt{\frac{2h}{g}}$, *where h = 1368 ft(417 m), g (gravity)* = 32.2 ft/s^2 (9.81 m/s^2), and time, or "t," is *t = 9.218* seconds, or *t = 9.22 s* for ball dropping from roof to ground in a vacuum.

NIST does not correlate with the seismic data recorded by the Seismology Group at Palisades, NY, shown in Table 2. That data shows a seismic event lasting less than 9.22 seconds. Thus, the issue of "collapse time" for WTC1 is not addressed by

NIST, whose analysis is therefore incomplete and inconsistent with the actual length of time the ground shook. The NIST analysis is incomplete.

The seismic signal for the destruction of WTC7 is discussed later in this chapter, in section L, "Seismic Disturbance from WTC7."

D. Compared with Other Earthquakes (January 17, 2001)

New York is not located in a major earthquake zone (see Figure 62), so designers would not anticipate designing and building with the likelihood of surviving major earthquakes. To analyze the shaking of the WTC Towers, let us compare the recorded ground shaking of a similar sized earthquake from January 17, 2001, in Midtown Manhattan.

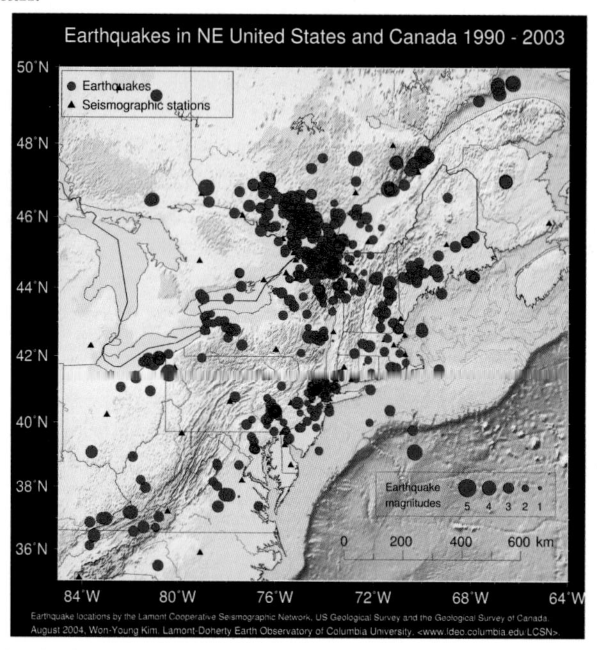

Figure 62. Earthquakes by location and magnitude, indicated by circles; locations of seismographic stations shown by triangles.
http://www.ldeo.columbia.edu/LCSN/Report/NE_Seismicity_1990-2003_color.pdf

Figure 62 is a map of earthquakes in the northeastern United States and southeastern Canada over the period 1990-2003. The small numbers refer to the events given in Table B-4 of the NIST report.[10] The Fox Islands (52.46 N, -169.28 E) are part of the Aleutian Islands of Alaska. Further, see graphics of the relative distance of the measuring sites for the WTC ground shaking as well as for earthquakes in the region.

Figure 64 indicates the locations of the seismic recording sites relative to the WTC and Midtown Manhattan for both the Manhattan quake and for the ground-shaking caused by the destruction of the buildings at the WTC. Figure 64a shows the amount of ground movement from a 2.4 Richter scale earthquake that hit NYC in January of 2001. The data appear to be raw and unfiltered. For example, the amplitude of the earth's movement for the earthquake is nearly double the 8 micrometers shown in the diagram for WTC1.

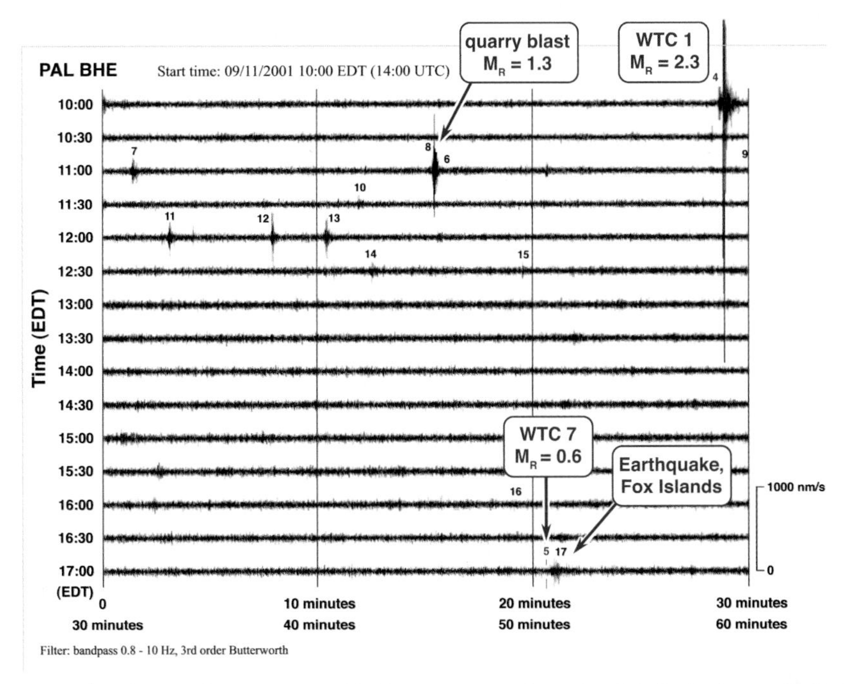

Figure 63. (9/11/01) E-W component, PAL seismic recordings (nm/s), beginning at 10:00 EDT.
Adapted from Figure B-2, page 652 (pdf 314 of 382), http://wtc.nist.gov/media/NIST_NCSTAR_1-9_vol2_for_public_comment.pdf

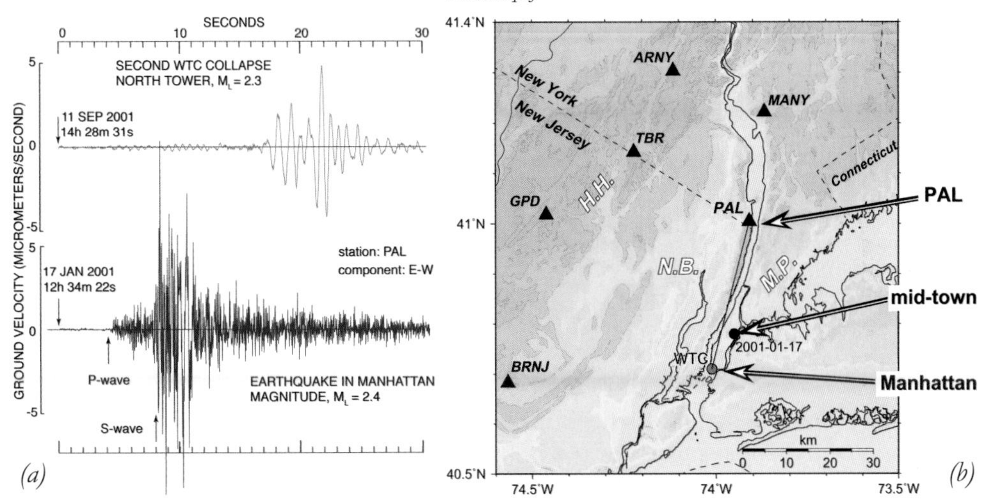

Figure 64. (a) Comparison of Palisades seismograms for collapse 2 and earthquake of 17 January 2001. Arrows at left indicate computed origin times.[11] *(b) Location of seismic recording stations in the New York City area.*[12]

The data for WTC1 appear very different from those for the January 2001 quake—smoother, fewer spikes, less complexity, and lacking distinctive S and P waves. There would also have been a delay between the P and S waves if it had been an earthquake (see Figure 68a) or even a conventional controlled demolition (see Figures 71, 78 and 79).

E. A Closer Look at Seismic Signals

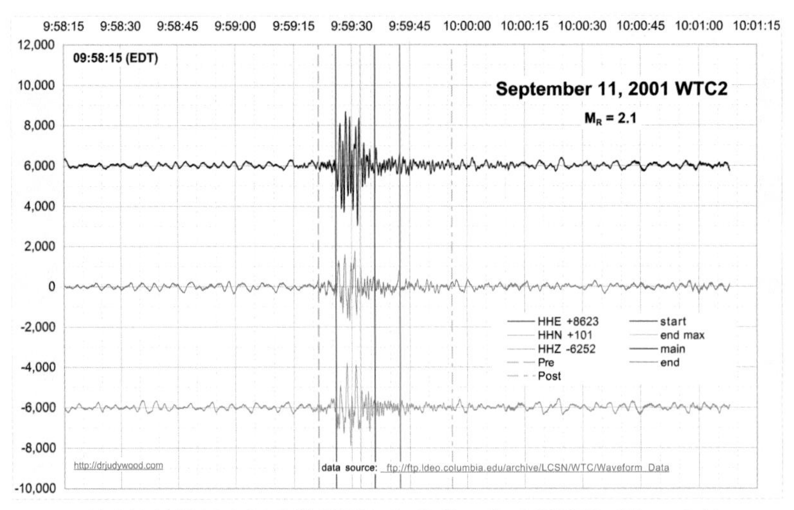

(a) 9/11/2001 9:59:04[13] EDT, the "collapse" of WTC2. (M_R = 2.1)

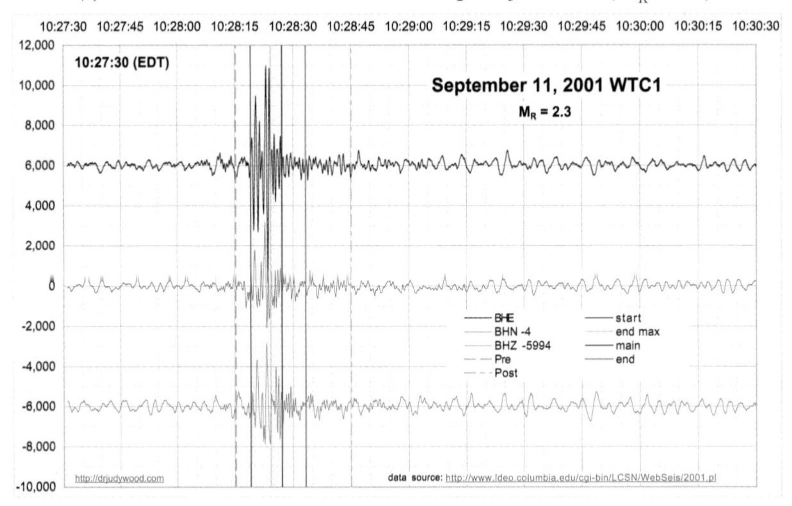

(b) 9/11/2001 10:28:31[14] EDT, the "collapse" of WTC1. (M_R = 2.3)

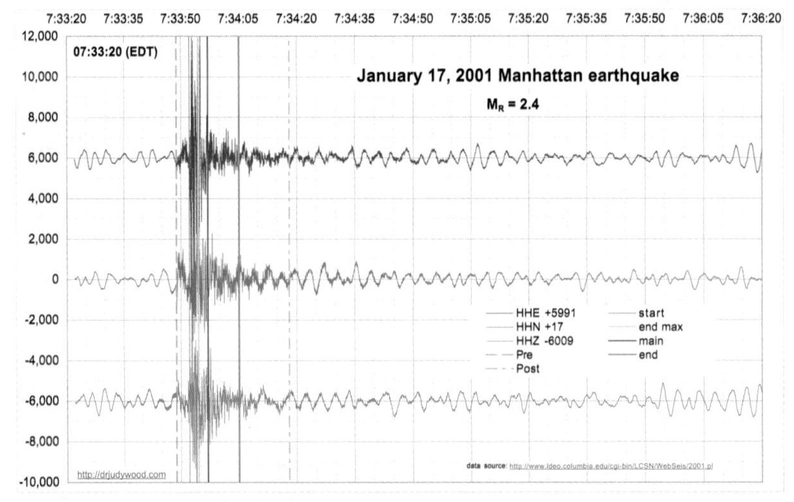

(c) (1/17/01) Manhattan Earthquake. (M_R = 2.4)

Figure 65. Three-minutes of seismic data recorded in Palisades, NY, for three events.[15]

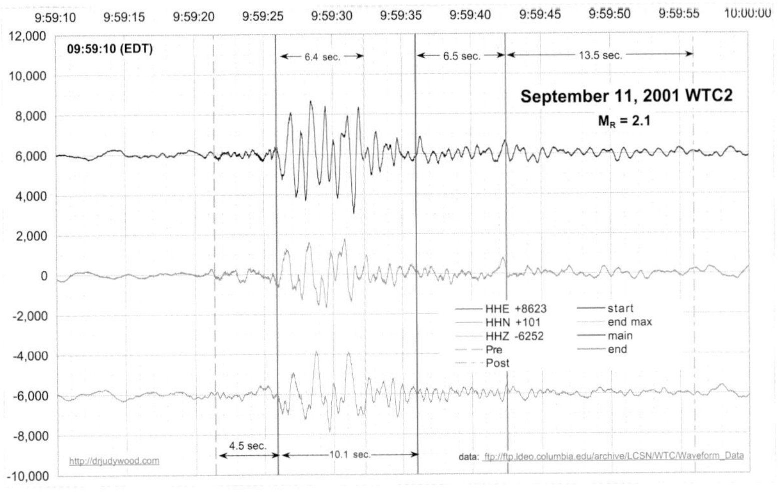

(a) 9/11/2001 9:59:04[16] EDT, the "collapse" of WTC2. ($M_R = 2.1$)

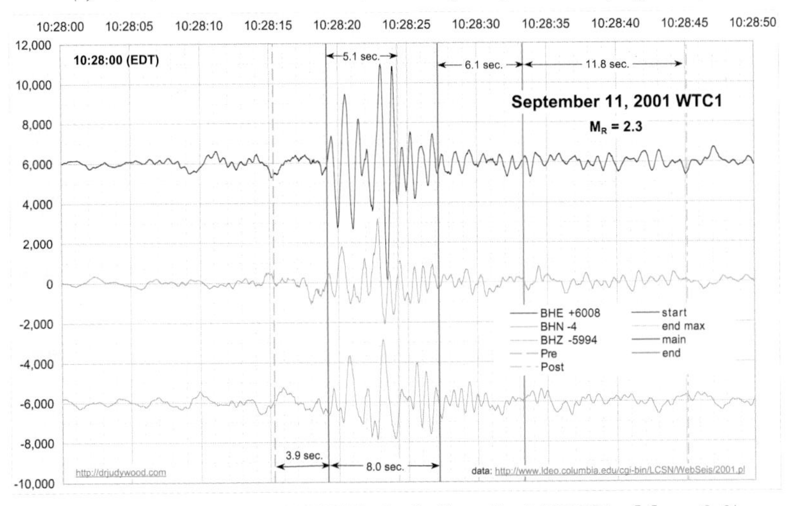

(b) 9/11/2001 10:28:31[17] EDT, the "collapse" of WTC1. ($M_R = 2.3$)

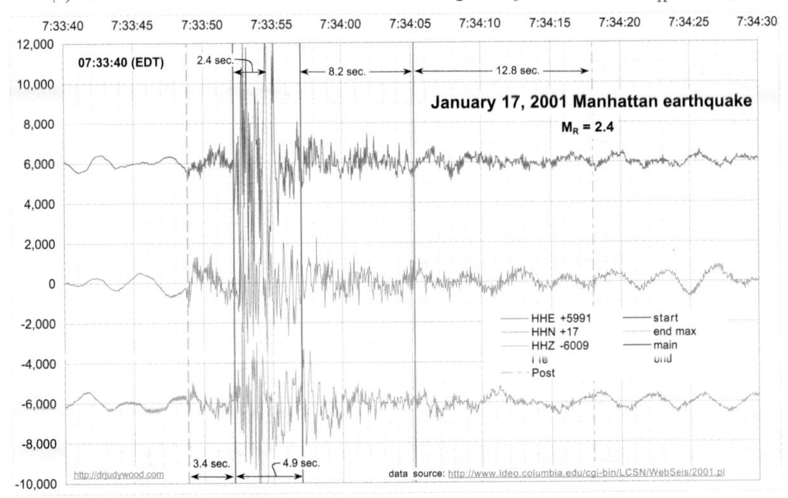

(c) (1/17/01) Manhattan Earthquake. ($M_R = 2.4$)

Figure 66. Fifty seconds of seismic data recorded in Palisades, NY, for three events.[18]

A comparison of seismic signals for the Manhattan earthquake and for the destruction of both WTC towers is shown in Figures 65 and 66. Figure 65 shows a three-minute time span surrounding each event and Figure 66 shows a 50-second time span surrounding each event with each signal duration for comparison. (Note that the signals are offset by 6,000 in order to plot all three signals for each event on the same chart [BHE, BHN, BHZ]).

The seismic signal from the January 17, 2001 earthquake (Figure 65c and Figure 66c) shows a striking difference from the seismic signals recorded during the final destruction of each of the WTC towers (Figures 65(a and b) and Figure 66(a and b)). The signal from the January earthquake shows a lower amplitude signal arriving approximately 3.4 seconds before the main signal. This is the P wave (Primary wave), which arrives before the S wave (Secondary wave). The NIST report notes that the seismic records from the five major WTC events (see Table 2) show no clear P or S wave arrivals. Short-period seismic surface waves, mainly of the Rayleigh type, are the predominant seismic waves on records at regional distances. [19]

Figure 67. (9/11/01) This type of debris doesn't make a thud.
Fulton Street, photo by Terry Schmitt http://ken.ipl31.net/gallery/albums/wtc/img_1480_001.jpg

The World Trade Center was 34km away from the station at Palisades, New York.

> For station PAL (Palisades, NY), data are recorded with 80 samples/second and are recorded by a broadband seismometer (STS-2, T_0=120 seconds), which has a flat response to the input ground velocity from 0.0083 - 50 Hz. Instrument gain is 629.145 counts/[micrometer/sec]. Hence, if you divide amplitude in integer counts by this gain constant, then you get signal amplitude in [micrometer/sec]. If you want to obtain displacement signal, then you must deconvolve instrument response (five poles and three zeros for displacement response). For data in the passband, you may

> simply integrate the velocity data and filter out long period and high frequency ends. It is the same, if you divide velocity signals in each frequencies with omega (= *2 x pi x frequency*).[20]

Emergency Medical Technician Michael Ober does not understand why he cannot remember the sound of a building hitting the ground.

> I don't remember the sound of the building hitting the ground. Somebody told me that it was measured on the Richter scale, I don't know how true that is. If the building is hitting the ground that hard, *how do I not remember the sound of it?* [21] [emphasis added]

This is a significant statement. Each of the WTC towers was 500,000 tons, or the equivalent of 50,000 full ten-ton dump trucks. If 50,000 full dump trucks crashed to the ground, some dropping from over 1/4-mile above the earth, certainly there would be a *very* loud crashing sound that would shake the ground. An empty dump truck speeding down a bumpy road makes a lot of noise, and in that case the road bumps cause drops of only a few inches.

An article describes the various waveforms involved in earthquakes.

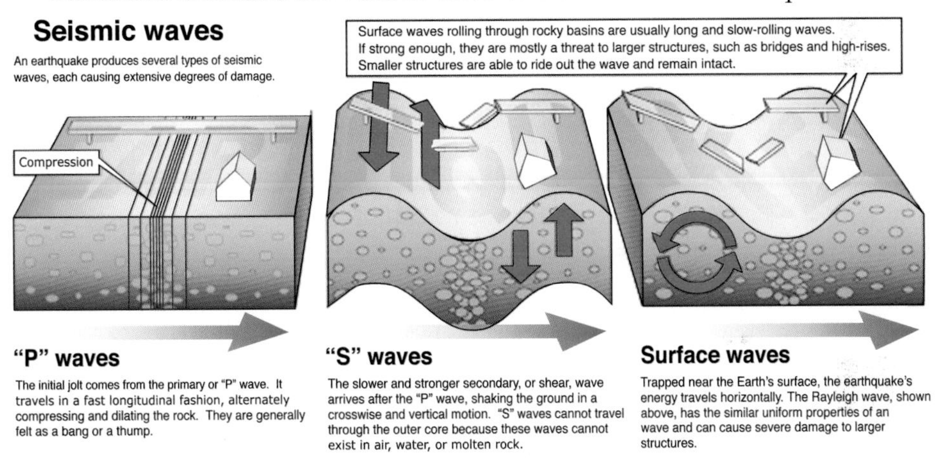

Seismic waves
An earthquake produces several types of seismic waves, each causing extensive degrees of damage.
Surface waves rolling through rocky basins are usually long and slow-rolling waves. If strong enough, they are mostly a threat to larger structures, such as bridges and high-rises. Smaller structures are able to ride out the wave and remain intact.
Compression
"P" waves
The initial jolt comes from the primary or "P" wave. It travels in a fast longitudinal fashion, alternately compressing and dilating the rock. They are generally felt as a bang or a thump.
"S" waves
The slower and stronger secondary, or shear, wave arrives after the "P" wave, shaking the ground in a crosswise and vertical motion. "S" waves cannot travel through the outer core because these waves cannot exist in air, water, or molten rock.
Surface waves
Trapped near the Earth's surface, the earthquake's energy travels horizontally. The Rayleigh wave, shown above, has the similar uniform properties of an wave and can cause severe damage to larger structures.

Figure 68. P-waves and S-waves.[22]

The article continues by explaining the various waveforms involved in earthquakes:

> **Magnitude:** Magnitude is a measure of the amount of energy released during an earthquake. All magnitude scales are calibrated to the original magnitude scale defined by Richter.
>
> **P wave:** Also called primary, longitudinal, irrotational, push, pressure, dilatational, compressional, or push-pull wave. P waves are the fastest body waves and arrive at stations before the S waves, or secondary waves. Their velocity in the crust varies between 5.0 and 7.0 km/s. The waves carry energy through the Earth as longitudinal waves, moving particles in the same line as the direction of the wave. P waves can travel through all layers of the Earth. P waves are generally felt by humans as a bang or thump.
>
> **S wave:** Also called shear, secondary, rotational, tangential, equivoluminal, distortional, transverse, or shake wave. These waves carry energy through the Earth in very complex patterns of transverse (crosswise) waves. These waves

move more slowly than P waves, but in an earthquake they are usually bigger. S waves cannot travel through fluids, such as air, water or molten rock.

Surface wave: Waves that move along the surface of the Earth, Rayleigh and Love waves are surface waves.[23]

Most importantly, *the amplitude of the 9/11 disturbance is less than half that of the January earthquake,* despite a similar peak Richter reading. It is almost as if the data from 9/11 have been attenuated, as if peak movements have been reduced by some kind of filtering process, as though some frequencies were not transmitted. Does this difference reflect real data, that is, differences in real phenomena accurately recorded? Or have the data been filtered asymmetrically or differently? Or have the data been completely manufactured? We do not know for sure, but for the sake of the analysis we use the Richter values reported. Could they have been *lower* than reported? Yes.

1. Kingdome

Maximum Richter readings of 2.3 and 2.1 for the Twin Tower "collapses" were no greater than the controlled demolition of the Seattle Kingdome on March 26, 2000, which "created the equivalent of a magnitude 2.3 earthquake."[24] It would be reasonable to infer that the potential energy from all three collapses was similar in kind, yet each of the towers had as much as 30 times the potential energy of the Kingdome.

2. Kingdome vs. Twin Towers

The Seattle Kingdome was demolished on March 26, 2000. Built of reinforced concrete, it had a 660-foot diameter, a footprint of 342,120 square feet, stood 250 feet tall and weighed an estimated 130,000 tons, equaling about 760 pounds per square foot in ground pressure (at rest). The implosion "created the equivalent of a magnitude 2.3 earthquake, with no vibration damage to adjacent structures." [25]

$A_{Kingdome} = \pi(660/2)^2 = 342{,}119 \text{ ft}^2$
$A_{WTC} = (208)^2 = 43{,}264 \text{ ft}^2$

Each twin tower, by contrast, had a footprint of 43,000 square feet, about a tenth of the Kingdome footprint, and weighed an estimated 500,000 tons, equaling about 23,114 pounds per square foot in ground pressure (at rest), or about 30.41 times that of the Kingdome. Yet the Lamont-Dougherty station at Columbia University only reported a peak of 2.3 Richter scale reading for the "collapse" WTC1 and 2.1 for WTC2, about the same as the Kingdome. (See Figure 60.)

A reading similar to the Kingdome's would be impossible if the twin towers were destroyed by conventional means, that is, from the bottom up (smashing to the ground), because much greater weight would have slammed into a much smaller chunk of land and therefore should have shaken the ground far more than the Kingdome did. In addition, the Kingdome was built on a soft soil basin while the towers were built directly onto bedrock. Each tower's "collapse" should have registered nearly 3.5 on the Richter scale, given the more than an order of magnitude difference between the twin towers' and Kingdome's potential energy and dimensions. (See Figure 73.) The apparent fact that the Richter reading peaked at 2.3 and the disturbance lasted only 8 seconds is our first indicator that the mechanism of destruction could not be a conventional one. An *energy technology of some kind had to have been used to eliminate the towers, while preserving the bathtub and surrounding structures. Conventional technology, which uses kinetic energy (bombs, gravity), cannot explain* the elimination of the towers without there being significant damage to the bathtub below or to the structures immediately surrounding the towers.

F. A Comparison with the Kingdome's Seismic Signal

We may now compare the destruction of the WTC towers with the controlled demolition signal of the Seattle Kingdome.

> The implosion of the Kingdome on March 26, 2001 [sic] generated seismograms at many seismographs in the Seattle area and farther. Detectable signals can be seen as far away as Mount Rainier (station FMW) and in the north Cascades (station RPW). The signals generated are more similar to those generated by a large rockfall than to those generated by an earthquake. Thus determining an earthquake magnitude for this event is not valid. However, using the same techniques for determining magnitude we can estimate an "equivalent" magnitude of 2.3, which matches fairly well with how the event was perceived by people in the area and how far away the seismic waves were recorded. [The seismic] source appears to start at 08:32 PST. [26]

Kingdome vs. Twin Towers

Demolished on March 26, 2000, the Seattle Kingdome was built of reinforced concrete, had a 720-foot outer diameter, a footprint of 407,000 square feet (including the outer ramps), or a 660-foot diameter for the main cylinder, with a 342,120 square-feet footprint. As noted, it stood 250 feet tall and weighed an estimated 130,000 tons. The implosion "created the equivalent of a magnitude 2.3 earthquake, with no vibration damage to adjacent structures."[27]

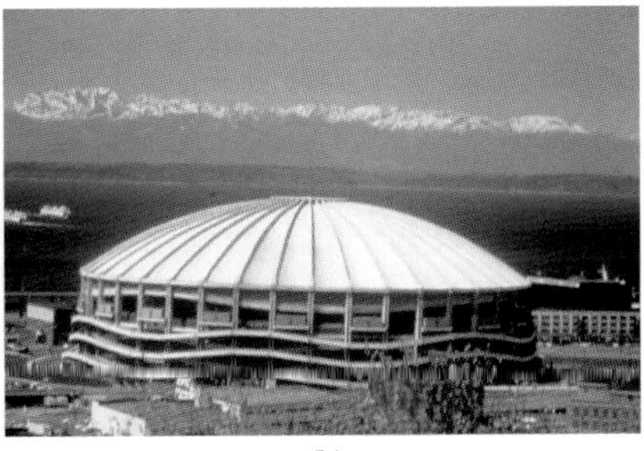

(a) *(b)*

Figure 69. (a) Finishing the interior of the Kingdome.
(b) Kingdome with the Olympic Mountains across Puget Sound.
(a)http://www.kingdome.org/images/img0030.jpg, (b)http://www.kingdome.org/images/img0045.jpg

Figure 70. (a) Demolition of the Kingdome (b) left a debris pile 30 feet high, (12% of its former height).
(a)http://seattletimes.nwsource.com/kingdome/gallery/photos/photo_09.jpg,
(b)http://seattletimes.nwsource.com/kingdome/gallery/photos/photo_05.jpg

Each twin tower, by contrast, had a footprint of 43,000 square feet, just over a tenth of the Kingdome footprint, and weighed an estimated 500,000 tons, resulting in a ground pressure in tons per square of approximately 30.61 times greater than that of the Kingdome. *Yet the Lamont-Dougherty station at Columbia University reported only a peak of 2.3 Richter scale reading for WTC1 and 2.1 for WTC2, about the same as the Kingdome.* The seismic signal from the Seattle Kingdome demolition lasted much longer than the destruction of the WTC, as shown in Figures 71, 78 and 79. The Seward Park seismic disturbance in the vertical direction was about 52 seconds, shown in Figure 71.

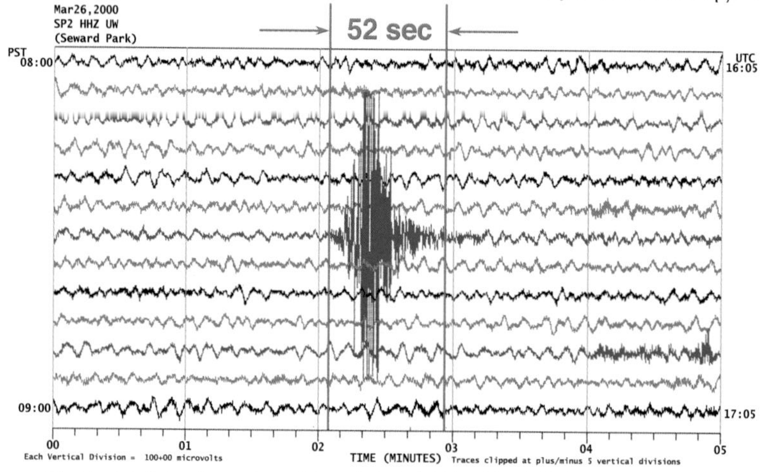

Figure 71. Seattle Kingdome demolition (3/26/00), SP2 HHZ UW (Seward Park).
http://www.geophys.washington.edu/SEIS/PNSN/WEBICORDER/KINGDOME/SP2.webi.kd.gif

Careful data on the Kingdome demolition[29] allow us to estimate what the earthquake-equivalent impact of the Twin Tower destruction should have been. The Kingdome data are pre-9/11 and unlikely to be politically corrupted.

There are, however, other factors to be considered in our analysis. Bedrock conditions are important in affecting earthquake-equivalent Richter readings. If a structure is anchored directly into bedrock, its demolition will yield a higher Richter than if it were not anchored this way. Why? Because if a structure is not anchored into bedrock, the energy released by its demolition is dissipated via the earth's "cushioning" materials. However, if it is anchored into bedrock, the released energy directly impacts that rock, "pinging" the earth directly *without* any dampening, thus allowing the signal to carry more efficiently to recording stations. It is like hitting a hammer against your mattress instead of against a tuning fork.

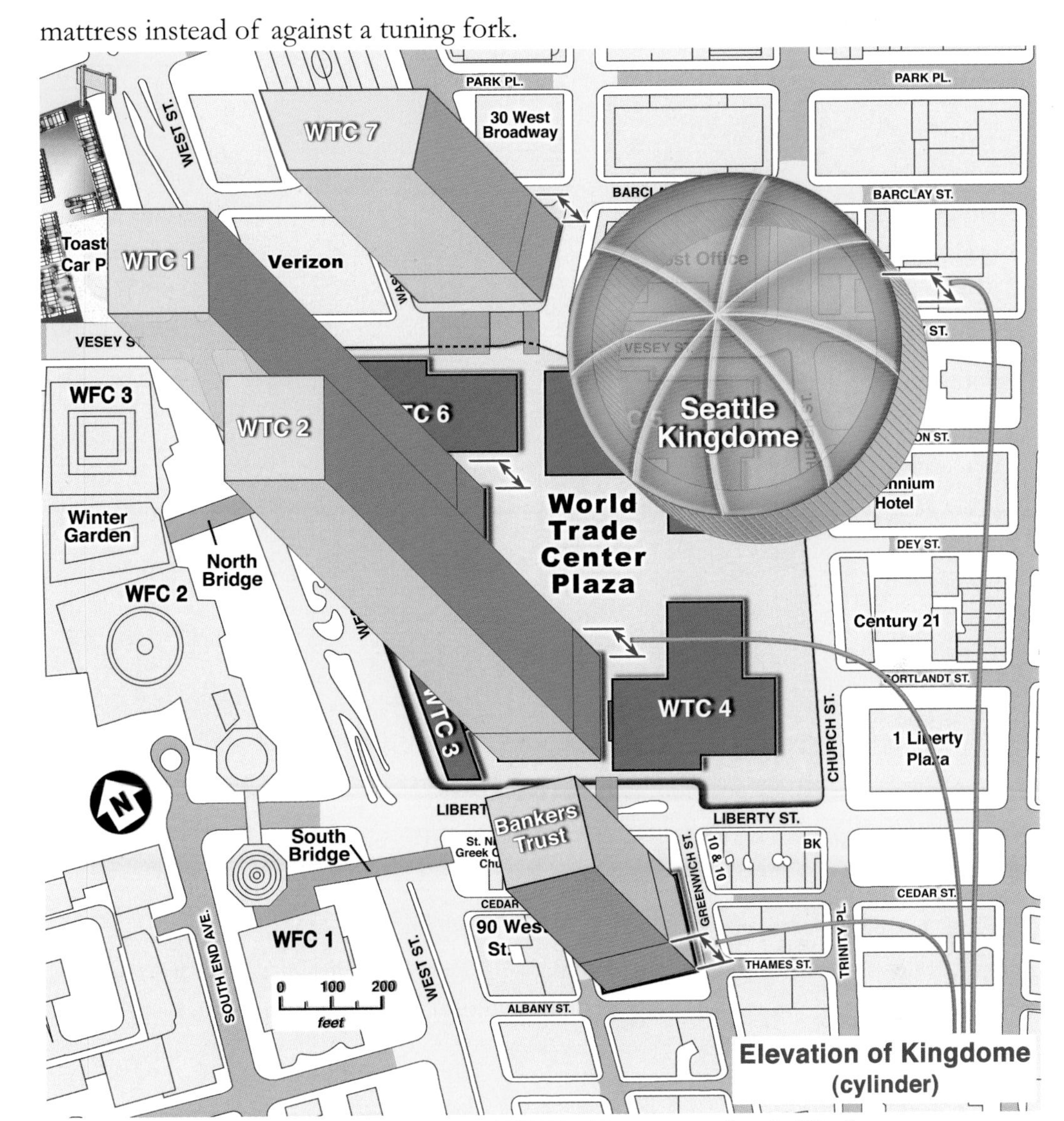

Figure 72. Relative size of the WTC buildings compared to the Kingdome.

The Kingdome was not anchored in bedrock. Consequently, if the Kingdome Richter value was a 2.3 reading transferred through soft material, a building with 30 times the potential energy anchored directly in bedrock should have transferred a *much higher* signal to earthquake-monitoring instruments.

Amazingly, however, the south tower reading of 2.1 was *lower* than the Kingdome's 2.3 *despite* the tower having 30 times the potential energy and being anchored in bedrock. *The difference between these Richter readings implies that the Tower had only 60% of the potential energy of the Kingdome instead of the real range of 3,000%.* The fact that the Towers were anchored in bedrock means that the energy release *should* have rung through to recording instruments loud and clear. But, notably, it did not. *To put it succinctly, once again the physical evidence, in this case* seismic *evidence, is inconsistent with the official explanation of a pancaking building of approximately 500,000 tons slamming into the ground, and,* similarly, it is inconsistent with any theory of a controlled demolition.

	WTC (each tower)	WTC7	Kingdome
steel(tons)	100,000[30]		443[33,34]
concrete(yd^3)	212,500[30]		52,800[33,34]
Concrete(tons)	315,562.5[34]		108,326
Steel+Concrete	415,562.5[34]		108,769
windows	21,800[30]	-	-
electric cables(mi)	6,000[30]	-	-
heating ducts(mi)	198[30]		
floors	110[30]	-	-
dimensions(ft)	208 by 208[30]		dia. =660[31,32,33]
base (ft^2)	43,264[34]	41,538[34]	342,119[34]
Height (ft)	WTC1: 1368[30] WTC2: 1362[30]	650[35]	total: 250[31,32,33] 133.5[31]+116.5[31]
Centroid(ft)	WTC1: 684[34] WTC2: 681[34]	325[34]	180.57[34, 36]
Volume(ft^3)	WTC1: 59,185,152[34] WTC2: 58,925,568[34]	26,999,655[34]	20,756,355+45,672,945 = 66,429,300
Weight (tons)	500,000[30]	228,596[34]	130,000[31,33]
elevators	97+6=103[30]		
Opening-Termination dates	WTC1:12/1970-9/2001[30] WTC2:1/1972-9/2001[30]	3/1987-9/2001	3/27/1976-3/26/2000
	Had **own zip code**, 10048		

Table 3. Data used in calculations for WTC Towers, WTC7, and Kingdome.

Following are some of the calculations I used to derive the comparison of potential energies in the WTC towers' destruction as opposed to the Kingdome's controlled demolition:

WTC1 and WTC2:

Note: 425,000 yd^3 x 3^3 (ft^3/yd^3)x (110)lb/ft^3 x (ton/2,000 lbs.)= 631,125 tons
Assuming this value is for both towers, one tower would be 316,000 tons.
Footprint: 208(ft) x 208(ft) = 43,264 (ft^2)
Volume WTC1: 43,264(ft^2) x 1368(ft) = 59,185,152 (ft^3)
Volume WTC2: 43,264(ft^2) x 1362(ft) = 58,925,568 (ft^3)
Volume WTC1,2: [59,185,152 (ft^3) + 58,925,568 (ft^3)]/2 = 59,055,360(ft^3)

KINGDOME:

Note: 52,800 yd^3 x 3^3 (ft^3/yd^3) x (152)lb/ft^3 x (ton/2,000 lbs.)= 108,346 tons
108,346 tons + 443 tons = 109,000 tons
Footprint: $\pi r^2 = \pi$ x 330(ft) x 330(ft) = 342,119 (ft^2)
Volume (top + cylinder): 20,756,355 + 45,672,945 = 66,429,300 (ft^3)

WTC7:

Footprint: (246.6+ 329)/2= 287.8(ft)
287.8(ft) x 144.3(ft) = 41,538(ft^2)
Volume: 41,538(ft^2) x 650(ft) = 26,999,655 (ft^3)
Approx. wt.: (26,999,655/59,055,360)* 500,000(tons) = 228,596 (tons)

The following article from *The Seattle Times* is further worth noting:

> Dust choked downtown for nearly 20 minutes, blocking out the sun and leaving a layer of film on cars, streets and storefronts. The dust cloud reached nearly as high as the top of the Bank of America Tower and drifted northwest about **8 miles an hour.**
>
> [...]
>
> Carefully placed explosives—4,461 pounds in all—collapsed the 25,000-ton roof like a cake taken out of the oven too soon. More than 21 miles of detonating cord exploded in a flash. The Dome's roof ribs and columns looked like they had been electrified with lightning.
>
> Rapid puffs of smoke followed, and the massive roof ribs that formed the Dome's 20 arches buckled first in three pie-shaped wedges. Then came the remaining three roof wedges, followed instantly by explosions in the support columns and in the roof's tension ring, which had held the roof together by exerting 8 million pounds of force around its base.
>
> While nearly the entire Dome, which once weighed about 130,000 tons, collapsed in on its own "footprint," chunks of concrete flew onto rooftops. The force of the blasts broke windows at the Salvation Army and Turner Construction buildings on Fourth Avenue South, and at F.X. McRory's steakhouse on South King Street. Residents of the nearby Florentine Condominiums had been taken to the restaurant earlier that morning, but no one was injured.
>
> A small army of street sweepers went into action moments after the blast. Businesses around the Dome were quick to reopen, with little damage reported. Engineers will survey adjacent buildings and structures over the next few days to assess any damage.
>
> The implosion registered a magnitude 2.3 on the Richter scale—a barely detectable ground motion that naturally visits the region once or twice a month. Scientists will use ground-vibration data from the implosion to learn more about the Seattle fault, which runs a few blocks south of the Kingdome.
>
> By afternoon, the job of pulverizing and hauling away the Kingdome was under way, with hydraulic jackhammers breaking columns into chunks. A couple hundred people gathered close to the site, taking pictures and searching for bits of the building to take home.
>
> The rubble is flatter than expected, only reaching about 30 feet high near the perimeter of the 9-acre Dome site. The Dome once stood 250 feet high.[37]

In other words, if the WTC had 425,000 cubic yards of lightweight concrete (72% the weight of normal concrete), then there were approximately 631,000 tons of concrete in the complex. This is a crude cross-check on the weight of the towers and the WTC, and it suggests that 500,000 tons is not an exaggeration.[38]

The *Seattle Times* account of the Kingdome demolition contrasts sharply with the destruction of the Twin Towers, as shown below:

1) "Dust choked downtown [Seattle] for nearly 20 minutes," yet ultra-fine dust plagued lower Manhattan for weeks and months;
2) Dust "drifted northwest about 8 miles an hour," the pace of stragglers at the end of a 26-mile marathon, yet people running full speed could not outrun the dust cloud from the Twin Tower destruction; and finally,
3) "Carefully placed explosives—4,461 pounds in all—collapsed" the Kingdome, which would imply the need for 17,158 pounds just to bring down a tower over a quarter-mile-high but not pulverize it or guarantee its falling within its own footprint. If we adjust for the tower's height of center of mass, potential energy from 67 *tons* of explosives would be required, resulting in an implied 3.5 Richter reading, far above the 2.3 reported for WTC1. Yet even this quantity of explosives would fail to pulverize the concrete and would leave an enormous rubble pile to jackhammer into smaller pieces, *none* of which were in evidence at the WTC sites.

G. Using the magnitude scale

The magnitude scale, patterned after the Richter scale, is a logarithmic scale that measures the amount of force or energy released by an earthquake. The scale is adjusted for different regions of the earth. Table 4 compares magnitude to the seismic energy yield of quantities of the explosive TNT.

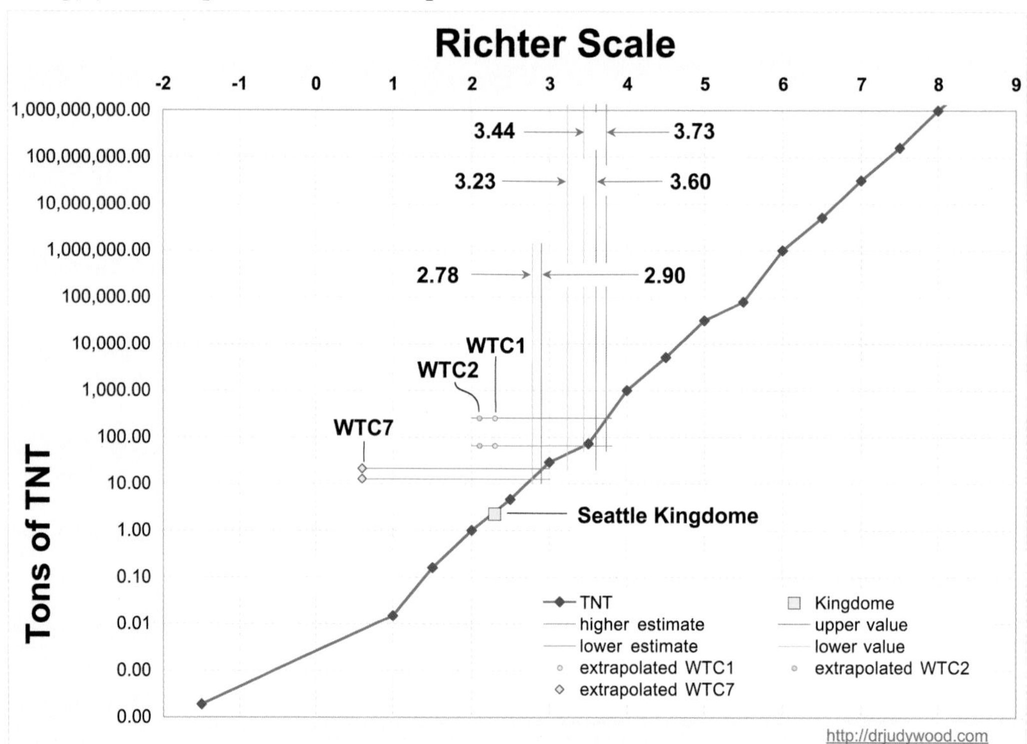

Figure 73. Correlation between Richter Scale and Tons of TNT.[39]
Adopted from: *http://www.lvrj.com/lvrj_home/1999/Apr-11-Sun-1999/photos/quakebig.jpg,* Dept. of Energy, USGS.

Magnitude	TNT for energy yield (tons)	Example (approximate)
-0.5	*6 ounces* 0.0001875	Breaking a rock on a lab table
1.0	*30 pounds* 0.015	Large blast at a construction site
1.5	*320 pounds* 0.16	-
2.0	1	Large quarry or mine blast
2.5	4.6	-
3.0	29	-
3.5	73	-
4.0	1,000	Small nuclear weapon
4.5	5,100	Average tornado
5.0	32,000	-
5.5	80,000	Little Skull Mountain, Nev., quake, 1992
6.0	1 million	Double Spring Flat, Nev., quake, 1994
6.5	5 million	Northridge, Calif., quake, 1994
7.0	32 million	Ryogo-Ken Nanbu, Japan, quake, 1995
7.5	160 million	Landers, Calif., quake, 1992
8.0	1 Billion	San Francisco, Calif., quake, 1906
8.5	5 Billion	Anchorage, Alaska, quake, 1964
9.0	32 Billion	Chilean quake, 1960
10.0	1 Trillion	San Andreas-type fault circling the Earth
12.0	160 Trillion	Fault dividing Earth in half through center or Earth's daily receipt of solar energy.

Table 4. Seismic Magnitude equivalence of energy yield of TNT.[40]

Extrapolation

Note again that in Seattle "Dust choked downtown for nearly 20 minutes" (not days or months).

And the dust following the Kingdome demolition drifted at about 8 miles an hour, while in NYC on 9/11, no one could out-run the rapidly expanding dust cloud. This information allows us to do some further calculations for the Kingdome:

The rubble height was 30 out of the original 250 feet height. 30ft/250ft = 12%
110 x 12% = 13.2 stories for the WTC
4,461 lbs. x (500,000/130,000) = 17,158 lbs. = (40 people) x (10 lbs. each trip) x (43 trips). That is, 40 people, each making 43 trips, carrying 10 pounds each trip.

But note: the Seattle Kingdome is not pulverized, nor is it controlled into its own footprint. Explosives only get the chunks down on the ground where they can be broken up and hauled away.

A little more calculating reveals more interesting things:

21 miles of detonating cord x (1368/250) = 115 miles of detonating cord needed for

WTC1, extrapolating from relative height.

That article in the Seattle paper also says that "hydraulic jackhammers (were) breaking columns into chunks," that is, the concrete of the Kingdome *was not pulverized.* And again, the article states clearly that "chunks of concrete flew onto rooftops." In contrast, the roof of Bankers Trust, directly across the street from WTC2, had only a few pieces of aluminum cladding on it and no wheatchex or other debris of significant mass (see Figure 195 in Chapter 10).

Now let's do a *Comparison of Potential Energy* of the Kingdome as opposed to the twin towers.

If each tower was constituted of 100,000 tons of steel and had a total weight of 500,000 tons, that means the steel is only 20% of the mass. So, if all the steel *except* that in the lower 36 floors were pulverized, then the lower 36 floors would be fairly light. I went through those numbers and discovered only the steel of the lower 36 floors are equivalent to the Kingdome's potential energy: That is, the bottom 36 floors of a 110 floor-building (where the entire 110 floors weigh 100,000-tons) has the same potential energy as the Kingdome.

The Kingdome did not have its weight evenly distributed. There was more density lower down, so one would expect the center of gravity to be lower than the geometric center. This would produce a lower potential energy than I actually used. But, on the other hand, the WTC was heavier on the lower floors than the upper floors, which would also produce a slightly lower center of gravity as well as a slightly lower potential energy. So the ratio of the WTC's potential energy to the Kingdome's potential energy is a reasonable approximation.

We know that each WTC tower did not slam to the earth and register as a 3.8 Magnitude earthquake. We also know that a lot of the building came down as dust.

So, if we assume that every floor contained 1/110th of the building's total mass, the bottom 20 floors of WTC1 alone would hold the same potential energy as the Kingdome. But, when the event on 9/11 was all over, we did *not* see the lower floors stacked up like pancakes that had slammed to the ground. What happened to all the concrete and marble? What happened to all the glass? What happened to all the desks? What we *did* see was a bunch of steel beams. So, if we were left with only steel beams, how many floors' worth of steel would have the same potential energy as the Kingdome?

The weight of all the structural steel in the building was 100,000-tons, 20% of the weight of the entire building. Again, if we assume that every floor contained 1/110th of the building's total mass of structural steel, just the steel in the bottom 36 floors of WTC2 had the same potential energy as the Kingdome.

So, as an approximation, the structural steel of WTC2 made up 36/110th of 1/5th the total mass of the building, or 6.5% of the building's mass. If this mass were evenly distributed over 36 floors, it would have the same proportional potential energy, relative to the Kingdome, that could be expected to cause the equivalent of a 2.1 earthquake when it slammed to the ground. But is this reasonable, considering the debris remaining after the event?

H. Compare WTC1 and WTC7 with the Kingdome's Seismic Strength —What Does That Say?

Here's the actual calculation: We know that the WTC towers must have had approximately 30 times the Kingdome's potential energy, given their mass and smaller footprints. The log of 30 yields 1.5, which must be added to 2.3 Richter for the Kingdome to yield 3.8 Richter for the towers. *But this result is not what the seismic data of the towers' destruction recorded at all, as we have previously seen.* Thus, *the seismic data does not support a gravity-driven or controlled demolition-driven "collapse."* The towers did not *collapse. They were destroyed where they stood, from the top down, by another mechanism entirely than collapse or controlled demolition.*

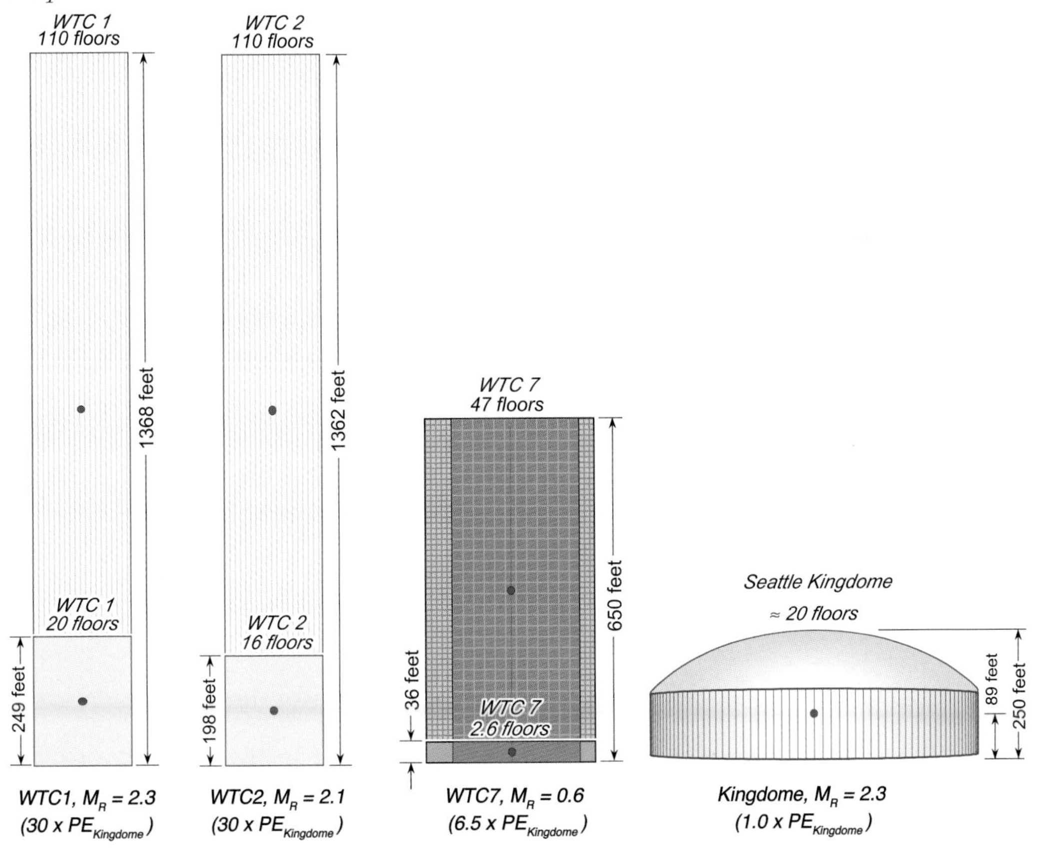

Figure 74. Physical size and approximate seismic size of WTC 1, 2, and 7.
The corresponding potential energy (PE) of each building is given relative to the Kingdome.

A seismographic chart similar to that describing the Kingdome demolition would be impossible if the twin towers had been destroyed by conventional means from the bottom up, since their far greater weight would have slammed into a far smaller chunk of land and would have shaken the ground much more severely than the Kingdome did. Each tower's collapse should have registered at least 4 on the Richter scale, given two orders of magnitude difference between the twin towers and Kingdome dimensions. *We conclude therefore that the apparent fact that the Richter reading peaked at 2.3 and the disturbance lasted only 8 seconds is an indicator that unconventional means were deployed to destroy the towers.* In the next section of this book we will discuss why no conventional means of destruction can explain the evidence, and therefore why it must have been an unconventional means, and more specifically, an unconventional energy weapon, that was used above ground to destroy the towers while preserving the bathtub and surrounding structures. The charts in Figures 74 and 75 help illustrate the calculations that brought me to these conclusions.

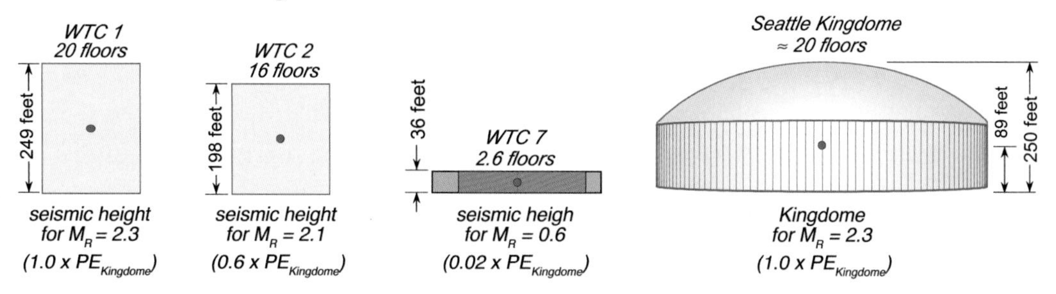

Figure 75. The approximate seismic height of WTC 1, 2, and 7 is shown relative to the Kingdome.

The log of 1 is zero, so, for WTC1, the mass of the lower 20 stories yields the equivalent magnitude earthquake as the Kingdome. For a building the size of WTC1 to have the same potential energy as the Kingdome, it would need to be 3.4% of its original density. That is, WTC1 would need to be 33.7/1,000 (33.7 thousandths) of its original density to explain the seismic signal. Similarly, a building the size of WTC7 would need to be 0.31% its original density to produce a seismic signal equivalent to a magnitude of M_R=0.6. That is, WTC7 would need to be 3.07/1,000 (3.1 thousandths) of its original density to explain the seismic signal. How can this be?

From Chapter 2, we know it would take over 9.22 seconds for a billiard ball dropped from the roof of WTC1 to hit the ground, without adjusting for air resistance. In addition, the seismic impact that did result was equivalent to what would be expected from just the lower 16-20 stories of each tower, not 110 stories weighting 500,000 tons.

I. Spatter Damage Height on Adjacent Buildings

There is one more important data set to consider in the refutation of the "collapse" and "controlled demolition" models, and this is the evident height of splatter damage to adjacent buildings. That is, we must ask whether this damage is consistent with projectiles of concrete-chunks and other debris being hurled against them as each tower undergoes a "controlled demolition"?

Consider figures 76 and 77.

Figure 76. (a) The tower is being peeled downward. Dark clouds shoot up, while light ones expand outward. The building has vanished above the white "snowball" while the lower part awaits termination. (b) Solid debris appears to have hit only the lower half of this 40-story building (Bankers Trust). However, the top few floors appear to have had their windows blown out.
(a)http://hereisnewyork.org//jpegs/photos/5245.jpg, (b)http://parrhesia.com/wtc/wtc034.jpg

Figure 77. WFC3 does not appear it was hit by any large objects above the 18th floor.
(a)http://parrhesia.com/wtc/wtc054.jpg, (b)http://parrhesia.com/wtc/wtc042.jpg

J. Kingdome's Seismic Signature

We've seen what the actual calculations for a collapse of the WTC towers indicate *should* have been seismically recorded. But now look at the actual charts themselves.

The Seattle Kingdome was made of a cylindrical base with a domed roof. The top of the dome was 250 feet from the ground, but most of the weight was in the cylindrical base, which was 133.5-feet high (see Table 3).

Once again, some simple calculations are quite revealing:

$t = \sqrt{\frac{2h}{g}}$, *where h = 250 ft (76 m), g (gravity)* = 32.2 ft/s^2 (9.81 m/s^2),

t = 3.94 seconds, or *t = 3.9 s*econds, and, t = 2.9 seconds for the 133.5-foot (41 m) height, where most of the mass was.

The seismic signal recorded during the demolition of the Seattle Kingdome is shown in Figures 71, 78 and 79.

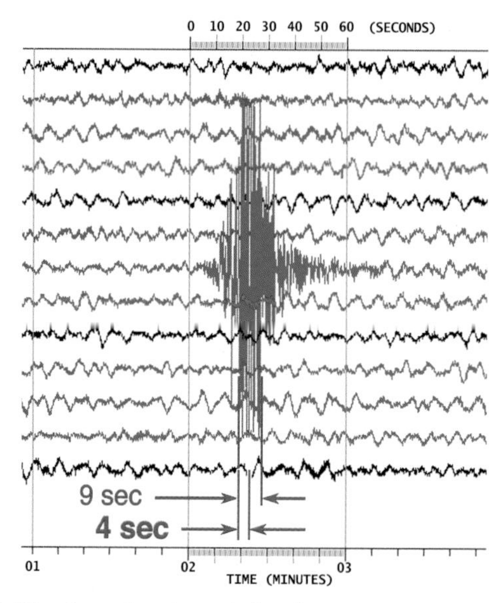

Figure 78. Seattle Kingdome demolition (3/26/00), SP2 HHZ UW (Seward Park).
http://www.geophys.washington.edu/SEIS/PNSN/WEBICORDER/KINGDOME/SP2.webi.kd.gif

Although conventional controlled demolition is known to bring buildings down very quickly, the ground will shake for much longer than the time it takes for something dropped from the roof to freely fall to the ground. The free-fall time of about four seconds is superimposed on the seismic chart from the Seattle Kingdome demolition and shown in Figures 78. In this recording, peak values are clipped at plus/minus 5 vertical divisions.[41] The duration of this clipping of peak signals is approximately 9 seconds, well beyond the calculated fall times of three and four seconds. The duration of the Seward Park seismic disturbance in the vertical direction (SP2 HHZ UW) was about 52 seconds, which was shown in Figure 71. The main Queen Anne seismic disturbance in the horizontal direction (QAW, ELN and QAW, ELE) was about 24 seconds, as shown in Figure 79. So it is clear that falling debris takes much longer to settle to the ground than something free-falling from the roof.

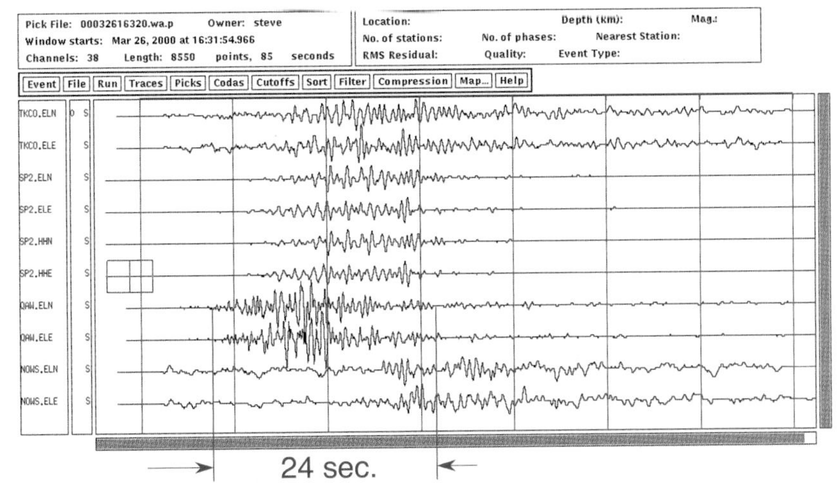

Figure 79. Seismographs of the Kingdome demolition recorded at various locations.
http://www.geophys.washington.edu/SEIS/PNSN/WEBICORDER/KINGDOME/kd.gif

In conventional controlled demolition, explosives are used to destroy the building's supports so that large sections of it will slam to the ground. The detonation of the explosives is heard and the chunks of building slamming to the ground is heard. Such evidence was not documented by video cameras recording the events on 9/11.

K. Sound of WTC7

As discussed before, EMT Michael Ober, a first responder, also mentions the *absence of hearing* the supposedly pancaking buildings—with all the terrible noise and roar that shearing steal and concrete should make in breaking away and hitting the ground. It is assumed that he is referring to not hearing a building "collapse" for the entire day.

> I don't remember the sound of the building hitting the ground. Somebody told me that it was measured on the Richter scale, I don't know how true that is. If the building is hitting the ground that hard, *how do I not remember the sound of it?* [42] [emphasis added]

A closer look at the *sound* of the "collapse" of WTC7 is in order:

> *5.7.5 Audio Characteristics Based on Video Soundtracks:*
>
> Three videos in the database included soundtracks that were used to investigate the audio signature associated with the period immediately prior to and during the collapse of WTC 7. All of these cameras were located at street level at least 640 m (2100 ft) from the building. Also, there were numerous other buildings between the cameras and WTC 7.
>
> The most usable soundtrack was recorded by Camera 3, with its West Street location. This video ran for many minutes *prior to and during* the collapse. *Even though sound was recorded by the camera, no interviews or commentary were recorded, and the microphone tended to pick up low level street sounds, such as sirens, traffic, and distant conversations.* Occasionally, the camera operators located nearby were recorded at a much louder level. Since the collapse was recorded on the video, it was possible to coordinate the sound recording with the actual WTC 7 collapse.

> A careful review of the audio clip *did not reveal any sounds that could be associated with WTC 7 until the global collapse began.* A low level waveform for the audio signal using Aftereffects software [was employed]. This video also did not reveal any features *that could be associated with the collapse until after the global collapse began.* In the analysis, the roughly 2 s delay in sound transmission between WTC 7 and the camera was accounted for. The amplitude of the sound signal increased while the global collapse was taking place, *but there were no loud, explosive sounds when the collapse began.*
>
> The response of the camera operators provides another indication of the audio environment. *Even though the east penthouse began to descend into the building 6.9 s prior to initiation of global collapse, there was no verbal response from the camera operators until 2.5 s after the global collapse began, when a loud shout of whoa, whoa, whoa, whoa was heard.* There is no evidence that the operators heard something that attracted their attention prior to this time.
>
> At the same time the Camera 3 video was being shot, a recorded street interview was being conducted a short distance away on West Street. In this video clip of the interview, WTC 7 is visible in the upper left hand corner of the frame. Even though *the east penthouse can be seen disappearing into the building, neither the camera operator, interviewer, nor interviewee responded in any way until just over 3 s after the global collapse began.* Again, there was no indication that sounds loud enough to attract attention or cause alarm were heard by people at the interview location prior to global collapse initiation. [43] [emphases added]

NIST stated that the soundtracks from the videos recording the destruction of WTC7 did not contain any sound as intense as would have accompanied such a blast, yet there is no descriptive analysis of what sound levels *should* accompany the sudden gravity-driven collapse model proposed. Sound is used as one of the criteria in eliminating the consideration of a blast event as cause of the destruction of WTC7. But the theory of a gravity-caused collapse has not been tested by the same sound-measurement criteria.

WTC7 weighed approximately 200,000 tons. That's equivalent in mass to about 10,000 to 20,000 dump trucks distributed in space throughout the height of the building. If those "trucks" suddenly collapsed to the ground, the sound should be audible, should register seismically, and should be included in NIST's analysis.

NIST acknowledges that it did not do an analysis of the soundtracks in order to verify its collapse hypothesis but, instead, used soundtrack analysis only to confirm that there was *no loud sound* that would have been expected from a blast event of the kind hypothesized here. NIST is aware that its work in this respect may be challenged as fraudulent.

L. Seismic Disturbance from WTC7

Now let's apply the same calculations to WTC7 that we applied to WTC1 and WTC2 in section C, "It Didn't Last Long Enough."

$t = \sqrt{\frac{2h}{g}}$, *where h = 650 ft (198 m), g (gravity)* = 32.2 ft/s^2 (9.81 m/s^2),

and t = 6.355 seconds, or *t = 6.4 s* for ball dropping from roof to ground in a vacuum.

So, t = sqrt ((2*h)/g) = sqrt ((2*650)/32.2) = 6.3539 seconds = ~6.4 seconds.
The seismic signal during the destruction of WTC7 was recorded at five seismographic stations in the area, shown in Figure 64.

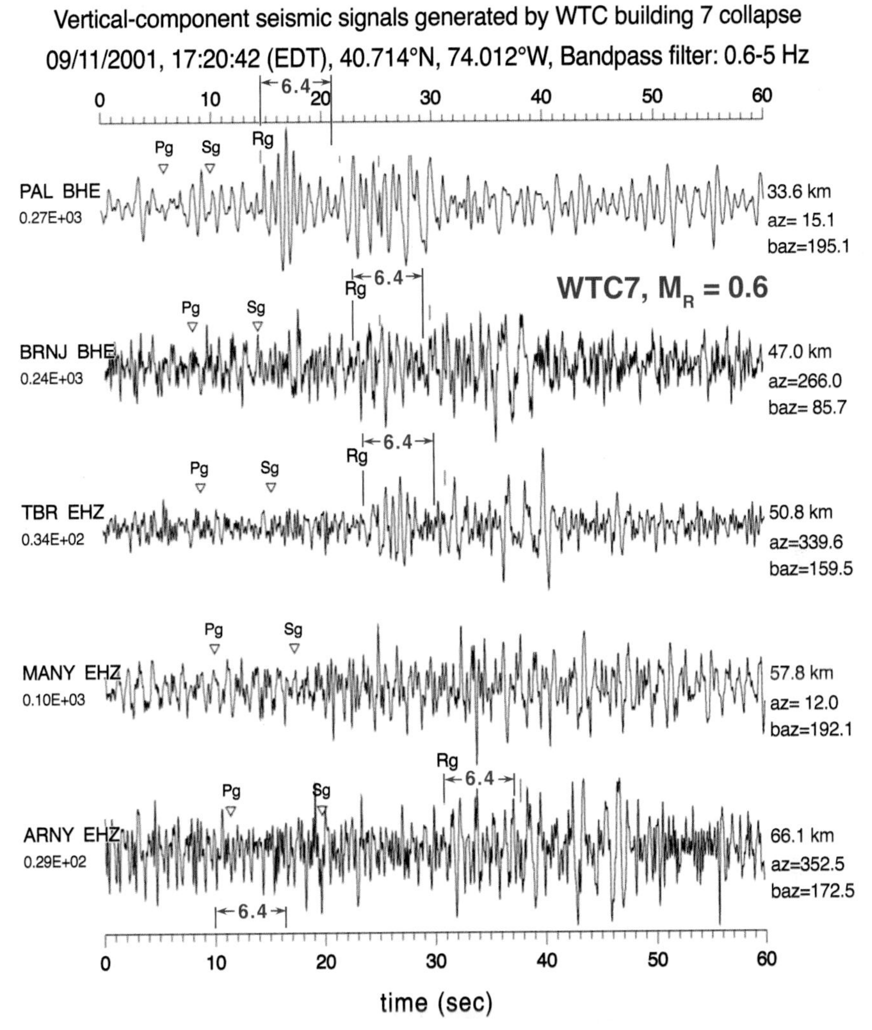

Figure 80. Seismic signals at various locations WTC7.
page 658 (pdf p. 320 of 382), http://wtc.nist.gov/media/NIST_NCSTAR_1-9_vol2_for_public_comment.pdf

The east-west and vertical component seismic records, filtered to pass signals below 5 Hz[44], are shown in Figure 80 with the time span of 6.4 seconds superimposed on each one having a recognized arrival time, R_g.

> R_g indicates the arrival of the seismic surface wave of the Rayleigh type which propagated with a speed of 2.4 km/s. P_g and S_g indicate P wave and S-wave estimated arrival times from the WTC site using a velocity model for the region (these wave arrivals were not detectable). The Rayleigh waves were quite weak and difficult to discern clearly at most of the stations. Nevertheless, R_g waves at PAL, BRNJ, TBR and ARNY were stronger than the background noise in Figure B5 [Figure 80]. [45]

In other words, there's basically no seismic event for WTC7's demise.

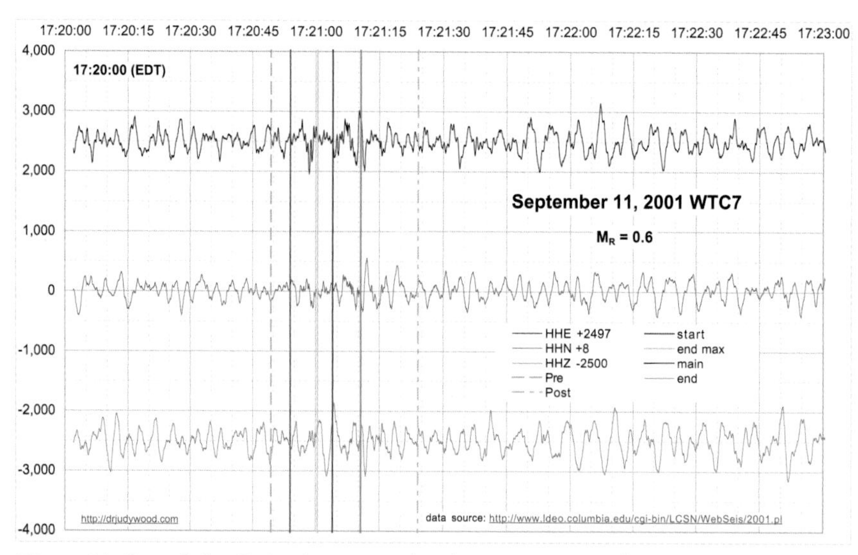

Figure 81. Recorded in Palisades, NY, 9/11/2001 17:20:33,[46] *the WTC7 "collapse."*
Data: http://www.ldeo.columbia.edu/cgi-bin/LCSN/WebSeis/2001.pl

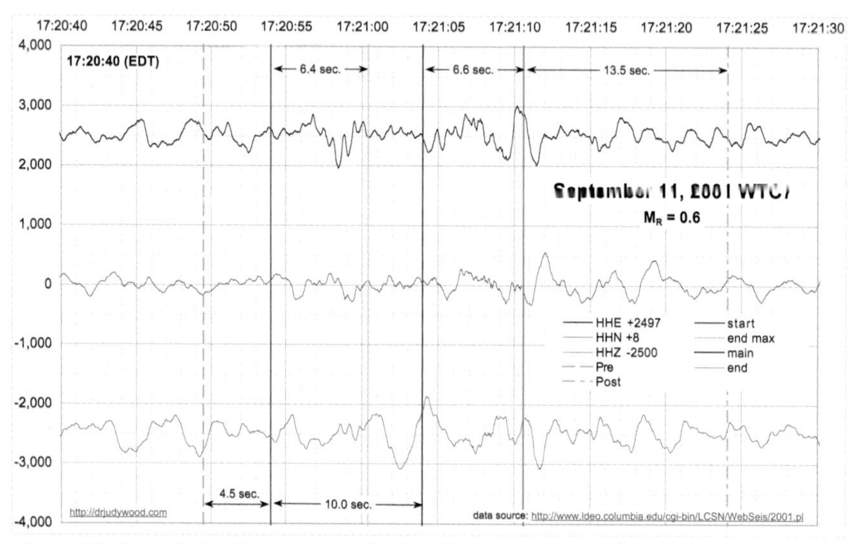

Figure 82. Recorded in Palisades, NY, 9/11/2001 17:20:33,[47] *the WTC7 "collapse."*
Data source: http://www.ldeo.columbia.edu/cgi-bin/LCSN/WebSeis/2001.pl

Let us review the three types of waves shown in Figure 68.

> [**P waves**]: carry energy through the Earth as longitudinal waves, moving particles in the same line as the direction of the wave. P waves can travel through all layers of the Earth. P waves are generally felt by humans as a bang or thump.[48]
>
> [**S waves**]: carry energy through the Earth in very complex patterns of transverse (crosswise) waves. These waves move more slowly than P waves, but in an earthquake they are usually bigger. S waves cannot travel through fluids, such as air, water or molten rock.[49]
>
> [**Surface waves**]: move along the surface of the Earth, Rayleigh and Love waves are surface waves.[50]

Figure 83. (pre 9/11/01) Looking north from Vesey Street.
(a)http://digitalgallery.nypl.org/nypldigital/dgkeysearchdetail.cfm?strucID=311582&imageID=505925, (b)http://digitalgallery.nypl.org/nypldigital/dgkeysearchdetail.cfm?strucID=311582&imageID=505924

Figure 84. (pre and post 9/11/01) The red arrow corresponds with the same location in Figure 85. Debris from WTC7 did not reach the sidewalk adjacent to the Postal Building.
(a)http://i56.photobucket.com/albums/g171/boloboffin2/911/WTC7TopfromWTC2.jpg, (b)USGS, (c)http://forums.therandirhodesshow.com/index.php?act=Attach&type=post&id=19357

NIST states that the surface waves were "quite weak and difficult to discern clearly at most of the stations"[51] and that "[s]eismic records from the five major WTC events show no clear P or S wave arrivals."[52] This indicates that the signal did not travel *through* the earth, which is consistent with the buildings *not crashing* to the earth. In other words, the seismic signals recorded are not consistent with a pancaking collapse, something hitting the ground, or debris being hammered into the basement. And debris was not hammered into the basement, as demonstrated by almost no damage to the bathtub.

Figure 85. (9/13/01) Looking south on West Broadway shows that the debris from WTC7 didn't even reach across the street to the Post Office. The red arrow corresponds with the same location in Figure 84.
http://www.hybrideb.com/source/eyewitness/complex/081.jpg

M. Bathtub Protection, or, "But What about the Bathtub?"

But why be concerned about keeping the bathtub intact?

We have seen before, and can see again in Figure 86, that the bathtub, remarkably, was left virtually intact, suffering almost no damage from the "collapse" of the two 110 story towers that stood above it. The bathtub could survive the collapse of the Twin Towers only if the majority of the buildings' mass was turned into powder before it landed. The reason for bringing about this result is quite simple: It avoided severely impacting the foundation of the towers, an event that would have caused subsequent and most likely massive and highly destructive flooding.

Figure 86. Clearly, the bathtub survived. This is a view from the footprint of WTC1. Notice that the parking structure that was under WTC6 also survived.
http://911research.wtc7.net/wtc/arch/docs/bathtub_wall1.jpg

N. Conclusions

What may be concluded from the seismic evidence? As outlined in this chapter, a comparison of the controlled demolition of the Seattle Kingdome and the destruction of the WTC towers reveals the following:

1) The smaller footprints of the WTC towers, plus their greater mass, as compared to the Seattle Kingdome, would result in a greater potential energy;
2) This greater potential energy would in turn require a commensurately larger Richter scale signature of the WTC towers' collapse, as compared to the Seattle Kingdome, although this was *not* evidenced by the actual seismic readings for the WTC towers' destruction;
3) The controlled demolition model, further, does not account for the lack of projectile damage on adjacent buildings at levels higher than the 20th story of those buildings;
4) The controlled demolition model does not account for the lack of significant damage to the bathtub beneath the WTC towers;
5) The controlled demolition model does not account for the "dustification" of the concrete within the towers;
6) It is also likely that if terrorists were responsible for the WTC towers' destruction, a controlled demolition would have been designed to *maximize* collateral damage to adjacent buildings, rather than *minimize* it, as is the actual case;
7) Based on the inadequacy of either model to explain the evidence, we conclude that an *unconventional* method was deployed to destroy the WTC towers. The conclusion that an unconventional means was a directed energy technology of some sort will be discussed later in this book.

But before we can examine the case for the deployment of unconventional directed energy technology, we must look more closely at the "controlled demolition" and "bombs in the building" models, in the next chapter.

---

[1] *http://images.google.com/hosted/life/f?q=World+Trade+Center+source:life&imgurl=4860fd0beb71abb4,* Date taken: 1971, by Henry Groskinsky, *http://images.google.com/hosted/life/l?imgurl=4860fd0beb71abb4&q=World+Trade+Center+source:life&usg=__AKIUx7-F3FCsE5Kz9ZDV7VWVXQk=&prev=/images%3Fq%3DWorld%2BTrade%2BCenter%2Bsource:life%26start%3D20%26hl%3Den%26sa%3DN%26ie%3DUTF-8*

[2] *http://images.google.com/hosted/life/f?q=World+Trade+Center+source:life&imgurl=744fbfaa572cd0e2,* Date taken: 1971, by Henry Groskinsky, *http://images.google.com/hosted/life/l?imgurl=744fbfaa572cd0e2&q=World+Trade+Center+source:life&usg=__Tc4l-m2mKJccH-GoEVMz1iUJMyI=&prev=/images%3Fq%3DWorld%2BTrade%2BCenter%2Bsource:life%26hl%3Den%26ie%3DUTF-8*

[3] *http://img.timeinc.net/time/photoessays/groundzero/zero03.jpg, http://www.time.com/time/photoessays/groundzero/2.html*

[4] *http://drjudywood.com/media/WTCSeismicSignatureNCSTAd.mp3, http://wtc.nist.gov/media/NCSTACmeetingDec06.htm*

[5] UTC = Coordinated Universal Time, also known as Greenwich Mean Time (GMT) where 8:40 AM EDT = 12:40 UTC.

[6] "Seismic Waves Generated by Aircraft Impacts and Building Collapses at World Trade Center, New York City," *http://www.ldeo.columbia.edu/LCSN/Eq /20010911_WTC/WTC_LDEO_KIM.pdf*

[7] "Fact Sheet," *Answers to Frequently Asked Questions,* The National Institutes of Standards and Technology (NIST*), http://wtc.nist.gov/pubs/ factsheets/faqs_8_2006.htm.*

[8] The height of the South Tower (WTC2) is 1362 feet, and the height of the North Tower (WTC1) is 1368 feet. *http://www.infoplease.com/spot/wtc1.html.*

[9] Seismology Group, Lamont-Doherty Earth Observatory, Columbia University. *http://www.ldeo.columbia.edu/LCSN/Eq/20010911_wtc.html*

[10] Table B-4, page 667 (pdf 329 of 382), *http://wtc.nist.gov/media/NIST_NCSTAR_1-9_vol2_for_public_comment.pdf*

[11] Won-Young Kim, L. R. Sykes1, J.H.Armitage, J. K.Xie, K.H. Jacob, P.G.Richards1, M. West1, F. Waldhauser, J. Armbruster, L. Seeber, W. X. Du1 and A. Lerner-Lam1, "Seismic Waves Generated by Aircraft Impacts and Building Collapses at World Trade Center, New York City," *Lamont-Doherty Earth Observatory of Columbia University*, Palisades, N.Y. 10964, USA; also *Dept. Earth and Environmental Sciences, Columbia University. http://www.ldeo.columbia.edu/LCSN/Eq/20010911_WTC/WTC_LDEO_KIM.pdf*

[12] *http://www.ldeo.columbia.edu/LCSN/Eq/20010911_WTC/WTC_LDEO_KIM.pdf*

[13] Seismology Group, Lamont-Doherty Earth Observatory, Columbia University. *http://www.ldeo.columbia.edu/LCSN/Eq/20010911_wtc.html*

[14] Seismology Group, Lamont-Doherty Earth Observatory, Columbia University. *http://www.ldeo.columbia.edu/LCSN/Eq/20010911_wtc.html*

[15] Data: *http://www.ldeo.columbia.edu/cgi-bin/LCSN/WebSeis/2001.pl*

[16] Seismology Group, Lamont-Doherty Earth Observatory, Columbia University. *http://www.ldeo.columbia.edu/LCSN/Eq/20010911_wtc.html*

[17] Seismology Group, Lamont-Doherty Earth Observatory, Columbia University. *http://www.ldeo.columbia.edu/LCSN/Eq/20010911_wtc.html*

[18] Data: *http://www.ldeo.columbia.edu/cgi-bin/LCSN/WebSeis/2001.pl*

[19] page 654 (pdf p. 316 of 382), *http://wtc.nist.gov/media/NIST_NCSTAR_1-9_vol2_for_public_comment.pdf*

[20] *ftp://ftp.ldeo.columbia.edu/archive/LCSN/WTC/Waveform_Data/Readme*

[21] File No. 9110093, *World Trade Center Task Force Interview,* "EMT Michael Ober," October 16, 2001, p. 10, *http://graphics8,nytimes.com/packages/html/nyregion/20050812_WTC_GRAPHIC/9110093.PDF*

[22] Redrawn from: *http://www.lvrj.com/lvrj_home/1999/Apr-11-Sun-1999/photos/quakebig.jpg,* Sources: Dept. of Energy, USGS.

[23] M.Lamontagne, S.Halchuk, J.F.Cassidy, and G.C.Rogers, "Significant Canadian Earthquakes 1600-2006," *Geological Survey of Canada, Open File 5539, http://earthquakescanada.nrcan.gc.ca/historic_eq/GSCOF5539/GSCOF5539_Lamontagne_etal.pdf*

[24] Byles, Jeff, *Rubble: Unearthing the History of Demolition*, Crown Publishing, New York (2005) p. 73.

[25] Byles, Jeff, *Rubble: Unearthing the History of Demolition*, Crown Publishing, New York (2005) p. 73.

[26] *http://www.pnsn.org/WEBICORDER/KINGDOME/welcome.html*

[27] Byles, Jeff, *Rubble: Unearthing the History of Demolition*, Crown Publishing, New York (2005) p. 73.

[28] For those interested, there is a video of the Kingdome demolition at these sites: *http://drjudywood.com/media/Kingdome_319180.mpg, http://portland.indymedia.org/media/media/2005/06/319180.mpg*

[29] The Seattle Kingdome was demolished on March 26, 2000, *http://seattletimes.nwsource.com/special/kingdome/k_implosion.html*

[30] (a) *http://www.infoplease.com/spot/wtc1.html*

[31] (b) *http://mlb.mlb.com/NASApp/mlb/sea/history/ballparks.jsp*

[32] (c) *http://www.kingdome.org/*

[33] (d) *http://seattletimes.nwsource.com/kingdome/k_gravity.html*

[34] (e) calculated

[35] (g) FEMA, NIST

[36] (f) The centroid is the center of cross-sectional area and not the center of mass. For the towers, the centroid and the center of volume are the same. For the Kingdome, the center of volume would be lower than the center of cross-sectional area and correspond to a lower potential energy. Also, most of the weight of the Kingdome was located in the base, so the center of mass would be lower than the center of volume, meaning the potential energy calculations for the Kingdome would be lower, using the center of mass.

[37] *http://archives.seattletimes.nwsorce.com/cgi-bin/texis.cgi/web/vortex/display?slug=4012219&date=20000327.*

[38] Note: 425,000 yd^3 x 3^3 (ft^3/yd^3)x (110)lb/ft^3 x (ton/2,000 lbs)= 631,125 tons. Assuming this value is for both towers, one tower would be 316,000 tons.

If the Kingdome had 52,800 cubic yards of normal concrete, then there were 109,000 tons of concrete in the dome. Therefore, the 130,000 ton estimate of the Kingdome's weight seems reasonable.

Note: 52,800 yd^3 x 3^3 (ft^3/yd^3)x (152)lb/ft^3 x (ton/2,000 lbs)= 108,346 tons, 108,346 tons + 443 tons ≈ 109,000 tons

[39] *http://www.lvrj.com/lvrj_home/1999/Apr-11-Sun-1999/photos/quakebig.jpg,* Dept. of Energy, USGS.

[40] Adopted from source: *http://www.lvrj.com/lvrj_home/1999/Apr-11-Sun-1999/photos/quakebig.jpg,* Mike Johnson, Review-Journal. Sources: Dept. of Energy, USGS.

[41] *http://www.geophys.washington.edu/SEIS/PNSN/WEBICORDER/KINGDOME/SP2.webi.kd.gif*

[42] File No. 9110093, "EMT Michael Ober," October 16, 2001, p. 10, *http://graphics8.nytimes.com/packages/pdf/nyregion/20050812_WTC_GRAPHIC/9110093.PDF*

[43] Page 333-334 of 404 of pdf, (labeled page 289-290 of report) *http://wtc.nist.gov/media/NIST_NCSTAR_1-9_Vol1_for_public_comment.pdf*

[44] page 657 (pdf p. 319 of 382), *http://wtc.nist.gov/media/NIST_NCSTAR_1-9_vol2_for_public_comment.pdf*

[45] page 657 (pdf p. 319 of 382), *http://wtc.nist.gov/media/NIST_NCSTAR_1-9_vol2_for_public_comment.pdf*

[46] Seismology Group, Lamont-Doherty Earth Observatory, Columbia University. *http://www.ldeo.columbia.edu/LCSN/Eq/20010911_wtc.html*

[47] Seismology Group, Lamont-Doherty Earth Observatory, Columbia University. *http://www.ldeo.columbia.edu/LCSN/Eq/20010911_wtc.html*

[48] M.Lamontagne, S.Halchuk, J.F.Cassidy, and G.C.Rogers, "Significant Canadian Earthquakes 1600-2006," *Geological Survey of Canada, Open File 5539, http://earthquakescanada.nrcan.gc.ca/historic_eq/GSCOF5539/GSCOF5539_Lamontagne_etal.pdf*

[49] M.Lamontagne, S.Halchuk, J.F.Cassidy, and G.C.Rogers, "Significant Canadian Earthquakes 1600-2006," *Geological Survey of Canada, Open File 5539, http://earthquakescanada.nrcan.gc.ca/historic_eq/GSCOF5539/GSCOF5539_Lamontagne_etal.pdf*

[50] M.Lamontagne, S.Halchuk, J.F.Cassidy, and G.C.Rogers, "Significant Canadian Earthquakes 1600-2006," *Geological Survey of Canada, Open File 5539, http://earthquakescanada.nrcan.gc.ca/historic_eq/GSCOF5539/GSCOF5539_Lamontagne_etal.pdf*

[51] page 657 (pdf p. 319 of 382), *http://wtc.nist.gov/media/NIST_NCSTAR_1-9_vol2_for_public_comment.pdf*

[52] page 657 (pdf p. 319 of 382), *http://wtc.nist.gov/media/NIST_NCSTAR_1-9_vol2_for_public_comment.pdf*

7.

Conventional Controlled Demolition: "Bombs in the Building"

To go against the dominant thinking of your friends, of most of the people you see every day, is perhaps the most difficult act of heroism you can have. —T.H.White

Whenever you find yourself on the side of the majority, it's time to pause and reflect. —Mark Twain

Even if you are only a minority of one the truth is still the truth. —Gandhi

On 9/11/01, the World Trade Center towers went away faster than they could have in a gravity-driven collapse. According to the seismographic evidence, WTC1 was destroyed more quickly than a billiard ball could free-fall from the roof to the ground in a vacuum. It has been suggested that thermite was used to destroy the buildings. It has also been suggested that the buildings were destroyed by Conventional Controlled Demolition (CCD) or even that "mini-nukes" blew up the buildings. The danger in grabbing on to quick *answers* of these kinds before evaluating all of the available evidence is that such positions, by definition, will lead to a bias favoring those explanations. As a result, data that contradicts a firmly held belief may be or is even likely to be overlooked or disregarded. Therefore, instead of grabbing on to a theory, let us grab on to the evidence. The evidence is the truth that the *correct* theory must mimic. Empirical evidence is the key. If we listen to the evidence carefully enough, it will teach us exactly what happened. The key is being ready to hear the evidence when it reveals itself.

We do not need to *know* what the answer is in order to *find* that answer, because we are capable of learning. That is, we need not be familiar with the cause in order to determine that cause, so long as we are capable of learning. If we are capable, we can then test our knowledge base against the evidence and eliminate explanations that contradict the evidence. For example, the official story that WTC1 "collapsed" in 8 seconds contradicts the laws of physics for free-falling bodies.

Consider the example in Figure 87. The roofline of Block A will take at least 9.22 seconds to hit the ground, not accounting for air resistance. If Block A (eight stories, or 36,000 tons) were dropped from a quarter mile above the ground in free-fall, it would hit the ground 9 seconds later, traveling at 200 mph. The impact would be the equivalent of 4000 dump trucks slamming to the ground from 1/4 mile above. One can imagine the seismic impact that would make.

But we did not have just Block A freely dropping to the ground at the World Trade Center. The case of Block B's path to the ground may give us a better approximation. That is, for Block B to hit the ground, 500,000 tons of building must slam to the ground with it—but in the same amount of time. However, if Block B encounters the resistance of the building under it, including compressing and ejecting all material on each floor, it will certainly take longer than 9 seconds.

In addition, we know that the seismic impact was less than would be expected if even just the top two floors alone had fallen (this invites the reader to wonder what the two-floor force of impact value is and may leave a question). So the value of 8 seconds contradicts the official story of a gravity-driven collapse, and the seismic evidence contradicts the assertion that a 500,000-ton building (WTC1 or 2) or a 230,000-ton building (WTC7) actually did slam to the ground.

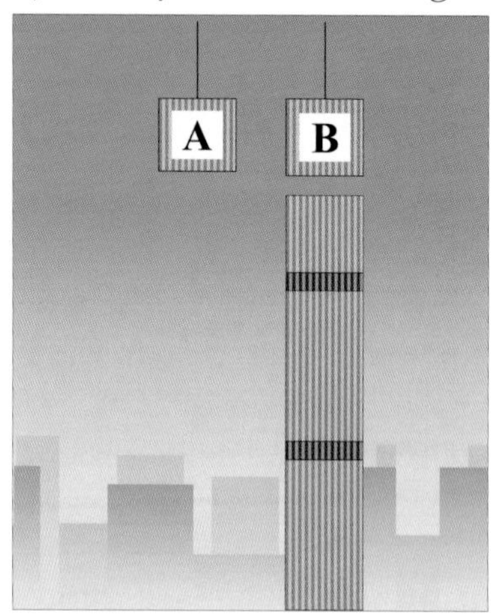

Figure 87. Diagram illustrating the concept of the building's resistance.

The official explanation cannot be the right explanation, as the example of Block A makes clear. The length of time is too short. "Bombs in the building" might help an argument in the case of Block B. Bombs might at least have blown out some of the supports, thus shortening the longer "collapse" time somewhat, but there is still the problem of 500,000 tons of building slamming to the ground with great force. That just didn't happen.

Two facts, a 2.3 Richter reading and the intact bathtub, contradict 500,000 tons of building material slamming to the ground in 8 to 10 seconds for whatever reason. Therefore, on these grounds alone we must rule out both the official explanation and the explanation of conventional controlled demolition ("bombs in the building"), whether with dynamite, RDX, or other kinds of explosives. Additionally, we can see that a number of other details preclude the possibility of "bombs in the building."

A. Conventional Controlled Demolition (CCD)

1. Chunks? (But we have Powder)

When buildings are demolished using Conventional Controlled Demolition (CCD), preparation must be made. Anything that can be removed must be removed, including non load-bearing walls. Glass windows must also be removed so that they do not become deadly projectiles, and so must anything else that could become a projectile. When things blow up, chunks go flying, and they stay as chunks until they hit something. Chunks of concrete are shown in Figure 88. So in planned controlled-demolition projects, the problem of projectiles is reduced to a minimum before the actual demolition. If the WTC buildings had been demolished with explosives, such preparation could not have taken place, for reasons explained below. As a result, there would have been a great number of projectiles ejected from the buildings. But this is not what we saw. Instead, there is virtually no evidence of projectiles having damaged the surrounding buildings—*and none at all above the 18th to 20th floors. In addition, there are virtually no chunks of debris, but only powder (see Chapter 8, "Dustification").*

Figure 88. Explosives testing facility.
http://www.dtra.mil/images/newsservices/full_photos/cp30l.JPG (now removed)

2. Preparing for Conventional Controlled Demolition (CCD)

In short, conventional Controlled Demolition requires considerable preparation in time and equipment. Once again, we will use details from the Kingdome to get an idea of what might be involved. The Kingdome used 4,461 pounds of carefully placed explosives and over 21 miles of detonating cord. Preparation in the case of the Kingdome began more than a month before the actual demolition. Careful planning took place as to the location, timing, and the quantity of explosions required for the job, as explained in this *Seattle Post-Intelligencer* article by Robert L. Jamieson from February 24, 2000.

> People at downtown's Columbia Tower even heard it. "I was at my desk," said Bev Devlin, a traffic reporter on the 66th floor. "I was surprised."
>
> What they, and many other early risers, heard at 7 a.m. was the blast of explosives at the Kingdome, mere blocks away. The planned mini-detonation, which triggered car alarms, was a test to see how Kingdome concrete would react to a small punch of gelatin dynamite.
>
> Information from the blast will help demolition experts calculate the fire power needed to implode the 110,000-ton stadium on March 19 or March 26.
>
> If not enough explosives are used, for example, the Dome may not collapse with ease; if too much explosives are used, parts of the Dome could hurl beyond expectation.
>
> [...]
>
> "This test shot is like a diagnosis a doctor gives before the operation," explained Mark Loizeaux, president of Controlled Demolition Inc., the Maryland-based firm hired to do the implosion. "If you get the wrong diagnosis, the operation is not going to go well."
>
> [...]
>
> "If we weaken parts of it," Gerlach said of the Dome, "gravity will do the rest."

> His words tidily summed up "implosion"—a word that contrary to the pyrotechnic imagery it invokes, is actually more like high-tech acupuncture; crews ply dynamite sticks instead of needles to bring down a structure, with safety and precision.
>
> In addition to information gleaned from yesterday's test blast, CDI and the contractors have plenty of blueprints and structural plans of the stadium from which to draw. So they know the building—its strengths and weaknesses—as if it were an old spouse.
>
> Crews today will continue to ready the Dome for implosion; the Public Stadium Authority board, will meet with city officials this afternoon to discuss related demolition issues, including preliminary street closures and outreach to homes and businesses affected by the implosion.
>
> In terms of public spectacle, today's proceedings will probably not compete with yesterday's baby boom, which left a puff of smoke and a bang.
>
> But yesterday's noise, Loizeaux said, will be dwarfed by Boomsday, which will probably break a few windows. But Loizeaux added the concussive force, or air blasts, from the big implosion would be lessened by drawing the detonation out over 15 or so seconds, thus lowering the air pressure.[1]

Each WTC tower comprised 4 times the mass of the Kingdome, but each would require more than 4 times the explosives because the mass of the Kingdome was not as concentrated, but was distributed over a quarter mile. It also follows that much more detonating cord would be required as well. If the equipment needed was so much larger, we can project that the time involved with the Kingdome set-up—at least a month—would have been concomitantly more extensive, *especially because it would have had to be done in secret.* One might speculate that the risk of exposure would have been too great for such a venture in any case. However, had each tower been blown up with conventional explosives, 500,000 tons of building material would have slammed down onto the foundation and would almost certainly have destroyed or at least significantly damaged the protective bathtub and flooded lower Manhattan as well as the subway tunnels and the basements of all buildings connected to these tunnels.

Some individuals have argued that the buildings were pre-wired for demolition when they were built. Among the problems with this hypothesis is that explosives have a shelf life, just like firecrackers, such that their reliability diminishes with time. Also, think of the safety hazard of having a building loaded with bombs. ("Oops, boss, I hit the wrong switch. Sorry about your building.") Because of the danger involved, it is illegal for vehicles to carry explosives (even bottled gas) through submerged tunnels. This is true of the Baltimore Harbor Tunnel (I-895), which opened in 1957, and the Fort McHenry Tunnel (I-95), which opened in 1985.[2] The Maryland Transportation Authority states:

> Vehicles carrying bottled propane gas in excess of 10 pounds per container (maximum of 10 containers), bulk gasoline, *explosives*, significant amounts of radioactive materials, and other hazardous materials are *prohibited* from using both the Fort McHenry and Baltimore Harbor Tunnels. In addition, vehicles in excess of 13 feet, 6 inches, in height, or 96 inches (8 feet) in width; and all double trailers are prohibited from using the Baltimore Harbor Tunnel. Vehicles carrying Class

> 1 explosives and radioactive materials require an escort at the Francis Scott Key Bridge. For additional information regarding these restrictions, please call 410-537-1374. [emphasis added] [3]

So how could one possibly justify pre-wiring a quarter-mile-high building? What would happen if there were a fire on two to three floors, such as the fire in 1975? It has also been reported that the towers were patrolled by bomb-sniffing dogs.

> The World Trade Center was destroyed just days after a heightened security alert was lifted at the landmark 110-story towers, security personnel said yesterday.
>
> Daria Coard, 37, a guard at Tower One, said the security detail had been working 12-hour shifts for the past two weeks because of numerous phone threats. But on Thursday, bomb-sniffing dogs were abruptly removed.
>
> "Today was the first day there was not the extra security," Coard said. "We were protecting below. We had the ground covered. We didn't figure they would do it with planes. There is no way anyone could have stopped that."[4]

Police K9 Sirius, one of the bomb-sniffing dogs, apparently had not detected any bombs in the building up until an explosion was heard on the morning of 9/11/01.

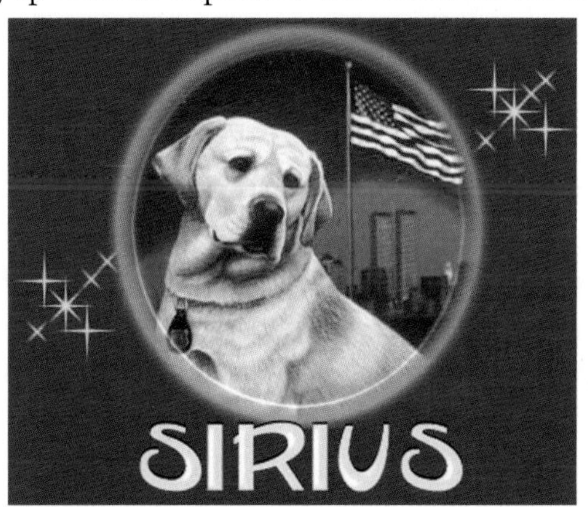

Figure 89. Police K9 Sirius[5], Badge Number 17.
Title image of Sirius adapted from a portrait by Debbie Stonebraker and used with the artist's permission provided credit is given and a link provided to Debbie's site.
http://www.novareinna.com/bridge/sirius.jpg, http://www.stonebrakerart.com/sirius.html

> Police K9 Sirius, Badge Number 17...a four-and-a-half-year old, ninety pound, easygoing, yellow Labrador Retriever...was an Explosive Detection Dog with the Port Authority of New York and New Jersey Police Department. Sirius, along with his partner, Police Officer David Lim, were assigned to the World Trade Center in New York, where their primary duty was to check vehicles entering the Complex, clear unattended bags and sweep areas for VIP safety. Sirius, who began work at the World Trade Center on July 4, 2000, was the only police dog to perish during the attack on the Twin Towers.
>
> On the morning of September 11, 2001, Sirius and Officer Lim were at their Station located in the basement of Tower Two. When Officer Lim heard the explosion, he thought at first that a bomb had been detonated inside the building. Believing he would be more effective alone, Officer Lim left Sirius locked in his six-foot by ten-foot crate, telling him, "I'll be back to get you," as

> he rushed to help with the rescue effort. At that time, Officer Lim could think of no safer place for his canine companion other than the basement. However, Officer Lim failed to return to Sirius. Becoming trapped in the falling debris of Tower One [sic], he wasn't rescued until some five hours or more later. Sadly, in the meantime, Sirius had perished when Tower Two collapsed. The remains of the loyal Sirius were recovered on January 22, 2002. Thankfully, it is believed that he died instantly when his kennel caved-in.[6]

What would be the purpose in having bomb-sniffing dogs patrol the building while placing explosives in the building? There is no record showing that these dogs had detected such explosives either during their careers or in patrolling the buildings immediately prior to 9/11/01. In addition, it would be inconceivably unwise to design a pre-planned demolition at the time of construction without knowing what buildings would be erected in the neighborhood during the lifetime of the towers, surrounding structures not having yet been built.

3. Ignition Temperature

(a) Ignition Temperatures of Office Material

A further piece of evidence that we must consider in our examination of the "controlled conventional demolition with bombs in the building" is the problem of ignition temperatures and standard office materials:

> The burning of a solid fuel often depends on the form of the fuel. The ignition temperatures of fuels differ. For a solid or liquid fuel to ignite, some of the fuel must first be heated to the temperature at which it vaporizes (turns to a gas). Solids generally have higher ignition temperatures than liquids because they vaporize at higher temperatures. For example, the ignition temperatures of most woods and plastics range from about 500 to 900 °F (260 to 480 °C). A liquid fuel such as gasoline can ignite at a temperature as low as -36 °F (-38 °C).[7]

(b) Ignition Temperatures of Commercial Fluids

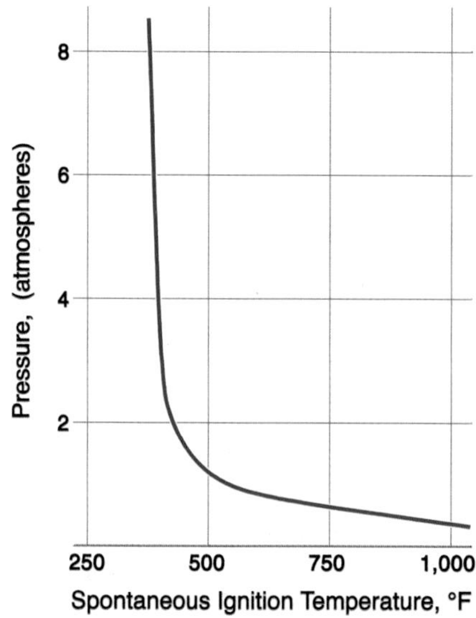

Figure 90. Spontaneous ignition temperature of JP-4 versus pressure.[8]

Pressure atm.	Ignition Temperature °F	
	JP-4	JP-5
1	484°F	477°F
5	376°F	415°F
9	378°F	408°F

Table 5. Variation of ignition temperature with pressure. [reproduced[9]]

Combustible	Pressure	
	1/2 atm.	1 atm.
JP-4	831°F	468°F
JP-3	840°F	460°F
JP-1	864°F	442°F
Av. Gas 100/130	1027°F	824°F
Av. Gas 115/145	1063°F	880°F
n-hexane	927°F	453°F
n-octane	869°F	428°F
n-decane	856°F	406°F
Hydraulic Fluid AN-0-366	838°F	437°F

Table 6. Ignition temperatures of commercial fluids at 2 pressures.[10]

(c) Ignition Temperatures of Commercial Explosives and Incendiaries

RDX: The boiling point of RDX is 234°C[11] [453°F]

C4: Ignition temperature: 440 – 465 °C[12] [824-869°F]

Blasting Powder: The ignition temperature of blasting powder is 518°F., and of rifle powder 528.8°F. The finest grades of sporting powder ignite at about 600°F.[13]

Properties of Nitroglycerine: Its ignition temperature, or, more properly, firing point, is 356°F. Exposed to a temperature of 365°F. it boils with the evolution of vapors. At 381.2°F. it volatilizes slowly. At 392°F. it evaporates rapidly. At 422.6°F. it detonates violently.[14]

TNT Ignition temp: 240°C[15] [464°F]

Dynamite: Dynamite Ignition temperature: min. 180°C[16] [356°F]

Thermite Ignition temp: Conventional thermite reactions require very high temperatures for initiation. These cannot be reached with conventional black-powder fuses, nitrocellulose rods, detonators, or other common igniting substances. Even when the thermite is hot enough to glow bright red, it will not ignite as it must be at or near white-hot to initiate the reaction. It is possible to start the reaction using a propane torch if done right, but this should never be attempted for safety reasons. The torch can preheat the entire pile of thermite which will make it explode instead of burning slowly when it finally reaches ignition temperature.

Often, strips of magnesium metal are used as fuses. Magnesium burns at approximately the temperature at which thermite reacts, around 2500 Kelvin

(4000 °F). This method is notoriously unreliable: magnesium itself is hard to ignite, and in windy or wet conditions the strip may be extinguished. Also, magnesium strips do not contain their own oxygen source so ignition cannot occur through a small hole. A significant danger of magnesium ignition is the fact that the metal is an excellent conductor of heat; heating one end of the ribbon may cause the other end to transfer enough heat to the thermite to cause premature ignition. Despite these issues, magnesium ignition remains popular amongst amateur thermite users.

The reaction between potassium permanganate and glycerine[sic] is used as an alternative to the magnesium method. When these two substances mix, a spontaneous reaction will begin, slowly increasing the temperature of the mixture until flames are produced. The heat released by the oxidation of glycerine[sic] is sufficient to initiate a thermite reaction. However, this method can also be unreliable and the delay between mixing and ignition can vary greatly due to factors such as particle size and ambient temperature.

Another method of igniting is to use a common sparkler to ignite the mix. These reach the necessary temperatures and provide a sufficient amount of time before the burning point reaches the sample.

A stoichiometric mixture of finely powdered Fe(III) oxide and aluminum may be ignited using ordinary red-tipped book matches by partially embedding one match head in the mixture, and igniting that match head with another match, preferably held with tongs in gloves to prevent flash burns.[17]

Material	Ignition °F	Ignition °C	Speed
Thermite*[18]	4000°F	2500°K	N/A
HMX[19]	997°F	536°C	9,100 m/s
C4[20]	824-869°F	440 - 465°C	8,050 m/s
Sporting powder[21]	600°F	316°C	
Rifle powder[22]	528.8°F	276°C	
Blasting Powder[23]	518°F	270°C	
Office material*[24] (woods and plastics)	500 - 900°F	260 - 480°C	
JP4 (jet fuel)*[25]	484°F	251°C	
JP4 (jet fuel)*[26]	468°F	242°C	
TNT[27]	464°F	240°C	6,900 m/s
RDX[28]	453°F	234°C	8,750 m/s
PETN[29]	286°F	141.3°C	8,400 m/s
Nitroglycerine[30]	50-60°F	122-140°C	7,700 m/s
Dynamite[31]	356°F	180°C	
gasoline[32]	-36 °F	-38 °C	

*Table 7. Ignition temperatures for various materials. (*Not an explosive.)*

If jet fuel (JP4) had been ignited in a building where controlled-demolition explosives were set to go off, the burning fuel would have ignited almost all conventional explosives. If burning jet fuel had ignited office material, the burning office material would have ignited almost any of the conventional explosives listed. Jet fuel (JP4) is kerosene and is not an explosive—but it does burn at temperatures above 484°F. JP4 burns at lower temperatures under pressure, as shown in Figure 90, but the pressure of a torn-open fuel tank would not be much greater than atmospheric pressure.

Therefore, *if* a fire had been ignited from JP4 jet fuel, the temperature of that fire could be anticipated to be about 468-484°F and greater if office materials began to burn. These temperatures are greater than those needed to ignite TNT, RDX, nitroglycerine, dynamite, or gasoline, *yet no such explosions were reported shortly after the initial fireball.* If such bombs had been placed in this area of each building at some time, they were no longer viable explosives. If there were no bombs in the top part of the building, there could be no top-down controlled demolition with conventional explosives. One exception is the C4 explosive, since it burns at a higher temperature and could have survived the initial fireball. However, C4 explosives blow material apart into chunks and launch these chunks at high rates of speed.

> To make C-4 blocks, explosives manufacturers take RDX in powder form and mix it with water to form a slurry. They then add the binder material, dissolved in a solvent, and mix the materials with an agitator. They remove the solvent through distillation, and remove the water through drying and filtering. The result is a relatively stable, solid explosive with a consistency similar to modeling clay.
>
> Just as with other explosives, you need to apply some energy to C-4 to kick off the chemical reaction. Because of the stabilizer elements, it takes a considerable shock to set off this reaction; lighting the C-4 with a match will just make it burn slowly, like a piece of wood (in Vietnam, soldiers actually burned C-4 as an improvised cooking fire). Even shooting the explosive with a rifle won't trigger the reaction. Only a detonator, or blasting cap will do the job properly.

Figure 91. (9/01) The façade of WFC2 appeared undamaged beyond the broken windows and the missing marble around the entrance.
http://img.photobucket.com/albums/v214/shadow-ace/sept%2011/windows.jpg.

> A detonator is just a smaller explosive that's relatively easy to set off. An electrical detonator, for example, uses a brief charge to set off a small amount of explosive material. When somebody triggers the detonator (by transmitting

> the charge through detonator cord to a blasting cap, for example), the explosion applies a powerful shock that triggers the C-4 explosive material.
>
> When the chemical reaction begins, the C-4 decomposes to release a variety of gases (notably, nitrogen and carbon oxides). *The gases initially expand at about 26,400 feet per second* (8,050 meters per second), applying a huge amount of force to everything in the surrounding area. At this expansion rate, it is totally impossible to outrun the explosion like they do in dozens of action movies. To the observer, the explosion is nearly instantaneous—one second, everything's normal, and the next it's totally destroyed.[33]

As said, the gases from a C-4 explosion initially expand at about 26,400 feet per second, which is about 18,000 miles per hour, or 23.5 times the speed of sound. So detonating such an explosive in quantities great enough to destroy the Twin Towers would have produced significant blast waves.

Figure 92. (approx. 9/13/01) The windows of WFC1 have rounded holes in them. These are double-paned windows, but the two panes do not show the same damage. In some cases, the inner window is intact while the majority of the outer window is missing. The marble-facade around the doorway is completely missing. The remaining building facade does not appear damaged.
http://img.photobucket.com/albums/v214/shadow-ace/sept%2011/aftermast25b.jpg,

But nothing of the sort was reported by any witness on 9/11/01, nor was there any photographic evidence consistent with explosive forces of this nature. The World Financial Center building 2 (WFC2), directly across West Street from WTC1, was missing its decorative marble around the lobby door, but the metal façade of the building surrounding the windows on floors 1 through 18 showed no significant damage or pock marks. Other than damage to the marble, there was no apparent harm done to the façade.

4. Cell Phone Issue

In the hypothetical case of controlled demolition, the detonations (by definition) must be controlled. Typically in such cases, explosives are detonated remotely. With the number and precise sequencing of detonations required for such a massive project, the number of independent signals would necessarily be extremely high.

Clearly, there is also a need to protect a prospective demolition site from random signal interference that could have disastrous consequences, which is why we are familiar with the following types of signs.

(a) (b)

Figure 93. How can all cell phones be turned off in New York City?
(a)http://www.mhd.state.ma.us/downloads/trafficMgmt/signs/warning/w22-1.jpg, (b)http://www.mhd.state.ma.us/downloads/trafficMgmt/signs/warning/w22-2.jpg

We are also told when it is safe to use cell phones again:

Figure 94. End of Blasting Zone sign signifies cell phone use can resume.
http://www.oce.oregon.gov/images/products/details/W22-3.jpg

The signs are to avoid premature detonation by stray signals from electronic devices during the preparation of the site.

Think of the implications of massive cell phone use by New Yorkers in Lower Manhattan prior to 9/11—surely including the vast majority of people who worked in the towers.

We also know that people trapped in the buildings used their cell phones to call loved ones shortly before the towers were destroyed. Thus the cell phones could not have functioned as detonators.

5. Has This Been Done Before?

Figure 95. (10/24/1998) J.L. Hudson Department Store, 5:47 PM.
Photos courtesy of Controlled-Demolition, Inc.[34]

Prior to 9/11, the tallest building ever demolished by controlled demolition was the J.L. Hudson Department Store in Detroit.[35] At 1368 feet, WTC1 was over 3 times taller than the Hudson's 439 feet. The largest building by volume was the Seattle Kingdome, which was approximately the same volume as one of the WTC towers, though the towers were 5.5 times the height and 3.85 times the mass. The Kingdome was essentially an empty shell.

Figure 96. (03/26/2000) Seattle Kingdome.
Seattle Kingdome, Seattle, Washington, USA, Record: The world's largest structure, by volume (19.821 million cu. m.), to be demolished by explosives. *Photos courtesy of Controlled-Demolition, Inc.*[36]

In effect, the largest demolition projects prior to 9/11 involved buildings that could not provide adequate models for a controlled demolition of the WTC complex, including its two quarter-mile-high buildings. On this subject, a question arises: If the WTC destruction were done by terrorists intent upon doing maximum harm, why would they not have tipped the buildings over and destroyed much of Lower Manhattan? In fact, given that such a massive demolition project had never been undertaken before, this calamitous result might be a danger even under the best planning.

6. Dust Clouds

Figure 97. Biltmore Hotel, Oklahoma City, Oklahoma.
Photos courtesy of Controlled-Demolition, Inc.[37]

Figure 98. Beirut Hilton - Beirut, Lebanon.
Photos courtesy of Controlled-Demolition, Inc.[38]

In conventional controlled demolitions, when the supports are blown out with explosives, the mass of the building drops to the ground. Big chunks, small chunks, and in-between-sized chunks of debris get created, along with some dust, but all of it, debris *and* dust, goes down rather quickly. Figures 97 and 98 show that the highest point of the dust cloud doesn't go much above the height of the original building. The two pictures in Figure 99 are here to make this point very clear.

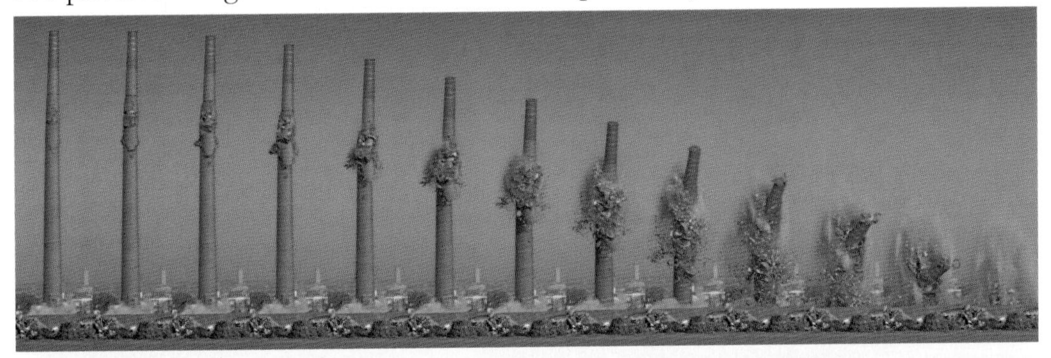

Figure 99. Demolition of a chimney at the former brewery "Henninger" in Frankfurt am Main, Germany, December 2, 2006.[39]

The picture in Figure 100 illustrates that the dust clouds produced by the destruction of WTC1 went *up*—in marked contrast to the results of conventional controlled demolition.

In the satellite picture (Figure 101) taken on September 11th, we have an even more striking image of large amounts of fumes from the WTC event being distributed far beyond the area where dust clouds would be found in a conventional controlled demolition—and lasting far longer. In the case of the Kingdome, the *Seattle Times* reported that "[D]ust choked downtown for nearly 20 minutes."[40] This gives us an idea of the time required in a conventional controlled demolition for coarse dust to *settle out* of the air. In contrast, the dust-cloud/fumes shown in the satellite picture from the WTC event *rose* into the upper atmosphere for many days.

Another aspect of conventional demolition observable in Figures 95 through 99 for the J. L. Hudson building, the Kingdome, the Biltmore, the Beirut Hilton, and the Frankfurt chimney is that the bulk of the dust clouds produced occurred when these structures slammed to the ground.

The postponement of the appearance of dust is another characteristic of conventional controlled demolition by virtue of the fact that the method employed involves getting the rigid ground to do the bulk of the work of breaking up the building into manageable pieces. As shown earlier, the majority of material composing the WTC towers did not slam to the ground.

Figure 100. (9/11/01) Fingers of lather emerge upward.[41]
http://www.studyof911.com/gallery/albums/userpics/10002/3%7E0.JPG

Figure 101. Light fumes move up and south, dark fumes dissipate west.[42, 43]
http://archive.spaceimaging.com/ikonos/2/kpms/2001/09//browse.108668.crss_sat.0.0.jpg

Figure 102. Remains of the Seattle Kingdome.
(a) http://www.seattlepi.com/kingdome/art/boom/400boom1.jpg, (b)http://lh5.ggpht.com/_2Ke6phA7W8k/R8sIkA6lsPI/AAAAAAAAA30/W-wdchW_hP8/400flagpole.jpg

Figure 103. An earthquake-induced collapse in Pakistan suggests how much rubble and how little dust should have been at Ground Zero if the "official" gravitational collapse story told to us were true.
(a)Source of photo: Rolling Stone[44] (b)http://www.post-gazette.com/images3/20051009wp_quake2_450.jpg

Figure 104. Steel core columns from WTC1 disintegrate into steel dust. WTC7 and water tower are in the foreground.[45]

7. The Sound of Explosions

Many witnesses reported hearing explosions. But the sound of an explosion does not necessarily mean that a bomb was detonated. Everything that goes "boom" is not necessarily a bomb. Consider what happens to a raw egg when cooked in a microwave oven. As the egg cooks it explodes due to a steam expansion. When water is heated and becomes steam, it expands in volume by 1,600 times.[46]

This leads us to the "Testimony of Exploding Scott Paks."

Figure 105. A "SCOTT air-pak."
http://memory.loc.gov/service/pnp/ppmsca/02100/02121/0211v.jpg

Firefighter Patrick Sullivan

A. There was a Deputy Chief's rig on fire that was extended to 113's rig. There was a big ambulance, like a rescue company truck, but it wasn't a rescue company truck. It was a huge ambulance. *It must have had Scott bottles or oxygen bottles on it. These were going off. You would hear the air go SSS boom and they were exploding.* So we stretched a line and tried to put that out. He could only use booster water.[47]

Firefighter Todd Heaney

A. I remember getting a drink of water out of their cooler there, and then we just started to put out the car fires, and the rigs were going, ambulances. I mean, there must have been 50 of these things burning heavily. *The Scott cylinders and the oxygen cylinders were all letting go. They were all blowing up left and right.*[48]

Emergency Medical Technician, Lieutenant René Davila

Q. At this point was your vehicle lost?

A. Basically all we [sic] to do is go around the building, came around. But it took longer than usual because you're walking in like this shit. Like you move and it's this soot like heavy dust.

While we're walking I realize that we only have two people. I see my vehicle. The seats are covered. I've still got my bag. I hold it like a trophy. Like people collect basketballs. *I haven't touched—whatever the force was, it was so strong that it went inside of the bag.*

But we were there. Vehicle 219 was destroyed.

Q. Was it on fire?

A. What?

Q. Was it on fire?

A. Fire? We saw the sucker blow up. We heard "Boom!" We were walking up Fulton Street. I don't know how far we made it up when someone says, "The building's coming down." By the time I realized, it's a repeat. [emphasis added] [49]

And there were exploding cars.

EMT Michael D'Angelo

I remember too, the cars started to explode inside the parking lot. I mean, the cars started cooking off, they started going off, boom, boom, boom, boom. I remember that. [50]

8. Squibs?

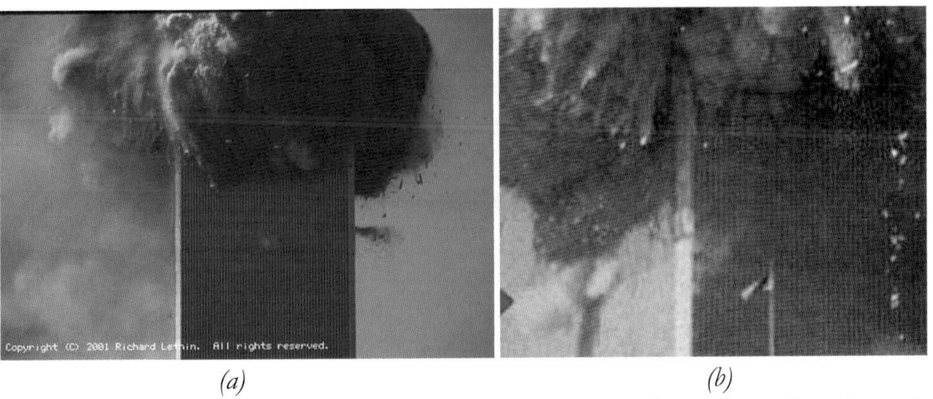

(a) (b)

Figure 106. Horizontal plumes below the "collapse wave" in WTC1 during the top-down destruction.
(a)http://www.reservoir.com/extra/wtc/wtc-small.1056.jpg, (b)http://911research.wtc7.net/wtc/analysis/collapses/docs/tower2_exp1c.jpg,

Air tanks and water tanks were not located on every floor but were spaced consistently throughout the height of each tower. During the destruction of the towers, air tanks and water tanks were ruptured, which would send their contents squirting outward at speeds relative to the pressures in the tanks.

How fast can you launch a pea from a pea shooter? How far will the pea go, having nothing but wind resistance to slow it down? In order for WTC1 to undergo a *pancake collapse* to ground level in 8 seconds or even at free-fall speed (9.22 seconds), all of the building's contents, including air, must be expelled outward at a rapid rate. The contents of each floor must be expelled as if being shot out of a pea shooter through every window on every floor. But at what speed would this volume need to be ejected?

For simplicity, let us assume the building contains only air between the floors.

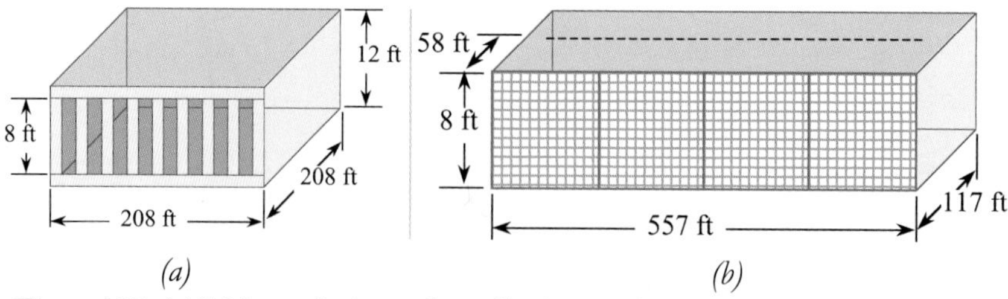

Figure 107. (a) Volume of air per floor. (b) Average distance air must travel to escape.

Length of window opening:
208.ft - (59 columns)(14.in)(1.ft/12.in) = 139.1667 ft
139.1667 x 4 = 556.667 ft

Area of window opening, per floor:
Total opening = (8 ft)(556.667 ft) = 4,453.33 ft^2

Volume of air per floor:
Volume of air = (208 ft)(208 ft)(12 ft) = 519,168 ft^3

Equivalent thickness of air.
(519,168 ft^3)/(4,453 ft^2) = 116.6 ft

Average distance air must travel
(116.6 ft)/2 = 58.3 ft

Average time for each floor to collapse down.
8 sec./110 stories = 0.07273 sec./floor

Average speed air must exit
(58.3 ft)/(0.07273 sec.) = 801.485 ft/sec
801.48503 ft/sec (3600)/(5280) = 546.47 miles/hr, or 550 mph

Now, considering actual fall rate:

(a) Assuming Free-fall speeds

$t = \sqrt{\frac{2x}{g}}$ =SQRT(2*1368/32.2)=9.21786 sec

v *(speed of "collapse")* = at = (32.2)(9.21786) = 296.815 ft/sec
one floor is (1368/110) = 12.436 ft

time to squash one floor at the bottom
t_s = 12.436 ft/(296.815 ft/sec) = 0.0418993 sec

So, air must escape at an average speed of:
(58.3 ft)/(0.041899 sec.) = 1391.186 ft/sec
1391.186 ft/sec (3600)/(5280) = **949 mph**

If the speed of sound is 1125 ft/sec (767 mph),
(1391.186)/(1125) = **1.237, or mach 1.2**

The *average* air ejection speed reaches mach 1 at floor 38.1, or 894.5 feet from the top or 473.5 feet above the ground.

The center of the building is 104 ft from a window. So, the speed that air from the

center must move is
(104 ft)/(0.041899 sec.) = 2482.159 ft/sec
2482.159 ft/sec (3600)/(5280) = **1,692 mph**
(2482.159)/(1125) = **mach 2.21**

Against the pile-driver argument:

(b) Assuming an 8-second "collapse" time

the necessary downward acceleration is 1368*2/(t^2) = 42.75 ft/s^2

$t = \sqrt{\frac{2x}{a}}$ =SQRT(2*1368/42.75)=8 sec

v *(speed of "collapse")* = at = (42.75)(8) = 342 ft/sec

one floor is (1368/110) = 1243.6 ft

time to squash one floor at the bottom
t_s= 12.436 ft/(342 ft/sec) = 0.03636 sec

So, air must escape at an average speed of:
(58.3 ft)/(0.03636 sec.) = 1602.97 ft/sec
1602.97 ft/sec (3600)/(5280) = **1093 mph**

If the speed of sound is 1125 ft/sec (767 mph),
(1602.97)/(1125) = **1.425**

Indeed, how far can you shoot a pea with a pea shooter?
The *average* air ejection speed reaches mach 1 at floor 55.8, or 673.6 feet from the top and 693.4 feet from the ground. Again, the center of the building is 104 ft from a window. So, the speed that air from the center must move is
(104 ft)/(0. 03636 sec.)= 2860.003 ft/sec
2860.003 ft/sec (3600)/(5280) = **1,950 mph**
(2860.003)/(1125) = **mach 2.54**

	Average ejection speed becomes greater than	*At location*
(with 8 second total "fall" time)	*a category 5 hurricane*	*108th floor*
	mach1	*56th floor*
(with 9.22 second total "fall" time)	*a category 5 hurricane*	*107th floor*
	mach1	*38th floor*

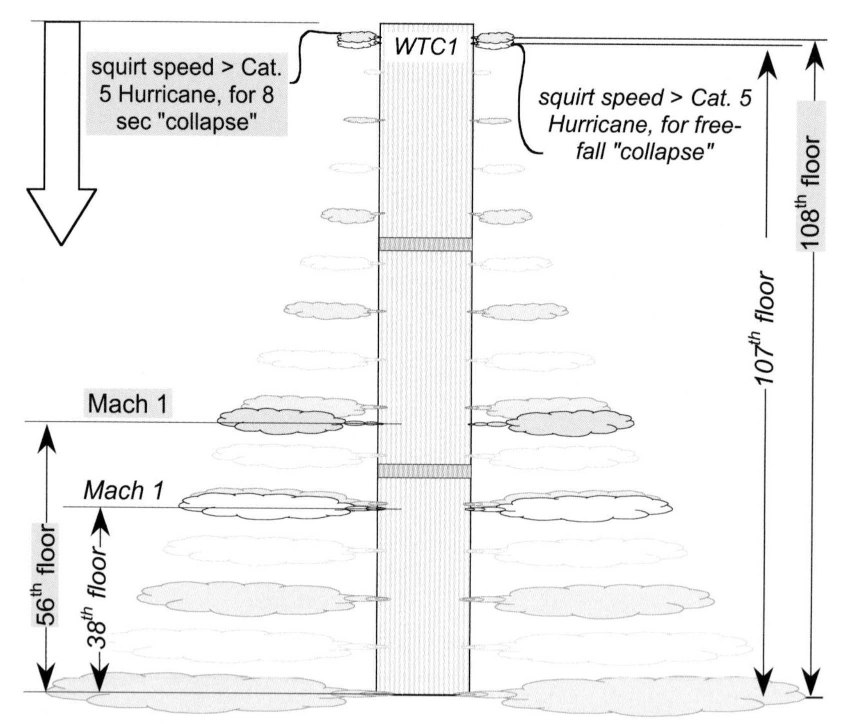

Figure 108. WTC1, average speed air "squirts" out windows for a free-fall "collapse" and for a "pile-driver" eight-second "collapse" consistent with the time the ground shook.

Figure 109. This ambulance was parked in front of WTC1.
http://www.fema.gov/photodata/original/5316.jpg

Does this scene look like building contents flew past the ambulance at speeds greater than mach 1? The ambulance in Figure 109 was parked in front of WTC1 during the destruction of the tower and yet appears to be unharmed. The vehicle was not blown away in the wind nor was the open door blown off. If air and debris blowing out of WTC1 at the wind speed of a category 5 hurricane, it is unlikely the ambulance would have remained there or remained there unharmed.

Obviously the WTC1 windows behind that ambulance were blocked (ambulance is still there), which would require the air and other contents being expelled out the other windows to move even faster. If velocity of that degree had happened, let us consider the type of damage that could have been expected.

Jet engine thrust values are used to determine specific requirements for dedicated engine run-up areas.

> **7. Analysis.** Past guidance was based on both mechanistic air velocity–air pressure relationships, as defined by the Bernoulli equation, and empirical observation. Based on the following Bernoulli model, the critical air velocity would be limited to 218 kilometers per hour (kph) (136 miles per hour [mph] or 199.8 feet per second [fps]):
>
> [. . .]
>
> However, empirical observation has indicated that the typical 51-millimeter (2-inch) -thick edge pavement can withstand velocities up to 362 kph (225 mph). This higher observed velocity was accepted as a valid basis for criteria development because the simple Bernoulli model ignored other forces which are difficult to model, such as friction, shear, and adhesion. Without being able to further refine the mechanistic model, guidance was issued based on empirical observations, with a safety factor of two applied. The active uplift force is a function of the velocity squared. Dividing the observed velocity of 362 kph (225 mph) by the square root of this safety factor yielded a threshold velocity of 257 kph (160 mph). This velocity was issued as criteria for establishing standoff distances.[51]

Boeing Commercial Airplanes provided an example of damage done to the pavement and aircraft when an engine run-up was performed on a taxiway instead of the designated run-up area.

> **Boeing Guidance for Design of Run-up Areas**
>
> The tail (vertical and left horizontal stabilizer) of this 767 [Figure 110] was heavily damaged as shown below, by pavement dislodged during a high-power engine run. The run-up was performed on a taxiway rather than on a dedicated run-up area. The debris field, shown below [Figure 111], was caused by the jet blast.[52]

Figure 110. (11/21/05) Boeing 767 (a) vertical and (b) left horizontal stabilizer.[53]

Figure 111. Pavement dislodged during a high-power engine run-up.[54]

Damage to the roadway and buildings adjacent to the WTC did not look like this. Compare the damage shown in Figure 111 with the nearly undamaged facades of WFC1, WFC2, and WFC3, shown in Figures 91, 92, and 113. There is also a false rumor that all 110 stories were squashed into the 7-story sub-basement. If this had actually happened, there would have been an even bigger problem with the pressure from ejected air and debris. If all of the air and contents did not squirt out above ground, they would have needed to squirt out below ground and would certainly have blown out the bathtub and sent debris all the way to New Jersey through the PATH tunnels.

Figure 112. Illustration of the WTC underground with PATH entry and exit.
http://www.terrorize.dk/911/maps/3d.map-217.jpg, Reference: http://911research.wtc7.net/wtc/arch/foundation.html

Compressing a 110-story building into the seven-story sub-basements would be like an engine piston, compressing everything into a pressure vessel. Certainly the bathtub was not designed as a pressure vessel or a designated "engine run-up area."

Figure 113. Notice how straight the holes are that cut down through WTC6.
http://www.pbase.com/watson/image/536207/original

In Figure 113, note how little "spatter" there is on adjacent buildings:

9. How Far Was Material Ejected?

Figure 114, as viewed from aerial photographs like the one in Figure 113, shows the approximate range of the debris that was ejected from each building. Here is an estimate of the average speed of debris ejection.

y = height of building
x = horizontal distance solid object traveled
a = vertical acceleration,

Free-fall $t = \sqrt{\frac{2 \times y}{a}}$

Or, assuming a "collapse" time, average acceleration, $a = \frac{2 \times y}{t^2}$

ejection velocity, $v = \frac{x}{t}$

where,
y is the height of the building,
x is the farthest horizontal distance debris landed (not carried by wind),
t_0 *(sec)* is the time for free-fall through the height of the building in a vacuum, which is the fastest time possible without being propelled downward,
v_0 *(mph)* is the average horizontal ejection speed in order to land x distance from the building

t *(sec)* is the longest estimated fall time of the object,

v *(mph)* is the average horizontal ejection speed in order to land x distance from the building at time t

t_t *(sec)* is the time for the total building height, while

t_f *(sec)* is the time to fall at the same downward *acceleration* the building "falls" at t_t

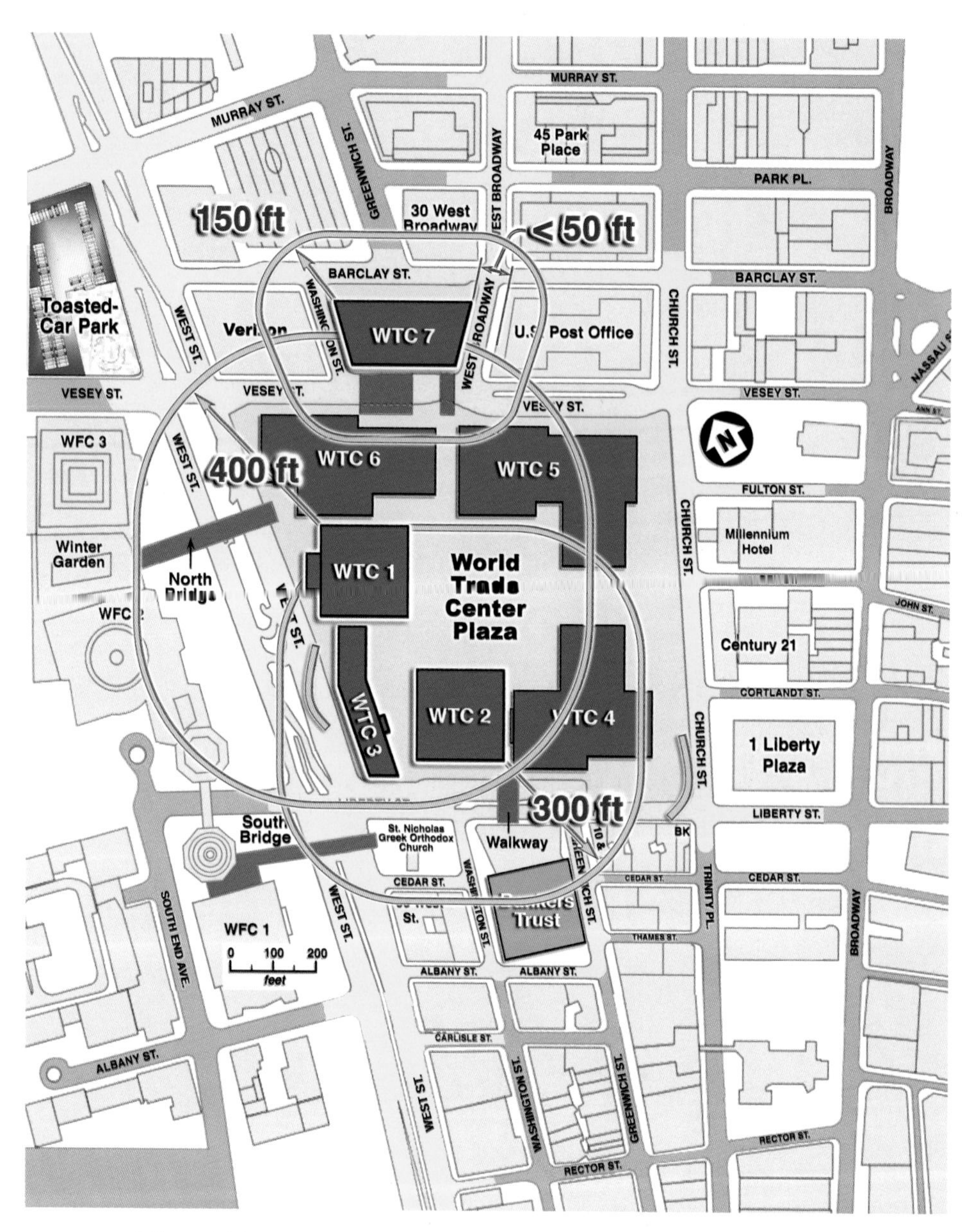

Figure 114. Approximate range of solid debris (omitting dust and paper).
Edit from p. 3 (pdf p. 53 of 298), http://wtc.nist.gov/NISTNCSTAR1CollapseofTowers.pdf

	y *(ft)*	x *(ft)*	t_o *(sec)*	*"sinking"* v_o *(mph)*	t_t *(sec)*	t_f *(sec)*	*"floating"* v *(mph)*
WTC1	1368	400	9.22	30	15	15	18
WTC2	1362	300	9.20	22	12	12	17
WTC7	650	100	6.35	11	18	18	4
WTC7	650	150	6.35	16	18	18	6

Table 8. Ejection speeds from the top, "sinking" down and "floating" down.

	y *(ft)*	x *(ft)*	t_o *(sec)*	*"sinking"* v_o *(mph)*	t_t *(sec)*	t_f *(sec)*	*"floating"* v *(mph)*
WTC1	684	400	6.52	42	15	10.6	26
WTC2	681	300	6.50	31	12	8.5	24
WTC7	325	100	4.49	15	18	12.7	5
WTC7	325	150	4.49	23	18	12.7	8

Table 9. Ejection speeds from half the height, "sinking" down and "floating" down.

* The last three columns are for those who say the building "fell" at less than free-fall speed, taking 15 seconds to "float" to the ground. Noting that the debris outside of the footprint fell at about the same speed as the destruction wave, a reduced acceleration of gravity was used, based on the slower "collapse" times.

Note that the *Toasted-Car Park* (see Chapter 11) is about $\sqrt{(600^2)+(400^2)} \approx 720$ feet from WTC1.

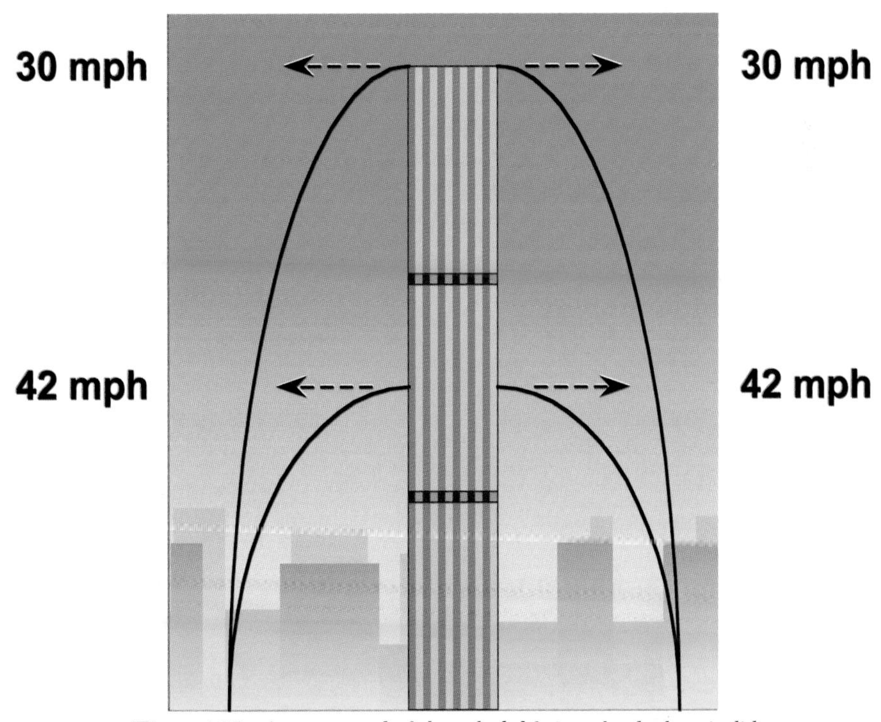

Figure 115. Average speed of launched debris to land where it did.

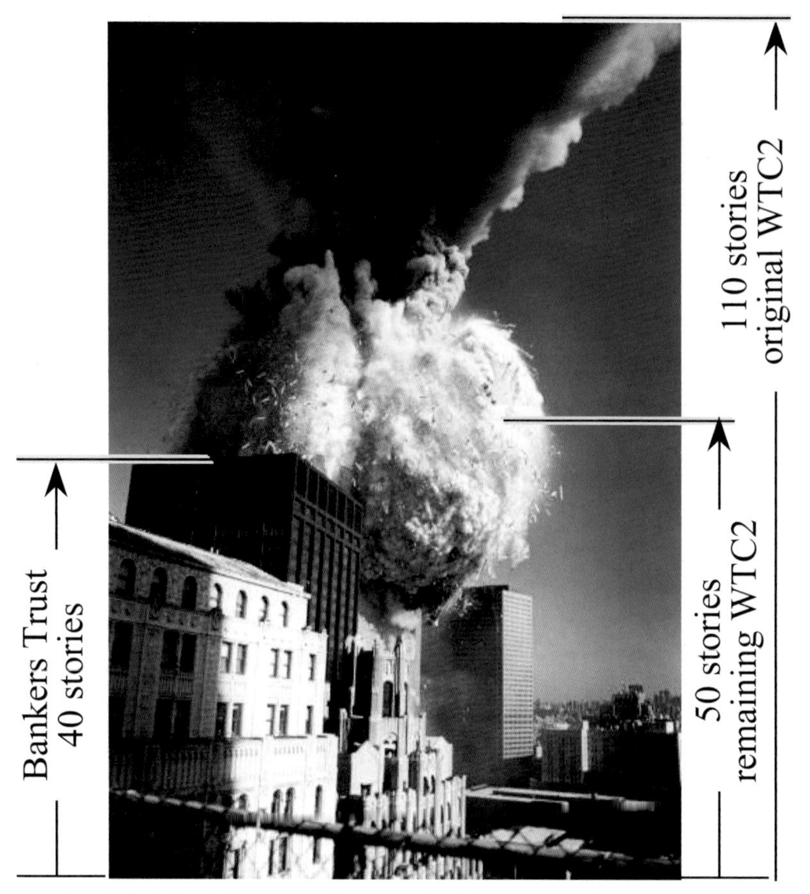

Figure 116. (9/11/01) The upper 60 stories of WTC2 are no longer visible, only powder.
http://hereisnewyork.org//jpegs/photos/5245.jpg

Figure 117. Little to no damage to bathtub.
http://911research.wtc7.net/wtc/arch/docs/bathtub_wall1.jpg

B. Claims for Controlled Demolition (CCD)

Here are some of the reasons that cause people to consider the collapse as having been the result of a controlled demolition.

1) Rapid onset of "collapse" But this is not a characteristic unique to CCD.
2) Sounds of explosions. But we have pointed out alternative explanations for these sounds, discussed in Section (7).
3) "Squibs." Again, there are alternative explanations discussed in Section 8 earlier in this chapter, in Figures 106 through 111, and elsewhere.
4) Dust clouds, but again, these are not unique to CCD. The actual magnitude of the dust clouds is not consistent with CCD.
5) "Vertical progression." Again, these are not unique to CCD.
6) "Symmetrical" progression through the path of greatest resistance. Yet again, this is not a characteristic unique to CCD.
7) How much energy? Enough: the buildings are no longer there. Many feel it is important to calculate how much energy is required to destroy the buildings. But what is the purpose of that? To determine they are still there? Or to determine that they are gone? It is simpler just to look. Besides, if you wanted to determine the amount of energy required, you would first need to know what process was used. Without that knowledge, any calculation of hypothetical energy requirements is arbitrary and without meaning.
8) Orange stuff dripping from window. (See Chapter 13, Section C, "The Rumor of High Heat" on page 262.)
9) Angle-cut columns. If you have a technology that can turn a building into powder in mid-air, why would thermite be used to cut a few beams? To make the job more difficult?

C. "Mini nukes": The Problems

Yet another explanation of the collapse offered in some circles is that "mini-nukes" were used. This raises its own set of difficulties:

1) Do they exist? But the problem is, were they used, and would they show these *types* of signatures?
2) Nuclear weapons *explode,* but the towers did not explode. They were pulverized and peeled to the ground, almost like bananas being peeled.
3) Moreover, the site wasn't "hot," that is, no significant ionizing radiation.
4) The bathtub survived, as has been seen, making it highly unlikely that nukes of any kind were used.
5) Additionally, the Richter reading did not show the use of a nuclear weapon. If a nuke large enough to destroy the WTC had been used, it would have registered a seismic signal greater than if the building had

fallen to the ground. What the seismograph showed was that the majority of the WTC did not hit the ground. (See Table 4, page 79.)

6) There was no evidence of high heat. Nuclear bombs produce heat and a lot of it, no matter if they are big nukes or small nukes.
7) The site showed massive amounts of unburned paper, an impossibility if nuclear weapons had been detonated. The "unburned paper" evidence will be discussed subsequently in conjunction with other factors.
8) There were fourteen survivors in Stairway B of WTC1 who were not burned or blown up but actually walked out on their own (see Chapter 9). There were two others from higher up in Stairway B of WTC1 who survived even though the stairway at that level did not.
9) Finally, multiple nukes or mini-nukes detonated throughout the building in sequence would each produce a shock wave that would interfere with subsequent detonations. In fact, digital cameras recorded the destruction of the buildings, while such cameras may cease to function in the presence of an electromagnetic pulse (EMP) caused by a nuclear detonation.

D. Thermite

1. What Thermite Is Used For.

Finally, we have the "nano-thermite" theory. But, as we will now discover, there are problems with that theory as well.

Thermite is a welding material, not an explosive. The military uses it to weld the guns and cannons of enemy artillery to render those weapons useless. Nano-thermite, like almost any fine powder, can be explosive. Flour can explode as well. Thermite grenades are used by the military to burn critical papers fast to protect secret documents from the enemy in case of an invasion.

Residues of burnt thermite like these, left by railway workers, can be found along railroad tracks.

The violent effects of thermite

Thermite reactions have many uses. Thermite is not an explosive but instead operates by exposing a very small area of metal to extremely high temperatures. Intense heat focused on a small spot can be used to cut through metal or weld metal components together by melting a very thin film where the components meet.

Thermite may be used for repair by the welding in-place of thick steel sections such as locomotive axle-frames where the repair can take place without removing the part from its installed location. It can also be used for quickly cutting or welding steel such as rail tracks, without requiring complex or heavy equipment. However, defects such as slag inclusions and holes are often present in such welded junctions and great care is needed to operate the process successfully.

A thermite reaction, when used to purify the ores of some metals, is called the Thermite process, or aluminothermic reaction. An adaptation of the reaction, used to obtain pure uranium, was developed as part of the Manhattan Project at

Ames Laboratory under the direction of Frank Spedding. It is sometimes called the Ames process.

Copper thermite is used for welding together thick copper wires for the purpose of electrical connections. It is used extensively by the electrical utilities and telecommunications industries.[55,56]

Figure 118. Thermite reaction proceeding for a railway welding. Shortly after this, the liquid iron flows into the mould around the rail gap.
http://upload.wikimedia.org/wikipedia/commons/8/82/Velp-thermitewelding-1.jpg

Military uses

Thermite hand grenades and charges are typically used by armed forces in both an anti-material role and in the partial destruction of equipment, the latter being common when time is not available for safer or more thorough methods. Because standard iron-thermite is difficult to ignite, burns with practically no flame and has a small radius of action, standard thermite is rarely used on its own as an incendiary composition. It is more usually employed with other ingredients added to enhance its incendiary effects. Thermate-TH3 is a mixture of thermite and pyrotechnic additives which have been found to be superior to standard thermite for incendiary purposes. Its composition by weight is generally 68.7% thermite, 29.0% barium nitrate, 2.0% sulfur and 0.3% binder (such as PBAN). The addition of barium nitrate to thermite increases its thermal effect, produces a larger flame, and significantly reduces the ignition temperature. Although the primary purpose of Thermate-TH3 by the armed forces is as an incendiary anti-material weapon, it also has uses in welding metal components.

A classic military use for thermite is disabling artillery pieces, and has been used commonly for this purpose since the Second World War. Thermite can permanently disable artillery pieces without the use of explosive charges and therefore can be used when silence is necessary to an operation. There are several ways to do this. By far the most destructive method is to weld the weapon shut by inserting one or more armed thermite grenades into the breech and then quickly closing it. This makes the weapon impossible to load. An alternative method is to insert an armed thermite grenade down the muzzle of the artillery piece, fouling the barrel. This makes the piece very dangerous to fire. Yet another method is to use thermite to weld the traversing and elevation mechanism of the weapon, making it impossible to aim properly.

Thermite was also used in both German and Allied incendiary bombs during World War II. Incendiary bombs usually consisted of dozens of thin thermite-filled canisters (bomblets) ignited by a magnesium fuse. Incendiary bombs destroyed entire cities due to the raging fires that resulted from their use. Cities that primarily consisted of wooden buildings were especially susceptible. These incendiary bombs were utilized primarily during nighttime air raids. Bomb sights could not be used at night, creating the need to use munitions that could destroy targets without the need for precision placement.[57,58]

2. Proof of Concept

There is no evidence that thermite, thermate, super thermite, or nano-enhanced thermite have ever been used to bring down major buildings in controlled demolition (as opposed to simply cleaning up debris). Thermite has never been used to bring down skyscrapers.

There is no proof of concept for the thermite hypothesis. Wikipedia Encyclopedia defines "Proof of concept" as "a short and/or incomplete realization (or synopsis) of a certain method or idea(s) to demonstrate its feasibility, or a demonstration in principle, whose purpose is to verify that some concept or theory is probably capable of exploitation in a useful manner. The proof of concept is usually considered a milestone on the way of a fully functioning prototype." In the case of thermite and the WTC towers, such a proof of concept has never been laid out.

3. "Thermitic Material" in the Dust

A report has come out stating that "thermitic material"[59] was found in dust samples from lower Manhattan after 9/11/01. The authors of the report did not say they found thermite, but only that they found "thermitic material." What is "Thermitic Material?" Presumably, the term refers to the ingredients of thermite, which is a substance made of aluminum powder and iron oxide (rust). The Twin Towers were steel structures with aluminum cladding. "Steel is a term used for iron to which between 0.02 to 1.7% carbon has been added.[60]" Typical low-carbon steel (e.g. ASTM A36) contains 99% iron. We know that a large portion of the towers was turned to dust (see Chapter 9, Dustification). And iron dust in atmospheric conditions will immediately rust. So it is natural and to be expected that materials the buildings were made of would be found in the nano-dust of their remains. The surprising thing would be if this nano-dust from the buildings did *not* contain "thermitic material." That the article is identified as having been "peer reviewed" is intended to imply the validity of the study. However, "peer reviewed" no longer means what it once did. Shortly after Bentham published the article about "thermitic material," *The Boston Globe* ran a news story about a hoax submission (an article made up of nothing but computer-generated nonsense) to another Bentham journal. This second article also passed "peer review."[61,62] Others have written about this deliberately fraudulent study.[63]

The buildings were turned to dust, and therefore the dust would be expected to contain traces of all materials that were in the buildings. Finding traces of chocolate, sugar, and nano-wheat (flour) in the dust would not prove that chocolate chip cookies turned the buildings to dust. It would not prove these ingredients had been combined as chocolate chip cookies in the buildings nor that such cookies were capable of turning buildings to dust. The same is true for thermite. Finding constituents of the building in the dust does not give a proof of what happened to the building. In this case, focusing on particular constituents found in the dust and claiming that those constituents support a particular theory may actually distract attention away from the central and incontrovertible fact that most of the building and its contents were, quite simply, turned to dust.

4. Pulverization

Where is the proof that thermite has *ever* been used to completely pulverize buildings in controlled demolition (not simply cleaning up debris)? The mechanisms of cutting and pulverization are mutually exclusive, and thermite cuts and melts but is not explosive. "Cutting requires action in one direction," says Jeff Strahl, a 9/11 researcher, "while pulverization requires action in all directions."

Where is the proof, either, that nano-*enhanced* thermite has *ever* been used to completely pulverize buildings in controlled demolition (not simply cleaning up debris)? Could thermite have been used to turn the upper 80-plus floors of the Twin Towers to ultra-fine dust?

Above all, how do angle-cut columns relate to pulverizing a building? What is the connection? We fail to see it.

E. Ignition and Control

How would the thermite have been ignited? Haven't we seen that thermite is difficult to ignite? Has thermite ever been ignited by remote control? Have multiple thermite ignitions ever been set off with exact timing by remote control? And haven't we seen that it takes a long time for thermite to cut through thin steel, far longer than the 8 to 11 seconds, the reported time it took the towers to "collapse." (See page 10.) Thermite heats the surface, so time is required for the heat to conduct into the material. The only contribution thermite could have made to the demise of the buildings would be the added weight to the buildings.

There is a similar challenge with timing explosives. How would ignition have been accurately controlled in a building of this size? How many remote control radio frequencies would be required to do this? How many ignition devices would be needed to cut 236 outer columns and 47 core columns on each of the 110 floors in each of the two towers? An ignition device on each column on each floor would total 31,130 ignitions. None of this would cut floor trusses or pulverize the concrete floors or any of the WTC contents, much less steel beams.

And what about the seismic signal? If most of the material from the Twin Towers crashed to the ground, there should have been a significant seismic event. Yet a NIST scientist says that "the collapse of the towers were not of any magnitude that was seismically significant." Here is the complete quotation made by Dr. Shyam

Sunder at The National Construction Safety Team (NCST) Advisory Committee meeting via teleconference on Thursday, December 14, 2006, regarding the seismic signals associated with the destruction of WTC1 and WTC2:

> "The signals' strength due to the collapse of the towers were not of any magnitude that was seismically significant from an earthquake design standpoint or from the design or a failure of a structural component or of I would say of a piping system that might be used in a structure, so ah there wasn't anything that gave us pause in terms of that being a significant seismic event to have ruptured the pipeline."[64]

F. The Kitchen Sink

Some may find it tempting to propose the idea that multiple methods of destruction were used, assuming that may have been a "fail safe" plan in case one or more methods failed. There is also the *false assumption* that if one promotes "everything including the kitchen sink" as the method of destruction, they're bound to be correct. However, when the physical requirements are considered, the complexity of implementing multiple methods, if even possible, would only add difficulty with no benefit. For example, consider thermite plus explosives. As discussed in the previous section, the ignition and timing of thermite is a challenge. Also, the ignition and timing of explosives (bombs) in the building is a challenge. Coordinating the timing of two such events would greatly increase the complexity, if not render it impossible. Even if it were possible, the fact remains that the physical evidence contradicts the use of thermite and/or "bombs in the building."

G. Occam's Razor

The concept of "Occam's Razor" is often brought up as an aid or test in deciding whether or not to reject a particular explanation. But it should be noted that Occam's Razor is not intended to be used as *proof* of any concept, but much more simply as a guide. Concisely stated, "Occam says minimize assumptions, not minimize stuff [evidence]."[65] In other words, the fewer assumptions that need to be made, the fewer opportunities there are to fall into errors. So it follows that the more evidence addressed in analyzing a problem, the slighter the need for assumptions. If *everything* is known, there is no need to make any assumptions or to indulge in speculation at all. With all this in mind, consider what assumptions must be made in arguing the case for Conventional Controlled Demolition. One must assume a sufficient quantity of explosives were placed ahead of time; one must assume that it was possible for the quantities of required explosives to be carried into the buildings without being detected; that workers would not notice the walls of their offices missing while the explosives were being placed; that there was a reason why bomb-sniffing dogs didn't detect it; that there really was a practical means of igniting it; that there exist silent explosives that are capable of the behavior we saw; and that there really must be a reason why the concept of Conventional Controlled Demolition is inconsistent with nearly all of the observable evidence remaining after the destruction.

H. Conclusions

What are we to conclude from the evidence presented here? Here are the essential points:

1) Conventional Controlled Demolition is not an adequate explanation for the destruction of the twin towers because:
   a) such a demolition requires careful preparation, and in this instance, would have to have been done secretly and over a very long period of time;
   b) because such a demolition would run the risk of exposure by bomb-sniffing dogs, none of which ever found such explosives, even on 9/11;
   c) because such controlled demolitions would leave collateral damage on adjacent buildings above the 20th floor as well as below that level, a condition not met at the WTC site;
   d) because such controlled demolitions would leave large chunks of debris;
   e) because such controlled demolitions do not adequately account for the large amount of "dustification" that occurred (see Chapter 8); and
   f) because such controlled demolitions would have to eject air and other material sideways from each floor at tremendous speeds in order to make the "pancake collapse" model work, leaving collateral ejecta on adjacent buildings and doing projectile damage to them.
2) The thermite model has essentially the same problems as the Conventional Controlled Demolition one. These are actually compounded by the presence of so much *paper*, a substance that could not conceivably have survived the intensely high temperatures created by thermite.
3) The nuclear model has the same problems as the Conventional Controlled Demolition and thermite models. In addition, fourteen people walked out of Stairway B without having been burned or baked or blown up or crushed.

In short, neither the conventional controlled demolition nor the bombs-in-the-building model offers a sufficient basis for explaining all the evidence.[66]

---

[1] *Implosion experts will study test explosion at Kingdome*, February 24, 2000, by ROBERT L. JAMIESON Jr., SEATTLE POST-INTELLIGENCER REPORTER, *http://seattlepi.nwsource.com/local/boom24.shtml*

[2] *http://www.sha.state.md.us/SHAServices/mapsBrochures/maps/oppe/trucker_back.pdf,*

[3] *http://www.mdta.state.md.us/mdta/servlet/dispatchServlet?url=/TollRates/rates.jsp*

[4] Curtis L. Taylor and Sean Gardiner, STAFF WRITERS, "Heightened Security Alert Had Just Been Lifted," September 12, 2001. (acc. 12/01/08) *http://www.newsday.com/news/nationworld/nation/ny-nyaler122362178sep12,0,1255660.story*

[5] Sirius is the brightest star in the sky and appears about twice as bright as the next brightest. (*http://en.wikipedia.org/wiki/List_of_brightest_stars*). Sirius is also known colloquially as the "Dog Star", reflecting its prominence in its constellation, Canis Major (Big Dog). (Hinckley, Richard Allen (1899), Star-names and Their Meanings. New York: G. E. Stechert. p. 117).

[6] *http://www.novareinna.com/bridge/sirius.html*

[7] *http://www.suryafireservice.com/training.htm*

[8] [Redrawn], *http://www.tpub.com/content/nasa2000/NASA-2000-tm210077/NASA-2000-*

tm2100770021im.jpg, http://www.tpub.com/content/nasa2000/NASA-2000-tm210077/NASA-2000-tm2100770021.htm; NASA-2000-tm2100770021im.jpg

[9] *http://www.tpub.com/content/nasa2000/NASA-2000-tm210077/NASA-2000-tm2100770022im.jpg*

[10] [Reproduced], *http://www.tpub.com/content/nasa2000/NASA-2000-tm210077 /NASA-2000-tm2100770022im.jpg*

[11] *http://en.wikipedia.org/wiki/RDX*

[12] MATERIAL SAFETY DATA SHEET, MIXTURE C4 HYDROCARBONS, SYNTHOS Kralupy a.s., issued: Nov. 13, 2007, Revision date: 1.6.2008, Page 3, *http://www.unipetrol.cz/docs/Rafinat%201%20aj.pdf*

[13] *Manual of explosives*, by De Kalb, Courtenay, b. 1861; Ontario. Bureau of Mines, Toronto : Ontario Bureau of Mines, Possible copyright status: NOT_IN_COPYRIGHT, page 15, *http://www.archive.org/stream/manualofexplosiv00dekarich/manualofexplosiv00dekarich_djvu.txt, http://www.archive.org/download/manualofexplosiv00dekarich/manualofexplosiv00dekarich.pdf*

[14]*Manual of explosives*, by De Kalb, Courtenay, b. 1861; Ontario. Bureau of Mines, Toronto : Ontario Bureau of Mines, Possible copyright status: NOT_IN_COPYRIGHT, page 25, *http://www.archive.org/stream/manualofexplosiv00dekarich/manualofexplosiv00dekarich_djvu.txt, http://www.archive.org/download/manualofexplosiv00dekarich/manualofexplosiv00dekarich.pdf*

[15]*The Chemistry of Explosives*, by Jacqueline Akhavan, Royal Society of Chemistry (Great Britain), Published by Royal Society of Chemistry, 2004, Page 143, *http://books.google.com/books?id=9tIQDn2uZz4C&ie=ISO-8859-1&output=html*

[16] *http://www.mbnamenska.com/dynamite.html*

[17] *http://en.wikipedia.org/wiki/Thermite*

[18] *http://en.wikipedia.org/wiki/Thermite*

[19] Encyclopedia, *http://www.nationmaster.com/encyclopedia/HMX*

[20] *http://science.howstuffworks.com/c-42.htm*

[21] MATERIAL SAFETY DATA SHEET, MIXTURE C4 HYDROCARBONS, SYNTHOS Kralupy a.s., issued: Nov. 13, 2007, Revision date: 1.6.2008, Page 3, *http://www.unipetrol.cz/docs/Rafinat%201%20aj.pdf*

[22] ibid.

[23] ibid.

[24] *http://www.suryafireservice.com/training.htm*

[25] *http://www.tpub.com/content/nasa2000/NASA-2000-tm210077/NASA-2000-tm2100770022im.jpg*

[26] ibid.

[27] *The Chemistry of Explosives*, by Jacqueline Akhavan, Royal Society of Chemistry (Great Britain), Published by Royal Society of Chemistry, 2004, Page 143, *http://books.google.com/books?id=9tIQDn2uZz4C&ie=ISO-8859-1&output=html, http://www.nationmaster.com/encyclopedia/Trinitrotoluene*

[28] *http://en.wikipedia.org/wiki/RDX*

[29] Encyclopedia, *http://www.nationmaster.com/encyclopedia/PETN*

[30] Encylopedia, *http://www.nationmaster.com/encyclopedia/Nitroglycerin*

[31] *http://www.mbnamenska.com/dynamite.html*

[32] *http://www.suryafireservice.com/training.htm*

[33] *http://science.howstuffworks.com/c-42.htm*

[34] J.L. Hudson Department Store, Detroit, Michigan, USA, Records: At 439 ft. tall Hudson's is the tallest building & the tallest structural steel building ever imploded. At 2.2 million square feet, Hudson's is the largest single building ever imploded, *http://www.controlled-demolition.com/default.asp?reqLocId=7&reqItemId=20030225133807*

[35] J. L. Hudson Department Store, 5:47 PM on October 24, 1998, *http://www.controlled-demolition.com/default.asp?reqLocId=7&reqItemId=20030225133807*, J. L. Hudson Department Store, Detroit, Michigan, USA, 10/24/1998, Records: At 439 ft. tall Hudson's is the tallest building & the tallest structural steel building ever imploded. At 2.2 million square feet, Hudson's is the largest single building ever imploded.

[36] *http://www.controlled-demolition.com/default.asp?reqLocId=7&reqItemId=20030317140323*

[37] h*ttp://www.controlled-demolition.com/default.asp?reqLocId=7&reqItemId=20030226162334*

[38] *http://www.controlled-demolition.com/default.asp?reqLocId=7&reqItemId=20030324142951*

[39] *http://upload.wikimedia.org/wikipedia/commons/4/49/Blasting_frankfurt.jpg, http://en.wikipedia.org/wiki/Demolition*

[40] Dome's Final Roar, by Jeff Hodson, Eric Sorensen, Alex Fryer, Beth Kaiman, Dionne Searcey, Sara Jean Green, John Zebrowski, Phil Loubere, Seattle Times staff reporter, Monday, March 27, 2000 - Page

updated at 12:00 AM, *http://archives.seattletimes.nwsource.com/cgi-bin/texis.cgi/ web/vortex/ display?slug=4012219 &date=20000327*

[41] *http://www.amanzafar.com/WTC/wtc67.JPG*

[42] (b) is contrast adjusted

[43]*http://archive.spaceimaging.com/ikonos/2/kpms/2001/09//browse.108668.crss_sat.0.0.jpg,* space imaging

[44] *Rolling Stone*, Issue 988, December 1, 2005, p. 80.

[45] [Cropped from originals.] *http://img156.imageshack.us/img156/2044/p9111200ms2.jpg, http:// img156.imageshack.us/img156/7799/p9111202fy8.jpg, http://img156.imageshack.us/img156/482/p9111203ac3.jpg, http://img152.imageshack.us/img152/605/p9111204pd0.jpg, http://img152.imageshack.us/img152/2684/ p9111205pg6.jpg,*

[46] *http://en.wikipedia.org/wiki/Steam*

[47] WORLD TRADE CENTER TASK FORCE, FIREFIGHTER PATRICK SULLIVAN INTERVIEW No. 9110235, p. 8, December 5, 2001. *http://graphics8.nytimes.com/packages/pdf/nyregion/ 20050812_WTC_GRAPHIC/9110235.PDF*

[48] WORLD TRADE CENTER TASK FORCE, FIREFIGHTER TODD HEANEY INTERVIEW No. 9110255, p. 13, December6, 2001. *http://graphics8.nytimes.com/packages/pdf/nyregion/20050812_WTC_ GRAPHIC/9110255.PDF*

[49] WORLD TRADE CENTER TASK FORCE, EMT RENÉ DAVILA INTERVIEW No. 9110075, pp. 27-28, October 12, 2001. *http://graphics8.nytimes.com/packages/pdf/nyregion/20050812_WTC_GRAPHIC/ 9110075.PDF*

[50] File No. 9110148, "EMT Michael D'Angelo," October 24, 2001, p. 11, *http://graphics8.nytimes.com/ packages/pdf/nyregion/20050812_WTC_GRAPHIC/9110148.PDF*

[51] ETL 07-3 Jet Engine Thrust Standoff Requirements for Airfield Asphalt Edge Pavements (02-14-2007) PDF 65 KB, 7 pgs, *http://www.wbdg.org/ccb/AF/AFETL/etl_07_3.pdf*

[52] 11/21/2005, Airport Technology, Boeing Commercial Airplanes, *http://www.boeing.com/commercial/ airports/faqs/ guidance _design_run.pdf*

[53] 11/21/2005, Airport Technology, Boeing Commercial Airplancs, (11/21/2005) "Boeing Guidance for Design of Run-up Areas,"Airport Technology, Boeing Commercial Airplanes, *http://www.boeing.com/ commercial/airports/faqs/guidance_design_run.pdf*

[54] 11/21/2005, Airport Technology, Boeing Commercial Airplanes, (11/21/2005) "Boeing Guidance for Design of Run-up Areas,"Airport Technology, Boeing Commercial Airplanes, *http://www.boeing.com/ commercial/airports/faqs/guidance_design_run.pdf*

[55] *http://en.wikipedia.org/wiki/Thermite*

[56] *http://www.springerlink.com/content/k57101130128rt78/, http://www.springerlink.com/content/ k57101130128rt78/fulltext.pdf,* Thermite reactions: their utilization in the synthesis and processing of materials, Journal of Materials Science, L. L. Wang1, Z. A. Munir1 and Y. M. Maximov1, 2, 12 November 1992, Springer Netherlands, Volume 28, Number 14 / January, 1993, pp. 3693-3708, Subject Collection: Chemistry and Materials Science, Springer, Link Date: Thursday, November 04, 2004

[57] *ibid.*

[58] *ibid*

[59] "Active Thermitic Material Discovered in Dust from the 9/11 World Trade Center Catastrophe," by Niels H. Harrit, Jeffrey Farrer, Steven E. Jones, Kevin R. Ryan, Frank M. Legge, Daniel Farnsworth, Gregg Roberts, James R. Gourley, Bradley R. Larsen. *The Open Chemical Physics Journal,* Volume 2, pp. 7-31, *http: //www.bentham-open.org/pages/content.php?TOCPJ/2009/00000002/00000001/7TOCPJ.SGM*

[60] *Online Metals.com, http://www.onlinemetals.com/steelguide.cfm*

[61] Fake paper tests peer review at open-access journal, by Elizabeth Cooney June 12, 2009, *http:// www.boston.com/news/health/blog/2009/06/phony_paper_tes.html*

[62] Oder, Norman, "Hoax Article Accepted by 'Peer-Reviewed' OA Bentham Journal." *http:// www.libraryjournal.com/index.asp?layout=talkBackCommentsFull&articleid=CA6664637&talk_back_header_ id=6605401* (Retrieved 2009-10-01.)

[63] "Questioning 'Active Thermitic Material Discovered in dust from the 9/11 World Trade Center Catastrophe'" by Andrew Johnson, April 8, 2009, *http://www.checktheevidence.co.uk/cms/index.php?option=com_c ontent&task=view&id=224&Itemid=60*

[64] NCST Advisory Committee Met December 14, 2006, 9:00-11:00AM, Webcast Entire session: (mp3); Segment: WTC Seismic Signature NCSTAd (mp3); See http://drjudywood.com/towers

[65] *http://www.physicsforums.com/archive/index.php/t-8259.htmlbr%2520/t-146601.html*

[66]For a recent article about the thermite theory, see: *http://www.opednews.com/articles/911-NanoTech-Thermite-Publ-by-John-R-Moffett-090616-456.html*

8.
Dustification

I close my eyes, only for a moment, and the moment's gone
All my dreams, pass before my eyes, a curiosity
Dust in the wind, all they are is dust in the wind
Same old song, just a drop of water in an endless sea
All we do, crumbles to the ground, though we refuse to see
— Kansas—*Dust In The Wind* [1]

The World Trade Center (WTC) towers did not "collapse" on 9/11/01. They were already turned to dust before a gravity-driven collapse was a possibility. It is time to discuss how this was done.

Figure 119. (9/11/01) Mostly unburned paper mixes with the remains of the Twin Towers. As seen a block away, a portion of the towers remains suspended in air.
This dust looks deeper than one inch. Most of the curb looks filled in. (East of camera view F, Figure 127)
Photo credited to Terry Schmidt (intensity adjusted)
http://ken.ipl31.net/gallery/albums/wtc/img_1479_001.jpg

A. Pulverized to Dust?

The common use of the word, "pulverize," refers to pounding or crushing a material[2]. Merriam-Webster defines *pulverize*: "to reduce (as by crushing, beating, or grinding) to very small particles."[3] The Twin Towers were not crushed or ground up; they turned to dust in mid-air. They were not vaporized, either. Vaporization refers to the conversion of a liquid or solid to the vapor state by the addition of latent heat.[4] The buildings were not cooked; they were turned to dust in mid-air. This was a new process and a new process needs a new word to represent it. We will call this *dustification*, saying that the buildings were *dustified.*

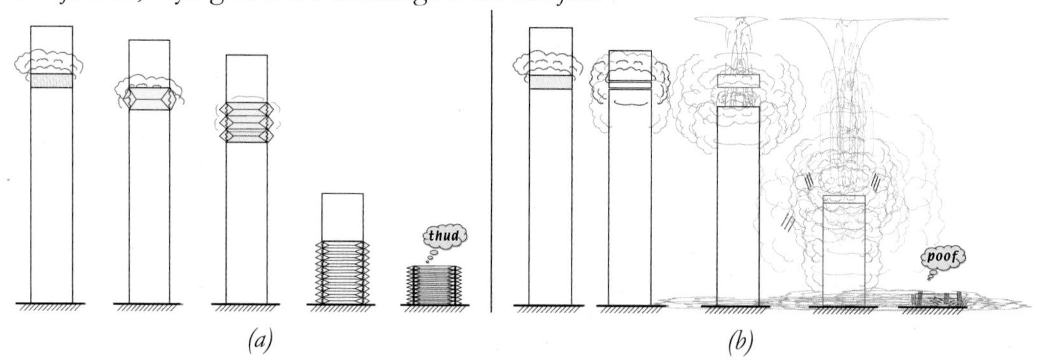

Figure 120. (a) An illustration of what a "pancake collapse" would look like. The rubble pile should be at least 1/8 of the original building height (12.5%). (b) An illustration of what the actual destruction looked like. Both towers went "poof." The rubble pile was no more than 2% of the original building height.

The defenders of the official story maintain that the Twin Towers were each hit by aircraft and that the subsequent fires weakened the steel in the upper stories, initiating a gravity-driven "pancake collapse" as illustrated in Figure 120a. As we have seen, there are many problems with this hypothesis. The most obvious is the near free-fall speed at which the buildings were destroyed (see Chapter 2). A second problem is the paucity of remaining material or debris. Where are the so-called "pancaked" concrete floors? Where is the office furniture? Where is the office machinery? Where are the filing cabinets? Where is the wallboard? Where are the bookcases? None of these were anywhere. Most of such material appears to have turned to dust, as illustrated in Figure 120b.

B. What Steel was Shipped to China?

It was widely reported that a substantial amount of WTC steel was sold as scrap, put on barges, and shipped to China to be melted down. But Figure 121 shows how little steel there *was* on the ground shortly after destruction of the WTC towers. We have evidence that steel was transported to Fresh Kills Island to be stored, and this steel may or may not have been subsequently shipped to China. But it could not have been a large amount.

The photo in Figure 121a is dated 9-13-2001. The sun is from the east (right side of picture), so the photo appears to have been taken on the *morning* of 9/13. The photo in Figure 121b is dated 9-14-2001 by NOAA. While, as we've said, it has been reported that much of the steel was removed from the site, loaded onto barges, and sent to China to be melted down, the steel could not have been removed *this* fast. So if it was not shipped to China overnight, where did it go? Very little of it was on the ground at the disaster site. It had to have been dustified and suspended in the air.

Figure 121. (a) The remains of WTC2 are in the foreground. Immediately behind WTC2 is where WTC3 (Marriott Hotel) once stood. Where did it go? In the background (upper-left) the World Financial Center (WFC) buildings have blown-out windows but little other damage. The remains of WTC6, an 8-story building, tower over the remains of WTC1. Note the blown out windows in the WFC. (b) Viewed from the opposite direction.

(a)http://www.hybrideb.com/source/eyewitness/groundzero/groundzero_07.jpg, (b)http://www.noaanews.noaa.gov/stories/images/groundzero.jpg

C. Dustification of Material

During the final seconds of the destruction of WTC1, a section of core columns remained rigid and standing upright after the rest of the building was gone: Then they simply turned to dust.

Figure 122. Steel core columns disintegrate into steel dust with WTC7 and water tower in the foreground. [Cropped from originals, contrast adj.] [5]

The sequence of images in Figure 122 shows a very distinct point at which the sharp, dark outline of the column simply disappears—not just blurred but gone, replaced only by a wider zone of dust that continues to fall at a slower rate as it widens and drifts.

Figure 123. (9/11/01) (a) Building turns to dust. (b) A closer view shows steel beams disintegrating into dust. (Camera view D*, Figure 127)*
(a)http://911wtc.freehostia.com/gallery/originalimages/GJS-WTC30.jpg, (b)cropped:http://911wtc.freehostia.com/gallery/originalimages/GJS-WTC30.jpg

Both Figures 123 and 122 are views taken from the northeast. Figure 122 was taken from near ground level while Figure 123 was taken from a police helicopter very nearly above the same location. A view from the opposite direction is shown in Figure 124, a superimposition of two images taken from the southwest, across the Hudson River.

Figure 124. Two superimposed views from the southwest.
http://1.bp.blogspot.com/_aJeegFsC3nY/ScY5RdeL42I/AAAAAAAABHo/eAy4n7dk7b8/s1600-h/superimp.spire.wtc1.jpg

An image taken during the destruction of WTC1 is superimposed on an image taken before 9/11 to show where these (momentarily) remaining core columns were located within the building. Figure 124 also shows the relative height of these columns; they appear to reach above the upper mechanical floors (floors 75-76). Because of its appearance, this group of briefly-remaining core columns has been referred to as "the spire."

Some have tried to argue that these remnants of the core had dust on them that shook loose just before the columns dropped to the ground. If so, how did the dust get there? How long did it take for the dust to get there? And how did it stay there?

If this much dust had been there before the destruction, it would have taken up a significant volume of the building. (See Figure 123.) It is very difficult to imagine that the remnants of the core were still covered in a thick layer of loose dust in the instants following the total disintegration of the building around them.

Or, if the dust had come *from* the destruction of the building, how could it settle so quickly and yet be fine enough to stay aloft? The tallest column in the "spire" does not appear even to have a horizontal surface for the dust to accumulate *on*. This theory of dust seems most highly improbable, if not impossible.

Consider the importance of this matter. These core columns stood rigidly and without support *after* the total destruction of the building around them. *If they had fallen over, they would have destroyed buildings over several city blocks.* These columns stood well above the 47-story WTC7 and extended half again that building's height, to a level of about 71 stories. From Figure 124, as we noted, they appear to extend above the upper mechanical floors of WTC1, or to a height above 78 stories.

1368 (ft)*(71/110) = 883 feet, or 1368 (ft)*(78/110) = 970 feet.

Figure 125. The minimum estimated distance remaining core columns would have reached had they fallen over.[6]

If this remaining group of core columns had been positioned horizontally rather than vertically, it would be approximately the length of three football fields. The shorter distance (883 feet) is shown as the radius of the smallest circle in Figure 125. Yet, there is no evidence that these core columns landed on or damaged *any* buildings in this "fall zone." Where did they go? What could have happened to them other than that they turned to dust?

In Figure 125, again, the two larger circles show how far each of the twin towers would have reached (in any direction) if, undamaged in any other way, it had simply tipped over onto its side.

Instead of falling, however, both towers, like the "spire," turned to dust. Here is where the building debris shown in Figure 126 as if it were falling from WTC1 *should* have landed. Figure 126 shows a sequence of video frames with several pieces of falling debris identified by number. The debris marked "1" can be seen falling in front of the northwest corner of WTC1, beginning with the first frame. In subsequent frames it becomes a cloud of dust.

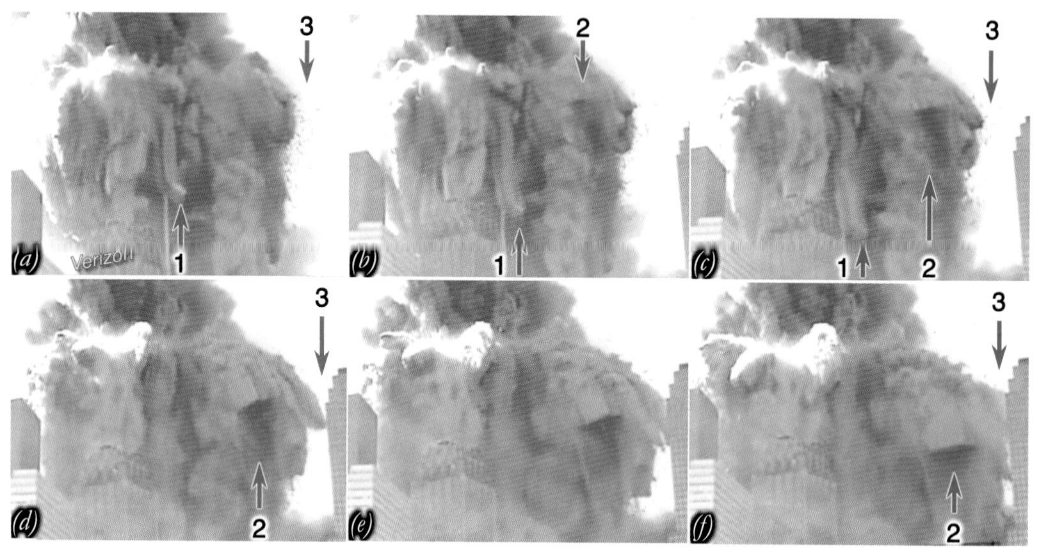

Figure 126. WTC1 turns to dust. (Camera view B, Figure 127)
[from YouTube video. Photographer is unknown. http://www.youtube.com/watch?v=5q7vz3ZEfBw]

If the buildings "fell" at free-fall speeds, the tops would have encountered no more resistance from the lower portions than they would have encountered from air. But, instead, the tops disintegrated while falling, as if they were encountering *very high* resistance. Here we have conditions that contradict each other. We do not see the steel beams striking anything other than air. *Then why are they being pulverized* as *they fall?*

View	Figure	Page	View	Figure	Page	View	Figure	Page
A	*128*	137	**C**	*170a*	180	**F**	*119*	131
B	*126*	136	**D**	*123*	134	**G**	*131*	138
B	*129*	138	**E**	*134*	140	**G**	*130*	138

Legend 1. Legend of Figure views for the map in Figure 127.

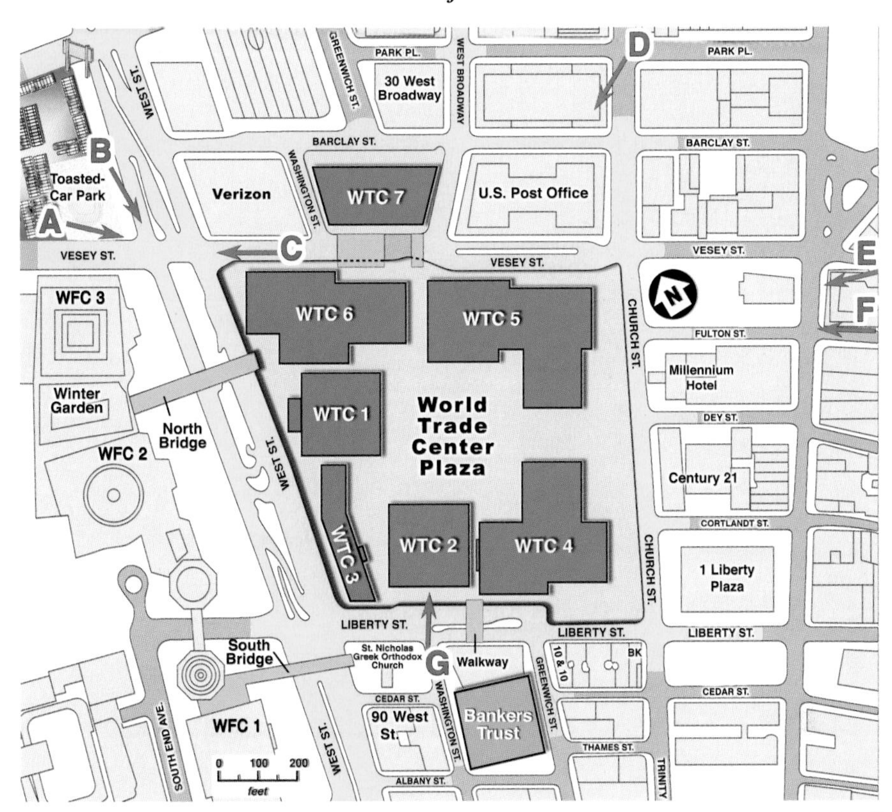

Figure 127. Letters indicate the approximate locations where the photos shown were taken.[7]

Figure 128. Photos taken in the afternoon of 9/11/01 before WTC7 was destroyed. (a) Looking southeast toward WTC6, across the West Street and Vesey Street intersection, (b) closer. (view A, Figure 127)
(a)http://911wtc.freehostia.com/gallery/originalimages/GJS-WTC131.jpg, (b)http://911wtc.freehostia.com/gallery/originalimages/GJS-WTC132.jpg,

Figure 129. Coming out of hiding after the destruction of WTC1. (view B, *Figure 127)*
(cropped, intensity adjusted) http://memory.loc.gov/service/pnp/ppmsca/02100/02121/0084v.jpg

Figure 130. Material turns to dust before it hits the ground during the demise of WTC2, viewed from the south. (Camera view G, *Figure 127)* YouTube video.[8]

Figure 131. Enlargement of frames (a) and (c) in Figure 130. (view G, *Figure 127)* YouTube video.[9]

Figure 132. (9/11/01): In-flight "pulverization" or dustification?
(a)http://hereisnewyork.org//jpegs/photos/1539.jpg, (b)Photo by Shannon Stapleton, Reuters[10]

D. Snowballs and Bullet Holes

Figure 132 shows clusters of falling wheatchex and debris trailing abundant volumes of dust. The volume of dust trailing from the falling material is sufficient to block all light. When we consider how much dust would need to have been placed on each column to leave such a trail, we can't conclude otherwise than that the steel itself is actually turning to dust.

Figures 133 and 134 show WTC2 dissolving into powder. An interesting detail is what appear to be "bullet holes," with dense perimeters, distributed throughout the dust cloud. These "bullet holes" look like donut holes within the denser-looking donut-shape of the dust cloud.

Figure 133. Above the white snowball-shaped dust cloud, the building is gone, while the lower part awaits termination. Dark material shoots up, while white material expands outward.
(WTC1 fumes in the background.) http://hereisnewyork.org//jpegs/photos/5245.jpg

Figure 134. The building appears to be dissolving into powder. No solid parts of a falling building can be seen here. (Camera view E, *Figure 127)*
http://hereisnewyork.org/jpegs/photos/5711.jpg

E. WFC3

Figure 135. As WTC2 is destroyed, disintegrating steel "wheatchex" fly over WTC3, the Marriott Hotel. (a)http://bocadigital.smugmug.com/photos/10746211-D.jpg, (b)http://image03.webshots.com/3/3/35/35/278633535003762706 2BOpdpJ_ph.jpg,[11] figure 3-6, page 3-6, www.fema.gov_pdf_library_fema40, (Richard Drew photo)

Some of the anomalies seen in Figure 135b are identified in Figure 136. Note that *the dust trails appear to shoot off of the steel beams in irregular patterns.* The white spots appear in patterns of arcs and circles. There is a shape that looks like a "seahorse." Just above this shape, the roofline becomes irregular, as though the building is coming apart *without* anything hitting it.

Soon after this photo was taken and after the destruction settled, the region of WTC3 where the irregular roofline and "seahorse" shape appeared was missing. (See Figure 138.)

Figure 136. Anomalies for identification in Figure 135.
http://bocadigital.smugmug.com/photos/10746211-D.jpg

F. Survivors in WTC3

Figure 137. (9/11/01) WTC3, (a) before (b) after.
(a)http://www.studyof911.com/gallery/albums/8000pics/WTC3456/2702.jpg, (b)http://memory.loc.gov/service/pnp/ppmsca/02100/02110v.jpg

An FDNY fire company was in the building during the collapses of both WTC 1 and WTC 2 and survived. The firefighters were near the top of the building in the process of making sure that there were no civilians present in the building, when the south tower collapsed. Firefighter Heinz Kothe is quoted as saying, "We had no idea what had happened. It just rocked the building. It blew the door to the stairwell open, and it blew the guys up near the door halfway down a flight of stairs. I got knocked down to the landing. The building shook like buildings just don't shake." Subsequently, the firefighters were in the lower portion of the southwest corner of the building when the north tower collapsed (Court 2001).[12]

The Chief Engineer of the Port Authority of New York and New Jersey, Frank Lombardi, was in the lobby of WTC 3 with other Port Authority executives during the collapse of WTC 2. They survived the collapse and were eventually able to leave the building (Rubin and Tuchman 2001).[13,14]

(a)

(b)

Figure 138. (9/11/01) Two of Bill Biggart's last pictures, following the destruction of WTC2. (a) Note the leafy trees. (b) Three wheatchext stand alone. They also withstood the WTC1 destruction. *(a)http://digitaljournalist.org/issue0111/images/Biggart1833.jpg, (b)http://digitaljournalist.org/issue0111/images/Biggart1836.jpg*

Figure 139. (9/13/01) The West Street laydown of wheatchex. WTC3 was reduced to this three to four story high narrow debris stack. A lower section of WTC1's west wall lies across the West Side Highway. http://www.photolibrary.fema.gov/photodata/original/3887.jpg

WTC3 was partly destroyed in a bizarre fashion during the destruction of WTC2. I would like to give special thanks[15] to Bill Biggart's family and friends for helping with this very valuable piece of the puzzle.

Most of WTC3 disappeared during the destruction of WTC1, as illustrated in Figure 140a. The pedestrian walkway over the West Side Highway had been connected to stairs now no longer there. The stairs were directly across Liberty Street from WTC3. The remains of WTC2 can be seen near the center of the photo and the remains of WTC1 are partly visible in the lower right corner.

Figure 140. (a) (9/27/01) Fuming continues 16 days later. (b) (9/12-13/01 or later) "Fumes" from the WTC site were visible for weeks.

(a)http://www.photolibrary.fema.gov/photodata/original/4228.jpg, (b)http://hereisnewyork.org//jpegs/photos/6823.jpg

Building fumes waft up from the WTC1 and WTC2 "piles" Where are they coming from? They resemble steam rising off of a manure pile. The fume or mist does not seem to originate from a single point but rises over a wide zone, like a haze, in a fairly uniform fashion. Smoke from a fire has an origin and continues to move away from the fire-source, dissipating along the way. Smog, such as in Los Angeles, may collect in a valley, but southern Manhattan is not a valley where smog would become trapped.

Figure 141. (a) Aluminum cladding and paper lay in the street, but where is the steel? And why isn't more of the paper on fire? (b) Why is this the only filing cabinet found?

(The intensity of Figure 141b has been adjusted for visibility, including the lettering.)

(a)http://www.september11news.com/AftermathReuters10.jpg, or http://jackmendes.com/wtc/wtc26.jpg, (b)http://thewebfairy.com/911/h-effect/image/filingcabinet_r1_c1.jpg

Destroyed is a 110-story building made of 100,000 tons of steel, yet the most abundant debris is *paper.* Why is it only *paper* that survived and not office furniture and office equipment? Most of this loose paper must have been in steel filing cabinets and bookshelves. Figure 141a shows what appears to be a fire, yet the paper nearby does not seem to catch fire at all easily. Is it a special kind of paper that does not burn well, or is it a special kind of "fire" that does not easily burn paper?

The twin towers together had an estimated 30,000 computers for nearly 50,000 workers. So 45,000 filing cabinets would not be an unreasonable estimate. It is reported that 200 complete bodies were recovered from among nearly 3,000 victims, about 1/15th. At the same ratio, we would expect that 3,000 complete filing cabinets out of the 45,000 should have survived. *Yet reportedly only one shrunken filing cabinet was found* (see Figure 141b).

G. Dust

Figure 142. WTC1 fumes obscure the WTC2 destruction as if there were a total eclipse of the sun.
http://911wtc.freehostia.com/gallery/originalimages/GJS-WTC21.jpg

WTC5 and WTC6 appear undamaged after the destruction of WTC2. Curiously, the fumes rising out of the top of WTC1 appear to flow horizontally, while the fumes from the destruction of WTC2 appear to drift upward along a diagonal path. In addition, the hazy fumes at the southern-most region of Manhattan (far right) appear to be drifting upward.

Figure 143. (9/13/01) (a) Scooping up the building. (b) Workers walk in thick dust atop the rubble pile, hardly higher than the lobby level.
(a)http://www.photolibrary.fema.gov/photodata/original/5455.jpg, (b)http://www.photolibrary.fema.gov/photodata/original/5453.jpg

The black building in the foreground of Figure 144 is the Bankers Trust Building (130 Liberty Street), which has a total volume that is approximately 28% of the total volume of just one WTC tower. The two WTC towers had seven times the volume of the Bankers Trust Building. How could seven buildings the size of Bankers Trust leave so little debris?

Buildings	**(W x D x floors)(12 ft/floor)**	**114,216,960 ft³**
WTC1 & WTC2	2x(208 x 208 x 110)(12 ft)	114,216,960 ft³
Bankers Trust	(182.5 x 182.5 x 40)(12 ft)	15,987,000 ft³
		ratio: 7.14

Table 10. Volume comparison.

Figure 144. (9/27/01) The black building in the foreground is the Bankers Trust Building. Bankers Trust is 130 Liberty Street. (Intensity adjusted and descriptions added.) *http://bocadigital.smugmug.com/photos/10948720_ejP7a-O.jpg, http://bocadigital.smugmug.com/gallery/275893_6MfUV/4/10948720_ejP7a/Original*

H. How Much Dust Would a Building Make?

Figure 145. Ground-level view of the enormous quantity of dust wafting skyward. Conventional demolition dust does not do this. *http://hereisnewyork.org//jpegs/photos/5717.jpg*

If a WTC tower were completely turned to dust, how much dust might be left? We can assume that dust is less dense than the solid material of the building and its contents. First, let us assume further that the building's materials were reduced to 10% of its original volume when turned to dust, then determine the volume of that dust.

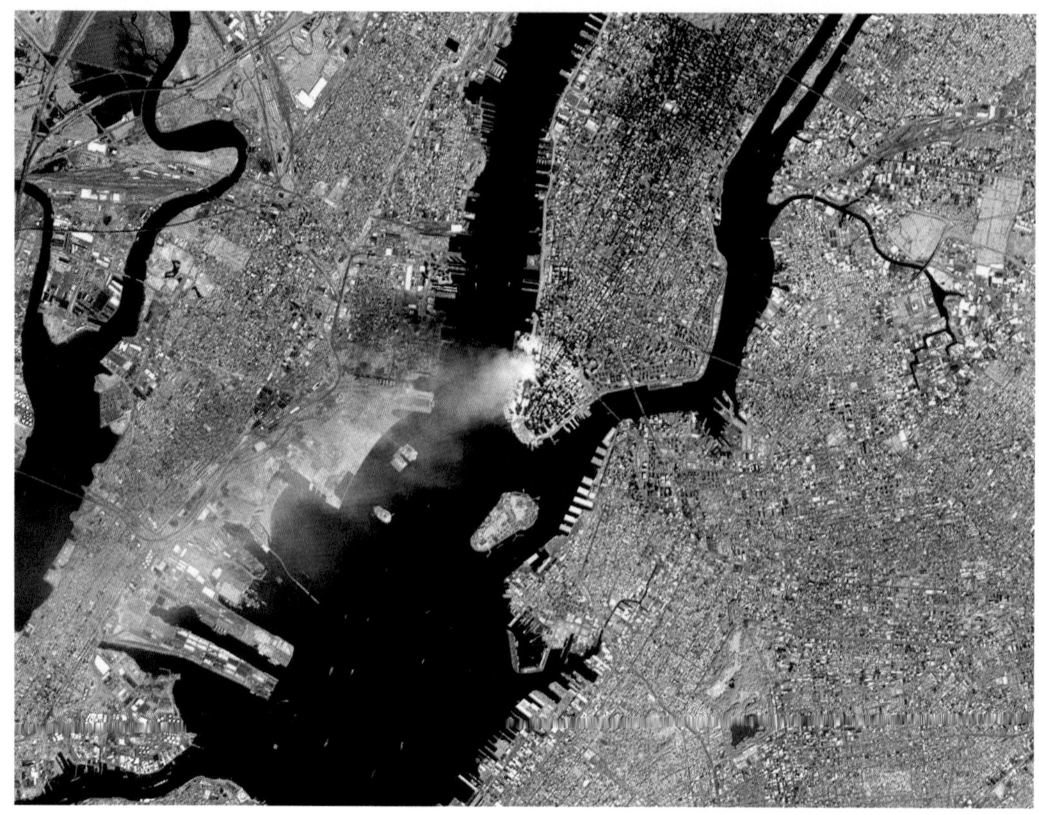

Figure 146. (9/12/01) This is a photograph taken from the International Space Station approximately 11:30 AM. (intensity adjusted)[16]
http://earthobservatory.nasa.gov/images/imagerecords/1000/1755/wtc-merge-321pan_lrg.jpg

Volume of one WTC tower = (208 *ft*)x(208 *ft*)x(1368 *ft*)
Dust Volume (one WTC tower) = (10%) x $\text{Volume}_{\text{tower}}$ (approx.)
Dust Volume for one WTC tower (approx.) = 5,918,515 ft^3
If spread over one square mile or (5280 *ft*) x (5280 *ft*),
5,918,515 $ft^3/(5280\ ft)^2$ = 0.212 *ft* deep or (0.212 *ft*)x(12 *in/ft*)
= 2.55 *inches* deep over one square mile,
or equivalent to one-*inch* deep over 2.55 square miles for one tower (or one-inch deep over 5.10 square miles for both towers).
An area of 2.55 square miles would be a radius of **0.901 miles**.

Note that this area would include both land and water.

If the building's materials and contents were reduced to only 5% of their original volume, that dust would be 1.27 *inches* deep over one square mile, or equivalent to one-*inch* deep over 1.27 square miles for one tower (or one-inch deep over 2.55 square miles for both towers). An area of 1.27 square miles would have a radius of **0.637 miles**.

These calculations suggest that the two towers had enough material to yield dust that, at about an inch deep, would cover approximately 2.5 to 5 square miles of lower Manhattan, not counting the dust carried over the Hudson River, the East River, Brooklyn, the Upper Bay, and into the upper atmosphere. Let us estimate the area where it appears that the dust did land.

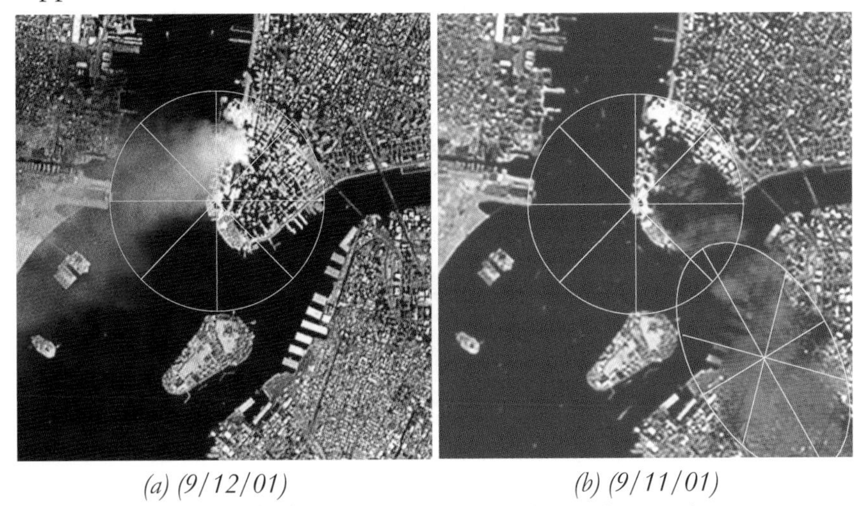

(a) (9/12/01) *(b) (9/11/01)*

Figure 147. Estimating the area covered by the heaviest dust.
(a)http://earthobservatory.nasa.gov/images/imagerecords/1000/1755/wtc-merge-321pan_lrg.jpg, http://upload.wikimedia.org/wikipedia/en/b/b8/September_11_from_space.jpg[17], (b)http://hybrideb.com/images/newyork/manhattan_temp.jpg

Because of wind and other factors, the zone of dust on the ground is not expected to be symmetrically arranged around where the towers stood. In the satellite image shown in Figure 147, the dust forms a circular boundary northeast of the WTC complex. Estimating the center of this arc, we can calculate the approximate area for this part of the dust zone.

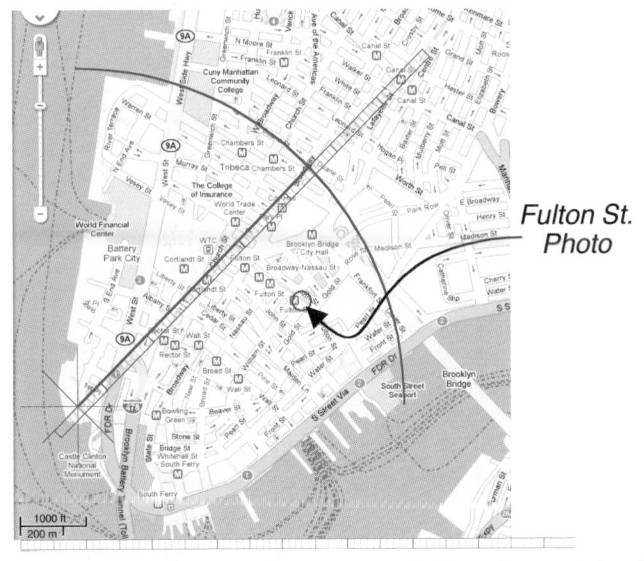

Figure 148. Arc of significant dust. Figure 119 shows dust on Fulton Street.
http://maps.google.com

But dust, smoke, and fumes were also blown southeast, so we will estimate that region to the southeast and beyond as approximately equal to the main circle (represented by the oval shape in the figure above). This dust would not be as dense, but in reality we know the plume continued down the eastern seaboard. (See Figure 146.)

An *estimation* for the main circle of dust may be calculated using a radius of 5,000 feet or (5,000/5,280 or 0.947 mile).

area = $\pi r^2 = \pi (0.947)^2 = 2.817$ sq. mi.

Figure 149. The destruction of WTC2 envelops lower Manhattan in a blizzard of ultra-fine dust.[18]
http://uploads.abovetopsecret.com/ats37434_wtc_2_60kb.jpg, (no longer online)

Based on the width of the dust cloud moving southeast over the East River and Brooklyn, an additional area equal to the circle will be used as an approximation of the down-wind area. That is, 2.817 sq. mi. x 2 = approximately 5.634 sq. mi. This value corresponds to our earlier calculations, when we assumed that the buildings' materials were reduced to 10% of their original volume when turned to dust.

Although these values are approximations, they provide a good indication that the majority of both towers had been turned to dust. Let us compare this amount of dust with the amount from the controlled demolition of the Seattle Kingdome.

I. Kingdome Debris

Typical dust and dirt settle out of the air fairly quickly. A lot of dirt and dust is kicked up when a vehicle drives down a dirt road, so we learn not to follow too close behind another vehicle, allowing time for the dust to settle. Typically, thirty seconds is sufficient. As discussed in the previous chapter, for the case of the Kingdome, the *Seattle Times* reported that "[D]ust choked downtown for nearly 20 minutes."[19] Conventional demolitions do not send dust far above the original building height, contrary to what we saw at the WTC. Figures 150 to 152 show the Seattle Kingdome demolition.

Figure 150. The Seattle Kingdome with the Olympic Mountains. (contrast adjusted)
http://www.kingdome.org/images/img0045.jpg

Figure 151. (a) The Kingdome implosion starts. (b) The right and left sides are "pulled" first to fold the building inward. (c) The dust expands. (d) The cylindrical wall collapses. (e) The dust barely goes higher than the original height of the Kingdome. (f) The pile of rubble that remains includes all of the beams and concrete.
Contrary to the WTC, all of the valuable furniture and fixtures had already been removed.
Source. http://seattletimes.nwsource.com[20]

Figure 152. The Kingdome was not pulverized. It was merely broken into manageable-size pieces. Note how tiny the people are relative to the rubble pile.
http://seattletimes.nwsource.com/kingdome/gallery/photos/photo_05.jpg

Visual evidence, along with the comments of experienced steelworkers, verify an unprecedentedly small debris-stack at the World Trade Center site.

J. WFC2 and WFC3
World Financial Center Two and Three

In contrast to the dramatic *dustification* of the WTC Towers, many windows in the World Financial Center (WFC) look as though they have been blown out from the inside, as seen in Figures 153a, 153b, and 155, or else shattered without having been blown either in or out—perhaps just turned to dust. The World Financial Center buildings (WFC) are on the west side of West Street, directly across from WTC1. The Winter Garden, located between WFC2 and WFC3, was damaged by steel debris but remained repairable. Surprisingly, WFC2 suffered no apparent structural damage to the façade other than blown windows.

World Financial Center Two (WFC2), on the left in Figure 153a, lost all of the marble immediately around the entrance and lost a lot of windows, yet the façade of the building in general suffered little to no damage.

Figure 153. (a) The all-glass Winter Garden had few broken windows. (b) (09/18/01) Damage to WFC3 was no higher than the 18th or 20th floor.
(a)http://bocadigital.smugmug.com/photos/10698505-D.jpg (b)http://www.fema.gov/photodata/original/3942.jpg,

As shown in Figure 153b, World Financial Center Three (WFC3) suffered damage to its southeast corner, and yet above the height of approximately its 18th to 20th floors, no steel beams hit the building at all. Inside WFC2, where the windows have been blown or shattered (Figure 155), the exercise equipment nevertheless appears undamaged.

In addition to the photographic evidence of what I am calling "dustification," we must consider the testimony of the firefighters, emergency medical responders, and law enforcement officials who came to the scene.

Figure 154. (1/17/08) The Winter Garden was repaired.
Personal photo, DSC04413

K. The First Responder Testimony

For example, here is Assistant Chief Fire Marshal Richard McCahey:

Assistant Chief Fire Marshal Richard McCahey

> Somebody, which actually helped me, I don't know who it was, after somebody said I can't breathe, somebody, I don't know which direction it came from, screamed out don't panic or relax, relax. It's not smoke. It's just dust. Just relax.[21]
>
> At that point, that's when I started to realize my mouth was filling up with like a sand ball. All of a sudden I realized when whoever said that, now I'm starting to pay attention to my surroundings. I realized there was no heat, you could breathe. *Stuff was going in your mouth* but it was like a cool air coming in *when you breathe it, so I said maybe he was right.* You couldn't see, it was gritty.[22]

And here is Deputy Chief Thomas Galvin

> I would say from that time maybe a minute later, two minutes later, is when the hotel just started shaking. Everything came down. I ran south down the lobby, and that's when I got caught in debris in the lobby there. There was just dust, debris coming down all over. *I got knocked to the floor with a lot of people.*
>
> You hear somebody say okay, the rumbling stopped, and you hear somebody say the rumbling stopped, we're all right, we're all right, everybody's okay. Let's just find our way out of here. You really could hardly see anything. [23]

Figure 155. (9/16/01) View from the WFC2 exercise room. (contrast adjusted)
http://www.photolibrary.fema.gov/photodata/original/3918.jpg

Emergency Medical Technician Immaculada Gattas

Or consider the statements of Emergency Medical Technician Immaculada Gattas, who describes finding no bodies, only dust:

> Just for a while, I don't know for how long, we stayed in that trailer in darkness, until one of us said okay, let's open the door. We have to open the door. We were scared to death to open the door. We didn't know what we were going to find. So we opened the door and we saw this oh, my god, everything covered by dust. *Everything, a cloud, everything was surrounded by dust.* People walking like zombies, covered in dust. Just like we were, but walking like zombies. [24]
>
> [...]
>
> *We started getting busses with hundreds of doctors and hundreds of nurses and members of the clergy from different denominations and medical equipment and all these areas were completely staffed and completely ready for what we were expecting we were going to get, which was patients, which we never got....* [25]
>
> [...]
>
> ...I got disoriented. I didn't—no idea, the slightest idea where I was, which was very scary. You couldn't see. You didn't know where you were, you didn't know if you were heading back because at the end you were just kind of walking in circles. I guess God was with me and I made it out of there and had to, you

know, deal with frustration and guilt because I had to go through that because you are trained to save lives, not to just witness a horrific destruction and no bodies found, nothing found, just dust. Just dust.

That dust, I'm telling you Christine, it was so freaking itchy. I was itching for hours and hours and hours. It was like glass stuck in my skin. I went—I don't know how many times I went to the bathroom to wash. It was like, like I had to get this thing out of my body and I cleaned my nose and my face and my arms and that thing, that dust, that stuff and when I went home, I took two baths and *I still couldn't get that stuff out of my skin* and I was coughing and spitting up stuff for days after that. [26]

Emergency Medical Technician Michael Ober

Other emergency medical technicians' testimony is remarkably similar. From Medical Technician Michael Ober, for example:

So when I got to the entrance, I turned around and saw the debris coming down. I didn't know if I was gonna die or get trapped. I was just like, I kept thinking, I don't want to be alone. I need to find somebody. I went into the parking garage, I made it about 5 feet. I saw some firefighter up against the wall, he was on the ground right close to the wall. I just jumped on this guy's back, I had no idea who he was. I said, are you all right? He was just like, yeah buddy, are you all right? Yeah, I'm fine. Just as he said that, a huge gust of wind just started coming down. It wasn't like huge chunks of debris, but it was a lot of dust, you know, just dust, it might have been ash too. But it was pitch black. We just laid [sic] there till it stopped coming down, then we finally got back up, it probably had to be inches of just stuff on top of everything. It had to be probably like six inches of dust and all kinds of stuff. After it stopped coming down, I finally got up, and thought we gotta get outta here. [27]

We got up. We're trying to figure out where we were. It seemed like a movie, it was just like guys in fire companies were screaming out their company numbers. I think it was 16, I'm not sure. But I remember just hearing someone in the background screaming 16, where are you? It didn't seem—I just felt like I was in a movie. So myself and this firefighter, I have no idea who it is, decide that we're gonna walk towards this entrance where we came in. We walked, and I'm like, we're stuck. He said, no we're not at the entrance yet. *Somebody says no, you're outside. I'm like we're not outside, it's dark out.* He said look at the air, there's so much stuff in the air, no sunlight's getting through. The amount of stuff that was in the air, you couldn't see it. It was like a curtain of just dust that blocked out all the sunlight. Breathing was horrible, it was *just ash* and *dust.* Every time I took a breath, I could feel more of it going into me. You try covering up your mouth, but it was still getting in. Cause it was just so thick in the air, it was hard to see 10 feet by you. It was horrible. There was just gear everywhere. I had no idea where I was, I had no idea how I was getting out. I somehow stumbled over a Scot Pak and picked it up, just in case. I never used it, but I just picked it up and brought it just in case I needed to use it. We found out we were outside. **At first you really couldn't see anything, it was just dark.** Then after that, we were just looking for people, anything around us that we would know, that could get us out. Someone that could help, it was just confusion at first. I didn't know what to do, where to go, who to help, who to look for or anything. Then the air started to settle down a

> little bit, and some of the air was coming through, and it felt like I was looking at a black and white TV. Everything was a shade of gray. Everybody around me was just covered in gray. I saw a fire engine that was just completely destroyed, and it was just gray. Every fire engine was just a different shade of gray. It was just crazy. Actually, with all the devastation that was going on, that was the real cool thing to see. I don't know why, it was just really interesting. It was just like I said, it was like a black and white TV. [28]

Clearly these emergency medical technicians are testifying uniformly to an extraordinary amount of dust. Note also that they are not describing sounds of explosions or of shearing metal or other sounds and sights one would expect to hear in a controlled demolition or in the pancaking floors of the official story. These first responders had dust land on them, not heavy chunks of debris. Michael Ober's statements describe what many other eyewitnesses also described, "It *was like a curtain of just dust that blocked out all the sunlight*" and *"it was pitch black,"* which is consistent with some of the photographs of the car fires. It looked darker than midnight. Let us think about this for a moment.

Dust has a high surface-to-weight ratio, so it is more affected by wind resistance than chunks of debris would be. Yet Michael Ober describes being covered with dust and not *"huge chunks of debris."* A *lot* of dust came down. The fact that it was pitch black indicates, also, that a substantial amount of dust went *up* in order to block out all of the sunlight.

Deputy Chief Medical Officer David Prezant

Deputy Chief Medical Officer David Prezant describes a similar scene:

> Unknown to me while this was happening but now obvious to me is that there were two large plywood sheets that had created a sort of roof above me. I was coughing tremendously. I was gagging. There was all sorts of particulate matter in my throat and in my eyes and *my eyes were burning. My throat was burning.* I was coughing. I was choking.
>
> Then I felt or saw or pushed these two plywood boards. Luckily, by moving them, I actually really did not have to spend much time digging myself out because by moving those two plywood boards I was able to have enough space so that I could get out of this debris and I could stand up. I was surprised that I could stand up. My right leg was hurting me a lot; my left leg a little bit. I had a lot of minor bruises on my back that were bothering me. I had like some bumps on the back of my head that I sort of felt.
>
> But the main thing was that, when I got up, *it was completely black.* It was blacker than midnight. *I could not see the sky. The air was like syrupy charcoal paste.* Again, coughing, gagging, eyes irritating, hard to breathe, and the only thing that I could think of at the time that could explain this was that I was still buried. I felt that the street had not collapsed underneath me, so I knew I wasn't subterranean. But I felt that the only way that *all this particulate matter could be creating this total blackness is that a roof had been created, so that maybe the collapse had created a tunnel above ground, a sort of mine structure above ground.* [29]

Michael Ober on the Unique Smell

But medical technician Michael Ober also described something else, something very interesting, and that was a unique *odor* in the air:

> The smell was just—it has a distinct smell. I've been to Manhattan many times since then, and that smell just brings back every single...I don't know if it's like World Trade Center cement. I don't know what exactly it is. It's just that like, the smell that we inhaled so many times with the rest of the *dust* and everything in the parking garage. It's just a nasty smell. I mean, it's not that bad, but I just don't like the smell. Um, I don't know...we were down there, there was rumors about the other stuff going on. But nobody had definite information. [30]

Ober Hears No Pancaking Collapse

Ober also mentions something else, something very interesting—an *absence.* That is, he does not remember *hearing* the supposedly pancaking building—with all the terrible noise and roar that shearing steal and concrete should make in breaking away and hitting the ground.

> I don't remember the sound of the building hitting the ground. Somebody told me that it was measured on the Richter scale, I don't know how true that is. If the building is hitting the ground that hard, how do I not remember the sound of it? [31]

Of course, in situations of shock, people often *do not* remember hearing or even feeling certain things. But in Ober's case his testimony is detailed and exact, and not what one would expect of someone in shock, as the following sections will demonstrate.

"Sprinkles"

Ober begins by noting something strange about the "collapse":

> You saw the towers right in front of you, and the first tower was hit. I couldn't tell at first if it was papers or birds, but something white was you know, like flickering up in the sky. It just seemed weird. [32]

But then he notices something else:

Jumper Parts

He provides one of the most significant of the statements by emergency medical personnel, about the horrible "rain of people" from the towers:

> So we left to go outside, and as we stepped out of the building, it *was like, raining people. People were just jumping from everywhere. Just all over it was bodies and parts just scattered.* We walked across, I believe it was West Street, and we set up the command post over there next to the fire command post, just trying to get everything in order. At the same time we're trying to watch to see what's going on. *The only thing that was going on was you could see the buildings burning and people just jumping.* You could watch them fall from like the 90th floor all the way down. It's like you go to school for so long to be able to take [care] of people and treat them

> and be able to fix them when there's something wrong with them, and there's nothing, they hit the ground, and that's it. [33]

Dep. Chief Thomas Galvin and EMT Christopher Attanasio's Testimonies

Deputy Chief Thomas Galvin's testimony is worth considering in some detail.

> Later on I know Ladder 5 was on the northbound side and Rescue 1 was right in front of them. Any other rigs, you know, I wasn't paying attention to the rigs. I was really looking up because there was just stuff falling out of the sky, debris falling down. [34]
>
> Q. It's the underground parking. Was there a lot of debris falling at the time?
>
> A. No, there was debris by the building across the street. You could see things coming down. It was parts of the building, maybe parts of the plane. And there was really no bodies on that side of West Street. [35]
>
> Q. At this point the buildings were still standing?
>
> A. Yeah, the buildings were still standing. Yeah, the buildings were intact.
>
> So then Chief Cassano came into the building, because he said to me, where's your aide? I said they're in there. I told them I would meet them in the south tower but I wanted to get the units to clear out the hotel, the units that we had here.
>
> Let me start moving to the south tower. And now hindsight's twenty twenty when you see the layouts of the buildings. I wasn't aware the Marriott directly interconnects—even this picture doesn't give it to you. It directly interconnects into the north tower.
>
> I remember saying to somebody how do I get into the tower, not realizing that this interconnected with both. And the guy actually sends me to the north tower. And I remember going through the atrium and hugging the wall, because when you come out—you had to go north in the lobby for about 100 feet and make a dog leg to get into this area. There was stuff falling down. I remember hugging a wall.
>
> When I got in there and I'm looking for the units and I'm saying where is everybody? So I grabbed somebody and said where's the fire command station, where's the chiefs? Oh, they already went outside. Why would they go outside already? What I didn't know is they had already made the decision to take the command post from the north tower out into the street.
>
> Now I'm making my way back into the thing. I don't think at that time I realized I was in the north tower. *I just said something's wrong here. It just doesn't seem right.* I remember coming back and there was more debris that's falling down into the atrium. I'm calling it an atrium, but it's just an open area. [36]

Now, consider the above statements in the context of emergency medical technician Christopher Attanasio's description of what *he* saw, just after the "collapse" of WTC2:

> We went back to where they were staging by Battery City and we went back, I went back to the tower that had come down, but a gentleman from OEM, who is a black male, that's all I remember, a very big black male, he had told us that his

> boss was in the building. *We proceeded to go back to the tower that was already down and when we pulled up, we saw* burnt vehicles, fire balls, smoke, debris, dust, bodies.[37]

Burnt vehicles, fireballs, smoke, debris, dust, and bodies?

This is hardly the type of list one would think of when describing the standard results of a controlled demolition, much less the government's official "airplane fuel caused the steel to melt and the building to pancake" theory. How does *either* theory account for the fact that vehicles *on the ground* caught fire? We will deal with this fascinating type of evidence—the burning ground vehicles—in a subsequent chapter. For now, we merely make note of the existence of this class of evidence and of the inability of either version of the "pancake" theory to account for it. Note also the description of "fire balls." This is an extremely important clue, as we will see.

EMT Roger Moore's Statement (Before the "End" of WTC2)

In addition to these statements, we must also consider those of emergency medical technician Roger Moore, prior to the "collapse" of WTC2:

> I got a helmet from his trunk and proceeded to walk to the command post, which was located at West and Vesey Street. While we were walking, we noticed three blocks away from the Trade Center that there were body parts scattered all over the roadway, and that there were police officers trying to safeguard the body pieces.
>
> Q. Were they civilians or you think they were from the aircraft?
>
> A. Unknown.
>
> Q. You don't know?
>
> A. I mean, you really had to look to see these—*these looked like hunks of meat. People were torn apart,* so we decided to go across the street on the water side and continued up to the command post.[38]

Note that Moore is describing the bodies—some of which we may assume were "jumpers"—as being "torn apart," yet another important clue, as we will eventually see.

Moore also describes what he saw *after* the "collapse" of WTC2:

> I was waiting for my assignment. Chief Nigro came over and was talking, along with First Commissioner Feehan, and a few minutes went by. They said, "We are going to give you an assignment." I said, "Okay, I'm here ready to go." The next thing I know, I heard Chief Nigro yell, "Look out, look up."
>
> We all looked up, and the building was on its way down, so we immediately dropped what we were doing and proceeded to run towards the water. That was there was a loading dock from the American Express building about 40 feet away.
>
> *As we were running, I managed to trip just outside of the loading dock and came down on the ground and hit the curb, taking out my knees and my left elbow.* I was now trying to crawl with one arm. I managed to roll a little bit, and a fireman grabbed and pulled me in. At the same time Lieutenant—Captain Sickles landed on top of me, but at this time—and the building was now hitting the ground outside—directly outside.

Essentially, I thought I wasn't in far enough, but at that point, you know, it was—you couldn't see any more. It was a black, black night, and dust, and dirt and everybody was choking.

After approximately ten minutes went by, you could start to see flashlights and people starting to talk and moving around. They came over to me, and asked, you know, let me help you up, and I couldn't get up. Both legs and the arm were shot. I figured they were broken. They said, "Don't go anywhere."

Commissioner Frank Gribbon came over and said, you know, he was dressed out in his uniform, in his turnout gear, and he said, "I'm going to get a crew, I'll be right back." Two firemen that were there said, you know, "We're not going to leave you," and they saw a stretcher over here, let's grab the stretcher. They put me on the stretcher, and said, you know, the best thing to do is probably go out through the loading dock into the building and out the back of the building where it's safe.

They said okay. So they wheeled me down to the loading dock, put me up on the loading dock into the building. Just as we are getting ready to get onto the elevator, we heard a high pitched whine and wind and heard thundering crashes. It turned out to be—we thought—we originally thought it was the building that we were in was coming down, but it turned out to be the other tower.

Q. That was the north tower?

A. Right.

At that point, the power went out in the building. They switched to their flashlights. After a couple of minutes, the *emergency generators*, I guess, kicked in or the emergency batteries. Some lights came back on, and they are, like, we need to get out of here, and I said, "Well, I can't walk." They said, "Well, we're not leaving you either," so they got on their radio and managed to get a couple of people that were upstairs. They came down the fire escape inside the stair building. I think it was Stairway F or G. The five of them carried myself and the hundred pound ambulance stretcher up four flights of stairs.

They had said, you know, "My name is Steve and my name is…" the other guy's name is Gary, and, you know, "We are going to help you." Listening to them talk on the radio, I was able to come to the conclusion that they were from Truck 6. They were saying Truck 6 to whoever, and they were, you know—those people came down.

They then proceeded to get me up the stairs. After about ten, fifteen minutes they got me up to the top, and they wheeled me out. There was a battalion chief there. I couldn't see his name.

They took me out the side door of the American Express Building and wheeled me down to the marina. The same time, there was a police launch pulling up, so they wheeled me down to the marina the rest of the way and put me on the police launch.[39]

More Burning Vehicles and Body Parts

Deputy Chief Thomas Galvin also reported the burning vehicles: "I saw all these ambulances turned over. Half of them are on fire." [40] Note, that only some of these vehicles, according to Galvin, were afire. This is, as we will see when we examine this type of evidence in a subsequent chapter, a very significant clue—and a very significant problem for the "pancake" explanation whether due to airplane fuel or to any type of controlled demolition.

Emergency Medical Technician Christopher Attanasio

Similarly, emergency medical technician Christopher Attanasio reports :

> Upon arrival, towers one and two were both ablaze. The second plane had hit the second tower already. Both towers were totally engulfed. *People were jumping out of the buildings.* There was airplane fuselage and landing gear around the site. *Body parts, victims' remains on the floor.* There were some injuries on the street. *Some cars were on fire.* [41]

Notably, Attanasio is one of the few "first responders" who mentions seeing airplane parts. But this is not, at the moment, what interests us. Note that he describes people jumping out of the towers, and that he also states that *some* cars were on fire.

Captain Michael Donovan

Finally, Captain Michael Donovan describes what he saw as WTC2 "collapsed":

> I saw one guy from 41 Engine, and he said we have to start searching this area. He was as disoriented as I was. The two of us, we went into what I believe was Three World Financial Center. All the windows were broken. We went in and we started searching in there. We didn't find anyone.
>
> Then he went off one way, and I continued searching the street area on West Street from just south of Vesey to the pedestrian bridge. I searched there for, I would say, 20 minutes to half an hour, and *I didn't see anyone.* I found a lot of masks and equipment all laying on the ground. Rigs were buried; rigs were destroyed. *Cars were on fire.* Buildings were on fire. [42]

Again, the testimony is that ground vehicles were on fire. Yet the airplanes slammed into the buildings, according to the "official story," several hundreds of feet *above* the ground. This would have meant that any airplane fuel catching fire would have done so several hundreds of feet above the ground. So what caused these vehicles on the *ground* to burn? And why, according to some of the testimony thus far presented, are only *some* of them burning? Note also that Donovan states that he didn't see anyone, a statement that, in the context of other first responders' statements attesting to the amount of dust in the air, makes sense.

Emergency Medical Technician Joseph Fortis

The testimony of Joseph Fortis is also of great importance, since certain statements within it allow a broad chronological sequence of his observations to be reconstructed. Accordingly, I have divided his testimony into chronological blocks. This is his statement for when WTC2 got its plane-shaped hole:

> I believe at that time—I can't even tell you the time frame. We actually heard like the engines I guess for the second plane coming in, and it just got louder and,

boom, the second plane hit the other tower. We stood there and we watched. I guess like a fireball cloud came down. We were all standing on the corner I guess by the pedestrian bridge off the corner there with our equipment. We were just amazed in awe. [43]

Everybody was running at us saying evacuate and we're under attack kind of thing. Then the light went on to run. We turned around and started running west on Vesey Street, and we made it just past the American Express building.

It was just coming too fast. We couldn't get away from that cloud per se, and we ducked into the lobby of a building there. I believe it was between—it might have even been the American Express building on the corner there.

Q. At this point it's the collapse of the first building now?

A. No, no. This is just when the second plane.

Q. Okay.

A. The buildings were still up. So it was just a blast from I guess the second plane, I think, came in, *and the dust cloud just came. We were on the corner there, and you just felt the heat. Our back and all our eyebrows were all singed and everything.* We had a little flash burn because we were right—I guess when it came we were right on West off the corner of I guess where six was, if that was it, if this is West Street here. [44] [emphasis added]

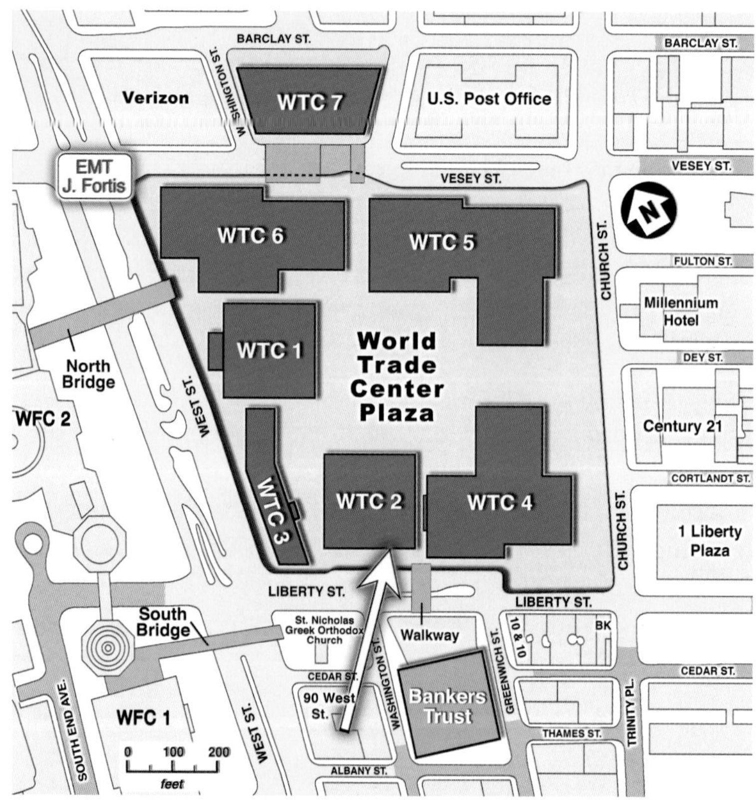

Figure 156. Location of EMT Joseph Fortis as WTC2 got its hole.[45]
Joseph Fortis was described standing at the Vesey-West Street intersection at the time WTC2 was "hit" by something. He reported that their eyebrows and backs were singed and bands on their jackets melted.

Joseph Fortis later continues

So we just backed the vehicle up there and, like I said, put our equipment in the front. Just as I was ripping tape to put 03 John on the windshield, I guess that's when we heard the turbines or whatever.

Q. The second plane?

A. The second plane per se. We didn't even know it was that until afterwards, until days after.

Q. Right.

A. The people you were with, it was Chris Attanasio and Roland. They were like, "Oh, no, dude, that was"—the first thing we heard was the plane coming in, because then we were there for the other collapses. I didn't even realize it because it was just chaos and everybody was just scared, to be honest with you.

Like I said, when that blast came out, everybody got this little singe. Even like the little bands on our jackets just melted right off, the reflective bands. We just dropped everything when the cloud came, composed ourselves, and went back to the site again.[46] [emphasis added]

Joseph Fortis describes being on the corner of West and Vesey Streets, across from WTC6 when WTC2 is said to have been hit by a plane, yet didn't know it was a plane "until days after." The corner he describes having been at is some distance from WTC2 and is blocked by WTC1. So it is curious that he said, *"We were on the corner there, and you just felt the heat. Our back and all our eyebrows were all singed and everything."* Joseph Fortis also describes what looked like a ticker tape parade or "sprinkles" of debris, similar to what Michael Ober described.

EMT Joseph Fortis' statements as WTC2 Goes Away

Like I said, we started ahead like halfway across West Street with our stuff, and the ground started shaking like a train was coming. You looked up, and I guess—I don't know, it was one that came down first or two? Which one?

Q. The first one to come down was the south tower, number two.

A. Two? We were standing on West Street, and the ground started to shake. You looked up, and it looked like a ticker tape parade off the back of the building, because *all this stuff started coming down. We thought it was just like all papers and everything. Like I said, there was pieces of body parts all over the place.*

We came halfway across the street, and the building was coming down. Everybody was running out of the same evac, the building's falling and ESU and everybody and everyone's screaming "Get back! Get back!"

We dropped all our stuff and started running again, west on Vesey, headed towards Vesey and West. Then we went west on Vesey. We just made the turn on I believe on North End Street. Just as we got to the corner, I guess the debris from the cloud came up Vesey and up Murray and then up and over that building that's there. So we actually stood up against the building here.

Q. Vesey and North End?

A. Right, right. The lobby is like right there. They didn't want us to go in the building. Actually everybody—it was just chaotic. *That cloud came, and we just leaned up against the building and it was just—no one could breathe or anything.*

Then after that we—an ambulance came by from I think it might have been Cabrini's ambulance, and they stopped right in front of that building on North

> End Street. We all jumped in the ambulance. Like I say, you could just see the cloud coming up and over the building. We waited and the ground was shaking and all that. We waited like two, three minutes. We got out of that vehicle and proceeded I guess east on Vesey towards the buildings again.[47]
>
> So we proceeded to the ambulance, put on our turnout gear, helmet and turnout coat, and as we were taking the equipment out of the ambulance, the second tower—the second tower, started to come down. As the tower was coming down, we ran. I ran, I guess it was west to the West Side Highway. The tower came down. I grabbed my partner, we ran.
>
> When the tower finally came down, *there was a white cloud of smoke that hit us, knocked us to our feet. It was very hard to breathe. We inhaled a lot of white powder, whatever it was, dust, concrete, whatever it was.*[48]

Notably, Fortis describes not only the choking cloud of dust, but also a "shock wave" from the cloud, and the shaking of the ground. So in other words, and bearing in mind what has been discussed in previous chapters, Fortis is not only corroborating the *dustification* of the buildings but is also corroborating the ground *vibrations* recorded by the seismographs in the area, *vibrations* which, as we saw, were *not* long enough in duration for the pancake model to be true. Thus, Fortis' testimony is an important clue as to what *other* types of methods might have been used to bring the towers down. In this regard, his statement that he was hit with some sort of "shock wave" as the dust cloud reaches him is also an important clue.

Supervising Fire Marshal Robert Byrnes

Another important bit of evidence is provided by Supervising Fire Marshal Robert Byrnes:

> As we were walking up there, I was around between Albany and I thought it was Carlisle or Cedar. We were right in the middle of the street and I happened to be looking up at tower number two and thinking to myself, *how are they ever going to put this fire out? It's probably just going to be a rescue operation until the fire burns itself out.*
>
> As I'm looking up at the building, I hear a loud noise and I see the south side of the building collapse. I see the south upper third of the tower start to pitch in my direction. At that point I yelled to Mike Kane, Mike, it's coming down. I turned around and I ran south on West Street.
>
> I actually ran towards the building line so that I could get adjacent to the building because I figured it would protect me from any falling debris because in my mind I thought the building was actually *toppling. I didn't realize that it actually tilted and then came down straight. My perception was it was toppling southward.*[49]
>
> Q. So you ran south you said?
>
> A. I ran south on West, but I ran adjacent to the buildings figuring it would protect me from any falling debris that may come this direction.
>
> Q. This is the south tower collapsing, so the first collapse?
>
> A. Right.
>
> I ran until I could no longer see and I had to slow my pace down. The cloud of smoke and the debris was coming around me. *There was a lot of dust. There were*

> *little pieces of debris coming down* that were bouncing around me, glass, *small pieces of concrete. Nothing hit me.*
>
> I don't know how far I went, but I was able to find a bus. I got into the bus and *we were able to breathe in the bus because the bus had fresh air.* There were several civilians in the bus. There were two or three windows that were open in the bus as well as the ceiling vents. I remember walking through the bus closing the windows and pulling down the ceiling vents. I got back to the front of the bus and the bus driver asked me, should I start the bus and try to drive away? I told him, no, don't start the bus. Let's just stay put and hope for the best.
>
> I stayed in there until it got dark. You couldn't see. It was pretty dark in there. A few moments later it lightened up. At that point I came out of the bus and I started to walk back up on West Street. As I'm walking back on West Street I'm thinking to myself that I'm going to see the bottom two-thirds of the building still standing there. *I'm thinking that just the top third came off. But you couldn't see up because the cloud of dust was still there.*
>
> As I'm walking back up West Street, I'm seeing lots of papers. There had to be maybe several inches of debris in the street, like *dust and powder*, a couple little fires I saw burning, like papers and stuff burning. I hear a rumbling again. I realize that this is the other tower coming down. At this time I turn around and I make my way back down West Street. I wind up down in Battery Park, where I ran into I believe it was Dr. Prezant. [50]

Again, Fire Marshal Byrnes testifies to the choking cloud of dust, and he adds a significant detail, namely, that the concrete debris is coming down in small pieces, not large chunks, as one would expect with a conventional demolition or a "pancaking" building. Moreover, Byrnes also thought that at least one of the towers was *toppling*. Finally, he expected *the fires to burn out normally, so that rescue operations could begin.* In other words, regardless of his suspicion that a portion of one tower was toppling, he fully expected to be able to enter the buildings at some point and begin rescue operations. He was *not* expecting that they would completely disappear, or "collapse."

Testimony of Chief Medical Officer David Prezant

Prezant's testimony is another significant indicator that some other mechanism than controlled demolition or a gravity-driven "pancaking" collapse was at work in the catastrophe of the twin towers:

> As they were getting their stuff ready and we were all sort of walking very slowly to the middle of the street, I noticed that everybody in front of me all of a sudden started to run away from the south tower. We were not looking at the south tower. We were looking towards the river now because we were walking to the middle of the street. But everybody in front of me all of a sudden started to run. I remember the first thought in my mind was what a bunch of wimps. What are they running from? There's been a little bit of noise ever since we've gotten here and there will be a little bit more noise and a little bit more debris and we've got a job to do.
>
> But within seconds, they were running and I started to run. To this day, no

matter how I stretch my mind and no matter how many firefighters I talk to, *what I think most about is a universal concept that there was not a lot of noise with this collapse. A little bit of noise. I don't know why that is. Maybe because it imploded inwards. Maybe because the noise was dampened by other buildings around it. It was not a lot of noise.* It was enough noise for all of these people to start running, but not enough noise for me to be all that concerned. I have to say I ran because they ran. [51]

Prezant, like many other first responders, describes "hearing a noise," but a noise that, according to his recollection, was strange in that it was not "a lot of noise." In other words, the sound was not commensurate either with a standard demolition or with a pancaking building. But it *could* be commensurate with "dustification."

"Levitation": The Statement of EMT Renae O'Carroll

Emergency medical technician Renae O'Carroll adds another significant clue:

> I'm running. I'm ahead of it. Everyone's running, and it's just a stampede. I'm about ten feet in front of it, running, actually sprinting because I'm an athlete and I'm running. What happened when I got to the corner, because I remember my feet hitting, coming off the sidewalk, another blob of stuff came around.
>
> Ash came around another building in front me, and it caught me in front of me and in back of me, and everything was pitch-black. Where it hit me from the front and the back, it actually lifted me off the ground and threw me. It was like someone picked me up and just threw me on the ground. [52]

O'Carroll seems to be describing the immense choking cloud of dust that came billowing towards her as she was trying to sprint away from it. But this is no ordinary "dust" cloud, for it appears to be driven by something that is actually able to pick her up and "throw" her to the ground.

O'Carroll was not the only person, or object, to experience this uncanny effect from the dust cloud.

Richard Drew on the Charlie Rose Show

In early October 2001, Charlie Rose interviewed David Handschuh and Richard Drew, two New York City Photographers, about what they experienced on 9/11.

> **0:50-0:57 Charlie Rose:** What do you remember most? What is the most searing image you remember?
>
> **0:57-1:28 Richard Drew:** The searing image is -- seeing the facade of the first building -- what I thought was the facade, it ended up being the whole building coming down. _(inaudible)____ All I saw was a chunk of it coming down. Sort of a crumbling, rumbling, of a creaking metal and like rock falling, and that kind of thing.
>
> And the people started walking out, quiet, ...a quiet that sort of invaded everything. And the people started walking out all covered with muck and injured and firemen....

[...]

01:50-2:00 Richard Drew: You thought you were watching a movie, [others agree] and it just sort of settled right down [hands indicating settling straight down].

02:51-2:36 Richard Drew: This is the second tower coming down. [image shown on screen] And I was standing a block and a half south of Stuyvesant High School. I'd been moved out of one position which was too close to the first tower, and then I sort of wanted to hide myself...standing in some bushes in a second median strip. I looked up and started to make a picture and all of a sudden the top [of the building] *just* poofed *out.*[53]

EMT Ronald Coyne and Flipped Cars

Ronald Coyne, another emergency medical technician, described what he saw as the WTC2 "collapsed," or, as I could just as accurately say from now on, "went 'poof'":

> At that point, I just heard a thunderous sound, and I looked up, and *I saw the building start to topple, start to sway, and it was swaying our way,* and we just yelled, "Run" and I tried to run as fast as I could, and I saw an SUV parked, and I figured that that would take some, you know, some of the hit, because I knew I couldn't out run the building, and by the time it took me to break the back window of the SUV, *my safety coat was already on fire. My socks were on fire.* I was already *covered with soot and all sorts of particles that were coming out of the building. I climbed into the truck, and that's when pieces of the building lifted the truck and came through the front window and* flipped *the truck over, and I was trapped in there for approximately 25 minutes to a half hour.*
>
> *I was falling asleep. I knew I was dying.* I just prayed that I wouldn't be found like that. I felt as if somebody was giving me fresh oxygen. I was able to dig my way out myself, out of the truck. I crawled into the street through ash and the fire, and I found a door, and I opened up the door, and it was a tavern. I ran over behind the bar, and I took the seltzer spray and started washing my eyes and my face, because it was burning, and washing my back. I just—I was just covered with burns and bruises, and I couldn't breathe at all.
>
> What little breath I had was just whatever I could get. After that it started to clear up again. I searched the entire building to make sure there was nobody in it. There was nobody found. Whatever apartment doors were open, I just took a peak in to see if anybody was on the floor, and then I automatically locked the doors and left. Then I left the building, and I went back to try and find my partner, *and I noticed that a lot of the ambulances were torched, flipped over, demolished, and I couldn't find my bus.*[54]

As shown elsewhere in this book, the speed at which material was ejected from the building was no more than 20 to 40 miles per hour, a velocity hardly sufficient to overturn vehicles. Even so, that's not all that Ronald Coyne saw that morning:

> We still couldn't find our ambulance. We didn't know where it was. I saw hands and legs, and *I saw a woman impaled into a wall across the street from the building.* I saw people jumping out of the windows when they were collapsing, going through

cars and hitting the pavement, and that was only early in the morning.[55]

Like many other emergency medical personnel that morning, Coyne expresses difficulty breathing when he was enshrouded in the choking cloud of dust. But note also that Coyne mentions seeing the tower start to *topple* and that he fully expected the toppling part to hit the ground. Yet, as we have noted already, very little collateral damage occurred to surrounding buildings, so the question remains: How did a toppling building right itself, turn into dust, and then go "poof," its remains falling into a small footprint with not enough rubble to cause damage to the bathtub? Also note that Coyne's own safety coat (made of fire-resistant material) was on fire even though he was at ground level and makes no mention of having been doused with airplane fuel. Finally, he expresses difficulty breathing in the dust, and the truck he climbed into was quickly flipped over. How does a dust cloud from a conventional controlled demolition, from thermite, or even from the airplane-fueled "pancaking floors" of the official story manage to do all this?

L. Conclusions

What are we to make of all this?

The amount of dust from the Twin Towers' destruction is clearly much too large for a "pancaking" theory—whether the official "airplane fuel"-based version of it or the conventional controlled-demolition version, or the "thermite-demolition" model preferred by many "9/11 Truthers." This vast, thick, choking cloud of dust was attested to not only by photographic evidence, but also by the statements and testimony of the "first responders," the law enforcement, medical, and fire-fighting personnel who were on the scene. A building in the process of being "dustified"—a building literally turning to dust before our eyes—would simply not have sufficient kinetic energy to dustify itself floor by floor, *and* to eject massive plumes of dust and debris, *and* to overturn cars and other vehicles on the ground. And yet, according to the testimony of some first responders, this strange cloud of dust did at least some of those things. Not only that, but a thousand feet *below* where we were told airplane-fueled fires were burning, vehicles were also ignited into flames.

Controlled demolition, thermite, mini-nukes, airplane fuel-based fires—none of them can do all of this.

Some other mechanism was at work. And we now turn to the evidence that more directly indicates what that mechanism was.

---

[1] *http://www.lyrics007.com/Kansas%20Lyrics/Dust%20In%20The%20Wind%20Lyrics.html*

[2] *http://www.answers.com/topic/pulverize*

[3] *http://www.merriam-webster.com/dictionary/pulverize*

[4] *http://encyclopedia2.thefreedictionary.com/vaporize*

[5] *http://img156.imageshack.us/img156/2044/p9111200ms2.jpg, http://img156.imageshack.us/img156/7799/p9111202fy8.jpg, http://img156.imageshack.us/img156/482/p9111203ac3.jpg, http://img152.imageshack.us/img152/605/p9111204pd0.jpg, http://img152.imageshack.us/img152/2684/p9111205pg6.jpg,*

[6] http://maps.google.com

[7] Map redrawn from p. 3 (pdf p. 53 of 298), *http://wtc.nist.gov/NISTNCSTAR1CollapseofTowers.pdf*

[8] *YouTube video: http://www.youtube.com/watch?v=kMr3ZSL6l-4*

[9] *YouTube video: http://www.youtube.com/watch?v=kMr3ZSL6l-4*

[10] *http://upload.wikimedia.org/wikipedia/en/c/c5/DustifiedWTC2.jpg, (now removed), http://hereisnewyork.org//jpegs/photos/1539.jpg,*

[11] Photo by Richard Drew, AP, World Trade Center Building Performance Study, Federal Emergency Management Agency (FEMA), Chapter 3, Figure 3-6, page 3-6 (pdf 6 of 9), *http://www.fema.gov/pdf/library/fema403_ch3.pdf*

[12] Court, Ben, eta1. 2001. "The Fire Fighters," *Men's Journal* Vol. 10, No. 10, pp. 70. November.

[13] Rubin, D., and Tuchman, J. 2001. "WTC Engineers Credit Design in Saving Thousands of Lives." *Engineering News Record.* Vol. 247, No. 16, *pp.* 12. October 15.

[14] William Baker, *Federal Emergency Management Agency* Report, p. 8, *http://www.fema.gov/pdf/library/fema403_ch3.pdf*

[15] I would like to thank a number of people for their help in solving this mystery. I especially want to thank Bill Biggart's widow, Wendy (*http://digitaljournalist.org/issue0111/biggart_intro.htm*), and his good friend Chip East (*http://digitaljournalist.org/issue0111/biggart01.htm*), for sharing with us those last photos taken by Bill Biggart, along with the story about them. (*http://digitaljournalist.org/issue0111/biggart21.htm*) Somehow, I know that Bill knew just how valuable those pictures would turn out to be. Thank you, Bill Biggart! And thank you, Wendy and Chip, for seeing that these images were made available to us.

[16] *http://earthobservatory.nasa.gov/IOTD/view.php?id=1755*

[17] *http://earthobservatory.nasa.gov/images/imagerecords/1000/1755/wtc-merge-321pan_lrg.jpg*

[18] Contrast adjusted.

[19] Dome's Final Roar, by Jeff Hodson, Eric Sorensen, Alex Fryer, Beth Kaiman, Dionne Searcey, Sara Jean Green, John Zebrowski, Phil Loubere, Seattle Times staff reporter, Monday, March 27, 2000 - Page updated at 12:00 AM, *http://archives.seattletimes.nwsource.com/cgi-bin/texis. cgi/ web/vortex/ display?slug=4012219&date=20000327*

[20] (a)*http://seattletimes.nwsource.com/kingdome/gallery/photos/photo_07.jpg,* (b)*http://seattletimes.nwsource.com/kingdome/gallery/photos/photo_08.jpg,* (c)*http://seattletimes.nwsource.com/kingdome/gallery/photos/photo_09.jpg,* (d)*http://seattletimes.nwsource.com /kingdome/gallery/photos/photo_10.jpg,* (e)*http://seattletimes.nwsource.com /kingdome/gallery/photos/photo_11.jpg,* (f)*http://seattletimes.nwsource.com /kingdome/gallery/photos/photo_12.jpg*

[21] File No. 9110191, "Assistant Chief Fire Marshal Richard McCahey," November 2, 2001, pp. 14, *http://graphics8.nytimes.com/packages/pdf/nyregion/20050812_WTC_GRAPHIC/9110191.PDF*

[22] File No. 9110191, "Assistant Chief Fire Marshal Richard McCahey," November 2, 2001, pp. 14-15, *http://graphics8.nytimes.com/packages/pdf/nyregion/20050812_WTC_GRAPHIC/9110191.PDF*

[23] File No. 9110197, "Deputy Chief Thomas Galvin," November 7, 2001, p. 11, *http://graphics8.nytimes.com/packages/pdf/nyregion/20050812_WTC_GRAPHIC/9110197.PDF*

[24] File No. 9110136, "EMT Immaculada Gattas," October 17, 2001, p. 8, *http://graphics8.nytimes.com/packages/pdf/nyregion/20050812_WTC_GRAPHIC/9110136.PDF*

[25] File No. 9110136, "EMT Immaculada Gattas," October 17, 2001, pp. 10-11, *http://graphics8.nytimes.com/packages/pdf/nyregion/20050812_WTC_GRAPHIC/9110136.PDF*

[26] File No. 9110146, "EMT Immaculada Gattas," October 17, 2001, pp. 11-12, *http://graphics8.nytimes.com/packages/pdf/nyregion/20050812_WTC_GRAPHIC/9110136.PDF*

[27] File No. 9110093, "EMT Michael Ober," October 16, 2001, pp. 5-6, *http://graphics8.nytimes.com/packages/pdf/nyregion/20050812_WTC_GRAPHIC/9110093.PDF*

[28] File No. 9110093, "EMT Michael Ober," October 16, 2001, pp. 6-7, *http://graphics8.nytimes.com/packages/pdf/nyregion/20050812_WTC_GRAPHIC/9110093.PDF*

[29] File No. 9110212, "Deputy Chief Medical Officer David Prezant," November 14, 2001, pp. 9-10, *http://graphics8.nytimes.com/packages/pdf/nyregion/20050812_WTC_GRAPHIC/9110212.PDF*

[30] File No. 9110093, "EMT Michael Ober," October 16, 2001, pp. 10-11, *http://graphics8.nytimes.com/packages/pdf/nyregion/20050812_WTC_GRAPHIC/9110093.PDF*

[31] File No. 9110093, "EMT Michael Ober," October 16, 2001, p. 10, *http://graphics8.nytimes.com/packages/pdf/nyregion/20050812_WTC_GRAPHIC/9110093.PDF*

[32] File No. 9110093, "EMT Michael Ober," October 16, 2001, p. 3, *http://graphics8.nytimes.com/packages/pdf/nyregion/20050812_WTC_GRAPHIC/9110093.PDF*

[33] File No. 9110093, "EMT Michael Ober," October 16, 2001, p. 4, *http://graphics8.nytimes.com/packages/*

pdf/nyregion/20050812_WTC_GRAPHIC/9110093.PDF

[34] File No. 9110197, "Deputy Chief Thomas Galvin," November 7, 2001, p. 4, *http://graphics8.nytimes.com/packages/pdf/nyregion/20050812_WTC_GRAPHIC/9110197.PDF*

[35] File No. 9110197, "Deputy Chief Thomas Galvin," November 7, 2001, p. 5, *http://graphics8.nytimes.com/packages/pdf/nyregion/20050812_WTC_GRAPHIC/9110197.PDF*

[36] File No. 9110197, "Deputy Chief Thomas Galvin," November 7, 2001, pp. 8-9, *http://graphics8.nytimes.com/packages/pdf/nyregion/20050812_WTC_GRAPHIC/9110197.PDF*

[37] File No. 9110204, "EMT-D Christopher Attanasio," November 9, 2001, p.4, *http://graphics8.nytimes.com/packages/pdf/nyregion/20050812_WTC_GRAPHIC/9110204.PDF*

[38] File No. 9110214, "EMS Roger Moore," November 29, 2001, pp. 3-4, *http://graphics8.nytimes.com/packages/pdf/nyregion/20050812_WTC_GRAPHIC/9110214.PDF*

[39] File No. 9110214, "EMS Roger Moore," November 29, 2001, pp. 4-7, *http://graphics8.nytimes.com/packages/pdf/nyregion/20050812_WTC_GRAPHIC/9110214.PDF*

[40] File No. 9110197, "Deputy Chief Thomas Galvin," November 7, 2001, p. 14, *http://graphics8.nytimes.com/packages/pdf/nyregion/20050812_WTC_GRAPHIC/9110197.PDF*

[41] File No. 9110204, "EMT-D Christopher Attanasio," November 9, 2001, pp. 2-3, *http://graphics8.nytimes.com/packages/pdf/nyregion/20050812_WTC_GRAPHIC/9110204.PDF*

[42] File No. 9110204, "Captain Michael Donovan," November 9, 2001, p. 20, *http://graphics8.nytimes.com/packages/pdf/nyregion/20050812_WTC_GRAPHIC/9110204.PDF*

[43] File No. 9110200, "EMT Joseph Fortis," November 9, 2001, p. 4, *http://graphics8.nytimes.com/packages/pdf/nyregion/20050812_WTC_GRAPHIC/9110200.PDF*

[44] File No. 9110200, "EMT Joseph Fortis," November 9, 2001, pp. 4-5, *http://graphics8.nytimes.com/packages/pdf/nyregion/20050812_WTC_GRAPHIC/9110200.PDF*

[45] Map redrawn from p. 3 (pdf p. 53 of 298), *http://wtc.nist.gov/NISTNCSTAR1CollapseofTowers.pdf*

[46] File No. 9110200, "EMT Joseph Fortis," November 9, 2001, p. 22, *http://graphics8.nytimes.com/packages/pdf/nyregion/20050812_WTC_GRAPHIC/9110200.PDF*

[47] File No. 9110200, "EMT Joseph Fortis," November 9, 2001, pp. 7-9, *http://graphics8.nytimes.com/packages/pdf/nyregion/20050812_WTC_GRAPHIC/9110200.PDF*

[48] File No. 9110204, "EMT-D Christopher Attanasio," November 9, 2001, pp.3-4, *http://graphics8.nytimes.com/packages/pdf/nyregion/20050812_WTC_GRAPHIC/9110204.PDF*

[49] File No. 9110206, "Supervising Fire Marshal Robert Byrnes," November 14, 2001, p. 5, *http://graphics8.nytimes.com/packages/pdf/nyregion/20050812_WTC_GRAPHIC/9110206.PDF*

[50] File No. 9110206, "Supervising Fire Marshal Robert Byrnes," November 14, 2001, pp. 4-7, *http://graphics8.nytimes.com/packages/pdf/nyregion/20050812_WTC_GRAPHIC/9110206.PDF*

[51] File No. 9110212, "Deputy Chief Medical Officer David Prezant," November 14, 2001, pp. 5-6, *http://graphics8.nytimes.com/packages/pdf/nyregion/20050812_WTC_GRAPHIC/9110212.PDF*

[52] File No. 9110116, "EMT Renae O'Carroll," October 18, 2001, pp. 6-7, http://graphics8.nytimes.com/packages/pdf/nyregion/20050812_WTC_GRAPHIC/9110116.PDF

[53] In early October 2001, Charlie Rose interviewed David Handschuh and Richard Drew, two New York City Photographers, about what they experienced on 9/11, *http://www.youtube.com/watch?v=ulE9OiZqQwg*

[54] File No. 9110395, "EMT Ronald Coyne," December 28, 2001, pp. 7-8, *http://graphics8.nytimes.com/packages/pdf/nyregion/20050812_WTC_GRAPHIC/9110395.PDF*

[55] File No. 9110395, "EMT Ronald Coyne," December 28, 2001, pp. 10-11, *http://graphics8.nytimes.com/packages/pdf/nyregion/20050812_WTC_GRAPHIC/9110395.PDF*

9.
Where Did the Buildings Go?

I looked and said, "Guys, there used to be 106 floors above us and now I'm seeing sunshine."..."There's nothing above us. That big building doesn't exist."...These are the biggest office buildings in the world and I didn't see one desk or one chair or one phone, nothing.[1] —Jay Jonas, (firefighter, survivor in stairwell B)

We just kept telling them we're in the B stairwell. ...I remember everybody had the same exasperation I did. We must have told them a hundred times: "B stairwell in the second floor, third floor, fourth floor of the north tower. I mean, B stairwell, second, third, fourth floor, north tower. B stairwell, north tower." "Where are you?" "North tower, stairway B, second, third floor." ...You could hear they didn't understand where we were. I'm keep going, "My God, why aren't you listening?" Then they said, "Where's the north tower?" *I was like, "What do you mean, where's the north tower??" ...It was unimaginable.*[2] —James McGlynn (firefighter, survivor in stairwell B)

Figure 157. Does this look like a collapse?
Cropped from: http://911wtc.freehostia.com/gallery/originalimages/GJS WTC28.jpg

A. What Debris Pile?

The WTC towers did not collapse. They did not collapse from fire nor did they collapse from "bombs in the buildings" (or conventional controlled demolition). They were turned to dust. They were turned to powder in mid-air. The majority of the building mass did not slam to the ground, as evidenced by the seismic data. Nearly all of each tower was turned to dust in mid-air and either floated to the ground or blew away. The majority of what remained of the towers was paper and dust. A gravity collapse (with or without bombs in the building) cannot turn a building into powder in mid-air.

Figure 158. The amount of steel barely covers the ground.
WTC6 supports the leaning remains of the north wall of WTC1.
http://farm1.static.flickr.com/47/132105299_a0ba16d412_o.jpg

Figure 157 should make it clear to anyone that the depicted conditions are inconsistent with a gravity-driven collapse. The destructive process seen here involves pulverization that is nearly instantaneous, not a collapse of steel upon steel. If the buildings had been blown up with explosives, 500,000 tons of debris would have slammed to the ground for each tower. Bombs do not turn buildings into powder; they only break them into chunks that in turn must slam to the ground. As we will see below, debris that actually slammed to the ground was almost non-existent. The seismic impact it created was far less than it otherwise would have been, as was discussed in Chapter 6.

When the air cleared, little to no significant debris remained, a situation shown in Figure 158. No substantial remains of WTC1 are visible in any of the photos immediately following the destruction of the WTC complex. This is not what a gravity-driven collapse looks like.

The size of the remaining rubble "pile" is superimposed in yellow on the base of WTC1 in Figure 159, showing its extraordinary tininess against the full height of the towers. Even WTC6, an eight-story building, towers over this remaining debris "pile," where the term "pile" itself is misleading.

Figure 159. The height of WTC6 is shown on the left.
(a)http://hereisnewyork.org//jpegs/photos/1593.jpg, (b)http://upload.wikimedia.org/wikipedia/commons/7/75/Wtc_arial_march2001.jpg

The photo in Figure 159 was taken just as the construction of the towers was being completed. In the foreground is new land being claimed by filling in the shoreline of the Hudson River. The World Financial Center (WFC) buildings were later constructed on this land.

Figure 160. (9/11/01) Where did the building go?
http://img.timeinc.net/time/photoessays/shattered/search2.jpg

The arrow in Figure 159, at the base of WTC1, shows the perspective of the photo as it was taken in Figure 160, where the remaining rubble "pile" is outlined in yellow. The lone column on the right of Figure 160 can also be seen in Figure 158.

Figure 160 looks to the north-northeast, where earlier that day a 110-story building stood between the camera and WTC7, the structure seen in the distance. This is how we know that the photograph in Figure 160 is from the day itself—9/11/01—since WTC7 disappeared only later in the afternoon. The shadows on the ground show the time to have been mid-day. The photograph is of an extreme wide-angle and may give the impression of a curved surface and an impression of height in its point of view. However, the ground-level pedestrian bridge can be seen at the left side of the photo just below the vertical center, confirming that the photograph was taken from ground level. So complete was the pulverization of the towers that the debris did not even rise to the height of a single story.

If the photographer of Figure 160 had walked about 100 feet north, then turned to his or her right, he or she would have been looking at the view shown in Figure 161. This photo has a view directly to the east, looking through what used to be the WTC complex. The ambulance in the photo is parked directly in front of what used to be the WTC1 entrance. This apparently undamaged ambulance is standing on street level and yet its roofline is higher than the remains of WTC1. It's true that there are sections of the aluminum cladding, but almost no pieces of steel structure are visible. Essentially the only recognizable objects remaining are the aluminum cladding and the ambulance.

Figure 161. (9/13/01) Looking east, directly in front of where WTC1 stood.
http://www.photolibrary.fema.gov/photodata/original/5316.jpg

In a collapse (for example, one caused by an earthquake), material either pancakes to the ground or shears over, but it is still nearly as recognizable on the ground as it was while standing. With explosives (kinetic energy devices), chunks go flying and remain in the form of chunks until they land. They do not "dissolve" into dust while traveling through the air. In the case of the WTC, almost all chunks that went flying dissolved into dust before they had time to slam to the ground in solid form.

With a pancake "collapse" or with an explosion, as we've said, a building is reduced to chunks that for the most part remain recognizable—items, for example, such as toilet fixtures. In the case of the WTC, however, not a single toilet fixture or even a recognizable portion of one was found anywhere in the buildings' remains. The two towers alone in all likelihood contained something near the number of 3,000 toilet fixtures. Not a single recognizable trace of any of them was found.

The truth is that WTC1 was *gone.* The minimal debris remaining could not conceivably represent the entirety of 110 floors' worth of steel and concrete and building contents. Even so, before I became involved in 9/11 research, of all the people looking at the above photo, none to my knowledge ever asked in any compelling way, "Where did the buildings go?" No one thought twice about calling this "the pile," so strong is what in Chapter 4 we called "Magic Shows," or "The Power of Suggestion."

On the LIDAR image in Figure 162, the colors correspond with the elevations in Table 11 relative to mean sea level.

COLOR	Value (meters)	Elevation Value (feet)
Dark Green	-9.272 to 0	-30.42 to 0
Green	0 to 30	0 to 98.43
Yellow	30 to 100	98.43 to 328.08
Magenta	100 to 150	328.08 to 492.12
Red	150 to 201.19	492.12 to 764.59

Table 11. Color legend for LIDAR image in Figure 162.
http://www.noaanews.noaa.gov/stories/s781.htm

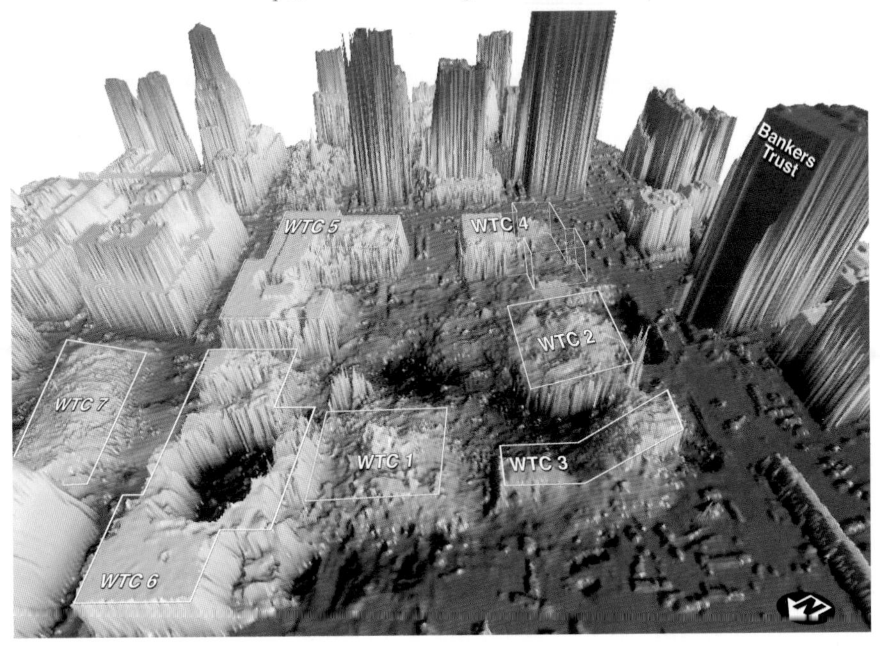

Figure 162. (9/16-23/01) Identification for each building.
http://www.noaanews.noaa.gov/stories/images/wtc2.jpg

The dark-green regions in the LIDAR image above represent regions below sea level. If there had been a breach in the bathtub, these regions would have been flooded.

Each WTC tower was 110 stories tall, but the amount of material that remained after their destruction was tiny even compared to the height of WTC6, which was only an eight-story building. The center of WTC6 was completely gone, down to the ground. The main part of WTC4 was completely gone, down to the ground, only the north wing remaining, as if it had been neatly sliced off. WTC3 was completely gone except for a few stories at the southern end. In actuality, there was no "pile."

> The flag-raising photo was made shortly after 5 p.m. on September 11, 2001. He was standing under a pedestrian walkway across the West Side Highway, which connected the World Trade Center to the World Financial Center *at the northwest corner.* Franklin said the firefighters were about 150 feet away from him and about 20 feet (6 m) off the ground, while the debris was about 90 feet beyond that. [3]

Figure 163. Two photos taken just after 5:00 p.m. on 9/11/01.
(a) Thomas E. Franklin, The Bergen Record. (b) Ricky Flores/The Journal News.
(color adjusted) (a)http://digitaljournalist.org/issue0110/original_images/Franklin_firemen.jpg, (b)http://digitaljournalist.org/issue0110/original_images/flag%20copy.jpg

Figure 164. (9/11/01) This locates the wheatchex.
(a)http://www.magnumphotos.com/CoreXDoc/MAG/Media/TR3/F/P/Y/G/NYC14401.jpg, (b)http://pubs.usgs.gov/of/2001/ofr-01-0429/wtc-r09.091601.usgs-thermal.jpg

View	*Figure*	Page	View	*Figure*	Page	View	*Figure*	Page
1	*158*	172	4	*164a*	176	3	*163a*	176
1	*161*	174	2	*165a*	177	3	*163b*	176
4	*160*	173	2	*166*	177	3	*165b*	177

Legend 2. Legend of Figure views for the map in Figure 165.

Figure 165. The ambulance in Figures 109 and 166 can be seen in the lower-left corner of Figure 165a, indicated by the arrow. Images are contrast enhanced to bring out the corner of WTC1.
(a)http://www.magnumphotos.com/CoreXDoc/MAG/Media/TR3/F/W/L/Y/NYC14148.jpg, (b)http://digitaljournalist.org/issue0110/original_images/flag%20copy.jpg

Scenes of carnage following the fall of the World Trade Center in lower Manhattan, NY
Photo © 2001 James Nachtwey / VII

Figure 166. (9/11/01) The wheatchex correspond with those in Figure 165b.
cropped from: http://digitaljournalist.org/issue0110/images/jn07.jpg

Language influences perception of events, and thus language can be used for perception management. If what remains of the WTC complex is referred to as a "pile," the assumption, conscious or not, becomes that it was indeed a "pile." The description of a rubble *pile* is quite different from a wide-open football *field* or rubble *field*. When Battalion Chief Richard Picciotto, one of the firefighters trapped in stairwell B, emerged from where he had been buried, he described coming out as if onto a football field: "I mean, *I'm looking out on the rubble field. I'm looking out at a big area, almost the size of a football field*, and, hallelujah, we're out of here."[4]

Because language influences perception, it may also lead to biased perceptions of events. For an independent evaluation of all the evidence, assumptions must be minimized to avoid such biases. For this reason, a new vocabulary is helpful. The use of new language, such as "*dustification*" to describe what is depicted in Figure 167, acknowledges that an observed process may not yet be well understood. This alternative terminology enables us to describe an observed process without unduly triggering pre-judgments of the observed events or of their results.

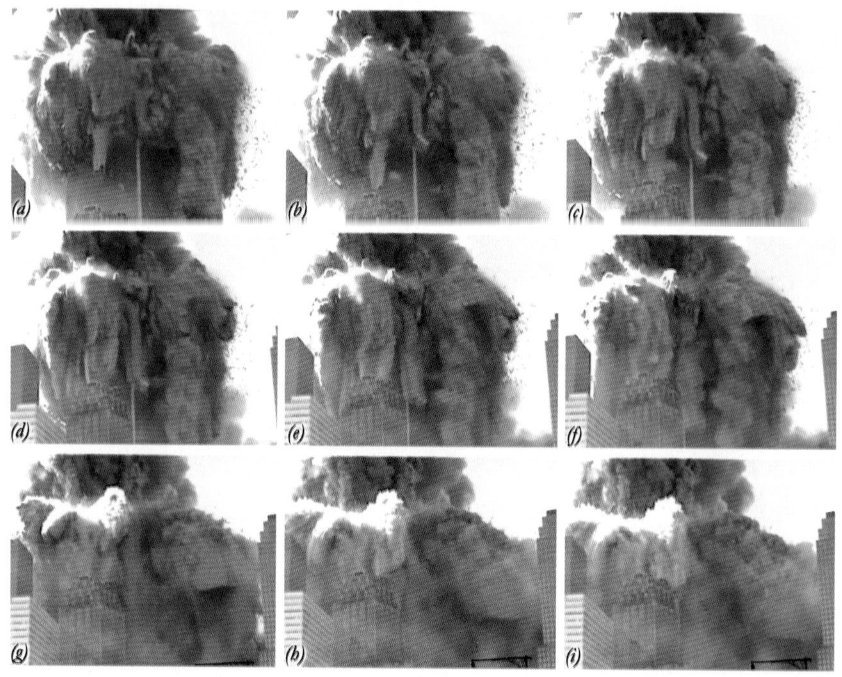

Figure 167. Dustification of WTC1, from the north-northwest.
YouTube video: http://www.youtube.com/watch?v=5q7vz3ZEfBw

In Figure 167, solid pieces of the building can be seen dissolving into dust as they fall

The people in Figure 169 show no apparent fear in going up to the missing building for a closer look. Instead, they are mystified and curious. Their body language suggests that they want to know what happened but not that they fear they are in danger of another building "falling" on them. Coming out of hiding, these people look amazed. From the postures, their jaws must be agape. Perhaps they are wondering if they are asleep and dreaming. Was this magic?

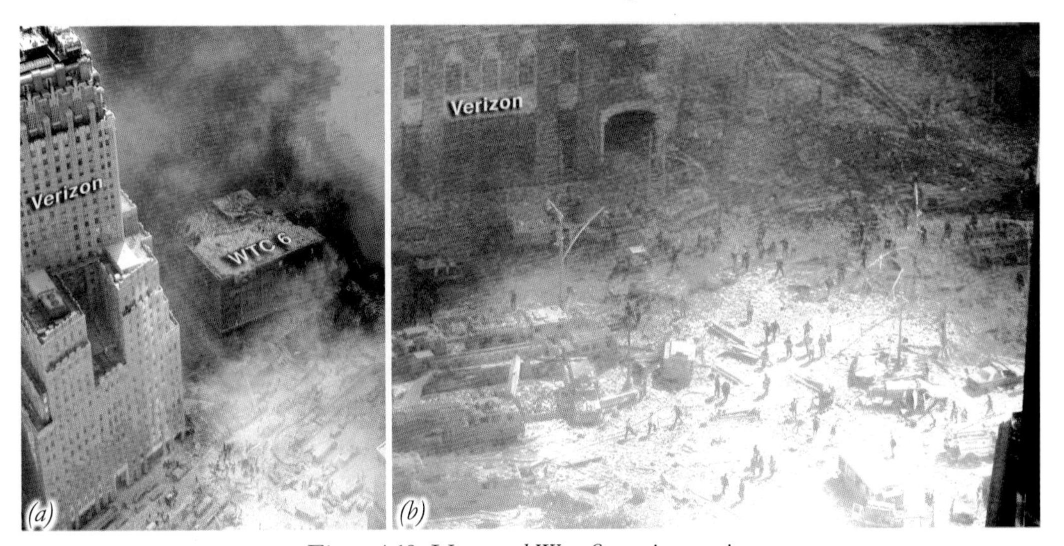

Figure 168. Vesey and West Street intersection.
(a) Looking southeast toward WTC6 at West Street and Vesey Street. (b) The minimal debris left by the destruction of both towers is remarkable. This was no collapse.
(a)http://911wtc.freehostia.com/gallery/originalimages/GJS-WTC131.jpg, (b)http://911wtc.freehostia.com/gallery/originalimages/GJS-WTC111.jpg

Figure 169. Coming out of hiding after the destruction of WTC1.
No significant rubble "pile" landed in this intersection.
http://memory.loc.gov/service/pnp/ppmsca/02100/02121/0084v.jpg

After all, there should be a 110-story building—or at least a "pile" of it—directly in front of them. Where did it go?

Figure 170. (9/11/01) (a) Looking west across West Street along Vesey Street after WTC1 was destroyed. (b) The toasted parking lot on the northwest corner of the Vesey and West Street intersection. *(a)http://memory.loc.gov/service/pnp/ppmsca/02100/02121/0087v.jpg, (b)http://911wtc.freehostia.com/gallery/originalimages/GJS-WTC105.jpg*

Tim McGinn, NYPD, said, "I was standing there for a couple of seconds thinking where the f**k is the tower? I simply couldn't comprehend it."[5] [*edited]

There is also a view west along Vesey Street across the West Street intersection, shown in Figure 170a. In the distance, on the right, is the *toasted parking lot* shown in Figure 170b. Sunlight is visible on WFC3 (upper-left). Paper in the foreground is not on fire. Why are vehicles in the parking lot on fire at some distance away? If a quarter-mile tall building just fell down here, why do we see so much paper and not much else?

View	Figure	Page	View	Figure	Page	View	Figure	Page	View	Figure	Page
A	157	171	**E**	161	174	**G**	163	176	**K**	175	186
B	170b	180	**E**	162	175	**G**	164	176	**L**	179	188
C	167	178	**F**	166	177	**G**	165	177	**M**	172	182
C	168	179	**F**	186	192	**H**	188	194	**N**	173	183
C	169	179	**F**	158	172	**I**	189	194	**O**	176	186
D	170a	180	**G**	160	173	**J**	174	185	**P**	178	187

Legend 3. Legend of Figure views for the map in Figure 171.

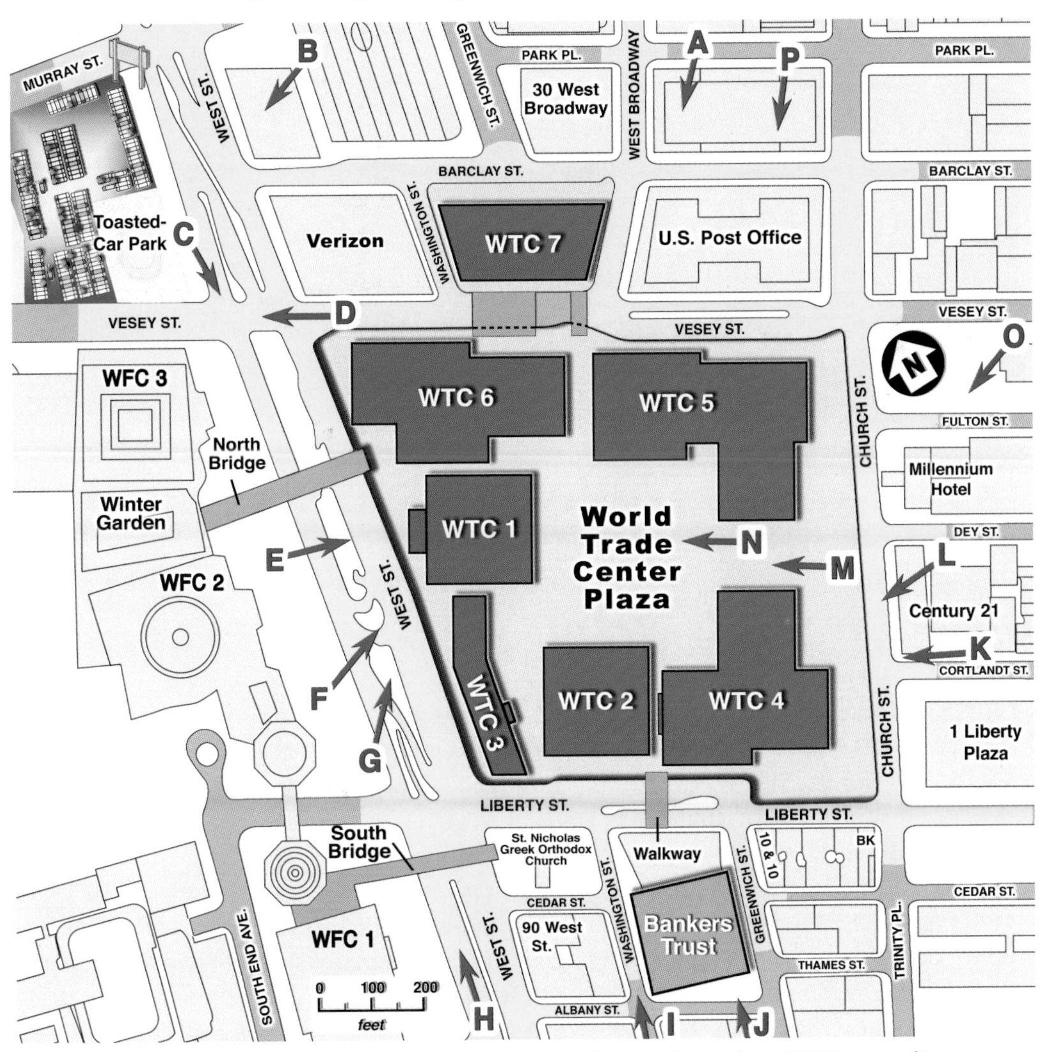

Figure 171. Location of photos in this chapter. (Map redrawn from NIST report[6])

B. Stairwell B

After the destruction of WTC1, all that remained was a small corner of the outer columns and a few stories of Stairwell B. Stairwell B was located in the center of the building. In Figure 172, as noted, Stairwell B can be seen in the remains of WTC1.

Figure 172. (9/13/01) Looking west through the remains at ground level.
http://www.photolibrary.fema.gov/photodata/original/5313.jpg

Another firefighter, Jay Jonas, trapped in stairwell B tells his story:

Firefighter Jay Jonas

And then the collapse stopped.

In a day of first experiences for everybody, well here's another one. I can't believe we just survived that. It was very quick and during the collapse you couldn't help but think that this is it. It's over. This is how it ends. I kept waiting for that big beam to hit or that big piece of concrete to come down and crush us.

It never came.

When it stopped, my first thought was oh, man, I can't believe I just survived that. But then we were in a battle with the dust and the smoke for a while. That big cloud of dust that was surrounding lower Manhattan, I was in the middle of that. I know exactly where Ground Zero was. It was the B stairway of the north tower, that's exactly Ground Zero. *That's the geographic center of that building. And I was in the middle of that.*

[...]

That's what was going on. And right around the three-hour mark, all of a sudden, a beam of sunshine hit the stairway. I looked and said, *"Guys, there used to be 106 floors above us and now I'm seeing sunshine."* They're like, "What?" I said, *"There's nothing above us. That big building doesn't exist."*

[...]

So we left. We left the stairway. Sal left the stairway. He had to kinda go down on the rope a little bit in order to get out because it was elevated and you had to make your way down across some debris. Then Tom Falco leaves the stairway. I wanted to be the last one of our group out of the stairway just because I don't want to have to go back, "Ah, where's Tommy? I gotta go back and see where he is."

Tommy Falco leaves the stairway and then he comes back in. He pokes his head in and he says, "Hey, Cap, wait until you get a load of this."

So I make my way up to the hole. I poke my head out and I couldn't believe what I saw. I couldn't believe it.

The first thing I saw was that corner facade that was still standing. And I was looking at it.

I said, "I can't believe this. This is unbelievable."... .[7]

Figure 173. Remains of Stairwell B
http://1.bp.blogspot.com/_aJeegFsC3nY/ScY52Dy4JCI/AAAAAAAABH4/imyW8Mb5fTo/s1600-h/4060.crop.jpg

Firefighter Jay Jonas continues

These are the biggest office buildings in the world and I didn't see one desk or one chair or one phone, nothing. The only thing you saw was steel, some reinforcing rods and this dust. That's all that was left. *There was nothing that was recognizable, no carpets, nothing like that.*

Out of all the hundreds and hundreds of firemen, police officers and civilians that were in that building when it collapsed, only 14 of us lived.

We just happened to be in the right spot. There was nothing magic about it. There was one pocket, one void and we happened to be in it.[8]

During the *World Trade Center Task Force Interview*, James McGlynn, who was also trapped in Stairwell B, describes how difficult it was to convey to others where they were. The searchers trying to rescue these firefighters could not find WTC1—as though there was nothing left to find.

> Q. What formed those gullies on the side? I mean, obviously there were eight basement levels below that, probably, so this is more or less like a collapse from the building parts—
> A. Right, all the stuff, right.
> Q. —just pushing everything down next to the building.
> A. Right.
> So basically we just stayed there and waited and tried to make contact. Finally we made contact. I think Jay Jonas—I don't know exactly who, but I know he made contact with Chief Visconti. The thing was we were trying to tell them where we were. We just kept telling them we're in the B stairwell.
> I remember everybody had the same exasperation I did. We must have told them a hundred times: "B stairwell in the second floor, third floor, fourth floor of the north tower. I mean, B stairwell, second, third, fourth floor, north tower. B stairwell, north tower." "Where are you?" "North tower, stairway B, second, third floor."
> You could hear they didn't understand where we were. I'm keep going, "My God, why aren't you listening?" Then they said, *"Where's the north tower?"* I was likc, "What do you mean, where's the north tower? The World Trade Center. There's Two World Trade Centers."
> Even though we knew there was a collapse, the idea that both of these buildings were totally down—
> Q. Just fallen.
> A. —and this was all that was left was something that was just—
> Q. You couldn't understand that?
> A. No. It was unimaginable.[9]

Battalion Chief Richard Picciotto was another of the firefighters trapped in stairwell B of WTC1. On the Montell Williams Show, September 17, 2001, he described coming out onto a "football field":

> I'm notifying one person. A lot of people aren't hearing this. So I start climbing up, and when I get to that light, I get out [of] that and then it opens up to a lot. I mean, *I'm looking out on the rubble field. I'm looking out at a big area, almost the size of a football field,* and, hallelujah, we're out of here. I mean, I got hurt people, you know. So I called Jay up, Jay, get up here, you've got to get up here, you know, we're out. We're out.
> We're out, but if you looked at it, it just looks like a pile of rubble. There's piles of rubble all over the place. We're just a pile of rubble, but we're in this pile, you know, there's piles all over the place. From the bottom of the rubble field, we're about, you know, it's a guesstimate, 30, 40 feet high. So now I bring my siren up again, my bullhorn, Mark, I'm out. Listen to this. Listen. So he hears it. I said come get us.
> Fifteen minutes later, no one is there. Over an hour, hour and a half later, we're in contact and, I mean, *I could see an area, like I said, the size of a football field.*[10]

C. Tipping Top of WTC2

At the beginning of the final destruction, the top section of WTC2 tips or twists to the east, but turns to dust before reaching the ground.

Figure 174. I call this image, "the exotic donut" for its appearance.
Figure 6-26 of NCSTAR1-6, page 183 of document, (265 of 470 of file), http://wtc.nist.gov/NISTNCSTAR1-6.pdf

The leaning top of the south tower falsifies the claim by the National Institute of Standards and Technology (NIST) of an "inevitable collapse."

> The physical condition of the tower had deteriorated seriously. The inward bowing of columns on the east wall spread along the east face. The east wall lost its ability to support gravity loads and, consequently, redistributed the loads to the weakened core through the hat truss and to the adjacent north and south walls through the spandrels. But the loads could not be supported by the weakened structure, and the entire section of the building above the impact zone began tilting as a rigid block to the east and south (Figure 3-5). Column failure continued from the east wall around the corners to the north and south faces. The top of the building continued to tilt to the east and south, as, at 9:58:59 A.M., WTC2 began to collapse.[11]

But then where did it go? NIST's explanation is misleading and deceptive. If the leaning top had fallen as a rigid block, it would have landed on WTC4. When we look, it is not there. Not only is it not there, but neither is the main body of WTC4 itself (Figure 175).

The top of WTC2 tipped to the side and then simply "went away." In addition, the main section of WTC 4 went away at the same time. The north wing of WTC4 appears to have been surgically sliced off from the main building while the main building has disappeared. Here we have a paradox having to do with resistance. The top section of the building turned to dust before hitting the ground, which implies it encountered high resistance. However, it turned to dust in mid-air, which is essentially *no* resistance. How can that be?

Figure 175. (a) (9/11/01) The remaining north wing of WTC4 (right). (b) The main body of WTC4 (left) is missing.
(a)http://www.september11news.com/JamesNachtweyTime_ash.jpg, (b)http://www.hybrideb.com/source/eyewitness/nyartlab/DSC08735.jpg

NIST acknowledges the tipping, but if there is *tipping,* the destruction cannot then be *symmetrical.* An asymmetrical loading cannot cause a symmetrical initiation process. The original damage was asymmetrical, the fires were asymmetrical, and the tipping created asymmetrical loading. The point is this: An *asymmetrical loading cannot cause a symmetrical "collapse."*

Figure 176. This series of photographs illustrates the tipping of the top of WTC2.
Cropped from: (a)http://www.studyof911.com/gallery/albums/userpics/10002/1%7E0.jpg, (b)http://www.studyof911.com/gallery/albums/userpics/10002/2%7E0.jpg, (c)http://www.studyof911.com/gallery/albums/userpics/10002/3%7E0.jpg,

In videos showing the destruction of WTC2, the top, as we've seen, begins tipping, then appears to stop tipping. Due to angular momentum, it is not possible for a rigid block to begin to rotate in space and then stop rotating, *unless it is either acted on by an external force or ceases to act as a rigid block.* The *latter* appears to be the primary explanation in this case, as can be seen in Figure 176. In this sequence of three photos, the top section stops rotating and turns to dust *before* the lower portion of the building begins to "collapse". But this presents another problem: *If the upper block turned to near-weightless dust, then there was no longer a force pressing down onto the lower portion of the building that could crush it.* And so we must ask this question: What caused *the lower portion* to turn to dust? The same argument exists for the tipping top, as illustrated in Figure

177. Once the tipping begins, the stress becomes less on the unloaded (western) side. So if it had not failed with the load that was on it, it certainly would not fail with that load removed. For that matter, once the tipping top has turned to powder, there is no longer any reason for the *remaining* structure to fail.

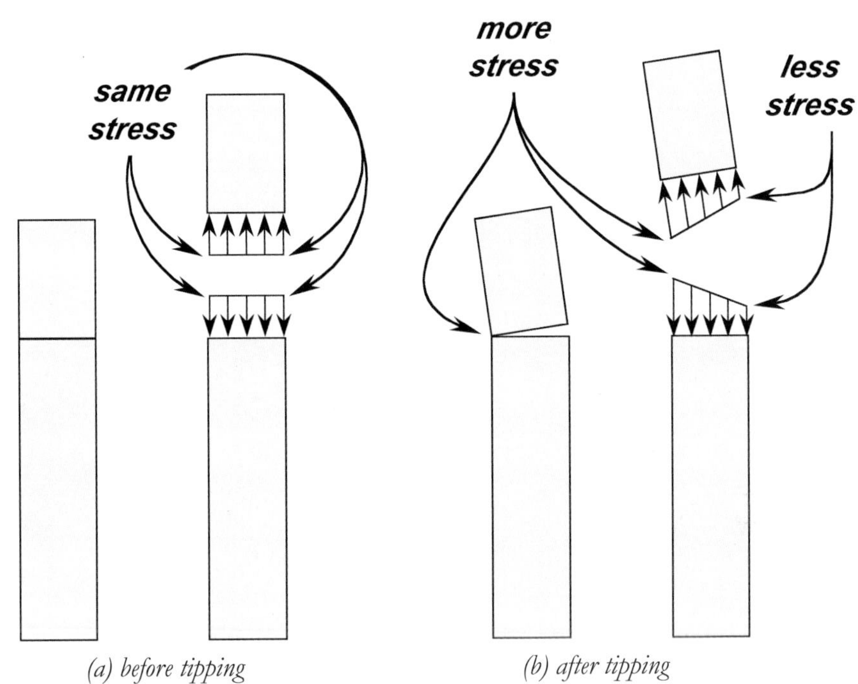

(a) before tipping *(b) after tipping*

Figure 177. A tipping top cannot collapse the lower building's back side.

Figure 178. WTC2 tipping eastward, as viewed from the north. The tipping top is disappearing. The arrow notes the mechanical floors.

http://sf.indymedia.org/uploads/const_in_foreground.jpg

Figure 179. The main part of WTC4 is gone. Below ground is still intact. (See Figure 181.)[12]
(WTC4 edge highlighted for clarity.) *http://bocadigital.smugmug.com/photos/10698660-D.jpg*

Figure 180. (9/19/01) Innovations Luggage, under WTC4 as viewed from Location C in Figure 183.[13]
(Also see Figure 181.) http://www.photolibrary.fema.gov/photodata/original/5345.jpg

Figure 181. (post 9/11/01) Under WTC4 (green) and WTC5 (purple) from Location D, Figure 183.[14] *The view is southward, under the WTC mall, where deliveries were made. Figures 54 and 180 are from directly above this green zone near the far end.*

Figure 182. (post 9/11/01) What remains of the Strawberry store, viewed from Warner Brothers.[15] *(Location A in Figure 183)*

Figure 178 shows the remaining portion of the 35-floor block, tipping to the east, while it disintegrates. The floors below this level appear to be intact. Using WTC1 as a guide for height, it appears that the disintegration of WTC2 involves the floors at and above the upper mechanical floor. (For each tower, as mentioned, floors 41-42 and 75-76 were mechanical floors.) The top of WT2 is visible, yet the top of WTC1 is above the top edge of the photo. From this comparison it appears that the top block of WTC2 is already less than half of its original height. Where did the material go?

View	*Figure*	Page	View	*Figure*	Page	View	*Figure*	Page
A	*182*	189	**C**	*54*	58	**air**	*179*	188
B	*51*	56	**C**	*180*	188	**D***	*181*	189

Legend 4. Legend of Figure views for the map in Figure 183.

* *Location D is below the mall, while the others are from the mall level in the first sub-basement level.*

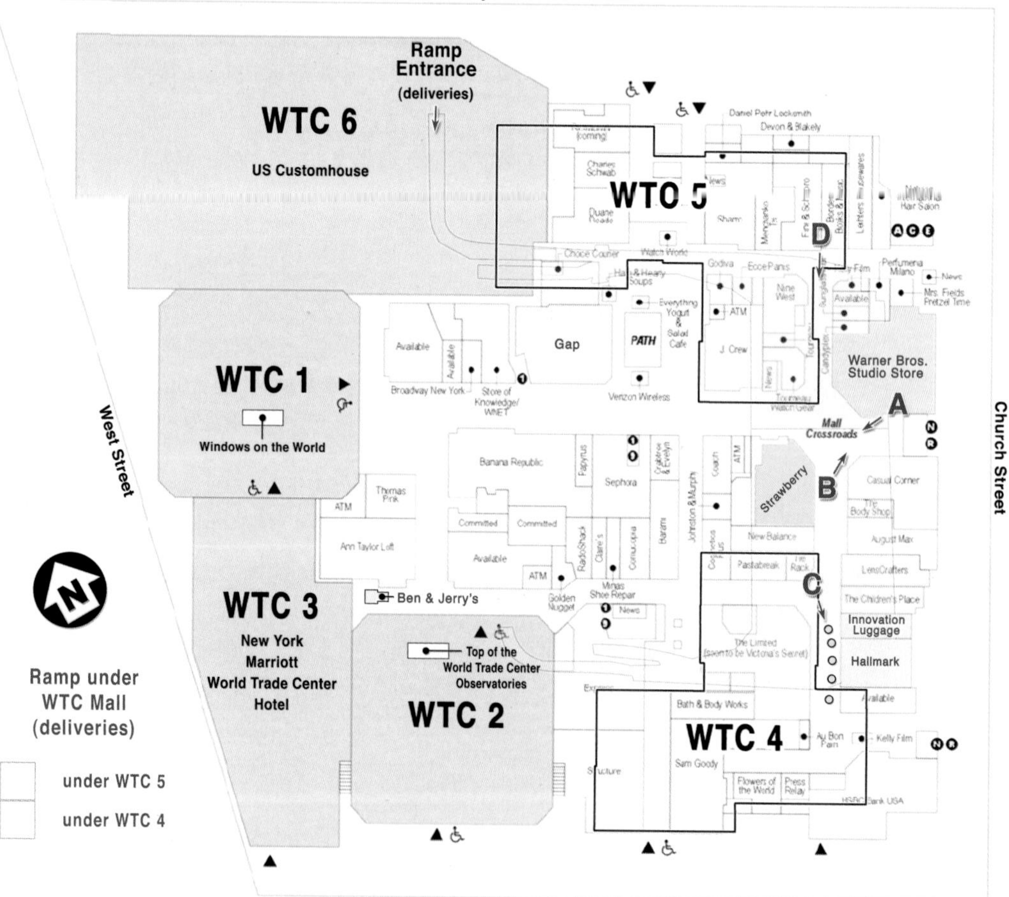

Figure 183. Map of underground mall.[16]

Images from positions A, B, C, and D, correspond with the figure numbers given in the Table above. The Ben & Jerry's ice cream store, shown in blue, between WTC1, WTC2, and WTC 3, is the location where the crumpled-up file cabinet was found, and thought to be the only file cabinet found in the remains.

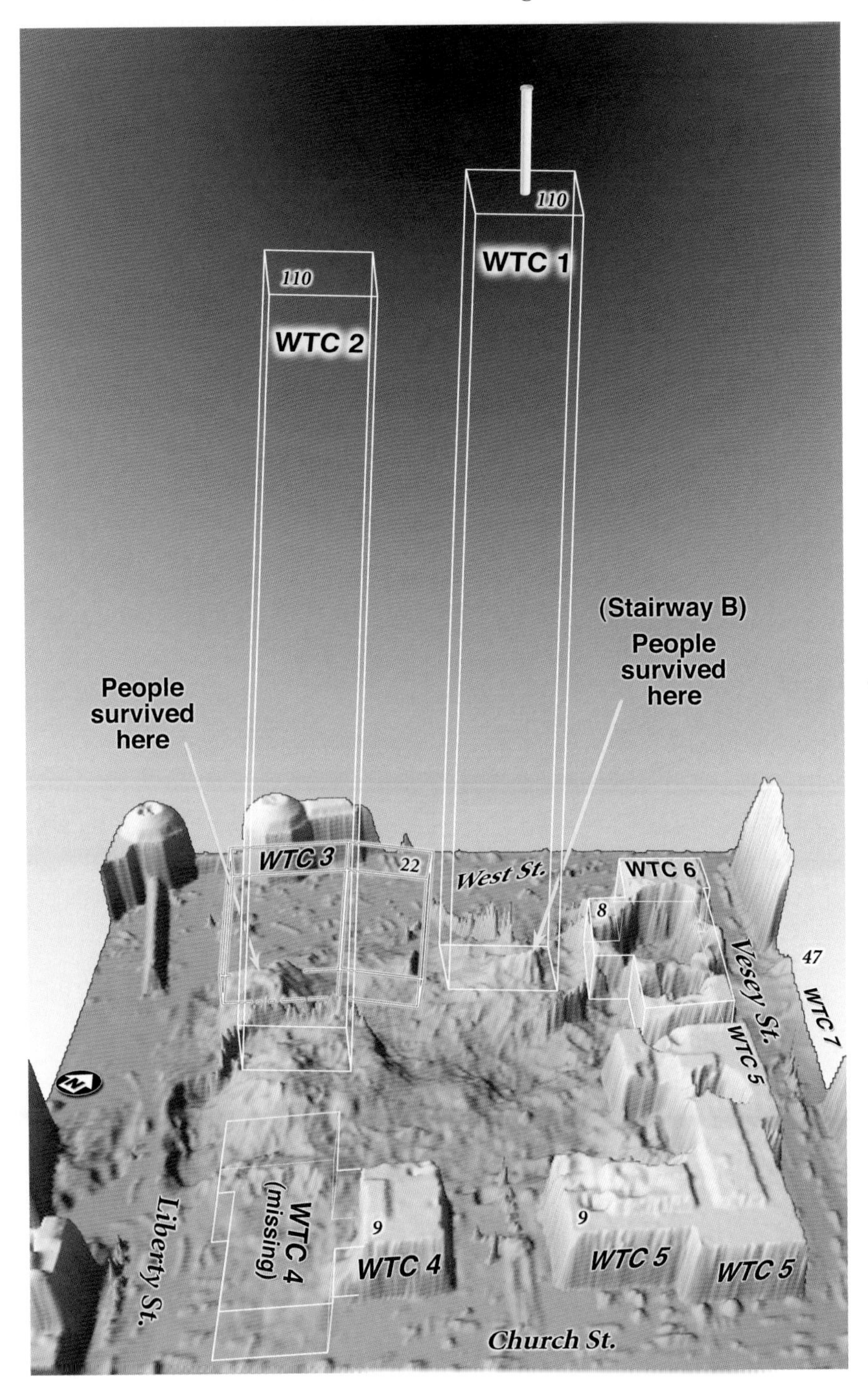

Figure 184. (9/23-27/01) Light Detection and Ranging (LIDAR[17]) image showing the topography of what remained with outlines indicating what was there.[18] The number of floors is shown for each building.

The main body of WTC4 disappeared during the destruction of WTC2. It looked as if it had been sliced off, where a viewing from south of where the main body stood, desktops in the north wing were visible, as shown in Figure 185b. The arrows in Figures 185a and 185b show the same viewing direction.

Figure 185. (9/01) WTC4 is sliced away.
The arrow is in the same approximate location for both (a) and (b).
A view out from the Century21 building (top of Figure 185a) is shown in Figure 405 on page 383.
(a)http://www.sharpprintinginc.com/911/images/photoalbum/12/Air_WTC0973.jpg, (b)http://hereisnewyork.org//jpegs/photos/3429.jpg

Similarly, the middle portion of WTC6 disappeared during the destruction of WTC1, as shown in Figure 186.

Figure 186. Remains of WTC6 tower over the footprint of WTC1.
http://www.studyof911.com/gallery/albums/userpics/10002/132105581_a75a50d39a_o%7E0.jpg

A view over the dome of WFC2 shows, in the center of the photo, the damage to WTC6. To the left are the remains of WTC7. Its debris stack is at least five stories high. To the right of WTC6 is the remaining north wall of WTC1, which leans toward WTC6. Where did the wall go? Where did the top 100 floors of the north wall go? They did not fall on WTC6 or WTC7 because there are no steel wheatchex there. A small portion of the core of WTC1 remains, but where is the rest of it?

D. Conclusions

The photo in Figure 187 was taken from about the 50th floor of WTC1 before 9/11/01 and shows the street below. The roof of WFC1, a 40-story building, can be seen in the lower right area of the photo and appears about ten floors below the photographer.

Figure 187. A look south along West Street from WTC1 before 9/11/01.
http://www.911digitalarchive.org/REPOSITORY/IMAGES/PHOTOS/2214.pjpeg

The photograph in Figure 188a was taken from the Albany Street and West Street intersection before noon on 9/11, as indicated by the sunlight on the face of WFC1 and the direction of the shadows.

WFC1 is on the left in Figure 188a. The pedestrian walkway crosses West Street just south of Liberty Street. Figure 188b was taken north of Liberty Street, directly in front of WTC3, at approximately position **H** on the map in Figure 171. Here, the remains of the building appear to be no more than dust and paper. Where did the buildings go? The answer now should be quite clear: The majority of the building material and its contents dissolved into dust before hitting the ground.

Ground Zero Photo © 2001 Allan Tannenbaum / Gamma James Nachtwey / Time

Figure 188. (9/11/01) A look north on West Street.
(a)http://www.digitaljournalist.org/issue0110/images/Tann3.jpg[19], *(b)http://img.timeinc.net/time/photoessays/shattered/search2.jpg*

Figure 189. WTC2 tipping eastward, as viewed from the north.
YouTube video: http://www.youtube.com/watch?v=kMr37SL6l4

An Interesting NY Times Article

There is an interesting article in *The New York Times of October 9, 2001, written by James Glanz* and entitled "Torn Steel, Cold Data of Salvage". Here are a few intriguing passages:

> So computer programs were designed, old blueprints were scoured and individual memories were ransacked to rebuild it on paper, to determine how much steel and concrete and gypsum and glass were there. What resulted was the most detailed accounting of just what the World Trade Center had been made of, down to the terrazzo flooring…
>
> For a more accurate assessment, the city and the Port Authority turned to *Leslie E. Robertson Associates*, one of the structural engineering firms involved in building the trade center. The firm took out the original plans and quickly started adding up the ingredients of the World Trade Center floor by floor, said William Faschan, a partner at the firm in Manhattan.
>
> The assessment came with a striking level of detail. It estimated that each of the twin towers contained 3,881 tons of steel reinforcing in the concrete floor slabs; 47,453 tons of vertical steel columns; 8,462 tons of aluminum and glass on the exterior walls; 2,531 tons of various ceiling materials; 4,218 tons of flooring; and 31,350 tons of partitions or walls.
>
> Added up, Mr. Lombardi of the Port Authority said, the total came to about 1,176,000 tons of debris, including about 285,000 tons of steel. *After some rounding, the very rough figure of 1.2 million tons of debris was born.*[20]

The point of all this, once again, is that there should be something close to 1.2 million tons of debris on the ground, and yet, the photographic evidence presented thus far is clearly not commensurate with that amount of debris. This statement, *"After some rounding, the very rough figure of 1.2 million tons of debris was born,"*[21] indicates that this number often quoted as the amount of debris removed was not measured, but *calculated*, as to what *should* be there.

Additionally, we have seen in this chapter that the leaning top of WTC2 gives lie to the collapsing "pancake" theory, whether pancaking caused by controlled demolition or, as in the official story, by structural weakening from fires. The physics is simply impossible either way :

1) If a top begins to topple, why is the subsequent "collapse" symmetrical?
2) If a top begins to topple, why is there no evidence of its having landed on top of neighboring buildings?

It must be that the top was dustified after it began toppling, a fact that implies something very significant, as we will see as we examine other types of evidence.

---

[1] The Entombed Man's Tale, transcript of Jay Jonas, http://archive.recordonline.com/adayinseptember/jonas.htm

[2] File No. 9110447, *World Trade Center Task Force Interview*, "Firefighter James McGlynn," January 2, 2002, pp. 22-24. *http://graphics8.nytimes.com/packages/pdf/nyregion/20050812_WTC_GRAPHIC/9110447.PDF*

[3] *http://en.wikipedia.org/wiki/Thomas_E._Franklin*

[4] Montel Williams show interview of Battalion Chief Richard Picciotto, September 17, 2001.

[5] Tim McGinn, NYPD, p. 76, Mitchell Fink and Lois Mathias, Never Forget: An Oral History of September 11, 2001, NYC: HarperCollins, 2002.

[6] Map redrawn from p. 3 (pdf p. 53 of 298), *http://wtc.nist.gov/NISTNCSTAR1CollapseofTowers.pdf*

[7] File No. 9110447, *World Trade Center Task Force Interview*, "Firefighter James McGlynn," January 2, 2002, pp. 22-24. *http://graphics8.nytimes.com/packages/pdf/nyregion/20050812_WTC_GRAPHIC/9110447.PDF*

[8] The Entombed Man's Tale, transcript of Jay Jonas, *http://archive.Recordonline.com/adayinseptember/jonas.htm*

[9] File No. 9110447, *World Trade Center Task Force Interview*, "Firefighter James McGlynn," January 2, 2002, pp. 22-24. *http://graphics8.nytimes.com/packages/pdf/nyregion/20050812_WTC_GRAPHIC/9110447.PDF*

[10] Montel Williams show interview of Battalion Chief Richard Picciotto on September 17, 2001

[11] NCSTAR1, Section 3.6, their page 44, file page 94.

[12] Brightness adjusted, building edge highlighted.

[13] This image as been lightened.

[14] *http://www.hudsoncity.net/tubes/tubesslide.4.jpg, http://www.nytimes.com/2005/07/25/nyregion/25tubes.html?8hpib*

[15] *http://upload.wikimedia.org/wikipedia/en/d/db/The_Mall_at_the_World_Trade_Center_after_collapse.jpg*

[16] Altered from: "This [map] is in the *public domain* in the United States because it is a work of the United States Federal Government under the terms of Title 17, Chapter 1, Section 105 of the US Code." *http://upload.wikimedia.org/wikipedia/en/3/35/WTCmall.png*, Details have been added from information in the NYTimes graphic, *http://graphics8.nytimes.com/images/2005/07/25/nyregion/20050725.tubes.graphic.gif*

[17] *http://www.ngs.noaa.gov/RESEARCH/RSD/main/lidar/lidar.shtml*

[18] NOAA CONDUCTS MORE FLIGHTS OVER WORLD TRADE CENTER SITE, LIDAR images of ground zero rendered Sept. 27, 2001 by the U.S. Army Joint Precision Strike Demonstration from data collected by NOAA flights. Credit "NOAA/U.S. Army JPSD." *http://www.noaanews.noaa.gov/stories/s798b.htm, http://www.noaanews.noaa.gov/stories/images/wtc-lidar092701-site.jpg*

[19] *http://www.digitaljournalist.org/issue0110/tannenbaum.htm*

[20] Presented as archived in *Torn Steel, Cold Data of Salvage*, New York Times, by James Glanz From

October 9, 2001, *http://911research.wtc7.net/cache/wtc/groundzero/NYT_steel_salvage.htm*

[21] Presented as archived in *Torn Steel, Cold Data of Salvage*, New York Times, by James Glanz From October 9, 2001, *http://911research.wtc7.net/cache/wtc/groundzero/NYT_steel_salvage.htm*

10.
Holes

We cannot live in a world that is interpreted for us by others. An interpreted world is not a home. Part of the terror is to take back our own listening, to use our own voice, to see our own light.
—Hildegard von Bingen

A. Introduction

Where did the buildings go? Why are there holes?

Figure 190. (9/23/01) NOAA[1] satellite image of the remains.
http://www.noaanews.noaa.gov/wtc/images/wtc-photo.jpg

Every destroyed building on 9/11 had a prefix of WTC. Surprisingly little collateral damage was suffered by the very nearby buildings that were not part of the WTC Complex. As for the WTC buildings *not* totally destroyed, multiple circular holes were visible in them—especially in buildings WTC5 and WTC6.

Buildings 5 and 6 had holes in them that were quite mysterious. Because of the *verticality* of these holes, they could not have been caused by conventional explosives. WTC6, an eight-story building, lost about half of its volume and yet there was remarkably little debris left at the bottom of the building. No one has attempted to explain these mysterious holes.

Where did Building 3 and Building 4 go? Half of WTC3 was missing after the destruction of WTC2, and the rest of it was gone after the destruction of WTC1 except for a small stub on the south end. Building 4 essentially disappeared except for its north wing. The missing material of WTC 3 and 4 and Bankers Trust shows patterns similar to the material that is missing in WTC 4 and 5.

Some have speculated that pieces of the Towers fell on these nearby buildings.

B. Locations

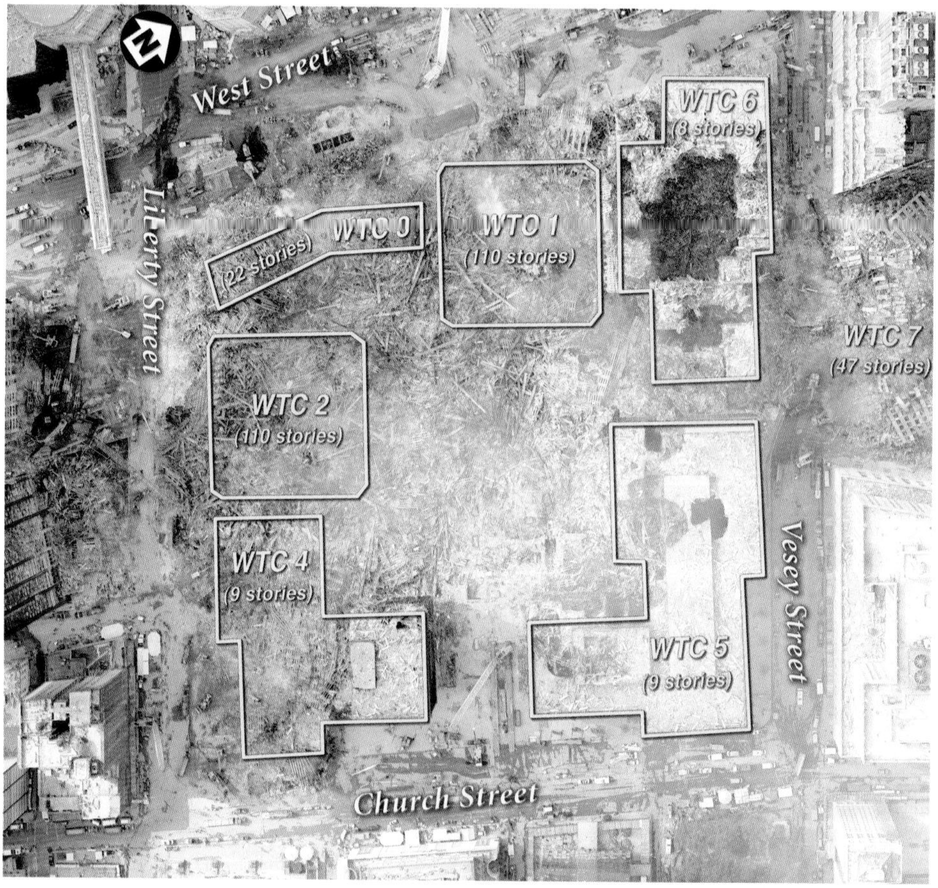

Figure 191. (9/23/01) A closer look. (NOAA)[1]
http://www.noaanews.noaa.gov/wtc/images/wtc-photo.jpg

In Figure 192, notice how straight the vertical holes are that cut down through WTC6. While there is abundance of aluminum cladding on the roofs of buildings 5 and 6, there is little or none in the holes.

Figure 192 looks west/northwest. Notice the substantial steel debris from the lower floors of WTC1 lying on West Street. This debris fell short of damaging the WFC2 façade, which suffered only broken windows.

Figure 192. (9/15?/01) Cylindrical holes were seen in WTC5 and WTC6. Notice how straight the vertical holes were that cut down through WTC6. http://www.hybrideb.com/source/official/hq/536207.351374.jpg

Figure 193. (9/15?/01) Cylindrical holes were seen in WTC5 and WTC6. This photo highlights the depth of the hole in WTC6. While there is abundance of aluminum cladding on the roofs of buildings 5 and 6, there is little or none in the holes. http://www.hybrideb.com/source/official/hq/536208.351380.jpg

The diagonal path of the steel "wheatchex" in the street look as if they had leaned back, jumped out of the bathtub, tried to do a pirouette, and plopped down in the middle of the West Side Highway, almost ready for loading onto trucks. The coherence of this wall from WTC1 laid out in West Street in single thickness strongly suggests that thermite was *not* used on it. Why? Because thermite would have cut steel and sent it tumbling down, creating a scattered trash pile and likely sending steel beams into adjacent buildings.

The north side of the outer wall of WTC1 is standing unsupported, leaning toward WTC6. The east side of the remaining WTC1 wall appears to be the highest standing section of unsupported wall that remains. A "pancake collapse" would have crushed (destroyed, buckled) these outer walls, yet they remain standing at attention like soldiers. The floors connected to these outer walls seem to have slipped away as if they had never existed. How could the floors be torn away from the walls while the walls stay erect, unbuckled, and unsupported? Once again, it appears that the floors were pulverized and simply disappeared. There is debris at the lobby level near the standing walls but *there are no floor "pancakes" stacked up.*

Figure 193 shows the remains of WTC6 just north of where WTC1 was standing. The vertical holes in WTC6 (U.S. Customs Building) have the shape of cylindrical core samples in soil. What could have done this? Explosives? Thermite? Mini-nukes? Beam weapon? Whatever it was, *it produced vertically straight holes while doing little apparent horizontal damage to the balance of the interior of WTC6.* In addition, the parking garage below WTC6 remained essentially undamaged, as Figure 194 shows.

Pulverized dust and some WTC aluminum cladding lay on the roof of the 40-story Bankers Trust building, but other damage was confined mostly to lower floors.

Figure 194. The western bathtub wall above continued to hold back the Hudson River after 9/11 with no significant damage. This is a view from within the WTC1 footprint.
http://911research.wtc7.net/wtc/arch/docs/bathtub_wall1.jpg

This same pattern prevails in other damaged buildings adjacent to the WTC complex. There is surprisingly little damage following destruction of two 110-story buildings directly across the street. Especially notable, why is there no serious damage to these adjacent buildings above their 20th floors? Twentieth-floor-level is at a height less than 20% of the height of a Twin Tower. What would explain this lack of damage at higher levels? Disintegration and pulverization into talcum-powder-sized dust *above* the 20th floors would explain it.

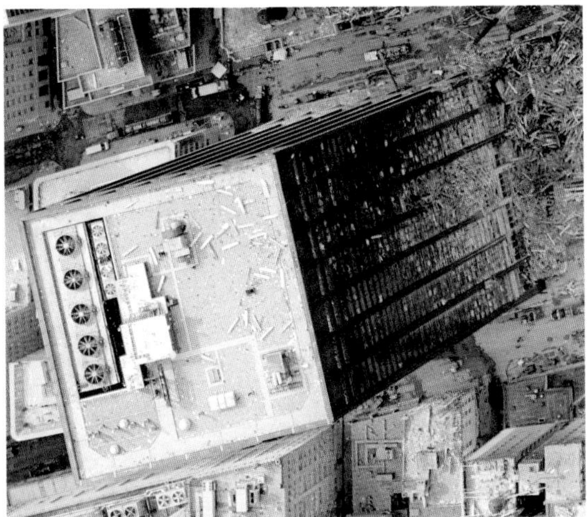

Figure 195. (9/23/01) Bankers Trust (Deutsche Bank) viewed from above. (NOAA)[1]
http://www.noaanews.noaa.gov/wtc/images/wtc-photo.jpg

C. Overhead

Figure 196. (9/23/01) A view of the WTC from above. (North is to the right.) (NOAA)[1]
http://www.noaanews.noaa.gov/wtc/images/wtc-photo.jpg

In Figure 196, some debris has been cleared, but the pulverized dust is still emerging. If most of the steel from the upper floors of WTC1 and WTC2 was pulverized, then how much steel was really shipped as scrap to China? Does anyone have these figures—or the receipts?

First responder testimony also corroborates the photographic evidence. For example, firefighter Tiernach Cassidy, in his testimony, repeatedly mentions finding a "hole" in the building, which he also described as a "gash" and, at one point, as a "hole sixty feet deep."[2]

D. Missing Wall

Figure 197. (9/12/01) Remains of WTC6 tower over the footprint of WTC1.
http://studyof911.com/gallery/albums/userpics/10002/132105581_a75a50d39a_o.jpg

In Figure 197, the view over the dome of WFC2 shows the damage to WTC6, in the center of the photo. To the left are the remains of WTC7. Its debris stack is several stories high. To the right is the remaining north wall of WTC1, which leans toward WTC6. Where did the wall go? Where did the top 100 floors of the north wall go? They *couldn't* have fallen on WTC6 or WTC7 because there are no metal wheatchex there. Some of the core of WTC1 remains, but where is the rest of the core? The amount of fallen debris barely covers the ground. These photos, as well, highlight the depth of the hole in WTC6.

(a) (9/13/01) *(b) (9/15/01)*

Figure 198. (a) The remarkable "lay down" of steel wheatchex from the lower stories of WTC1 on West Street (West Side Highway). The WTC3 debris pile is in the background, next to the unsupported WTC2 wall. (b) The three steel wheatchex stabbed into West Street in the foreground and the remains of WTC3 in the background, in front of the west wall of WTC2.

(a)http://www.photolibrary.fema.gov/photodata/original/3887.jpg, (b)http://www.photolibrary.fema.gov/photodata/original/3930.jpg

E. WTC6

Figure 199. (9/23/01) This shows the vertical cutouts in the center of WTC6. To the left of WTC6 are the remains of WTC1. Note the fairly consistent diameter of the holes. The holes are essentially empty: little debris visible inside the holes. (NOAA)[1]

http://www.noaanews.noaa.gov/wtc/images/wtc-photo.jpg

You can locate the "wheatchex" object above in the large photo (top center, near cranes) in Figure 199 and use it as a scale (Figure 200).

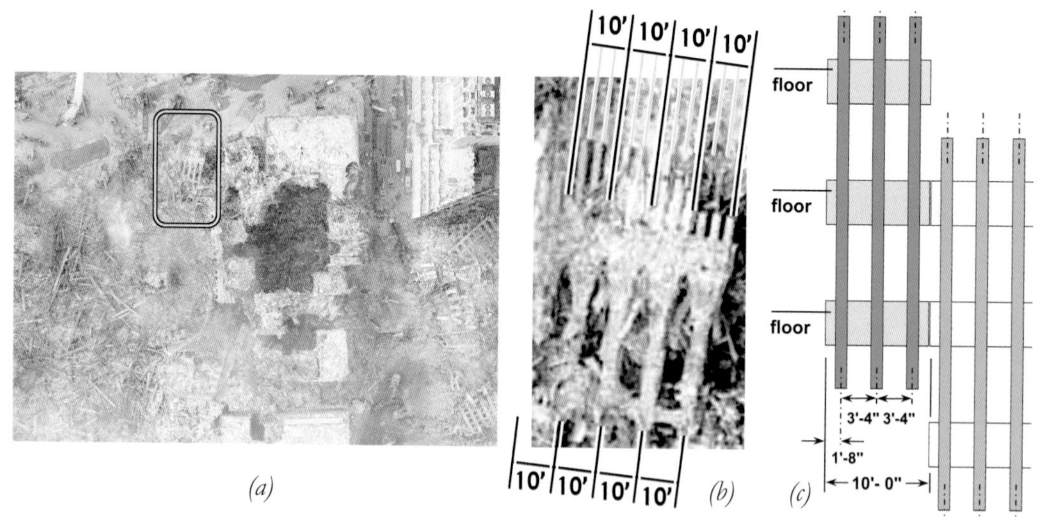

Figure 200. (a) Wheatchex as a scale. (b) "Pitchfork" dimensions with "wheatchex" as prongs. (NOAA)[1]
(c) Wheatchex dimensioned in assembled groups.
(a)(b)http://www.noaanews.noaa.gov/wtc/images/wtc-photo.jpg, (c)JW

Figures 199 and 200 both contain the same overhead view of WTC6. This photo shows not a single cutout but a cluster of vertical cutouts that coalesce together and form a scalloped border. *No collapsed floors are visible at the bottom of the hole, and the heart of the building is gone.* A bomb cannot do this. The debris inside the building is minimal and all at ground level, no deeper. The base of the north wall of WTC1 is to the left of WTC6, and you can see the remains of a cluster of core columns in the center of the WTC1 footprint. The debris from WTC1 also is almost non-existent, dwarfed by the remains of the 8-story WTC6.

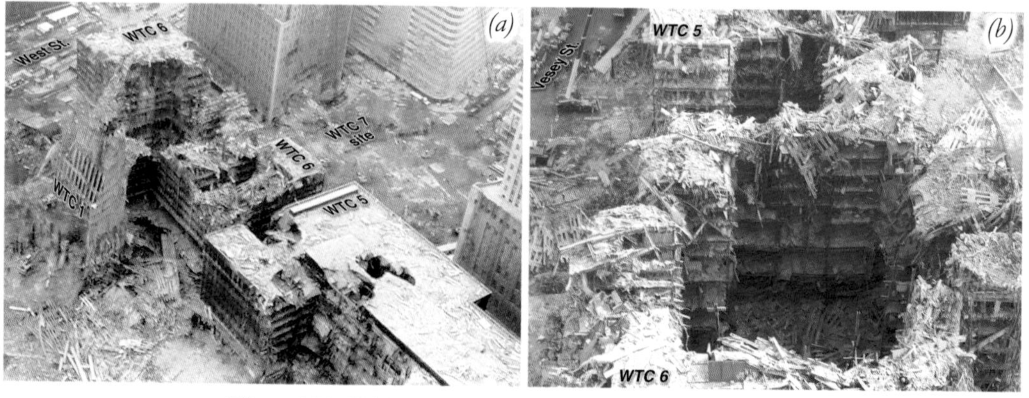

Figure 201. WTC6 (a) from the east and (b) from the west.
(a)http://thewebfairy.com/killtown/images/wtc6/wtc6-side-aerial-hr.jpg, (b)http://i24.photobucket.com/albums/c49/IgnoranceIsntbliss/911/buildingsthatshouldhavefallen/10220017.jpg

The vertical cutouts in WTC6 approximate the shape of circles (viewed from above), each with a diameter of approximately 24 feet, as measured and calculated from the known dimensions of the metal "wheatchex." Where did the *core* material of the building go? Was it disintegrated, pulverized into a fine dust? If so, how?

Figure 202. (9/27/01) (a) Two people view the inside of WTC6. (b) Inside the 8-story WTC6 hole.
(a)http://www.studyof911.com/gallery/albums/userpics/10002/wtc_5f21.jpg, (b)http://www.photolibrary.fema.gov/photodata/original/5395.jpg

Here is the first-responder testimony of Mark DeMarco, 49 Emergency Services Unit, NYPD:

> The people who had been with us by the lobby door, Sergeant Mike Curtin, D'Allara, and the firemen, were in the *whole* part of the building that had caved in. *It was like a massive crater.* I felt that they were more than likely gone. We just happened to be in a small pocket of the lobby, and we survived. *Three-fourths of the lobby was devastated.* When I went back the next day, I brought my supervisor back to show him exactly where I was when it happened, and where I thought our missing people were. And when the dust had settled and the sun was coming in, I just couldn't believe how much of building 6 was gone. At the time I felt there was no hope for them. It was just going to be a recovery.[3]

Once again, notably *absent* from DeMarco's description is anything remotely describing pancaked floors. Instead, he emphasizes "how much of building 6 was gone" and that it was like the whole building simply was not there: "It was," he states, "like a massive crater."

The photo in Figure 202 was taken from inside WTC6. The vertical cut-outs seen in Figure 199 do indeed appear to go completely to the ground floor, with relatively little debris remaining. The number of floors within the building can be counted and eight floors are visible (counting floors on the far side, from top down, appears easiest). Again, the evidence suggests that all 8 floors somehow were pulverized or "disappeared" Why? And how?

Figure 203. (10/20/01) The middle of WTC6.
http://www.photolibrary.fema.gov/photodata/original/5494.jpg

F. WTC3

Figure 204. One of Bill Biggart's last photos, showing a section missing from WTC3.
http://digitaljournalist.org/issue0111/images/Biggart1836.jpg

G. WTC5

Figure 205. Holes in WTC 5, after the destruction of WTC1, WTC2, and WTC7.
(a)(b)http://www.noaanews.noaa.gov/wtc/images/wtc-photo.jpg, (c)http://i24.photobucket.com/albums/c49/IgnoranceIsntbliss/911/buildingsthatshouldhavefallen/10220018.jpg

Figure 206. No holes in WTC6 or WTC5, yet. (a) From over the Hudson River. (b) A closer look.
http://911wtc.freehostia.com/gallery/originalimages/GJS-WTC21.jpg

H. Liberty Street Holes

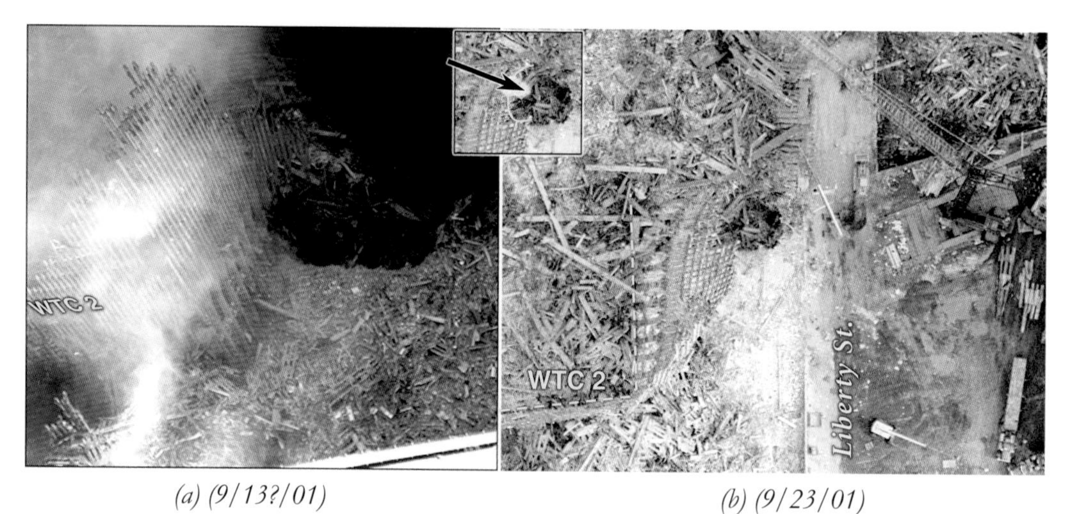

(a) (9/13?/01) (b) (9/23/01)

Figure 207. The cylindrical hole in Liberty Street.
(a)http://www.studyof911.com/gallery/albums/userpics/10002/gzd_019.jpg, (b)http://www.noaanews.noaa.gov/wtc/images/wtc-photo.jpg

Figure 208. (9/22/01) Holes, missing buildings, and the bathtub wall.
http://bocadigital.smugmug.com/photos/10698619-D.jpg, http://bocadigital.smugmug.com/

1. Liberty Street Hole #1

Figure 209a shows a hole outside the wall of WTC2 closest to Liberty Street and exhibits a number of anomalous characteristics. One is the missing lower portion of that beam on the right end. *The three outer columns in the center of the picture have a strange, flanged appearance as if they had unfolded, and they look cooked.* It looks as if a steel wheatchex dove into the hole. The fact that you can only see the tip of the wheatchex suggests how deep down the hole extends.

Figure 209. Liberty Street holes. (a) Hole #1 (9/21/01) (b) Hole #2 (9/18/01)
(a) This hole is adjacent to WTC2 (see Figure 208) and is through the sidewalk and pavement. This hole contains more debris than the hole discussed in Figure 211, but is still quite minimal. It looks as if the debris fell in the hole. (Note the scale, shown in Figure 210.) (b) Ground Zero workers near a stepladder in hole 2 in Liberty Street, shown in Figure 208. The remaining wall of WTC2 is in the background.
(a)http://www.photolibrary.fema.gov/photodata/original/4015.jpg, (b)http://www.photolibrary.fema.gov/photodata/original/3952.jpg

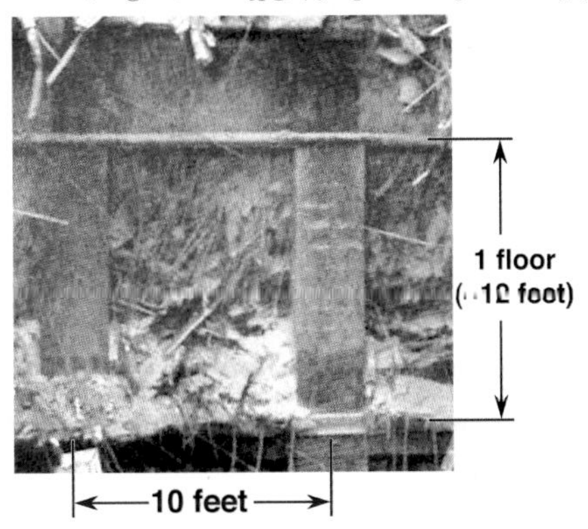

Figure 210. Spacing of columns underground and approximate floor height.
http://www.photolibrary.fema.gov/photodata/original/4015.jpg

The WTC2 columns are "pitch fork handles" at the lobby level, spaced on 10-foot centers. A six-foot person could lie down between the columns with his feet against one column and reach out and almost touch the adjacent column. Large cars and trucks can fit between these columns. Therefore the column's characteristic of "unwrapping" suggests the effect of extraordinary force.

2. Liberty Street Hole #2

As in Figure 209b, again there are "serpent-like" steel beam remnants hanging over the hole. They look as though they stopped short of complete disintegration. The metal in the lower right corner of the picture, with a camouflage appearance, looks deformed and dissolved as if attacked by acid. There is a large amount of material distributed throughout that looks as if it had been run through a paper shredder. In the lower levels, concrete rebar is exposed, apparently because the concrete attached to it was pulverized.

In Figure 209b there is some strange "toasted" steel. The worker by the ladder stares down at several pieces of discolored steel beams as if he may be saying, "What the heck is this?" There is a long red beam that rises from the lower-right corner that looks wavy like a serpent. Before the steel disintegrates, does it crinkle, become wavy, or shrivel? Near the tip of the "serpent," there is a vertical piece of material that has a gold-copper appearance. Below the ladder, there are two similar gold pieces that are highly reflective. The WTC2 shows a strange pattern where the lower portion of a steel column is missing, but not its upper portion. Was gravity playing tricks at that point? In the background, behind the ladder, you can see a section of intact sidewalk with almost no debris on it. This is adjacent to the base of the 110-story South Tower, yet the sidewalk was not crushed by falling debris.

Rising dust and vapor from zones in the rubble continued for weeks after 9/11. This phenomenon needs an explanation. Pools of molten metal appear to provide a possible explanation, but it is not a correct one.

Figure 211 looks down into the basement of WTC2. This is adjacent to the southeast corner of the bathtub wall, which was damaged at the top, and there seems to be a wet floor or puddle of water. But there is no steam rising, and there *clearly* is no molten metal visible in this section of the basement.

Figure 211. (9/18/01) Rescue workers descend into the subbasements below WTC2. (Hole #2) While there is extensive damage, there is little building debris at the bottom of the hole. There is no sign of molten metal. A worker in the distance walks along a massive core column.
http://www.photolibrary.fema.gov/photodata/original/3946.jpg

3. Holes #1 and #2 Coalesce

What may we make of this? Neither controlled demolition nor thermite (much less burning airplane fuel in the towers) can explain both of these anomalous holes and the fact that there is no *pancaked* debris at the bottom of them.

In short, something *else* must now be considered as the mechanism of destruction, something that, quite literally, bored holes down through the buildings, and something, also, that dustified the twin towers.

Figure 212. (9/13?/01) Deep empty holes under Liberty Street.
(a)http://bocadigital.smugmug.com/photos/10698496-D.jpg, http://bocadigital.smugmug.com/, (b)http://bocadigital.smugmug.com/photos/10698500-D.jpg, http://bocadigital.smugmug.com/

---

[1]NOAA's Aerial Photo of World Trade Center, Image taken by NOAA's Cessna Citation Jet on Sept. 23, 2001 from an altitude of 3,300 feet using a Leica/LH systems RC30 camera., *Image released Oct. 2, 2001 at 6:15 p.m. EDT, credit "NOAA" http://www.noaanews.noaa.gov/stories/s798b.htm*

[2] File No. 9110413, "Firefighter Tiernach Cassidy," December 30, 2001, pp. 10-18, *http://graphics8.nytimes.com/packages/pdf/nyregion/20050812_WTC_GRAPHIC/9110413.PDF*

[3] Never Forget: An Oral History of September 11, 2001, Mitchel Fink and Lois Mathias, NYC: HarperCollins, 2002, pp.203-205.

11.
Toasted Cars

I've lived to bury my desires,
And see my dreams corrode with rust
Now all that's left are fruitless fires
That burn my empty heart to dust.
—Alekandr Sergeyevick Pushkin

A. Introduction

One of the most mysterious and unexplained categories of phenomena that occurred around the WTC complex on 9/11 was that of the *toasted cars*. The term I'm using here refers to condition, rather than cause, and is derived from the casual phrase. "It's history, it's toast," meaning it is unsalvageable.

Figure 213. (9/13/01) Peculiar wilting of car doors and deformed window frames of this vehicle found under FDR Drive.
http://nyartlab.com/bombing/09-13/DSC07998.jpg

Although some of the cars did appear to have burned, paper right next to them did not. Some appeared to have wilted, as if subjected to high heat, while their plastic trim remained unaffected. Some vehicles appeared to have burned on the inside but not on the outside. Vehicles were reported to have exploded and burst into flames by spontaneous combustion,[1,2,3,4] and others were described as "half there and half disintegrated."[5] Some vehicles were flipped upside down yet appeared relatively undamaged.

A reported 1,400 vehicles were damaged on 9/11,[6] some of them as far away as the FDR Drive, about seven blocks from the WTC, along the East River, as witnessed by Emergency Medical Technician, Alan Cooke (see page 218). These vehicles showed peculiar patterns of damage. For example, some vehicles had missing door handles or blown out windows; the window frames of others were deformed, their engine blocks disintegrated, their steel-belted tires left with only the steel belts remaining, while in other cases the front ends of vehicles had been destroyed while little or no harm had been done to the back ends. Damage of these highly varied and extremely mysterious kinds cannot be explained by falling debris, jet fuel, conventional fires, or even thermite. What could have caused these kinds of damage?

B. NYPD Car 2723 - The Waxed Spot Car

The back end of the police car shown in Figure 214 appears to be in pristine condition. The shiny finish looks as if the car just came off the show-room floor with a new wax job. The right side of the police car (shown in Figure 214a) looks, literally, half *toasted*. That is, half of it looks toasted and half of it looks pristine, with an abrupt boundary between the two zones being the front and rear doors, shown by the green arrows. The front door is completely toasted while the back door is not at all toasted and appears shiny, with the "new wax job" appearance. Fires do not burn with an abrupt boundary in an all-or-nothing fashion. It does not seem possible for a fire to burn the front door so completely without leaving any soot on the back door, adjacent to it.

Figure 214. (9/13/01) The back end of this police car looks better than the front.
(a) There is extensive damage to the front of car 2723. (b) There is no door handle on the driver's door. There is an unusual, unburned circular area on the rear door.
(a)http://nyartlab.com/bombing/09-13/DSC08010.jpg, (b)http://nyartlab.com/bombing/09-13/DSC07993.jpg,[7]

In addition, the roof-mounted light bar (indicated by the red arrows in Figures 214a and 214b) does not appear to have melted or burned. The domes on police-car light bars are made of plastic (polycarbonate).[8] One manufacturer reports their polycarbonate has a glass transition temperature (T_g) of 145°C (293°F)[9] and a melting temperature (T_m) of 225°C (437°F)[9]. Another manufacturer reports a melting temperature (T_m) of 149°C (300°F)[10] and another source reports a melting temperature of 267°C.[11] The temperatures at which enamel paint burns or melts as well as the temperature at which steel significantly softens are much greater than this. A regular hydrocarbon fire in a fireplace, burning wood, will easily melt plastic but not the metal grill supporting the logs.

Car doors have rubber gaskets around them to protect the interior from the weather as well as to block out road noise. A rubber gasket is not a firewall and would have little effect in stopping a normal fire. But let us note that rubber is an excellent electrical insulator, having very low electrical conductivity. Copper[12], Aluminum[13], and iron[13] are about 10^{20} to 10^{21} times more conductive than rubber. In other words, the electrical conductivity of rubber is almost zero.

The left-rear door of the car, as seen in Figure 214b, appears to have had the upper half of it *toasted* except for a circular spot near the rear of the door, indicated by the yellow arrows. Appearing to have retained its wax job in this spot, it became known as "the waxed-spot car."

While analyzing the photograph in Figure 214b, I first realized that the mechanism of destruction of the WTC must have involved the interference of energy fields. This was my first "Aha!" moment. I was not at all familiar with energy weapons, nor did I know if such a thing existed. It wasn't something I had thought about or looked for. However, my scientific area of expertise has to do with a type of optical interferometry, so that I am very familiar with interference of energy in the optical range. Visible light constitutes only one small range of the wavelengths along the entire electromagnetic spectrum. The same principles of interference that exist for light waves can exist for other wavelengths as well. In regions where energy fields interfere with one other, there will be zones of constructive interference and zones of destructive interference. That is, you will have zones where effects are multiplied and zones where effects are canceled.

Figure 215. (9/13/01) The toasted interior of police car 2723.
(a)http://nyartlab.com/bombing/09-13/DSC08011.jpg, (b)http://nyartlab.com/bombing/09-13/DSC07994.jpg,[11]

As said, the photograph in Figure 214b shows an apparently unaffected circular zone on the back door, amidst a larger toasted area. It reminded me of a masked-out area when exposing photographic paper, the light being blocked from exposure to the paper. But the sheet metal, iron, aluminum, plastic and glass of an NYPD vehicle are not photographic film. The unburned circular area on the rear door of police car 2723 (a.k.a. "The Waxed Spot") in Figure 214b is very revealing. More examples of abrupt boundaries between affected zones and unaffected zones can be seen in photographs of other cars as well. (See Figures 245 and 247).

Because of the abrupt boundary between the toasted and non-toasted zones on the police car's rear door, we know that this damage cannot be the result of a regular fire. When fire burns, the temperatures between the hot regions and the cool regions smoothly transition into and between one another. If we assume the burnt-looking region to be the result of a very hot flame that had been held next to the car while something blocked or masked off that circular region, we would be assuming incorrectly. This is because the degree of heat needed to burn the "toasted" region would have been conducted into that adjacent circular region that appears unburned. Consequently, the *toasted* effect cannot be from conventional heat. Much other evidence supports the lack of high heat on 9/11—the great volume of unburned paper fluttering and lying amid the destruction, for example.

Also revealing are the *insides* of police car 2723. The photograph in Figure 215a is the inside of the same police car, number 2723, that is shown again in Figure 214b. (A closer view of Figure 215a is shown in Figure 257 on page 250.) The photographs of the car were taken on 9/13/01, just two days after it was damaged. It and other cars had been pushed off the roadway and under the FDR Drive so that they wouldn't block traffic. This car, then, in Figure 215, had been parked in the shade for two days—days without rain[15]—so that the condition of its interior cannot be explained by any normal environmental degradation. But notice that the steel transmission and drive shaft hump between the two front seats appears to be full of small holes, as if the steel has been partially eaten through.

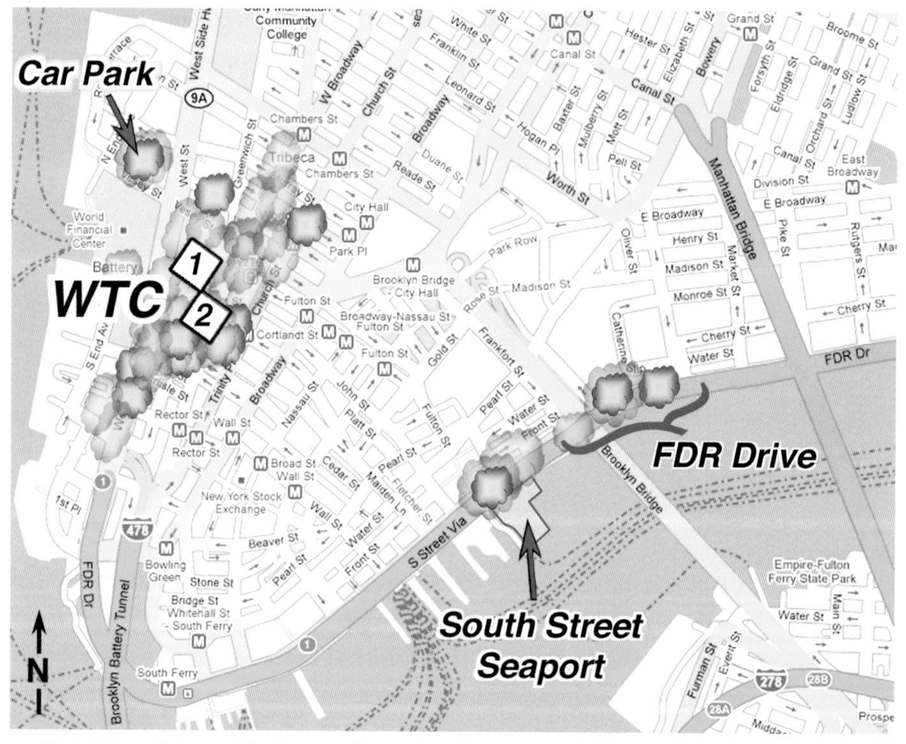

Figure 216. Map of lower Manhattan shows the WTC and FDR Drive a half mile or more apart.[16]

Note that the toasted-car lot is about $\sqrt{(600^2)+(400^2)} \approx 720$ feet
The map in Figure 216 locates the damaged vehicles on a map of lower Manhattan. The toasted cars and flipped vehicles are far outside of the possible extent of the debris fields of the Twin Towers, especially the toasted cars on FDR Drive near the South Street Seaport. The official story offers no explanation for this damage, for the unexpected location of the damage, nor does it even make any mention of these bizarrely damaged vehicles.

C. Toasted Cars under FDR Drive

Similar to the clear boundary-lines between "toasted" and "not-toasted" areas that we saw in the case of police car 2723 (a.k.a. "the wax spot," Figure 214a), many other examples of abrupt boundaries between damaged and undamaged materials are to be seen. In the case of the police car in Figures 214 and 215, that boundary runs through the entire car itself, with an almost razor-like precision, leaving the front half "toasted" and the back half relatively unscathed. Another example of this is shown in Figure 247.

Figure 217. (9/13/01) Vehicles under FDR drive were randomly toasted. This is at least 1/2 mile away from the WTC. Note the waviness of the tire tracks. What happened?
http://nyartlab.com/bombing/09-13/DSC07989.jpg

When the debate over toasted cars arose, some argued that the wrecked vehicles under the FDR Drive had been damaged at the WTC and were then loaded up and transported to the FDR Drive for storage. But there are problems with this theory. First, there is *no* evidence that the moving of vehicles was done. Second, it makes no sense to load up wrecks and transport them only to dump them by a busy thoroughfare for "storage." These wrecks would then have had to be picked up yet again and transported once more. If vehicles were truly moved from the WTC to the FDR Drive, we wonder why WTC steel beams were not stacked up under the FDR Drive as well, if it was such a good storage area. Third, governments may be stupid, but we doubt they could be this inefficient. Further, if reported, such tampering with evidence would have been declared felonious. Marks on the roadway suggest that some of these vehicles were pushed to the side of the roadway until they could later be removed.

Figure 218. (9/13/01) Disintegrated cars under FDR Drive.
(a)http://nyartlab.com/bombing/09-13/DSC07999.jpg, (b)http://nyartlab.com/bombing/09-13/DSC08000.jpg,

Figure 219. (9/13/01) This car did not burn and there is little debris inside.
(a)http://nyartlab.com/bombing/09-13/DSC07990.jpg, (b)http://nyartlab.com/bombing/09-13/DSC08013.jpg,

Regarding the distance between the WTC and the FDR Drive, the following is of interest. After the initial "impacts" at WTC1 and WTC2, emergency medical technician Alan Cooke reported what seemed to be an explosion, not at the WTC, but at the South Street Seaport, at Fulton Street and the East River. The South Street Seaport is just south of where the toasted cars were found under FDR Drive, as shown on the map in Figure 216 on page 216.

EMT Alan Cooke

> We got to the beginning of the FDR Drive, by the ferry, and I guess because of the way the streets channel everything, *one of the fire balls or whatever, had to have made it as far as the South Street Seaport, because what happened at that time, it seemed like an explosion was coming from there.* I thought an explosion was coming from there. That's when everybody started running towards us from the Seaport.

Now we had everybody running [to] us from the Seaport and running to us from the west side, so we couldn't go either way. That's when all the ash and everything started coming. We had a couple of people stop us because they were complaining *of chest pain...*

Nobody could breathe and everybody was trying to climb up on to the wall of the FDR Drive...I thought what happened was that there was an explosion at the World Trade Center. Then I thought there was another one at the Seaport. I thought that was a secondary and herding everybody towards the Brooklyn Bridge, because everybody was asking me where should we go, where should we go. I just told them to get on the highway and head north or towards Brooklyn.[17]

While an "explosion" could account for *some* of the vehicular damage in the vicinity of the toasted cars under the FDR Drive, it still does *not* explain why there was such an abrupt *boundary* between areas where damage occurred and areas where it did not. An "explosion" would not explain the damage to the vehicle in Figure 220, either, no matter where it had been parked. Towing a vehicle cannot account for this damage. This photograph was taken under the elevated FDR Drive. The underside of the highway can be seen in the upper-right corner.

Figure 220. (9/13/01) Vehicle under FDR Drive with unburnt seatbelts, plastic molding, and upholstery. http://nyartlab.com/bombing/09-13/DSC08002.jpg, [18]

The vehicle shown in Figure 220 appears to have unburnt seatbelts and upholstery. Also, the plastic molding around the passenger window opening appears undamaged as does the rubber gasket on the rear side window. But strangely, the exterior of the vehicle looks toasted, especially around the window openings. The brightwork (silvery trim) around the two window openings in the foreground is completely missing. In addition the exterior area that had been covered by the brightwork appears more affected than the other areas. This is counter to a normal car fire, where the brightwork would serve as a protective layer.

D. Flipped Cars and Levitation

Atypical burn patterns were not the only evidence of unconventional physical effects during the demolition of the World Trade Centers. Consider the words of New York Firefighters and a photographer who were on the scene when the buildings *poofed.*

Photographer David Handschuh

> I was almost being picked up by a tornado…it was like being picked up…and [it] just picked me up and tossed me about a block. Just…one second I was running and the next second I was airborne…[19]

Figure 221. A flipped car near the Liberty and West Street intersection.
http://i24.photobucket.com/albums/c49/IgnoranceIsntbliss/911/NonBurnedCars/a3.jpg

It is often assumed that vehicles like those in Figure 221 were flipped over by gusts of wind during the destruction of the WTC. However, the physical evidence does not support such an idea. Figure 221 shows flipped cars and flipped fire trucks (rigs) in the near vicinity of trees with full foliage still intact. If wind could overturn a fire rig, certainly it would strip leaves from a tree. In addition, a number of eyewitnesses have described being picked up and carried some distance when the Towers went away.

Firefighter Todd Heaney

> When I got to the front of the building, it tossed rigs down the street like it was—like they were toys. They were upside down, on fire. [20]

Firefighter Michael Macko (WTC2 Demolition/Poofing)

> We were making our way down West Street. We got just about south of the north overpass, about 50 feet past that, when the first collapse occurred. I looked up. I was awed by—I thought it exploded at the top. Everybody I guess at that point started running, and I luckily ran north where I came from to try to run out from under this—which happened to be a collapse, realized I couldn't. I was going to stay under the overpass.
>
> I realized I couldn't get out from under the collapse. *I dove under an ESU truck that was facing north on the west side of West Street.* I dove under that and waited for the building to come down.
>
> When the building did come down, I actually thought I was trapped, and the truck was blown off me, pushed off me, I guess. It was not there. *At that point I*

was just really shocked and didn't know what was going on at that point. I didn't know—I was really, really shocked...

At that point I encountered some I guess emergency personnel that were trying to help, you know, guys that were coming out of the collapse at that point. I sort of assessed myself and found that I didn't have any real physical damage.[21]

Figure 222. (9/12/01) Close up of flipped car in Figure 246.
This does not appear to have tumbled here. The underside is not scuffed.
http://hereisnewyork.org//jpegs/photos/3534.jpg

EMT Brian Gordon (WTC2 Demolition/Poofing)

So basically all the patients started getting up and running to the back and so did I, and the tower hit and it was like it picked me up and threw me.[22]

EMS Captain Mark Stone (WTC2 Demolition/Poofing)

I don't know whether I was in 10, 20, 30, 50, 100 feet. I know I wasn't in far, but all of a sudden just a whoosh and a thrush, just, I started getting hit by debris. I got picked up and started being thrown...I was actually in the air flying along...[23]

Brian Smith (WTC2 Demolition/Poofing)

It was like a bomb went off, you know, and it just hit me so suddenly, you know, and it picked me up, picked me right up off my feet and threw me a good 30 feet through the air, because now you can imagine where the ambulance is.[24]

EMT Renae O'Carroll (WTC2 poofing)

I tried to grab people. People was grabbing on to me. We were just running, running, running. I have never seen anything like that before. I understand what [sic] someone says I looked death in the face. That was death coming to me. That's all I know.

I'm running. I'm ahead of it. Everyone's running, and it's just a stampede. I'm about ten feet in front of it, running, actually sprinting because I'm an athlete and I'm running. What happened when I got to the corner, because I remember my feet hitting, coming off the sidewalk, another blob of stuff came around.

Ash came around another building in front [of] me, and it caught me in front of me and in back of me, and everything was pitch-black. Where it hit me from the front and the back, it actually lifted me off the ground and threw me. It was like someone picked me up and just threw me on the ground. [25]

Something told me—I looked to the left on the ground, and I saw a *red light.* I don't know what that was. I'm thinking it's another light.

I can laugh about this now. At that time I couldn't laugh about it. I couldn't laugh about it.

And that's when I put my hand to the left to see what that light was, and I felt glass. What happened to me was just a miracle. The glass door opened up. It was a door. It opened up. It opened up, *and it felt like someone put their hands under me just pulled me,* picked me up and pulled me.[26]

Figure 223. Flipped police car on Church St. at Cortlandt St.[27]
(Street names on the sign have been enhanced for clarity.) http://memory.loc.gov/service/pnp/ppmsca/02100/02102v.jpg

Firefighter Robert Salvador (WTC2 Demolition/Poofing)

By that time I noticed that it was mushrooming and I knew then it was collapsing, so I started to run and the concussion picked me up and threw me 30 feet and I fell on my back, right near two parked cars. I turned around toward the wall and the cloud was right on top of me. So I crawled underneath two parked cars and that was it for me. [28]

[...]

Either the cloud engulfed them or they ran towards Church Street, I don't know, but I ran, and the cloud just swept me and picked me up and threw me 20 feet, and I crawled under two parked cars, and I said my prayers.[29]

Firefighter John Wilson

Everybody just ran. We ran away from the noise. So the noise was to my east as I was standing there. We ran back basically the way we came in, I don't know how far. All of a sudden I was just picked up. I felt the wind from behind me *just pick me up and throw me.* So now I'm on the ground. [30]

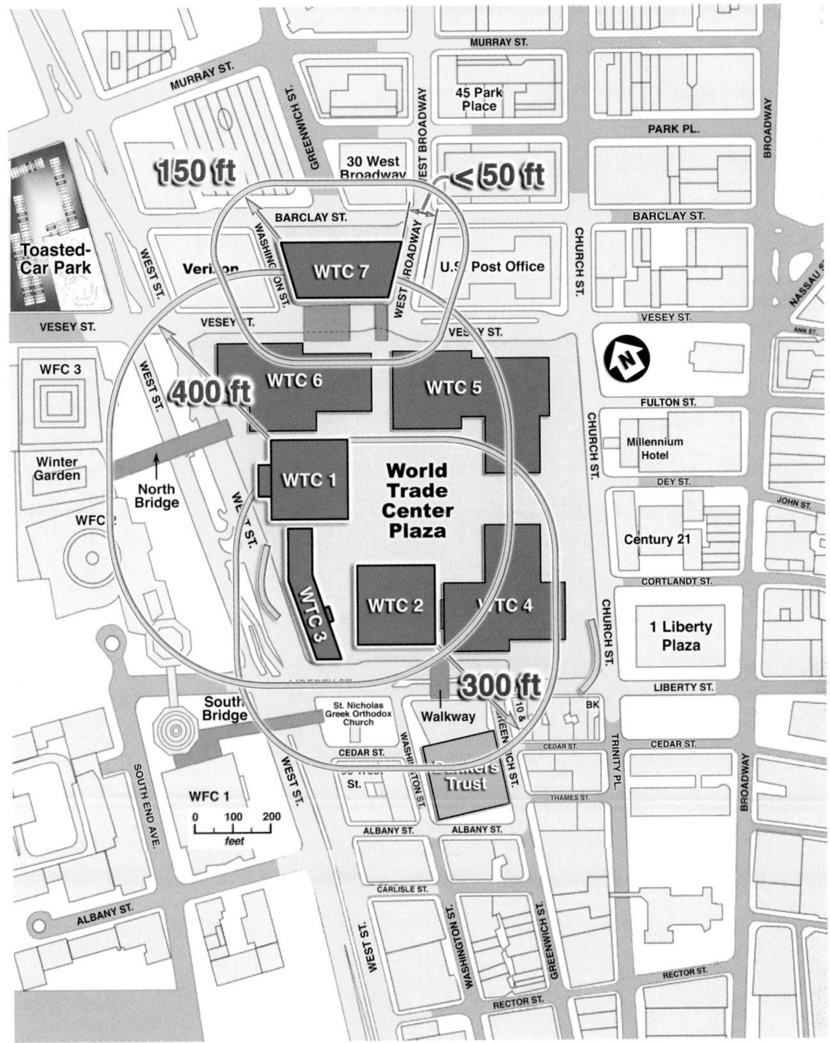

Figure 224. Estimated footprint of falling solid debris from the WTC.
The streets shown in yellow indicate areas where anomalies such as toasted cars were photographed.
Edit from p. 3 (pdf p. 53 of 298), http://wtc.nist.gov/NISTNCSTAR1CollapseofTowers.pdf

Paramedic Manuel Delgado

But as I ran, I got knocked down by *it seemed like even someone punched me in the back, like a blast it seemed.* It just kind of picked me up and knocked me down. I scraped my elbow, I twisted my ankle, my pants got all ripped, my glasses got blown off and the helmet came off. I get up quickly and all I can see now is just—it had to be on Fulton—the blast, this dark cloud coming at us, at me anyway, because I don't even know who's around me at this point. You kind of lose all—*I lost all track of time.* I lost basically all body movements and I was going on, and then we're engulfed in the smoke, which was horrendous.[31]

Firefighter First Grade Dan Walker (earlier)

I just said to myself, you know, I don't want to fall down. Like feet don't fail me now. I can remember going to jump over that guy and that was it. The wind just laid me down. I didn't get hurt. That's what struck me. It's not like I fell with a thud. Like I tripped over something. It was like the wind was blowing so hard that it just kind of picked me up and laid me down softly.[32]

Firefighter First Grade Dan Walker (later)

So I kind of got blown a little to the side and I wound up right in the corner of—there was this huge beam behind, I guess it was like a marble fascia and I can just remember getting down on my knees and covering up with whatever I could do. I'm holding on to my helmet and I'm watching this debris fly out this door and I mean every bit of it is going straight. It doesn't matter what it was. *Rocks or dust or whatever it was. It was just going straight because of the force of the wind.*[33]

Firefighter John Moribito

The building came down. **The rush of wind lifted me up off the ground, and threw me about** 30 feet back into the lobby of 1 World Trade Center.[34]

Does or can the official story of a pancake collapse possibly correspond to the actual highly confined nature of the debris field and to the extraordinary effects demonstrated on people, vehicles and property? (see Figure 115 in Chapter 7.)

E. Toasted Parking Lot and Cheeto Cars

Figure 225. (9/11/01) The toasted car lot in progress. Flaming vehicles can be seen. The air upwind of the WTC has visibly become clearer.
http://911wtc.freehostia.com/gallery/originalimages/GJS-WTC105.jpg

As the following photos show, several vehicles were on fire after the "dustification" of the towers. Figure 226 was the first photo of toasted-cars I came across. I imagined being there that day and diving under one of these cars. Then I wondered, how would I know which one to pick? The pattern was not clear. Note that in Figure 226, in a picture taken on or shortly after 9/11, there is little visible sign of rust on the toasted cars. Yet, as Figure 227 shows, in a picture taken on September 23, 2001, with dust still all around, there is already evidence of rust occurring. *And in Figures 231 and 232 below, more rust is visible, and again, one can plainly see anomalous damage on some vehicles, with one section damaged, and another, not.*

Figure 226. The toasted car park after apparent "fires" ended.
The toasted-car park is northwest of the WTC complex. (See Figure 224, page 223.) This view is facing away from the WTC. There is little visible rust, so the photo was probably within a few days of 9/11.
http://whatreallyhappened.com/IMAGES/cars_wtc2.jpg

Figure 227. (9/23/01) The rust is noticeable. (NOAA)
NOAA: http://www.noaanews.noaa.gov/wtc/images/wtc-photo.jpg

It appears that those who like vehicles with sunroofs also like to park in the same area. Many of the cars in the back row of the *toasted-car lot* are missing their sunroofs—or have had a rectangular hole eaten through the roof of their vehicle. But there are two cars in particular, marked A and B in Figure 228, that have even more than usual in common. They appear to be the same make, although slightly different models, and the engine compartments of both have been destroyed, as shown in Figure 229.

Figure 228. (9/23/01) There are a lot of missing sunroofs or eaten roofs. (NOAA)
Cropped: NOAA: http://www.noaanews.noaa.gov/wtc/images/wtc-photo.jpg

Car A and car B appear to be the only two vehicles in this parking area that have had the engines eaten out. These are similar vehicles by the shape of each, the angle to the A-pillars of each (structural member on the sides of the windshield), and the type of wheels of each, and they have similar damage.

Figure 229. (9/12/01) Toasted cars in a lot near the WTC.
http://static.flickr.com/119/297635136_93a1f7e6e7.jpg

The back row of cars, in Figure 230, appears to be *burning*, but where are the flames? Where is the dark smoke? Several vehicles in this back row are glowing orange, but they don't appear to be on fire. It appears to be at this time that the engine compartment of car A disintegrates. The inside of the car marked "C" is glowing throughout. In front of that car and up slightly toward car A, there is also an orange glow.

Figure 230. The glowing back row of the toasted car lot.[35]
http://img68.imageshack.us/img68/4764/19wtc108rj0.jpg[36]

It appears that the sunroof is missing from many of the cars in Figure 230, including the five in a row counting up from the bottom of the photograph.

Most of the 1,400 toasted cars had some, if not all, of their windows missing. What would cause this? Figure 229 shows a silver convertible with its top apparently intact but missing some of its window glass. If the windows were blown out from pressure inside, we would certainly expect the convertible top to have been torn or blown off. The windows may not have been blown out by internal pressure—but broken, shattered, or caused to go missing. This suggests that the windows were not blown out by internal pressure, but may have been shattered by some other means.

Figure 231. (9/01) The toasted-car park.
(a)http://inlinethumb61.webshots.com/959/1022378944031921077S600x600Q85.jpg, (b)http://inlinethumb32.webshots.com/1311/1022378952031921077S600x600Q85.jpg

The cars parked in the back of the "toasted car lot" suffered extensive damage. The cars on the left in Figure 232 and on the far right look like they have had their engines "eaten." The cars in the foreground experienced less damage, although many windows are missing. At the same time, partial and whole windows also remain.

Figure 232. (approximately 9/23/01 to 10/23/01) Towing finally begins.
http://inlinethumb32.webshots.com/1311/1022378952031921077S600x600Q85.jpg

Most of the 1,400 toasted cars had some, if not all, windows missing. What would cause the breakage or elimination of the glass?

F. Fuming

Figure 233. (9/11/01) Fuming begins ahead of the dust rollout.
http://911wtc.freehostia.com/gallery/originalimages/GJS-WTC80.jpg

Figure 234. (9/11/01) Fumes emerge from toasting cars.
(a)http://911wtc.freehostia.com/gallery/originalimages/GJS-WTC67.jpg, (b)http://studyof911.com/gallery/albums/userpics/10002/imdf11092001221934a.jpg

G. Flaming Cars

In the photos in Figures 235a and 235b, there is paper all around the vehicles, yet it isn't burning. In the photo in Figure 235b, the front end of the car is burnt and deformed. Aluminum cladding has fallen around the car, but these aluminum façade-pieces are not burning, nor do they show any signs of being burnt or melted.

Figure 235. (9/11/01) (a) Why doesn't the paper burn? (b) The front end of a toasted car can be seen on the other side of the "fire." (Location D, Figure 303, page 290.)
(a)http://debunking911.com/street1.jpg, (b)http://images6.fotki.com/v1/photos/1/115/34570/DSC_0193-vi.jpg

A plastic wastebasket can be seen lying on the ground near the left edge of the photo (Figure 235b) along with an Igloo cooler. It shows no signs of being toasted. This location is about a quarter mile east of WTC2, so it seems unlikely that these items traveled that distance. If they had, it is curious how they would survive when other commonplace office artifacts did not.

The damage on the car suggests that it flamed more vigorously before this photo was taken. The whole front half of the car looks burned, but, in this photo, the fire is confined to the wheel well. Similar patterns of front-end fires are seen, in progress, in Figure 236.

Figure 236. (9/11/01) Unburned paper adjacent to what appears to be fire.
(a)http://hereisnewyork.org//jpegs/photos/5244.jpg, (b)http://www.wtc911.us/911_photos/wtc11.jpg

H. West Broadway (or "The Swamp")

West Broadway is a one-way street (going south) from Beach Street to Vesey Street, ending next to WTC7 at the north edge of the WTC complex. See Figure 224 (page 223). On 9/11, nearly every vehicle along West Broadway lit up with flames for several blocks. A large amount of water was used on the fires and mixed with the dust and debris, giving West Broadway the appearance of a swamp, shown in Figures 237 through 245.

As if all the strange and anomalous burning of the *toasted cars* were not enough, at least one firefighter, Patrick Connolly, stated that many of these vehicles seemed to simply "light up spontaneously" as he walked up West Broadway.

Patrick Connolly had just walked under the Vesey Street pedestrian bridge shown in Figure 237 when WTC1 went poof (10:28:31 AM). He said he was hit with debris, *"But nothing hard. Like just, you know, considerable smacks and stuff like that, but nothing that you couldn't take. Nothing that was pulverizing me."*[37] [emphasis added] He then found his way around the corner to the door of WTC7 on West Broadway, indicated by the arrow in Figure 237. The photograph in Figure 237 was taken shortly after he came back out that door, which was apparently left open.

Firefighter Patrick Connolly

> ...Brian and Joe thought that we were in a fall out shelter, but it was getting dustier and dustier. The conditions were getting—the visibility was starting to deteriorate a lot.
>
> The only thing I knew, that I wanted to do was I wanted to go back out the door that we came in and make a left turn and walk straight up West Broadway out to safety. They were a little bit more controlled. They decided to stay put about five minutes. And think things through and they were gonna go down. There were some interior stairs and they were gonna go down to probably more like a bomb shelter area and I think when Joe looked, it looked like it was destroyed. So we decided that we would tie the search rope off to the doorway and the three of us hand in hand decided that we would walk out. *And as we came out we started to walk north and slowly but surely and up and over and under steel and cars were—cars with tires and cars were popping and they were just starting to light up spontaneously and there was near zero visibility at this stage.* It was better though than it was right after

> the collapse. And then we walked up two blocks. We walked up, there was a hot dog stand there. We broke the window in the hot dog stand, took bottles of water and we were washing our eyes, because our eyes were burning. We made a left on Park Place.[38] [emphasis added]

Patrick Connolly said that Brian and Joe thought they were in a fallout shelter, implying they were below ground level. He also said he thought that when Joe looked down for the bomb shelter, "it looked like it was destroyed."[39] Both towers were gone before Patrick Connolly entered WTC7, so his comments that "but it was getting dustier and dustier" and "[t]he conditions were getting—the visibility was starting to deteriorate a lot,"[40] indicate an ongoing process. This is consistent with the statements of Michael Hess, chief lawyer for the City of New York who recounted his experience in the stairwell of WTC7 just as WTC1 went poof.[41] Hess also described a sudden deterioration of visibility and the destruction of the building below floor six, which cannot be explained by chunks of WTC1 landing on WTC7.

Figure 237. (9/11/01) The West Broadway door to WTC7 after the demise of WTC 1 and 2. Firefighter Patrick Connolly came out this door and walked up West Broadway.
Figure 5-18. http://wtc.nist.gov/NCSTAR1/PDF/NCSTAR%201-8.pdf, p.104, file page 158 of 294,

Patrick Connolly said he came out of the door and walked north (to the right in Figure 237). Referring to the aerial view of West Broadway in Figure 238, Connolly walked north from intersection **A** to intersection **C**, then walked west along Park Place to intersection **D**, according to his description. The roadway north of the Murray Street intersection (below location **E** in the photo) has a matte appearance, consistent with a uniform coating of dust. The roadway south of Murray Street (above location **E** in the photo) appears shiny with dark patches, consistent with water-covered streets and remains of car fires. The boundary between matte and shiny is fairly linear, as indicated by the arrows on either side of the intersection at location **E** in Figure 238, marking the north end of the "swamp." Figures 239 through 245 are photos along

West Broadway between intersection **A** (Vesey Street) and intersection **E** (Murray Street). Also, a street-level photo of car fires at intersection B (in Figure 238) is shown in Figure 304 on page 291.

Figure 238. (9/11/01) An aerial view southward over West Broadway. The photograph's EXIF information indicates a time of 11:57:22, although the shadows imply that it could be earlier. WTC1 is no longer standing, indicating the photo was taken after 10:30 AM.
http://911wtc.freehostia.com/gallery/originalimages/GJS-WTC128.jpg

West Broadway between intersections **A** and **E** (Figure 238) is in a man-made *canyon* between the buildings. After both towers had been destroyed, there were no apparent chunks of debris on the roofs of these buildings, only dust. There is no evidence that flaming chunks of the buildings flew over 1,000 feet north during the destruction of WTC1 or WTC2. There were no apparent chunks of building in any of the ground-level images, either. And Patrick Connolly described walking up this street while cars *"were just starting to light up spontaneously,"*[42] and did not describe having been hit with anything. [emphasis added] So there is no evidence that falling debris started these fires, but the contrary. The distance between the toasted cars was too great for one car to ignite the next in sequence and the distance between these vehicles was covered with unburned paper. So something else had to have caused the spontaneous lighting up of these vehicles, some as far as 1,000 feet from the WTC.

Figure 239. (9/11/01) Toasted cars provide the only light when all sunlight is blocked. (a) Looking southeast across West Broadway at the Park Place intersection (location C, Figure 238). (b) Glowing "cheetos" on the ground don't ignite paper while fire is trapped inside a minivan. (See Figure 293.)
(a)http://.studyof911.com/images/wtc/cars/2431.jpg, (b)http://i24.photobucket.com/albums/c49/IgnoranceIsntbliss/911/BurnedCars/5772.jpg

Figure 240. A view south along West Broadway from Park Place (intersection C, Figure 238).
http://memory.loc.gov/service/pnp/ppmsca/02100/02114v.jpg

Figure 239b, appears to mainly have *internal* fires while the car in Figure 239a appears to mainly have *external* fires. The tires (and, it would appear, even the pavement under the car) are on fire, with a line of fire along the trunk lid. Also, the right front fender has turned white and the red van is unburned. Figure 240 shows the entire car turned white and the red van toasted.

Figure 241. The toasted tour bus does not appear to have burned (between B *and* C *in Figure 238).*
http://i24.photobucket.com/albums/c49/IgnoranceIsntbliss/911/BurnedCars/064.jpg

Later that evening, Diane Sawyer interviewed a volunteer[43] who witnessed these toasted cars and said, "where these cars were, there were no fires around." I believe he is talking about West Broadway. He mentions Building Seven, which is on West Broadway, where a toasted tour bus was located.

> **Diane Sawyer:** And I've got Don Dahler in here with me, we're dividing up all the duties, here, and he's been down to the scene, and also J. D. Halperin who says he's just a volunteer who came in here from California and was around the area.
>
> I just wanted to know how much fire is there. You said you were just at Ground Zero. How much fire is left?
>
> **J. D. Halperin:** Well at Building Seven there was no fire there whatsoever, but there was one truck putting water on the building. But, it's collapsed completely. And then, the other building, that there were some flames still coming up was in World Trade Center One, not a lot.
>
> **Diane Sawyer:** You said you saw melted tour busses? Melted cars?
>
> **J. D. Halperin:** The cars that were right down there ...it was just unbelievable. They were twisted and melted into nothing. The build... the debris is just unbelievable. And then you can see fire trucks and police vehicles that were down there early, that um, all their windows, their windshields, are completely blown out—from...it must have been from when debris dropped.[44]

The bus shown in Figure 241 *appears* to be melted, but there is no evidence of fire. There are very few, if any, charred areas, and the bus appears to have been uniformly affected by *something* from front to back. No rubber tires can be seen on any of the vehicles, only bare wheels. The car parked on the far right of the photograph does not appear to have any charred areas but has zones of discoloration that look like rust. We know this photo was taken before 5:20 PM on 9/11 because WTC7 is still standing. Curiously, the traffic lights are still working, as are the lights in Fiterman Hall, the building next to the bus. And it appears there was no shortage of water from the fire hydrants, because West Broadway has gotten a swamp-like appearance. Unburned paper and dust can be seen covering the sidewalk to the right as well as the roadway. So it seems likely that the wet debris in the street is composed of dust and unburned paper.

Figure 242. Looking (a) south, and (b) north, on West Broadway.
(a)http://image03.webshots.com/3/3/42/18/21934218NQChpEAhHb_ph.jpg, (b)http://image03.webshots.com/3/3/40/70/21934070XFlNSYUYrp_ph.jpg

Figure 243. West Broadway, looking north through intersection E (Figure 238 page 232).
http://images30.fotki.com/v478/photos/1/115/34570/DSC_0092-vi.jpg

Figure 243 shows a parked car whose front end was completely burned out. The car in front of it appears to have suffered more damage in the rear (going by the manner in which it is sagging). The front-end-burned car also appears to have an intact windshield but no side windows, while the car in front of it appears to have no windows at all. The car in front appears to be surrounded by fire-retardant foam lying on the road. Neither car looks like it was struck by flying or burning debris.

Figures 244 and 245 picture a mini-van, a Crown Victoria and a RAM. Both cars appear to have burned more vigorously at their engine (front) ends. However, note that on both of them the paint behind the engine compartments appears still to be intact. The amount of rusting that has occurred is noteworthy. The front tires of both the RAM and the Crown Victoria appear to have been burned away. The rear tires of the Crown Victoria remain intact, but the grill and the plastic/rubber front bumper are gone.

Figure 244. Looking southeast along W. Broadway at Park Place.
(a)http://images30.fotki.com/v478/photos/1/115/34570/DSC_0087-vi.jpg, (b)http://images30.fotki.com/v478/photos/1/115/34570/DSC_0096-vi.jpg

Both cars have gas tanks in the rear. However, I see no signs that the gas tanks caught fire in either of the vehicles. There are no signs of impact from heavy objects on any of these three cars. As for the engine compartments, other photos show firemen pointing their hoses into the engine compartments of various cars.

Figure 245. Looking southeast along W. Broadway at Park Place.
(a)http://images30.fotki.com/v478/photos/1/115/34570/DSC_0081-vi.jpg, (b)http://images30.fotki.com/v478/photos/1/115/34570/DSC_0083-vi.jpg

I. Various Other Anomalies

There were a reported 1,400 vehicles damaged on 9/11.[45] These vehicles had peculiar patterns of damage, as we have seen: Some had missing door handles, windows blown out, window frames deformed, melted or missing engine blocks, steel-belted tires with only the steel belts remaining, while other vehicle front ends were destroyed although little or no damage was visible on the back ends of the vehicles.

Portions of cars burned while paper nearby did not ignite. In addition to these anomalies, some photos show vehicles that have been completely flipped over while others in the same vicinity have not (see Figure 246). The pictures in Figure 247 document a wide variety of damage to vehicles, including total incineration, toasted and disappeared engine blocks, steel-belted tires with only steel belts remaining, deformed wheels, missing grills, broken and missing windows and mirrors, absent door handles, wilted doors, along with the great mystery of there being unburned paper right next to burning cars.

Figure 246. (9/12/01) Why is this car upside down? A burned out SUV with missing windows and toasted front end is parked in front of WFC2 on West Street but will not be moving under its own power. (Close up view is shown in Figure 222.)
http://911pictures.com/images/previews_lg/911-1392.jpg

The C-pillar is the part of the car between the rear passenger window and the back window. The C-pillar of the NYPD car in Figure 247a has a line of discoloration across it, indicated by the yellow arrow. The abrupt boundary between the toasted C-pillar with the non toasted window frame is similar to the door boundary on police car 2723 in Figure 214 on page 214. If this was an ordinary fire, why did the plastic lights on the top of the police car not melt? Similarly strange is the fire truck in Figure 248, whose front end appears to have wilted.

Figure 247. (9/11/01) (a) NYPD Police car along W. Broadway at Park Place. (b) A badly damaged fire truck. Where did its engine and radiator go?
The bottom of the tire has turned to goo below a distinct horizontal line in the tire.
(a)http://image03.webshots.com/3/3/41/59/21934159QBVXpGOZjb_ph.jpg, (b)http://worldfiredepartments.com/animations/images/firetruck-3.jpg

Figure 248. (9/12/01) Why would the front of this fire truck wilt?
(a)http://www.tarrif.net/wtc/4.jpg, (b)http://hereisnewyork.org//jpegs/photos/2941.jpg

J. Peeling appearance

Many vehicles had their outer surfaces peeled away as if they had been sardine cans. In some instances, a good deal more than peeling paint was involved. It appears in some cases that sheet metal delaminated as though something caused the material to melt or disintegrate from the inside, at a certain depth. In other cases there appears to have been more heat damage inside than outside and vice versa. For example, consider figures 249a and 249b. What happened to the axles, wheels, and tires of this car? On the doors, it appears that a heavy layer of paint peeled down from the top, yet the paint on the lower part of the doors has not been peeled.

(a) (9/11-13/01) *(b) (9/13/01)*

Figure 249. (a) The peeling on the side door appears to be delaminated. Sheet metal is not laminated, so why would it do this? (b) How did the wheel get under the rear suspension?
(a)http://hereisnewyork.org//jpegs/photos/1805.jpg, (b)http://nyartlab.com/bombing/09-13/DSC08111.jpg

K. Swiss Cheese Appearance

Figure 250. (a) (9/11/01) The FDNY Hazmat truck is in front of WTC6 on West Street. The remaining upper part of the truck has been peeled and evaporated in areas. The upper part of the cab is gone and the engine block seems to have disappeared. The photo was taken on 9/11 after WTC1 went away, but before WTC7 collapsed. (b) Why does this ambulance have melted inside doors? The inside looks to have suffered more heat damage than the outside. What would cause that?
(a)http://911pictures.com/images/previews_lg/911-1328.jpg, (b)http://amny.com/media/photo/2006-08/24912294.jpg

L. More Eyewitnesses

Paramedic Gary Smiley

Paramedic Gary Smiley testified that he hid under an emergency vehicle during the destruction of one of the towers.

> When it sounded as if the debris stopped, I couldn't - the sides were already covered as if it had snowed, so I started to dig and I cleared enough away that I could see a glow, which I knew was not - which was fire, because *there was fires burning around the vehicles.*[46] [emphasis added]

Additionally, Smiley describes taking refuge in a salad bar across from WFC4, near North End Avenue, which was across from what I refer to as "the toasted car park." (See Figure 224 on page 223) Gary Smiley reflects the lack of understanding about what was going on, stating, "We didn't know..." He does not explain the cause of the fires and explosions, only what others were saying at the time to explain what they were experiencing.

> ...there was a lot of fire around there. The parked cars that had been parked there *were all on fire and which wasn't on fire was exploding.* We didn't know at that point. *I know that one of the police officers said that he thought that they were bombs and maybe they rigged them to blow up.* Just secondary explosions...The explosions outside were too much, so we decided to get out of there...and we ended up coming out of there and we ended up on North End Avenue.[47]
>
> ...It was crazy, people thought that there was—*they were going to bomb everything.*
>
> People were running around. *FBI people saying that there were bombs in different parks* and just running from one area to the next. ...I remember we got to the point where we were actually running out of places to go. We were going to go into the water. I said maybe we should take off our stuff and go into the ocean. We didn't know where else to go.[48] [emphasis added]

Lieutenant René Davila corroborates the fact that some vehicles were simply exploding:

Lieutenant René Davila

> But we were there. Vehicle 219 was destroyed.
>
> Q. Was it on fire?
>
> A. What?
>
> Q. Was it on fire?
>
> A. Fire? We saw the sucker blow up. We heard "Boom!" We were walking up Fulton Street.[49]

Firefighter Steve Piccerill described a grizzly "war zone" in his statements:

Firefighter Steve Piccerill

> Meanwhile we thought the tunnel was clear at that time for emergency traffic. When we got in the tunnel, it wasn't clear. It was—we were stuck in there for maybe ten minutes. I don't know. That's when I started to get a little nervous, because I was afraid that someone in the cars just, you know, figuring in the tunnel, this would be a place to hit also.
>
> We finally got out of the tunnel. *We were driving out of the tunnel up West Street, and we're seeing body parts in the street, torsos, chunks of flesh, parts of the airplane, landing gear, car fires everywhere. It was like a war zone.*[50]

Many such statements testify to the fact that, inside ambulances and other emergency vehicles equipped with oxygen tanks, explosions could be heard. Similarly, there are many statements to the effect that gas tanks were also exploding. Yet these statements and mechanisms still do not explain the *causes* of the anomalous types of damage observed in the photos we have included here, these being but a very *small* sampling of all those that are available. It must be that we are looking at a cause that is not in accord with the official story, but instead as an unexpected, untypical, perhaps unknown mechanism that was put to work in creating the anomalous kinds of damage that we have seen, and also in destroying the towers themselves.

While there is a wealth of such testimony from firefighters, police, and medical personnel that day, the statements of firefighter Kevin McCabe are especially worth noting:

Firefighter Kevin McCabe

I remember when I came up Liberty to West Street, now I could see this devastation that I definitely didn't see to that magnitude after the first building came down, but now there was just mountains of steel. *I remember seeing what turned out to be 10 Truck flipped on its side, kind of twisted.* I remember walking across the bumper of a TAC unit, thinking that there were guys that I knew in TAC, who actually didn't make it. I remember walking across the bumper of one of the TAC units, climbing up on some steel and just figuring that if Kasey was around maybe I would see him.

I ran into some guys from other companies that I knew. What was left of the hotel was—what I had seen of the hotel was no longer standing. It was just one portion, I guess of what would be exposure 4 side still standing, I guess looking at it. I'm on the west side of West Street now, looking across at the hotel. I guess the one end of the hotel closest to Liberty that was, I guess, you know, the only remaining portion of the hotel that was still standing.[51]

Firefighter Michael Macko (WTC2 Demolition/Poofing)

Q. At this point the first tower was already down?

A. No, nothing was down at this point.

I made my way onto West Street, still looking for a flashlight, a radio, some equipment that I could use to maybe try and search or help out. I end up going south on West Street. I ran into a couple guys, one guy from Squad 18 and a detail that I've seen from the previous night.

We were making our way down West Street. We got just about south of the north overpass, about 50 feet past that, when the first collapse occurred. I looked up. I was awed by—I thought it exploded at the top. Everybody I guess at that point started running, and I luckily ran north where I came from to try to run out from under this—which happened to be a collapse, realized I couldn't. I was going to stay under the overpass.

I realized I couldn't get out from under the collapse. I dove under an ESU truck that was facing north on the west side of West Street. I dove under that and waited for the building to come down.

When the building did come down, I actually thought I was trapped, and the truck was blown off me, pushed off me, I guess. It was not there. At that point I was just really shocked and didn't know what was going on at that point. I didn't know—I was really, really shocked.

I ended up hooking up with a guy from 18 Engine or Squad 18. Scott, I think his name was. We made our way across West Street to the World Financial Center and broke through a lobby window in there and made our way out the other side of the lobby.

At that point I encountered some I guess emergency personnel that were trying to help, you know, guys that were coming out of the collapse at that point. I sort of assessed myself and found that I didn't have any real physical damage.[52]

Firefighter Todd Heaney

When I got to the front of the building, it tossed rigs down the street like it was—like they were toys. They were upside down, on fire. There was a large chunk of the facade basically where we were standing. I didn't know where the officer was or what happened to that chief, but I found Tommy Hansard, the guy who was caught outside with me. [53]

Briam Smith (WTC2 poofing)

I grabbed him by the KED that we had on his waist and his hip, grabbed his shirt, and I started to drag him and looked behind me, and then it was just like "bam," and it came so fast. It was like a bomb went off, you know, and it just hit me so suddenly, you know, and it picked me up, picked me right up off my feet and threw me a good 30 feet through the air, because now you can imagine where the ambulance is. I'm on the other end of the bay side. Ain't nothing blocking the—you know, between the street and me, and it just hit me and sent me flying and sent me closer to where the gear racks were. They moved them since, because I've been down there, but the gear racks were set up on the left-hand side behind where I set up all the chairs.

It sent me past that part, and I got to the back of the gear racks and hit down face first, and then I felt the rush of everything coming behind me. I felt things coming by, hitting me, and it was just a sound, like whoosh, and I liken it to being hit by a wave on the beach, you know, because that's kind of what it felt like. I mean, the stuff was so heavy I thought I was being buried.[54]

Figure 251. Flipped car and trees with full foliage.
http://i24.photobucket.com/albums/c49/IgnoranceIsntbliss/911/NonBurnedCars/a3.jpg

It is often assumed that these vehicles were flipped over by a gust of wind during the destruction of the WTC. However, the physical evidence does not support such an assumption. The photograph in Figure 251 shows flipped cars in the vicinity of trees with full foliage. Wind strong enough to flip cars and trucks would certainly strip leaves. In addition, eyewitnesses have described being picked up and carried.

EMT Brian Gordon (WTC2 poofing)

So basically all the patients started getting up and running to the back and so did I, and the tower hit and it was like it picked me up and threw me. [55]

EMS Captain Mark Stone (WTC2? Poofing, confused)

I don't know whether I was in 10, 20, 30, 50, 100 feet. I know I wasn't in far, but all of a sudden just a woosh and a thrush, just, I started getting hit by debris. I got picked up and started being thrown. [56]

EMT Renae O'Carroll (WTC2 poofing)

I tried to grab people. People was [sic] grabbing on to me. We were just running, running, running. I have never seen anything like that before. I understand what [sic] someone says I looked death in the face. That was death coming to me. That's all I know.

I'm running. I'm ahead of it. Everyone's running, and it's just a stampede. I'm about ten feet in front of it, running, actually sprinting because **I'm** an athlete and I'm running. What happened when I got to the corner, because I remember my feet hitting, coming off the sidewalk, another blob of stuff came around.

Ash came around another building in front me, and it caught me in front of me and in back of me, and everything was pitch-black. Where it hit me from the front and the back, it actually lifted me off the ground and threw me. It was like someone picked me up and just threw me on the ground. [57]

EMT Renae O'Carroll

Something told me—I looked to the left on the ground, and I saw a red light. I don't know what that was. I'm thinking it's another light.

I can laugh about this now. At that time I couldn't laugh about it. I couldn't laugh about it.

And that's when I put my hand to the left to see what that light was, and I felt glass. What happened to me was just a miracle. The glass door opened up. It was a door. It opened up. It opened up, and it felt like someone put their hands under me just pulled me, picked me up and pulled me.

I rolled down some stairs, and the door behind me closed. Down there it was a basement to somewhere. I remember there was a subway station that I ran past. I figured maybe it was the other side of the subway station. [58]

Firefighter Robert Salvador (WTC2 poofing)

By that time I noticed that it was mushrooming and I knew then it was collapsing, so I started to run and the concussion picked me up and threw me 30 feet and I fell on my back, right near two parked cars. I turned around toward the wall and the cloud was right on top of me. So I crawled underneath two parked cars and that was it for me. [59]

Either the cloud engulfed them or they ran towards Church Street, I don't know, but I ran, and the cloud just swept me and picked me up and threw me 20 feet, and I crawled under two parked cars, and I said my prayers.[60]

Firefighter John Wilson

Everybody just ran. We ran away from the noise. So the noise was to my east as I was standing the [sic] there. We ran back basically the way we came in, I don't know how far. All of a sudden I was just picked up. I felt the wind from behind me just pick me up and throw me. So now I'm on the ground. [61]

Paramedic Manuel Delgado

A. I remember seeing that and I remember saying I'm not going down there because I don't want to be suffocated, I don't want to be suffocated. But as I ran, I got knocked down by *it seemed like even someone punched me in the back, like a blast it seemed. It just kind of picked me up and knocked me down.* I scraped my elbow, I twisted my ankle, my pants got all ripped, my glasses got blown off and the helmet came off. I get up quickly and all I can see now is just—it had to be on Fulton—the blast, this dark cloud coming at us, at me anyway, because I don't even know who's around me at this point. You kind of lose all—I lost all track of time. I lost basically all body movements and I was going on, and then we're engulfed in the smoke, which was horrendous.[62]

Firefighter First Grade Dan Walker (4-9)

I just said to myself, you know, I don't want to fall down. Like feet don't fail me now. I can remember going to jump over that guy and that was it. The wind just laid me down. *I didn't get hurt. That's what struck me. It's not like I fell with a thud. Like I tripped over something.* It was like the wind was blowing so hard that it just kind of picked me up and laid me down softly.

Q. Really?

A. As far as I can think it over in my mind, you know.

Next thing I know I'm kind of rolling around on the floor and getting blown down the hallway. The windows are breaking, the lights are going. Lights went out, people are screaming. I mean it was horrible.

Now we are in complete darkness. I had a radio because I had control. I had the control position that day, so I had the radio. I had a pretty good flashlight with me.[63]

Firefighter First Grade Dan Walker (4-9)

So I kind of got blown a little to the side and I wound up right in the corner of—there was this huge beam behind, I guess it was like a marble facia and I can just remember getting down on my knees and covering up with whatever I could do. I'm holding on to my helmet and I'm watching this debris fly out this door and I mean every bit of it is going straight. It doesn't matter what it was. *Rocks or dust or whatever it was. It was just going straight because of the force of the wind.*[64]

Firefighter John Moribito

The building came down. *The rush of wind lifted me up off the ground, and threw me about* 30 feet back into the lobby of 1 World Trade Center. [65]

---

[1] File No. 9110048, "EMT Patricia Ondrovic," October 11, 2001, p. 4-8, *http://graphics8.nytimes.com/packages/pdf/nyregion/20050812_WTC_GRAPHIC/9110048.pdf*

[2] File No. 9110075, "Lieutenant René Davila," October 12, 2001, pp. 27-28, *http://graphics8.nytimes.com/packages/pdf/nyregion/20050812_WTC_GRAPHIC/9110075.pdf*

[3] File No. 9110039, "Paramedic Gary Smiley," October 10, 2001, pp. 10-13, *http://graphics8.nytimes.com/packages/pdf/nyregion/20050812_WTC_GRAPHIC/9110039.pdf*

[4] File No. 9110072, "EMT James McKinley," October 21, 2001, p. 5, *http://graphics8.nytimes.com/*

packages/pdf/nyregion/20050812_WTC_GRAPHIC/9110072.pdf

[5] File No. 9110025, "EMTD Richard L. Erdey," October 10, 2001, pp. 7, *http://graphics8.nytimes.com/packages/pdf/nyregion/20050812_WTC_GRAPHIC/9110025.pdf*

[6] *http://.apwa.net/Publications/Reporter/ReporterOnline/index.asp?DISPLAY=ISSUE&ISSUE_DATE=032004&ARTICLE_NUMBER=770*

[7] For other images, see *http://nyartlab.com/bombing/09-13/DSC07993.jpg; http://nyartlab.com/bombing/09-13/DSC08011.jpg; http://nyartlab.com/bombing/09-13/DSC08011.html; http://nyartlab.com/bombing/09-13/DSC07994.jpg.*

[8] *http://www.alibaba.com/product-gs/240323325/police_car_light.html*

[9] *http://www.polymerprocessing.com/polymers/PC.html*

[10] *http://www.dynalabcorp.com/technical_info_polycarbonate.asp*

[11] *http://en.wikipedia.org/wiki/Polycarbonate*

[12] Griffiths, David (1999) [1981]. "7. Electrodynamics" in Alison Reeves (ed.). *Introduction to Electrodynamics* (3rd edition ed.). Upper Saddle River, New Jersey: Prentice Hall. pp. 286. ISBN 0-13-805326-x. OCLC 40251748.

[13] Serway, Raymond A. (1998). Principles of Physics (2nd ed ed.). Fort Worth, Texas; London: Saunders College Pub. pp. 602. ISBN 0-03-020457-7.

[14] For other images, see *http://nyartlab.com/bombing/09-13/DSC07993.jpg; http://nyartlab.com/bombing/09-13/DSC08011.jpg, http://nyartlab.com/bombing/09-13/DSC08011.html; http://nyartlab.com/bombing/09-13/DSC07994.jpg.*

[15] The first rain after 9/11/01 was on Friday, 9/14/01. (See Glossary page 494.)

[16] Redrawn from original: maps.google.com

[17] File No. 9110040, "EMT Alan Cooke," October 10, 2001, pp. 5-6, *http://graphics8.nytimes.com/packages/pdf/nyregion/20050812_WTC_GRAPHIC/9110040.pdf*

[18] Original source: *http://nyartlab.com/bombing/09-13/DSC08002.jpg*, alternate: *http://www.hybrideb.com/source/eyewitness/nyartlab/DSC08002.jpg*

[19] Charlie Rose Show, 2001, *http://youtube.com/watch?v=ulE9OiZqQwg*, 03:32

[20] File No. 9110255, "Firefighter Todd Heaney," December 6, 2001, p. 9, *http://graphics8.nytimes.com/packages/pdf/nyregion/20050812_WTC_GRAPHIC/9110255.pdf*

[21] File No. 9110506, "Firefighter Michael Macko," January 25, 2002, pp. 3-5, *http://graphics8.nytimes.com/packages/pdf/nyregion/20050812_WTC_GRAPHIC/9110506.pdf*

[22] File No. 9110171, "EMT Brian Gordon," October 30, 2001, p. 11, *http://graphics8.nytimes.com/packages/pdf/nyregion/20050812_WTC_GRAPHIC/9110171.pdf*

[23] File No. 9110076, "EMS Captain Mark Stone," October 12, 2001, pp. 9-10, *http://graphics8.nytimes.com/packages/pdf/nyregion/20050812_WTC_GRAPHIC/9110076.pdf*

[24] File No. 9110136, "Briam Smith," October 23, 2001, pp. 27-28, *http://graphics8.nytimes.com/packages/pdf/nyregion/20050812_WTC_GRAPHIC/9110136.pdf*

[25] File No. 9110116, "EMT Renae O'Carroll," October 18, 2001, pp. 6-7, *http://graphics8.nytimes.com/packages/pdf/nyregion/20050812_WTC_GRAPHIC/9110116.pdf*

[26] *Ibid.*, pp. 9-10

[27] Photo trimmed, "Church St." enhanced for clarity, "Liberty St." added.

[28] File No. 9110032, "Firefighter Robert Salvador," October 10, 2001, pp. 2-10, *http://graphics8.nytimes.com/packages/pdf/nyregion/20050812_WTC_GRAPHIC/9110032.pdf*

[29] File No. 9110474, "Firefighter Robert Salvador," January 18, 2002, p. 5, *http://graphics8.nytimes.com/packages/pdf/nyregion/20050812_WTC_GRAPHIC/9110474.pdf*

[30] File No. 9110376, "Firefighter John Wilson," October XXX, 2001, p. 7, *http://graphics8.nytimes.com/packages/pdf/nyregion/20050812_WTC_GRAPHIC/9110376.pdf*

[31] File No. 9110004, "Paramedic Manuel Delgado," October 2, 2001, pp. 16-17, *http://graphics8.nytimes.com/packages/pdf/nyregion/20050812_WTC_GRAPHIC/9110004.pdf*

[32] File No. 9110341, "Firefighter First Grade Dan Walker," December 12, 2001, pp. 6-7, *http://graphics8.nytimes.com/packages/pdf/nyregion/20050812_WTC_GRAPHIC/9110341.pdf*

[33] *Ibid.*, p. 19

[34] File No. 9110354, "Firefighter John Moribito," December 12, 2001, p. 13, *http://graphics8.nytimes.com/packages/pdf/nyregion/20050812_WTC_GRAPHIC/9110354.pdf*

[35] Intensity adjusted. Photo enlarged.

[36] I believe this to be from a GJS photograph (NYPD)

[37] File No. 9110453, "Firefighter Patrick Connolly," January 13, 2002, p. 8, *http://graphics8.nytimes.com/packages/pdf/nyregion/20050812_WTC_GRAPHIC/9110453.pdf.*

[38] *Ibid.*, pp. 10-11,

[39] *Ibid.*, pp. 10-11

[40] *Ibid.*, p. 10

[41] *http://www.youtube.com/watch?v=hy5lpp6yADw*, original: *http://www.bbc.co.uk/blogs/theeditors/2008/10/caught_up_in_a_conspiracy_theo.html?page=12*, video included at *http://drjudywood.com/towers*

[42] File No. 9110453, "Firefighter Patrick Connolly," January 13, 2002, pp. 10-11, *http://graphics8.nytimes.com/packages/pdf/nyregion/20050812_WTC_GRAPHIC/9110453.pdf.*

[43] *http://liveleak.com/view?i=1c9_1184090191*

[44] *http://liveleak.com/view?i=1c9_1184090191*

[45] *http://.apwa.net/Publications/Reporter/ReporterOnline/index.asp?DISPLAY=ISSUE&ISSUE_DATE=032004&ARTICLE_NUMBER=770*

[46] File No. 9110039, "Paramedic Gary Smiley," October 10, 2001, pp. 10-11, *http://graphics8.nytimes.com/packages/pdf/nyregion/20050812_WTC_GRAPHIC/9110039.pdf*

[47] *Ibid.*, pp. 12-13

[48] *Ibid.*, p. 14

[49] File No. 9110075, "Lieutenant René Davila," October 12, 2001, pp. 27-28, *http://graphics8.nytimes.com/packages/pdf/nyregion/20050812_WTC_GRAPHIC/9110075.pdf*

[50] File No. 9110330, "Firefighter Steve Piccerill," December 12, 2001, p. 3, *http://graphics8.nytimes.com/packages/pdf/nyregion/20050812_WTC_GRAPHIC/9110330.pdf*

[51] File No. 9110344, "Firefighter Kevin McCabe," December 13, 2001, p. 25, *http://graphics8.nytimes.com/packages/pdf/nyregion/20050812_WTC_GRAPHIC/9110344.pdf*

[52] File No. 9110506, "Firefighter Michael Macko," January 25, 2002, pp. 3-5, *http://graphics8.nytimes.com/packages/pdf/nyregion/20050812_WTC_GRAPHIC/9110506.pdf*

[53] File No. 9110255, "Firefighter Todd Heaney," December 6, 2001, p. 9, *http://graphics8.nytimes.com/packages/pdf/nyregion/20050812_WTC_GRAPHIC/9110255.pdf*

[54] File No. 9110136, "Briam Smith," October 23, 2001, pp. 27-28, *http://graphics8.nytimes.com/packages/pdf/nyregion/20050812_WTC_GRAPHIC/9110136.pdf*

[55] File No. 9110171, "EMT Brian Gordon," October 30, 2001, p. 11, *http://graphics8.nytimes.com/packages/pdf/nyregion/20050812_WTC_GRAPHIC/9110171.pdf*

[56] File No. 9110076, "EMS Captain Mark Stone," October 12, 2001, pp. 9-10, *http://graphics8.nytimes.com/packages/pdf/nyregion/20050812_WTC_GRAPHIC/9110076.pdf*

[57] File No. 9110116, "EMT Renae O'Carroll," October 18, 2001, pp. 6-7, *http://graphics8.nytimes.com/packages/pdf/nyregion/20050812_WTC_GRAPHIC/9110116.pdf*

[58] *Ibid.*, pp. 9-10

[59] File No. 9110032, "Firefighter Robert Salvador," October 10, 2001, pp. 2-10, *http://graphics8.nytimes.com/packages/pdf/nyregion/20050812_WTC_GRAPHIC/9110032.pdf*

[60] File No. 9110474, "Firefighter Robert Salvador," January 18, 2002, p. 5, *http://graphics8.nytimes.com/packages/pdf/nyregion/20050812_WTC_GRAPHIC/9110474.pdf*

[61] File No. 9110376, "Firefighter John Wilson," October XXX, 2001, p. 7, *http://graphics8.nytimes.com/packages/pdf/nyregion/20050812_WTC_GRAPHIC/9110376.pdf*

[62] File No. 9110004, "Paramedic Manuel Delgado," October 2, 2001, pp. 16-17, *http://graphics8.nytimes.com/packages/pdf/nyregion/20050812_WTC_GRAPHIC/9110004.pdf*

[63] File No. 9110341, "Firefighter First Grade Dan Walker," December 12, 2001, pp. 6-7, *http://graphics8.nytimes.com/packages/pdf/nyregion/20050812_WTC_GRAPHIC/9110341.pdf*

[64] *Ibid.*, p. 19

[65] File No. 9110354, "Firefighter John Moribito," December 12, 2001, p. 13, *http://graphics8.nytimes.com/packages/pdf/nyregion/20050812_WTC_GRAPHIC/9110354.pdf*

12.
Tissue Beams and Tortilla Chips

Science is not about building a body of known 'facts'. It is a method for asking awkward questions and subjecting them to a reality-check, thus avoiding the human tendency to believe whatever makes us feel good.
— Terry Pratchett

A. Thinning Beams

The pictures below tell a story of extraordinary deterioration that can best be described as molecular dissociation. It is understood from the FEMA report[1] that these images and text pertained to WTC7.

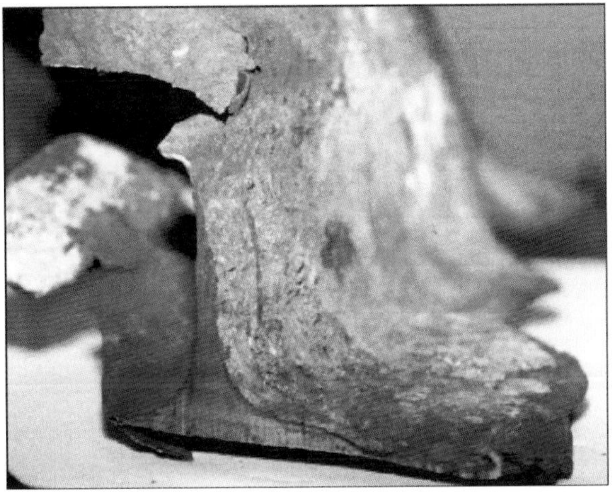

Figure 252. (2001-02) This looks like a hot-spicy tortilla chip, but it is a close up view of highly-eroded wide-flange beam section.
Figure C-2, http://www.fema.gov/pdf/library/fema403_apc.pdf

Figure 253. Swiss Cheese?
(a)http://i286.photobucket.com/albums/ll116/tjkb/NotMelted.jpg, (b)Figure 10, http://www.nistreview.org/WTC-ASTANEH.pdf

The "Deep Mystery" of Melted Steel[2]

Materials science professors Ronald R. Biederman and Richard D. Sisson Jr.

confirmed the presence of eutectic formations by examining steel samples under optical and scanning electron microscopes. A preliminary report was published in JOM, the journal of the Minerals, Metals & Materials Society. A more detailed analysis comprises Appendix C of the FEMA report. The New York Times called these findings "perhaps the deepest mystery uncovered in the investigation." *The significance of the work on a sample from Building 7 and a structural column from one of the twin towers* becomes apparent only when one sees these heavy chunks of damaged metal.[3]

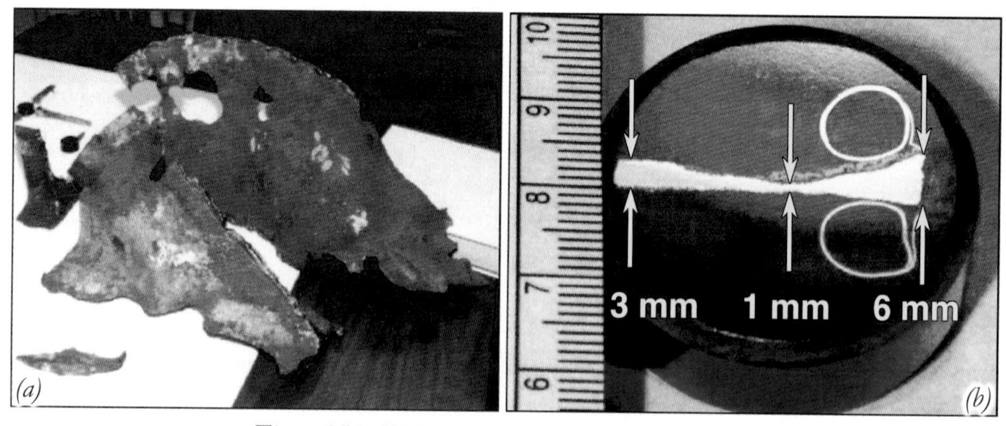

Figure 254. (2001-02) Thought to be from WTC7.
(a) Swiss Cheese? (Figure C-1. Eroded A36 wide-flange beam.) (b) (Figure C-3. Mounted and polished severely thinned section removed from the wide-flange beam shown in Figure C-1.)
(a)http://www.fema.gov/pdf/library/fema403_apc.pdf, (b)http://www.fema.gov/pdf/library/fema403_apc.pdf

What has instantly rusted this steel? It was collected in 2001.

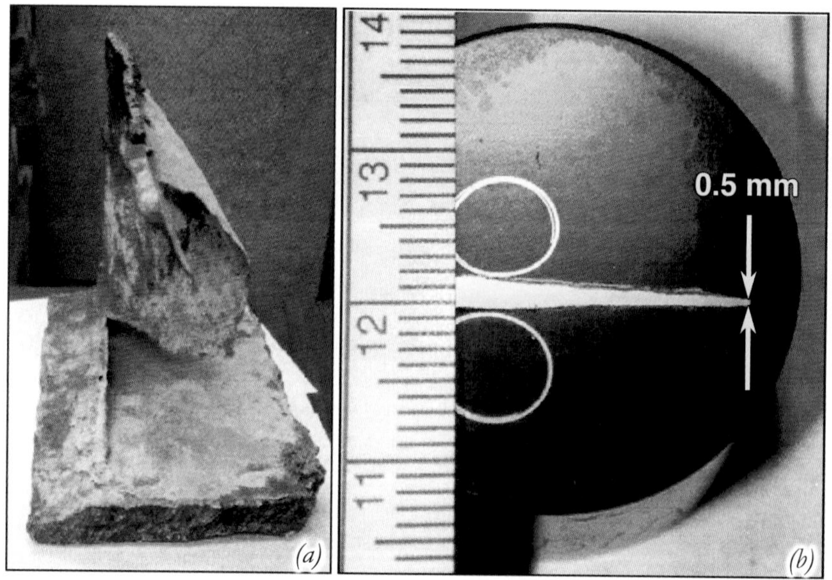

Figure 255. (2001-02) Thought to be from a WTC1 or WTC2 column.
(a) Why is this steel beam as thin as tissue paper and curled like a tortilla chip? (Figure C-9. Qualitative chemical analysis.) (b) Mounted and polished severely thinned section.
(a)http://www.fema.gov/pdf/library/fema403_apc.pdf, (b)http://www.fema.gov/pdf/library/fema403_apc.pdf

Figure 256. (9/21/01) (a) The Liberty Street hole. (b) Enlarged view of the lower-right corner of the large image. (c) Location map.

(a)http://www.photolibrary.fema.gov/photodata/original/4015.jpg, (b)http://www.photolibrary.fema.gov/photodata/original/4015.jpg, (c)http://www.noaanews.noaa.gov/wtc/images/wtc-photo.jpg

A team led by Abolhassan Astaneh-Asl, was supported by a grant from NSF to study the properties of the structural steel from the WTC.

> Contribution to Project: A. Astaneh was PI for this Small Grant for Exploratory Research (SGER) of NSF. One week after the tragic collapse of the World Trade Center, supported by this GSER, he travelled [sic] to New York and stayed for two weeks in Hotel Tribeaa [sic] which was few blocks from Ground Zero. First he met with Mr. Leslie Robertson and visited Ground Zero with him. Mr. Robertson is the structural designer of the collapsed World Trade Center Towers.
>
> ABSTRACT
> On September 11, 2001, World Trade Center towers collapsed because of a terrorist attack. The twin 110-story buildings had tube steel structures steel [sic] and were the fifth and sixth tallest buildings in the world with a height of about 1365 feet. The tragic collapse due to this criminal act will require comprehensive analyses of the structure subjected to initial impact of airplane as well as the intense fire that followed. Such analyses no-doubt will be conducted in the future to establish the cause(s) of the collapse and failure sequence in order to improve

> design of new buildings and retrofit existing structures. *The studies will need reliable data on mechanical properties of materials used in the towers, actual as-built steel connections and members as well as floor systems.*
>
> The objective of this SGER is to conduct post-disaster reconnaissance and collection of perishable data. Specifically, samples of material and structural members and connections that later can be used to establish their properties will be collected. *Of particular interest is to collect samples of steel from areas that were heavily affected by heat of the jet fuel and/or impact of the planes.*
>
> The PI is collaborating with David McCallen of the Lawrence Livermore National Laboratory, who is providing expertise in computational mechanics, modeling and analysis of large structural systems. He is also collaborating with Frederick Mowrer of the University of Maryland who is collecting perishable data on the fire protection engineering aspects of the WTC. Together, the data obtained in these projects will be of critical importance to future analyses of the WTC collapses.[4]

Reviewing this material evidence, one has to ask some questions: First, what consequence of controlled demolition or burning airplane fuel could account for these corrosive results? Moreover, what force or heat in either process could not only reduce these beams to "tissue-paper-thinness" but then furthermore *curl* them or make them like "Swiss cheese"? What type of energy fields could develop the extreme types of *vorticular* bending and damage found in these photos? Before we can begin trying to answer such questions as these, we have to consider the problem of "the fires without heat."

B. Rusty Cars

Figure 257. (9/12/01) This toasted interior of car 2723 was consumed.
http://www.hybrideb.com/source/eyewitness/nyartlab/DSC08011.jpg

Figure 257 shows the inside of police car 2723, which is "the waxed-spot car" shown in Figure 214 on page 214. Figure 214 shows the abrupt boundary between the front and rear door of police car 2723 from the outside, while Figure 257 shows the door boundary from the inside. How does a car rust that fast on the inside? If it had been on fire, shouldn't we see a blackened-burned appearance? Instead, we see a tremendous amount of rust on the inside of the car. Why is that?

C. Rustification

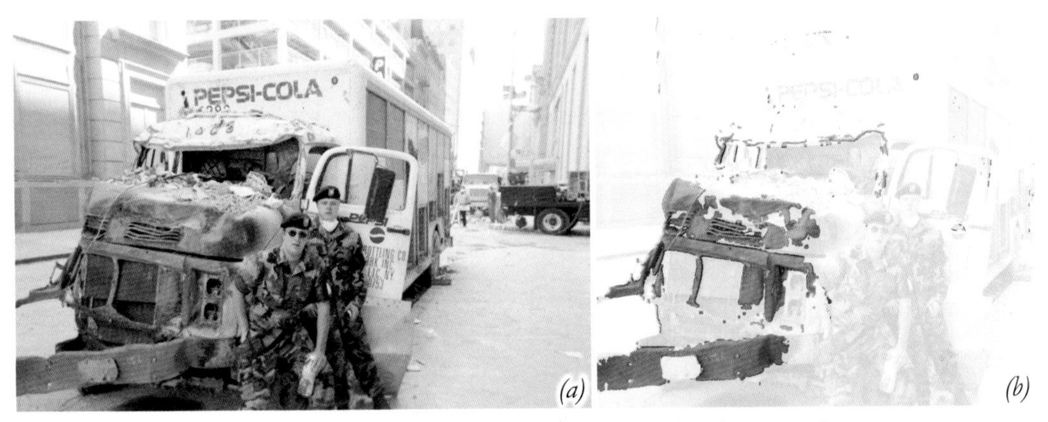

Figure 258. (9/12?/01) Not all of the exposed steel was rusted.
(a) Full picture. (b) Rusted regions lighted.
http://i24.photobucket.com/albums/c49/IgnoranceIsntbliss/911/BurnedCars/004_NR.jpg

The damage to the Pepsi truck in Figure 258 is amazing. The rust looks like a point-of-impact "splash" of something. Some areas look "burned" but show no surrounding smoke damage. Some areas appear melted but show no paint damage. The windshield glass is completely gone, as well as the rubber seal, yet the underlying paint is not even discolored. The surrounding frame appears as if melted yet shows no heat damage.

D. Rusty Beams and Pipes

Figure 259. (9/16/01) Uniformly and brilliantly rusted pipes.
(cropped from original, intensity adjusted)
http://www.photolibrary.fema.gov/photodata/original/5326.jpg

If you clean an iron skillet and leave water standing in it, you will find that it turns bright orange with rust rather quickly and very uniformly. If you were to do the same thing with a steel pipe, you would not have the same result. It might rust, but not uniformly. Over time, steel rusts "here and there" like "creeping crud." Structural steel is made of iron mixed with other materials, such as carbon and manganese, that improve its structural properties as well as its resistance to rust and corrosion. It appears that in the samples we're seeing here, the molecules have let go of these additional ingredients on the surface, effectively turning the material back into iron (if the iron molecules reconnected).

E. Musical Chairs

The mental model I began to visualize is analogous to the game of "musical chairs." When the music starts, everyone gets up and moves around. (Meanwhile, a chair is removed.) When the music stops, everyone is supposed to sit down, but someone will be left out because a chair has been removed. This model seems to work well for a lot of the anomalies at the WTC. While the "music" is playing, the molecules let go of their grip on each other. And when the music stops, they reach for something they recognize, so the "do-dads" might get left out, especially on the surface. This is also my model for "jellification" and resolidification. I think this is what remains because it's at the outskirts of the affected zone, as if the music had been playing very slowly there. The heart of the affected zone might be analogous to the music's playing very fast and very excitedly. So when the music stops, those molecules don't connect with each other.

Figure 260. (before 9/21/01) Dissolving steel with recent and extensive rust.
http://bocadigital.smugmug.com/photos/10697258-D.jpg

I think some of them connect with the oxygen in the air, but there are others that do something else. I think this "something else" may be related to the ongoing behavior that can be observed in the various materials we've looked at. That ongoing behavior within the materials does exist, and remains my working model, based, as always, on all the empirical evidence available to me. My current explanation, then: molecular dissociation and transmutation. The surface of the steel has become iron, and some of the iron on the very extreme surface may not be strongly attached, almost like powdered iron dust—so that with a light rain, it turns bright orange the way rusty iron tends to do—and it flows with the water as if rinsed by an acid bath.

F. Bankers Trust

The Bankers Trust Building (1 Bankers Trust Plaza) was built during 1973-1974, immediately after the Twin Towers were completed. It was a 41-story building at 130 Liberty Street connected to the World Trade Center by a pedestrian bridge that ran between WTC2 and WTC4. In 1999, Deutsche Bank A.G. bought Bankers Trust, and the name was changed from Bankers Trust to Deutsche Bank.[5]

(a) (9/27/01) *(b) (9/20/01)*

Figure 261. The Damage to Bankers Trust on 9/11/01.
(a) The north wing of WTC4 is visible in the foreground, but the main body of WTC4, which had blocked this view of Bankers Trust, is completely gone. (b) The wheatchex shown could not have caused this damage.
(a)http://www.photolibrary.fema.gov/photodata/original/5681.jpg, (b)http://www.photolibrary.fema.gov/photodata/original/3988.jpg,

On September 11, 2001, Bankers Trust was left with a large circular arc or gash down its front, which faced WTC2 and WTC 4. Surprisingly, only two bank employees were killed.[6] It was reported by FEMA[7] that "[t]here were no fires in this building," so it was assumed that all of the damage resulted from the "falling debris" from WTC2 shown in Figure 261. The wheatchex shown hanging from the gash in Figure 261 is only a fraction of the size of the missing building. Also, these wheatchex still have sharp ends, are not bent, and appear to be in near-pristine condition. The type of damage shown in the photographs is not consistent with mechanical loading. (See Figure 6-10 of the FEMA report[8] and Chapter 21 in this book.)

The authorities apparently thought Bankers Trust could be restored by replacing the damaged portions and adjacent beams. But it wasn't that simple.

(a) Missing floors. (11/3/03) *(b) Repaired. (2004)* *(c) Deconstruction begins. (07/28/06)*

Figure 262. Bankers Trust repaired, followed by deconstruction.

(a)http://i.pbase.com/u35/apmillard/upload/23202568.06AbandonedDeutscheBankbuilding.jpg, (b)http://img22.photobucket.com/albums/v65/RandySavage/100_0030.jpg, (c)http://i.pbase.com/o6/51/57151/1/80214224.iJeOlpzP.IMG_1701_edited.jpg

A Building's History[9]

February 26, 2004 — An agreement is made for the Lower Manhattan Development Corporation (LMDC) to buy Deutsche Bank (formerly Bankers Trust) and demolish it. April 13, 2004, the LMDC approves a $45 million contract with the Gilbane Building Company to dismantle the building by the end of 2005; the contract is canceled in May 2005. September 2004, an environmental study finds high levels of toxic material in the building. January 2005, the Environmental Protection Agency (EPA) criticizes the draft demolition plan, saying the plan to monitor air quality during the work is "not acceptable." April 11, 2005, the LMDC awards a two-year $75 million contract to Bovis Lend Lease to clean and dismantle the building. September 2006, the EPA approves the revised plan for the dismantling of the building.

> December 2006 — Workers for the John Galt Corporation, a subcontractor, walk off the job. Galt and Bovis request $30 million more ($75 + $30 = $105 million). Eliot Spitzer and Mayor Michael R. Bloomberg intervene in January to broker a settlement: Bovis will receive an additional $9.7 million, but could lose up to $29 million in additional payments if it does not complete the work in time.
>
> February 2007, demolition begins. Completion is expected by early 2008. May 17, 2007, a 22-foot metal pipe falls from the 35th floor and crashes through the roof of the Engine 10 firehouse across the street. Work is halted for about a week. August 18, 2007, a fire starts on the 17th floor; two firefighters die.

After repairs and midway through deconstruction of Bankers Trust, at the level where the damage had been repaired, rusty beams were exposed in approximate area located in Figure 263a and shown in Figure 264.

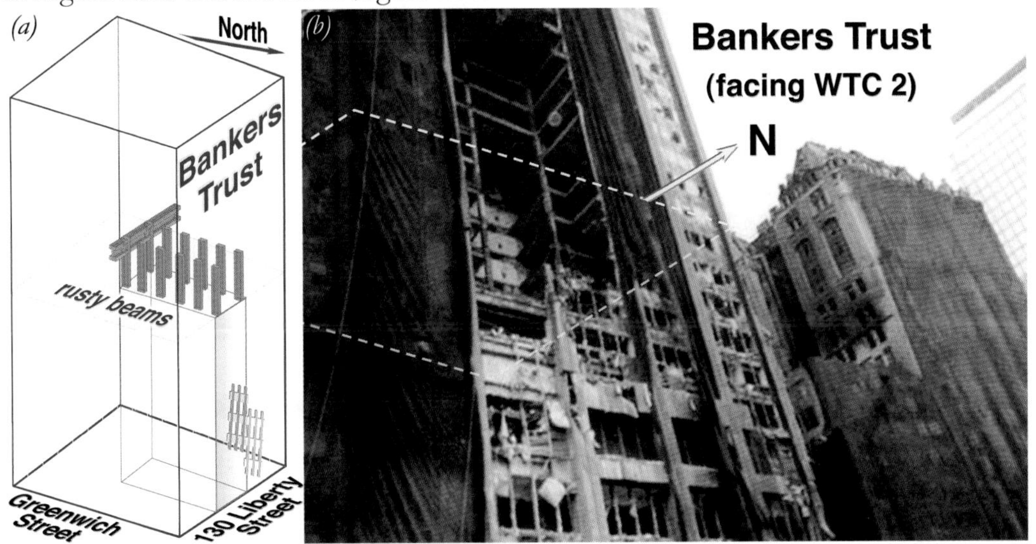

Figure 263. Bankers Trust damage zone opened up for repair.
(a) Diagram of Bankers Trust with location of rusty beams. (b) Photo after 9/11 and before 11/03/03.
(b)http://www.wtclivinghistory.org/images/gallery/wtc_24_big.jpg

Figure 264. (after 01/07) Bankers Trust "furry rust."
This is Bankers Trust being taken apart: (a) original, (b) rusty zones highlighted white.
http://graphics8.nytimes.com/images/2007/08/16/nyregion/17bank_CA07.jpg[10]

The rusting beams in Figure 264 look like they've been at the bottom of the ocean for years. There seems to be a trend for where the really furry-looking rust is. (WTC2 was across the street, to the right of where the first photo in Figure 264 was taken.) That red furry-looking beam in the center of the photo is almost incredible. How many years in ocean water would be required to form that extent of rust?
It appears that those in charge of repairing Bankers Trust were unaware of the non-self-quenching nature of the process of molecular dissociation that was initiated on 9/11/01.

---

[1] Ronald R. Biederman and Richard D. Sisson, Jr., World Trade Center Building Performance Study, Appendix C, *Limited Metallurgical* Examination, Federal Emergency Management Agency (FEMA).

[2] The "Deep Mystery" of Melted Steel, *http://www.wpi.edu/News/Transformations/2002Spring/steel.html*

[3] The "Deep Mystery" of Melted Steel, *http://www.wpi.edu/News/Transformations/2002Spring/steel.html*

[4] World Trade Center Post-Disaster Reconnaisance and Perishable. *http://www.nistreview.org/WTC-ASTANEH.pdf, http://nsf.gov/awardsearch/showAward.do?AwardNumber=0139542*, Structural Engineering Data Collection, Final Report: 0139542, Submitted on: 11/07/2005, Final Report for Period: 10/2001 - 09/2002, Principal Investigator: Astaneh-Asl, Abolhassan, Senior Personnel Name: Astaneh-Asl, Abolhassan, *http://www.nsf.gov/awardsearch/afSearch.do;jsessionid=7CB39E30A9617457307F3427153DDF23?QueryText=&StartDateTo=&ProgEleCode=1637&StartDateFrom=&StartDateOperator=&d-49653-e=2&COPIFirstName=&PILastName=&AwardNumberTo=&ProgOfficer=&SearchType=afSearch&AwardInstrument=&ExpDateFrom=&ProgRefCode=&ExpDateTo=&ProgProgram=&ExpDateOperator=&PIState=&ProgFoaCode=&Search=Search&CongDistCode=&6578706f7274=1&AwardAmount=&IncludeCOPI=&PIInstitution=&AwardNumberFrom=&PICountry=&PIFirstName=&ProgOrganization=&page=4&AwardNumberOperator=&PIZip=&COPILastName=*

[5] *http://graphics8.nytimes.com/images/2007/08/20/nyregion/BldgTimeFull.gif, August 20, 2007,*

[6] *http://graphics8.nytimes.com/images/2007/08/20/nyregion/BldgTimeFull.gif, August 20, 2007,*

[7] Page 6-1 (pdf page 1 of 16), *http://www.fema.gov/pdf/library/fema403_ch6.pdf*

[8] Page 6-9 (pdf page 9 of 16), *http://www.fema.gov/pdf/library/fema403_ch6.pdf*

[9] *http://graphics8.nytimes.com/images/2007/08/20/nyregion/BldgTimeFull.gif*

[10] *http://www.nytimes.com/slideshow/2007/08/16/nyregion/20070817_BANK_SLIDESHOW_6.html*

13.
WEIRD FIRES
FIRES WITHOUT HEAT, HEAT WITHOUT FIRES

We were putting out the car fires, or attempting to, and there was no—the water had no effect on the car fires at the time.[55]—Firefighter Armando Reno

An error does not become truth by reason of multiplied propagation, nor does truth become error because nobody sees it. —Mohandas Gandhi

A. Introduction

Let us begin approaching the problem of the "fires without heat" by considering the two pictures below.

Figure 265. (9/11/01) (a) A view north, across Pine Street, two blocks south of Liberty Street. (b) Vehicles burn, but paper does not (West Broadway).
(a)http://digitaljournalist.org/issue0110/original_images/Reuters/rtr08.jpg (b)http://www.debunking911.com/street1.jpg

Near the WTC complex, the dust clouds in Figure 265a moved too quickly for anyone to out-run them. Although the clouds caught up to and passed the people trying to flee the area, there were no piles of burned bodies left behind. No one reported having been burned by the dust cloud. *Vehicles burned, but people did not. Street signs did not burn, and neither did trees.*

Figure 266. West Broadway looking (a) south, and (b) north.
(a)http://image03.webshots.com/3/3/42/60/21934260YgQzEznyWJ_ph.jpg,
(b)http://images30.fotki.com/v478/photos/1/115/34570/DSC_0092-vi.jpg,

As we have also seen, paper strewn near burning vehicles did not burn. In fact, according to Chief Jerry Gombo,

> You had this soot coming down on you and then in lieu of snow on the ground you had *mounds and mounds of paper, I mean, an unbelievable amount of paper. I'm not talking about one. Just every time you took a step there were mounds of it, it was inches high, plus all of the soot and debris and stuff like that.* [1]

Figure 267. WTC3 from Church Street. The flags did not burn.
http://www.positiontoknow.com/S-11/img/WTC6.jpg

In the pictures in Figure 267, we see a building or a pile of debris on fire in close proximity to what appears to be paper, and yet the paper—whose ignition temperature is comparatively low—is not burning. Again, this is a phenomenon we encountered before, when examining the "toasted cars."

Now let us look at some more pictures.

(a) (9/11/01) *(b) (9/14/01)*

Figure 268. (a) WTC4, viewed from the east. (b) Unburned flags and shrubs.
(a)http://digitaljournalist.org/issue0110/original_images/AFP/11FLAME%20FLAG.jpg, (b)photographer1 [2]

The trees and the shrubs in Figure 267 do not show any indication of having been burned, despite the appearance of many fires around them.

Figure 269. (9/14/01) Unburned shrubs.
Photographer1 [2]

Note in Figure 270a that the building's cladding *itself* appears to be on fire, dissolving into smoke.

(a) (9/11/01) *(b) (9/14/01)*

Figure 270. WTC Tower and WTC 5 have similar patterns. This corner of WTC5 faces away from the towers.
(a)http://www.libertynews.org.nyud.net:8090/wtc/wtc59.jpg, (b) photographer1[2]

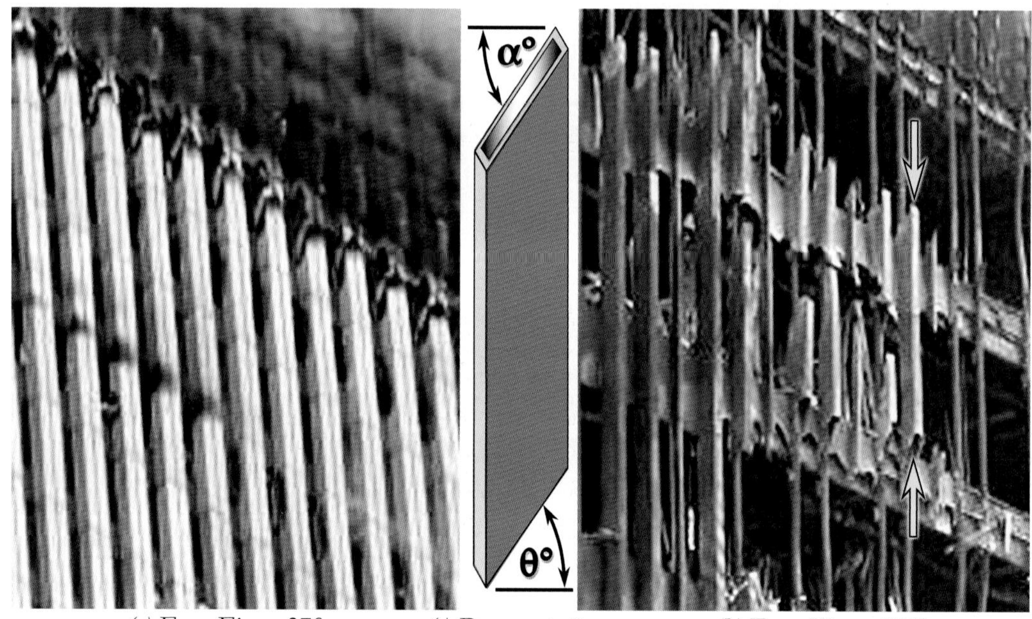

(a) From Figure 270a *(c) Damage pattern* *(b) From Figure 270b*

Figure 271. Enlargements of Figures 270a and 270b above.
(a)http://www.libertynews.org.nyud.net:8090/wtc/wtc59.jpg, (b)photographer1[2]

Now consider a 911 emergency call from Melissa Doi. I reproduce the transcript in full below. Doi's comments are in plain type, the 911 Dispatcher's in *italics*. Boldface is used for emphasis.[3]

B. Melissa Doi Transcript[4]

Doi: Well, there's no one here yet, and the floor's *completely* engulfed. We're on the floor and we can't breathe.

911 Dispatcher 8695: *(Unintelligible)*

Doi: And it's very, very, very hot.

911 Dispatcher 8695: *It's very...are the lights still on?*

Doi: The lights are on, but it's very hot.

911 Dispatcher 8695: *Ma'am, now m'am...*

Doi: **VERY** hot! We're all on the other side of Liberty, and it's very, very hot.

911 Dispatcher 8695: *Well the lights, can you turn the lights off?*

Doi: No, no the lights are off.

911 Dispatcher 8695: *Okay good, now everybody stay calm, you'd doing a good job...*

Doi: Please!

911 Dispatcher 8695: *Ma'am listen, everybody's coming, everybody knows, everybody knows what happened, okay...they have to take time to come up there, you know that. You got to be very careful.*

Doi: …very hot

911 Dispatcher 8695: *I understand. You've got to be very, very careful… how they approach you, okay? Now you stay calm. How many people where you're at right now?*

Doi: There's like, five people here with me.

911 Dispatcher 8695: *All up on 83rd floor?*

Doi: 83rd floor

911 Dispatcher 8695: *with five people...unintelligible, everybody's having trouble breathing?*

Doi: Everybody's having trouble breathing, some people are worse...than others.

911 Dispatcher 8695: *Anybody's unconscious, everybody's awake?*

Doi: So far yes, but it's...

911 Dispatcher 8695: *Listen, listen, everybody's awake?*

Doi: Yes, so far.

911 Dispatcher 8695: Conscious? And it's very hot there, but no fire, right?

Doi: **I can't see because it's too high.**

911 Dispatcher 8695: *No, no, very hot, no fire for now, and no smoke right? no smoke, right?*

Doi: **OF COURSE THERE'S SMOKE!**

911 Dispatcher 8695: *Ma'am, ma'am, you have to stay calm.*

Doi: There is smoke! I can't breathe!

911 Dispatcher 8695: *Okay, stay calm with me, okay? I understand...*

Doi: I **think** there is fire because it's very hot.

911 Dispatcher 8695: *Okay, it's...*

Doi: It's very hot, everywhere on the floor.

911 Dispatcher 8695: Okay, I know you don't see it and all, but I'm, I'm (...stumbles over words...) I'm gonna, I'm documenting everything you say, okay? And it's very hot, you see no fire, but you see smoke, right?

Doi: It's very hot, I see...I don't see, I don't see any *air* anymore!

911 Dispatcher 8695: *Okay...*

Doi: **All I see is smoke.**

911 Dispatcher 8695: *Okay dear, I'm so sorry, hold on for a sec, stay calm with me, stay calm, listen, listen, the call is in, I'm documenting, hold on one second please...*

Doi: I'm going to die, aren't I?

911 Dispatcher 8695: *No, no, no, no, no, no, no, say your, ma'am, say your prayers.*

Doi: I'm going to die.

911 Dispatcher 8695: *You gotta think positive, because you gotta help each other get off the floor.*

Doi: I'm going to die.

911 Dispatcher 8695: *Now look, stay calm, stay calm, stay calm, stay calm.*

Doi: Please God...

911 Dispatcher 8695: *You're doing a good job ma'am, you're doing a good job.*

Doi: **No, it's so hot, I'm burning up.**

911 Dispatcher 8695: *Okay ma'am, the floor is hot, everything is hot, is there, are there, are there desks there? You go up high, you get yourself away from the smoke.[2] Okay, I know you know, hold on a second.*

911 Dispatcher 8695: *(into radio) 83rd floor, three people trapped[3]*

unidentified: *(into radio) Where?*

911 Dispatcher 8695: *(into radio) Way out on the 83rd floor, they're trapped, very hot...that's what I'm doing, just letting you know.*

Doi: (*excited interrupting*) Wait! Wait! We hear voices! (*screaming to firefighters*) **HELLO! HELP!**

911 Dispatcher 8695: *Hello, ma'am? Ma'am?*

Doi: (screaming to firefighters) **HEEEEELP**

911 Dispatcher 8695: *Okay, stay calm, ma'am, ma'am.*

Doi: (screaming to firefighters) **HEEEEELP**

911 Dispatcher 8695: *Stay calm, stay calm, just don't move.*

There are several interesting things that emerge from the audio recording as well as from this transcript. It may be best to lay them out in a list:

1) Over and over again Melissa Doi emphasizes how *hot* it is;
2) She emphasizes that there is such a degree of "smoke" that she can barely "see air"; and finally,
3) In spite of stating there is extreme heat, she cannot see any *fire.*
4) Although she emphasizes how *hot* it is, Melissa Doi's speech is rapid and emotional, not lethargic. Someone who is physically hot (sauna, hot tub) is lethargic, not panicked.
5) On the audio recording, Melissa Doi states that everyone is having trouble breathing, yet we never hear her or anyone else cough or gasp.

All of these descriptions are similar to the types of effects one would be experiencing in an energy field, such as a microwave field. That said, we may now make some additional observations. If there was a large microwave field in use on that day, it would explain why so much paper was not catching fire in the close vicinity of other burning debris, for that debris is literally—like the chicken on the paper plate in your microwave oven—being cooked from the inside out. Perhaps the outer aluminum cladding on the towers might itself have acted as a kind of "microwave oven grating" to keep whatever kind of energy it was *locked inside of it* and building up an energy field, a kind of standing wave, that could—literally—turn the buildings to powder. Before we go too far, though, there is other evidence to consider.

C. The Rumor of High Heat

The image in Figure 272a has often been used by those challenging the official story of 9/11, claiming that it proves there was "high heat" in the rubble from the WTC.

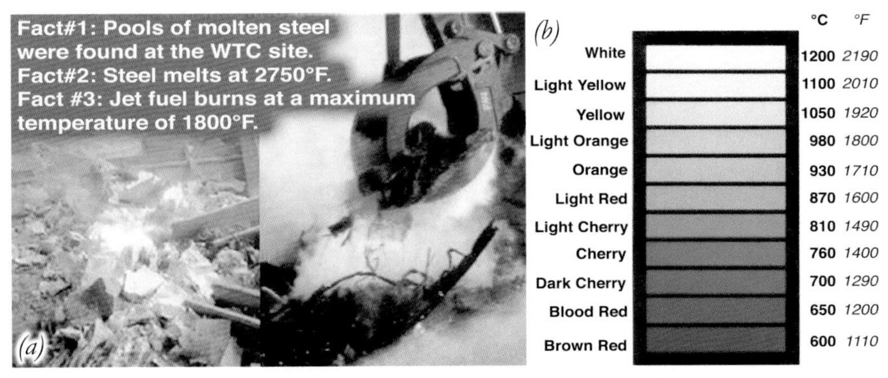

Figure 272. (a) This promoted image has been incorrectly interpreted. (b) Temperature chart. (The text of Figure 272a was enhanced for clarity.)
(a)http://i204.photobucket.com/albums/bb181/slimpimpinstb/WTC20Molten20Steel20Photos20Fact-1.jpg, (b)http://www.processassociates.com/process/heat/metcolor.htm

Hot things glow, but not everything that glows is hot. Consider a fluorescent light. It glows enough to light the room but does not put out light by means of heat (as in incandescence). So, having a photo of something glowing does not prove that it was hot. We need to consider more information in regard to the photo. When materials are heated to a particular temperature, they glow a particular color, as shown on the chart in Figure 272b. This applies to all metals, molten or not, indoors or out. The material on the right side (in the jaws of the grappler) of Figure 272a has a color that would imply that most of it is between 810°C and 1050°C. But we have to recognize that hydraulic systems are permanently damaged if operated at temperatures above 82°C.[5] This seeming impossibility or contradiction is addressed more fully in the next section.

Figure 273. Strange glowing appearance without burning paper or melting aluminum.
http://www.pixelpress.org/september11/sept11_pix/index_pix/alan2.jpg

The material on the left side of Figure 272a has a color that would imply that most of it is between 1050°C and 1200°C. But we also see a lot of unburned paper right next to the glowing material. It even appears that some of the paper might be glowing.

The full version of the images above is shown in Figure 273. A piece of aluminum cladding near the center of the photo appears to be glowing at one end.[6] But aluminum melts at 660°C, well before it becomes hot enough to glow. This piece of aluminum cladding has not melted and is still holding its shape. The glowing material on the far right (the material that looks like "Cheetos" on the ground) are spread out and don't have much mass. So it is unlikely that they are glowing due to heat. It is interesting that fumes are rising from the aluminum cladding but not from the material that is glowing.

Luminance without heating is a phenomenon that is seen in cold fusion as well as in the Hutchison Effect, which will be examined later.[7],[8] Regarding the Hutchison Effect, Richard Sparks stated:

> A less frequent phenomenon is the apparent heating to incandescence of iron and steel specimens having high length to width ratios. This event is not accompanied by the [apparent] heating, [apparent] charring or the burning of combustible materials in contact with the specimen throughout the duration of the event of about two minutes. The fracturing of certain iron and steel geometries accompanied by an anomalous residual magnetic field, permanent in nature, is not uncommon.[9]

The photograph in Figure 274a has often (and inappropriately) been used as evidence of "molten metal" and "proof of thermite." This image appears in the NIST report where the compilers acknowledge having adjusted the intensities in the photograph. Let us assume the color was not changed, only the brightness.

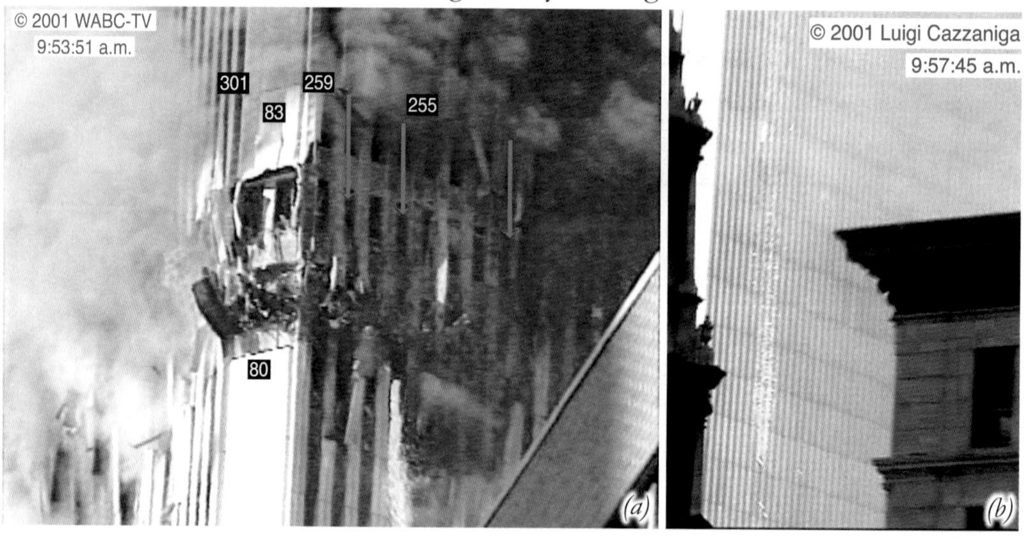

Figure 274. Mystery material emerging from WTC2: (a) 9:53:51, (b) 9:57:45.
(a)NISTNCSTAR1-5A, Figure 9-75, appendix C, page 382 (pdf page 86 of 268),http://wtc.nist.gov/NISTNCSTAR1-5A_chap_9-AppxC.pdf,
(b)NISTNCSTAR1-5A, appendix C, Figure 9-77, page 384 (pdf page 88 of 268),http://wtc.nist.gov/NISTNCSTAR1-5A_chap_9-AppxC.pdf

The material pouring out of the window on the 80th floor, directly below the arrow noting column 255, appears to be glowing a yellow-orange color. If the material is glowing because it is hot, then it would need to be approximately 1,100°C. It also appears to be splashing over the aluminum cladding below it, yet that aluminum cladding does not appear melted. Furthermore, the flow appears to be enormous,

coming from a reservoir of perhaps hundreds of gallons, all of it presumably 1,100°C. And yet nothing adjacent to it appears to be hot, melted, or deformed.
The photograph in Figure 274b was taken near ground level, below where the apparent liquid emerges from the building some sixty or so stories above. In this Figure, the material has been dispersed through the air yet maintains its color. The finer droplets greatly increase the surface area that is cooled by the air, and yet despite this dispersal, the color appears to remain yellow and uniform. This is not the behavior of hot molten metal. The sparks and spatter from a welding torch typically cool from a bright glow before falling to the ground from just a few feet up, changing color in flight as they cool. Remember, hot things glow, but not everything that glows is hot.

Figure 275. (9/21/01) This is called the "Cheeto" picture because the orange object in the lower left corner looks like a Cheeto.[10] Photo credited to Frank Silecchia.[11,12]

Now, keeping in mind all of the above, consider the image again in Figure 275, and note that the grappler is obviously making use of hydraulics. Also realize how conductive metal is. If this grappler had been digging around a hot rubble pile, the bucket cylinder would have been heated to a point of seizing up long before it could have lifted something glowing due to heat. The tolerance in a bucket cylinder is designed for a particular operating temperature, similar to a piston of a car engine. It has been suggested that water was sprayed on the equipment to keep it cool. Water is sprayed on an overheated car engine may likely crack the engine block. Hydraulic equipment, like a car engine, has a temperature range in which it is designed to operate.

Now let us take a closer look at hydraulic systems and their maximum operating temperatures:

> Hydraulic fluid temperatures above 82°C (180°F) damage most seal compounds and accelerate oil degradation. A *single* overtemperature event of sufficient magnitude can permanently damage all the seals in an entire hydraulic system, resulting in numerous leaks. The by-products of thermal degradation of the oil (soft particles) can cause reliability problems such as valve-spool stiction and filter clogging.[13]

Viscosity Value	**cSt**	**Temperature (VG68)**
Minimum Permissible	10	95°C
Minimum Optimum	16	78°C
Optimum Bearing Life	25	65°C
Maximum Optimum	36	55°C
Maximum Permissible	1,000	2°C

Table 12. Correlation of operating viscosity, fluid temperature, and grade.[14]

What does all this mean?

In David Ray Griffin's book, *Debunking 9/11 Debunking: An Answer to Popular Mechanics and Other Defenders of the Official Conspiracy Theory*, Griffin, with the following statements, attempts to convince his readers that there was molten metal at the World Trade Center site.

> It was, however, in relation to the issue of molten metal that (Jim) Dwyer (NYT) most fully displayed his ignorance of crucial facts. Pointing out that Steven Jones had argued that *"the molten [metal] found in the rubble was evidence of demolition explosives because an ordinary airplane fire would not generate enough heat,"* Dwyer gave the final word to the director of Protec, a demolition monitoring firm, who said that *"if there had been any molten steel in the rubble, it would have permanently damaged any excavation equipment encountering it."*[15]
>
> We have here an extreme example of the tendency to favor a priori arguments over empirical evidence. As we saw in Chapter 3, the testimony to the existence of molten metal in the rubble is so strong as to put the issue completely beyond doubt.[16]

In the article David Ray Griffin quotes from, Jim Dwyer writes,

> Jones also argues that the *molten steel* found in the rubble was evidence of demolition explosives because an ordinary airplane fire would not generate enough heat. He cited photographs of construction equipment removing debris that appeared to be *red.*
>
> In rebuttal, Blanchard of Protec said that if there had been any molten steel in the rubble, it would have permanently damaged any excavation equipment encountering it. "As a fundamental point, if an excavator or grapple [sic] ever dug into a pile of molten steel heated to excess of 2000 degrees Fahrenheit, it would completely lose its ability to function," Blanchard wrote. "At a minimum, the hydraulics would immediately fail and its moving parts would bond together or seize up."[17]

I agree with the wording Jim Dwyer used, describing this photograph as being an image of construction equipment removing debris that "*appeared* to be red." But a photo such as this, from an optical camera, not a thermal-imaging camera, captures the *color,* not the *temperature* of the objects in the photograph. David Ray Griffin then, instead of consulting a source on machine hydraulics, quotes second-hand information to refute the credibility of Jim Dwyer's information.

> In one of these statements, moreover, Greg Fuchek said, "sometimes when a worker would pull a steel beam from the wreckage, the end of the beam would be dripping molten steel." Evidently this worker's crane was not "permanently damaged." Dwyer was writing a story for the New York Times, which likes to think of itself as having the highest standards of excellence. But he apparently did not check to see if the evidence supported Jones rather than the man from Protec—even though a Google search for "molten metal at Ground Zero" would have turned up over 300,000 items, many of which contain the testimonies of the people quoted in Chapter 3, such as Peter Tully, Mark Loizeaux, and Leslie Robertson.[18]

A Google search for *"molten metal at Ground Zero"* turned up over 300,000 items, yet not one photograph of a solidified glob of molten steel. Further, since when have scientific questions come to be settled by the number of "items" a Google search turns up? As for Peter Tully, Mark Loizeaux, and Leslie Robertson, all were paid millions of our tax dollars. Are they unbiased witnesses?

> Leslie Robertson, a member of the engineering firm that designed the Twin Towers, said, "As of 21 days after the attack, the fires were still burning and molten steel was still running."[19]

But *where* was this molten steel "still running," and under what exact circumstances? We're given no concrete evidence or information at all.

Again, it appears that David Ray Griffin values the say-so of those on the payroll more than he does physical facts. This is very troubling. The molten metal story appears on the evidence to be exactly that—a story. Certainly, it is a very different thing from the abundant recorded and observable evidence of *glowing* metal. We saw the chart that shows the maximum temperatures at which hydraulic systems can be operated. That chart supports what NYT reporter Jim Dwyer said. Nevertheless, Griffin insults Dwyer by stating that he *"fully displayed his ignorance of crucial facts."*

The crucial facts, however, support Dwyer—namely, that high temperatures would have permanently damaged if not destroyed hydraulic equipment. The facts are well-established and available in mechanical engineering handbooks as well as on the internet.[20] Additionally, as of the date of this publication, no evidence has yet come to view that Peter Tully, Mark Loizeaux, or Leslie Robertson ever witnessed molten metal:

> Mr. Bryan:[21]
> I didn't personally see molten steel at the World Trade Center site. It was reported to me by contractors we had been working with. Molten steel was encountered primarily during excavation of debris around the South Tower when large hydraulic excavators were digging trenches 2 to 4 meters deep into the compacted/burning debris pile. There are both video tape and still photos of the molten steel being "dipped" out by the buckets of excavators. I'm not sure where you can get a copy.
> Sorry I cannot provide personal confirmation.
>
> Regards,
> ==========================

Mark Loizeaux, President
CONTROLLED DEMOLITION, INC.
2737 Merryman's Mill Road
Phoenix, Maryland USA 21131

[. . .]

www.controlled-demolition.com

Figure 276. (10/13/01) Fire does not cause instant and uniform rust. (The orange fumes are probably from oxy-fuel torch cutting.)
http://www.photolibrary.fema.gov/photodata/original/5446.jpg

Additionally, there is a YouTube video whose transcript, reproduced below, reveals some interesting discrepancies:

VIDEO[22] TRANSCRIPT:

GEORGE PATAKI: This was the South Tower. This was 104, 106 stories tall, and this is what is left. Up there was the North Tower. And you look, and you see there's no concrete, there's very little concrete. All you see is aluminum and steel.

REPORTER: What happened to the concrete?

PATAKI: *The concrete was pulverized*, and I was down here Tuesday, and it was like you were on a foreign planet. All of Lower Manhattan, not just this site. *From river to river there was dust, powder, 2-3 inches thick. The concrete was just, er—pulverized and…*

REPORTER: What's to explain Governor, the smoke that still comes out of the [inaudible (fire)]?

PATAKI: There is still fire down below, there's such an incredibly deep pile of rubble and the tower goes down 5, 6 stories underground that, er, there's still fire underneath.

FIREMAN: This is how it's been since Day One. Oh, it's unbelievable. This is six weeks later, almost six weeks later. And as we get closer, to the center of this it gets hotter and hotter. It's probably 1500 degrees. We've had some small windows into what we thought was a boardroom [war zone?] some point [?][inaudible], and it looked like an oven, you know—just roaring inside. It just had a *bright, bright, reddish orange* color.

Or is the bright reddish orange color he is speaking of actually red rust?

FIREMAN: See that stuff he's pulling out?
PERSON: What was that chief?
FIREMAN: Ya know we're going to hold off on the water. You see that stuff he's pulling out right now? [pointing to a grappler].
PERSON: Yeah
FIREMAN: It's red hot. If we hit it with too much steam he won't be able to see what he's doing.
PERSON: Great
FIRE CHIEF: You see how this debris is still smoking? That's from the fires that are still burning. Eight weeks later we still got fires burning. Every now and then, one of the pieces of equipment will dig in, will open up a small area. The oxygen will rush in and you'll get this plume of brown-black smoke coming up. That's because that fire just got more oxygen. So I mean these things are burning. At one point I think they were about 2800 degrees.

The fire chief, though, seems unaware that brown smoke does not normally come from "fire."

V/O: Underground fires, ignited by burning jet fuel, smoldered for months, *fed by molten steel* and buried carpeting, office furniture, wood paneling and paper.

Is steel considered a fuel? *Did paper burn? Do carpeting, office furniture, wood paneling, and paper fuel a fire hot enough to melt steel? If they did, why wouldn't the grill in a fireplace melt? Why wouldn't a common wood stove melt?*

Then we have the following statement:

WOMAN IN HARD HAT: Steel-toed boots is [sic] one of the biggest things. Out still on the rubble it's still, believe um, 1100 degrees. The guys' boots just melt within a few hours.

Let's think about this for a minute. Consider the steel-toed boots as an oven, only hotter. If you put a turkey in those boots it would be incinerated in just a few minutes. *Why didn't the feet inside those boots get incinerated?*

FIREMAN: You'd get down below and you'd see molten steel. Molten steel running down the channel rails, like you're in a foundry. Like lava in a volcano.
V/O: One of the more unusual artifacts to emerge from the rubble is this rock-like object that has come to be known as 'the meteorite.'
MAN IN SUIT: There's this fused element of—of steel, molten steel and concrete and all of these things, all fused by the heat into one single element.
MAN IN BLAZER: And almost like a chunk of lava from Kilowaya [sic] or Iceland where there are very sharp but—but breakable shards on the ends here.

Now please note, in *no way* am I saying or implying that the fireman was lying about "molten metal." On the other hand, memories are not like tape recorders and there is no question but that memories can be manipulated. False memory syndrome refers to the creation of memories that are not based on factual events. See references for more information.[23],[24] But consider the following statements:

> Subjectively, what are misinformation memories like? One attempt to explore this issue compared the memories of a yield sign that had actually been seen in a simulated traffic accident, to the memories of other subjects who had not seen the sign but had [had] it suggested to them (Schooler et al. 1986). The verbal descriptions of the "unreal" memories were longer, contained more verbal hedges (I think, I saw...), more references to cognitive operations (After seeing the sign the answer I gave was more of an immediate impression...), and fewer sensory details. Thus statistically a group of real memories might be different from a group of unreal ones. Of course, many of the unreal memory descriptions contained verbal hedges and sensory detail, making it extremely difficult to take a single memory report and reliably classify it as real or unreal…
>
> A different approach to the nature of misinformation memories came from the work of Zaragoza and Lane (1994) who asked this question: Do people confuse the misleading suggestions for their "real memories" of the witnessed event? They asked this question because of the real possibility that subjects could be reporting misinformation because they believed it was true, even if they had no specific memory of seeing it. After numerous experiments in which [subjects] were asked very specific questions about their memory for the source of suggested items that they were embracing, the investigators concluded that misled subjects definitely do sometimes come to remember seeing things that were merely suggested to them. They referred to the phenomenon as the "source misattribution effect." But they also noted that the size of the effect can vary, and emphasized that source misattributions are not inevitable after exposure to suggestive misinformation.…
>
> Taken together these studies show the power of this strong form of suggestion. It has led many subjects to believe or even remember in detail events that did not happen, that were completely manufactured with the help of family members, and that would have been traumatic had they actually happened…
>
> Misinformation can cause people to falsely believe that they saw details that were only suggested to them. Misinformation can even lead people to have very rich false memories. Once embraced, people can express these false memories with confidence and detail.[25]

We tend to adapt to what (even if unconsciously) we think we should remember. This is what allows humans to recover from brain injuries or strokes, by taking on the memories of those around them as their own. In addition, if people are unsure of their own memories and perceptions they often strengthen them with the thoughts of those nearby (even if unconsciously). This is why we ask for advice. The same would be true in high-stress and emergency situations. If someone on the street frantically yelled, "GET DOWN ON THE GROUND," most people would immediately follow the instruction, especially if they thought that everyone else also was getting down on the ground. If, as members of a group, our memory isn't perfect or our perception of an event isn't perfect, we can adapt—but this also means that we can be fooled through being members of that group.

Now consider that we may be talking here about the biggest psyop of all time. The designers of this psyop know how the human psyche operates. They know that immediately after an enormous and traumatic event like 9/11, humans will need supplemental information and will ask questions. And, as it is often said that "first impressions stick."[26] As soon as they're given an answer, they will latch onto it and won't want to look back. After all, to look back could be to relive the most horrible moments in a person's life. How many people would want to do that? And in addition, to look back would contradict our natural tendency to adapt to our environment. So the first answers people are given tend to be blindly repeated over and over, self-reinforcingly. This is probably what brainwashing looks like, though it is important to remember that this pattern of behavior seems to be a typical aspect of human nature—which I believe has something to do with why this world-wide psyop worked so well.

After all, in the news videos from September 12, 2001 (discussed on page 297), ABC news reporters Peter Jennings, George Stephanopolis, and Robert Krulwich essentially asked, "Where did the towers go?" Why did these reporters stop asking this question soon after that? And why did the rest of us stop asking this question (at least out loud) as well?

Back to the thermal evidence. Consider the thermal images of Ground Zero shown in Figure 277. One image is said to have been taken on Sept. 16, 2001, and the second one a few days later, on Sept. 23.

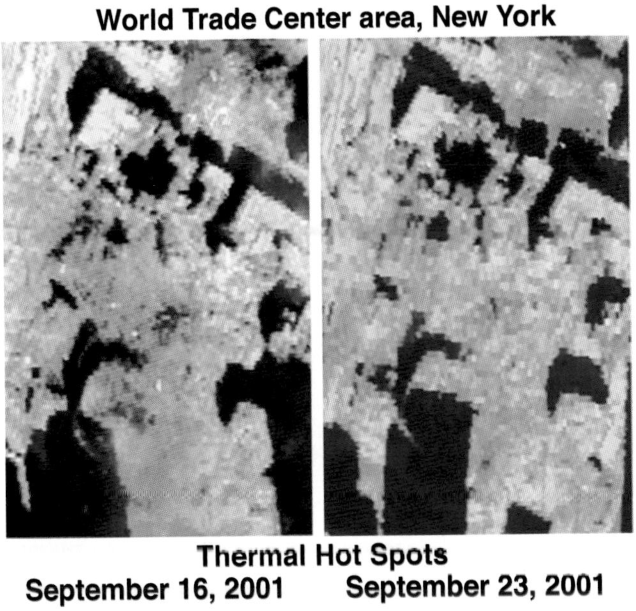

World Trade Center area, New York
Thermal Hot Spots
September 16, 2001
September 23, 2001

Figure 277. (9/16-23/01) What have been called hot spots show as orange and yellow areas. Dozens of hot spots are seen on September 16, but most had cooled or the fires had been put out by September 23.[27]
http://pubs.usgs.gov/of/2001/ofr-01-0429/hotspots-compare.jpg, http://pubs.usgs.gov/of/2001/ofr-01-0429/thermal.r09.html

If these hot spots at Ground Zero were "real," rescue workers, at 400-800 degrees Celsius, would have been instantly cooked, charred, and desiccated. A turkey cooks at 177 degrees Celsius (350 degrees F). How could anyone walk around on top of an 800-degree Celsius oven and live for more than a second or two?

Hot Spot	Kelvin	°C	°F
A	1000	727	1341
B	830	557	1035
C	900	627	1161
D	790	517	963
E	710	437	819
F	700	427	801
G	1020	747	1377
H	820	547	1017

Table 13. (9/16/01) Thermal Hot Spot Data.[28]

(a) (9/16/01) *(b) (9/11/01)*

Figure 278. (9/11/01) Hot spots under water with no boiling?
(a) Locations of some of the hot spots on 9/16/01.[29] *(b) West Street on 9/11, flooded from a broken water main.*
(a)http://pubs.usgs.gov/of/2001/ofr-01-0429/hotspot.key.tgif.gif, (b)http://digitaljournalist.org/issue0110/images/jn08.jpg

The photo in Figure 279 was taken at location "F" as shown in Figure 278a. There appears to be a puddle of water at the bottom of the debris-pit, and the people walking through it are obviously not cooked. It simply cannot be 800°F at this location.

Figure 279. (9/18/01) Basement of WTC2 two days after the September 16, 2001 thermal imagining.
http://www.photolibrary.fema.gov/photodata/original/3946.jpg

D. Steam Explosions

Figure 280. (7/18/07) Steam shot from the broken pipe in New York City. USA Today
http://i.usatoday.net/news/_photos/2007/07/18/steamx-large.jpg

Figure 280 is equally important, for if there *had* been molten metal on 9/11, there should have been a steam explosion with all of this water. Yet, as we see below, not only is there water, and little steam, but very live people walking around in the midst of it all.

Figure 281. (9/11/01) (a) Here is the southern "shore" after a water main broke. Note the "steam" appearance along the "northern shore" that cannot be the result of "hot water" because there are live people wading along that shore. (b) The person wading out into it does not look like a boiled chicken.
(a)http://www.magnumphotos.com/CoreXDoc/MAG/Media/TR3/F/P/Y/G/NYC14401.jpg,
(b)http://www.magnumphotos.com/CoreXDoc/MAG/Media/TR3/F/W/Q/1/NYC14164.jpg

Figure 282. (9/11/01) Wading through the "lake" over the reported hot spot.
(a) The wheatchex extending upward to the top of the picture appears to be the same one as is on the far right side of Figure 281a. (b) This photo was taken in the same area as Figure 282a but looking to the left, showing how far back the water goes.
(a)http://images30.fotki.com/v478/photos/1/115/34570/DSC_0287-vi.jpg,
(b)http://images30.fotki.com/v478/photos/1/115/34570/DSC_0290-vi.jpg

Again, in the pictures shown in Figure 283, if there really is molten metal present in the "raging fires" beneath the rubble, then where are the "steam explosions" resulting from water being applied? Similarly, if there were molten metal in the basements, more "steam" would be expected in wet weather than in dry weather. But that was not the case, as shown in Figure 283. Most everyone knows not to put water onto a grease or oil fire. Why not? If water is thrown onto hot oil, the water will instantly heat up and become steam, expanding in volume by about 1,600 times and producing an explosion. We need to ask why there are scientists and engineers who say that there was "molten metal at the WTC for weeks," declaring the molten metal also to be evidence of "a classic controlled demolition."[30]

The case is very clear. If there really were temperatures high enough to produce molten metal "for weeks," *especially* if water were added into the mix, it would have been deadly for *any* people to have been near these sites, let alone directly on them. And yet that is exactly what we see.

(a) (9/15/01) Dry *(b) (9/20/01) Wet*

Figure 283. (a) Dry weather. Note the appearance of "steam." (b) Wet weather. No "steam."
(a)http://www.photolibrary.fema.gov/photodata/original/3914.jpg, (b)http://www.photolibrary.fema.gov/photodata/original/3984.jpg

For a certainty, whatever this heat was, it was not ordinary heat. To see more evidence of why it can't have been "ordinary," let's take a look at the testimony of Renae O'Carroll, an emergency medical technician, who, along with her partner Eddie Rodriguez, arrived at the scene about 11 minutes before WTC2 went "poof." It is important to note that the two had driven over the Manhattan Bridge.

Emergency Medical Technician, Renae O'Carroll

> We went down Atlantic Avenue *going towards the Manhattan Bridge, and the heat was so intense, so intense, you could actually feel it while you were up on the bridge. I mean that intense you could feel the heat.* Cars were coming this way, and we were driving that way.
>
> They assigned us to go to Church and Vesey. We were going towards that area. The cars are coming this way. People are screaming and running, and we're going the opposite direction into the mess, into the belly of the beast of this thing. I still can't believe I had the nerve to do that to this day.
>
> We got down there, and the scene was just horrendous. I've never seen anything like this before in my life. You were kind of afraid, but you knew you had a job to do. You knew there were people in there.[31]

Now, note the *distance* between the Manhattan Bridge, where O'Carroll first noticed what was a "heat so intense, so intense, you could actually feel it while you were up on the bridge," and the World Trade Center, where the actual source of the heat is supposed to be.

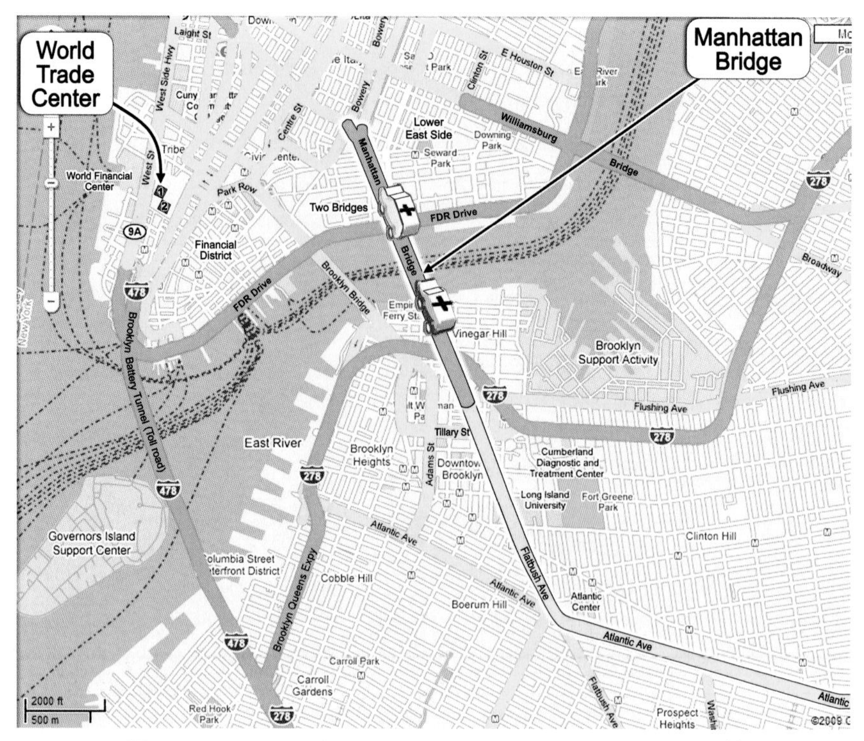

Figure 284. Route described by Renae O'Carroll. (maps.google.com)[32]

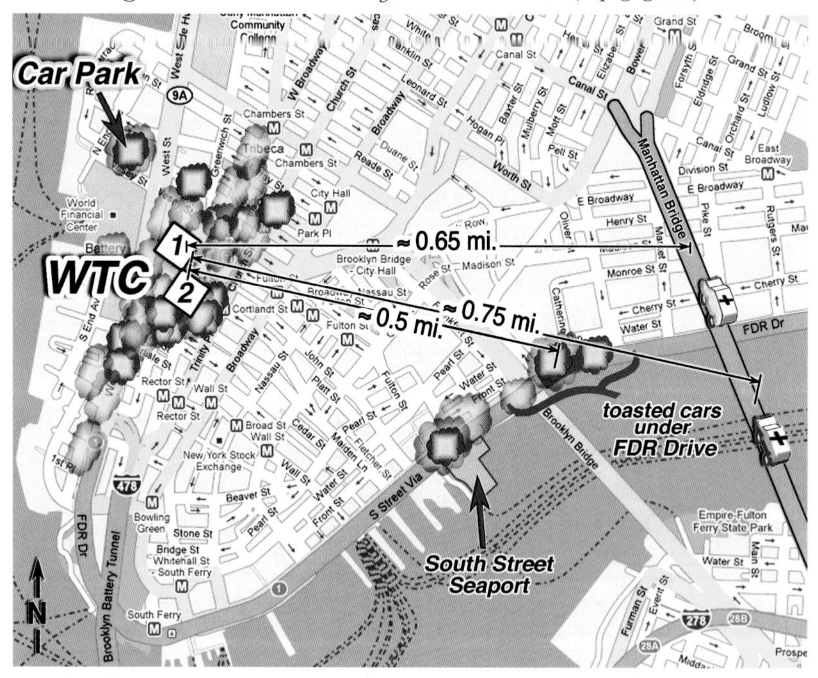

Figure 285. WTC to the Manhattan Bridge distance. (maps.google.com)[33]

Continuing with O'Carroll

Everything was pitch-black…

It was just basically dark. I had never been through anything—I thought I was dying. The only thing I could see was balls of fire, just balls of fire. *At one*

> *point I thought I was on fire because it was that close to me. I could feel the heat. I said to myself, wow, I'm on fire. This is what it feels like to be on fire. I don't know what it feels like to be on fire. I thought that's what it feels like to be on fire.*[34]

What was it that EMT Renae O'Carroll saw? If it really had been *balls of fire*, the paper in the streets would have ignited and the trees would have been burned. But there is no evidence of any burned trees in lower Manhattan that day. I have seen thousands of photographs taken of the area on September 11, 2001, and in the days that followed, and I have *yet* to see a photograph showing a burned tree. (Recall the earlier pictures we saw of "toasted cars" with unburned paper and trees nearby.)

Figure 286. Leafy green trees remained unburned along West Broadway.
(a)http://image03.webshots.com/3/3/40/70/21934070XFlNSYUYrp_ph.jpg,
(b)http://images30.fotki.com/v478/photos/1/115/34570/DSC_0092-vi.jpg,

Paramedic Robert Ruiz

Now consider paramedic Robert Ruiz's statements. Ruiz was on the ground directly in front of WTC2 when it went poof. Unable to get into any of the buildings from where he was, he crouched down in a doorway of a building.

> *But the biggest problem there was I couldn't breathe. The smoke was really, really hot, and I couldn't catch my breath. So I ended up taking off my* shirt and wrapping it around my face, trying to get some air, but I couldn't.
>
> What I remember is, oh, man, I'm going to die here and—who will protect my family. I was wondering how does it feel to be dead. That's what I remember thinking while I was stuck in that corner. I was just like I'm ready for it. I wasn't so calm like I am right now, but I remember thinking this while I was there.
>
> The only words I remember saying is, oh, my God oh, my God, over and over and over, and trying to pull the fence off so I could get inside, but there was no way. There was no way for me [to] open the door. It was locked. I was just trapped there. *So I was just waiting for the big impact that was just going to take me out.*
>
> *Miraculously the building stopped falling. The noise completely stopped. But the smoke kept coming. I was having a real, real bad time breathing. The shirt wrapped around my face wasn't helping at all, because it was saturated in that dust as well.*
>
> I'm saying, oh, my God, what do I do? Do I stay here? Do I try to run out? But then I thought if I run out of that little corner I'm going to get hit with the rest of the stuff that's flying. I was trapped there.
>
> *Like things weren't bad enough already, the car that's parked right on that corner catches*

on fire. I don't mean a little fire, the entire thing. Don't ask me how. *The entire* car *caught on fire. You would think maybe just a motor part or just the engine part. But this entire car just goes up in fire.*[35]

[. . .]

Now thanks to the car fire—*because it was so big, I could see now. Before then I couldn't see anything. It was literally like people say you stick your hand right in front of your face and you couldn't see it. Now with this* car fire, *I could see a few feet in front of me.*

I got out of the door well there, and I walked a little bit. I notice that there were windows there as well, but the windows had gates on them. *All the stuff that fell knocked the gates off the building. So the windows weren't there anymore, the gates weren't there.*

All I had to do was break a little bit of glass that was left there, and I ended up jumping inside the building, because you couldn't walk in the street. I was right along like the edge. It's really hard to describe. It was like a mountain of stuff.

I was walking right alongside the building, and I ended up going inside there. There was no gate. I tried to remember where were the gates, and I didn't see the gates at all. That whole corner was full of windows with gates.[36]

It is significant that Ruiz says that he was expecting the building to fall, yet, "miraculously," it did not. What *did* impact him was the spontaneous car fires and the lack of window gates that had prevented him from breaking in through the windows to find a safe refuge from what he thought would be a building falling on him.

Now look at Figure 287, of workers at Ground Zero.

Figure 287. (a) Steam? If this were as hot as a grill, these people would become something that looked more like a grilled-cheese sandwich. The Oxy-fuel torch, with the hose draped across the rubble from the pressurized tanks without exploding, is evidence contrary to a high-heat environment. (b) (9/21/01) An oxy-fuel line is draped across the ground. (Hose lightened to emphasize.)

(a)http://hereisnewyork.org//jpegs/photos/5103.jpg, (b)(9/21/01)http://www.photolibrary.fema.gov/photodata/original/4092.jpg,

As noted, if these fumes *were* steam from the "fires" raging below because of molten metal, then, first, we would see a steam *explosion* and not what appear to be simply non-scalding "fumes" of some sort wafting away. And, second, in either case, whether in the case of a steam explosion or in the simple case of workers standing in the *midst* of steam, they would be quite literally cooked and hydraulic grapplers would be rendered useless.

(a) (10/11/01) *(b) (10/13/01)*

Figure 288. (a) If this were steam, these workers would be cooked like steamed clams. (b) If this "pile" were 1,100°F, those grapplers would be fatally damaged before the operators had time to drive out there to the middle of the pile.

(a)http://img.timeinc.net/time/photoessays/groundzero/zero04.jpg, (b)http://www.photolibrary.fema.gov/photodata/original/5445.jpg

Fuel-torch hoses can be seen draped across the site in Figures 289 and 287, indicating a lack of dangerous heat. The WTC plaza level had been covered with cement blocks before 9/11, and the above photos dated 9/13/01 show it covered with dirt. A "collapse" does not cause a building to turn into dirt and neither does fire.

Figure 289. (9/13/01) (a) WTC plaza covered with dirt. (b) Fuel-torch hoses draped across the site. (Hose lightened to emphasize.)

(a)http://www.photolibrary.fema.gov/photodata/original/5313.jpg, (b)http://www.photolibrary.fema.gov/photodata/original/3885.jpg

We were told there were fires burning for 99 days and that fires burned underground. But do fires burn underwater and under wet dirt? Dirt was trucked in and watered down, yet fumes are still seen emerging from the saturated dirt shown in Figure 290.

Paramedic Joel Daniel Pierce

Paramedic Joel Daniel Pierce describes a scene that, in his own words, resembled Dante's *Inferno*, and yet he was not cooked:

> I remember at one point I was back down, I think I was down by—in front of Engine 10/Ladder 10—no. I was in front of Liberty Plaza and they said they needed morphine down there. They found somebody inside and they were going to have to take his legs off and they need morphine. They saw me and they said are you a medic? I go yeah. You got morphine? I go yup. The guy who was with

me, he said you're coming with me and they physically grabbed me. They said you're coming with us. They said you're coming with us, we need you now, we need your drugs.

At that point I found myself on the pile and I don't know if—it felt like I was hallucinating because just looking around, *I was up on top of that pile a good way in, it was hot and I was looking at all these holes down.* It was like one wrong step and I'm dead. If I fell into that little hole, I'm dead and that was it, because I could see the *flames below, you could see the redness.* I knew if I went in, it was like the pits of hell. It was like Dante's Inferno, I guess you could call it. I was in the remains of the south tower, between the south tower and near the hotel, and it was something, just being there, all the way in. I don't know how far in I was. I must have been a good 200 feet in when I was going up and down these piles of debris. Then I got called out and I learned that they managed to free the guy's legs up, so that made me happy. I got out of there. I climbed all the way back, assisted with the hose pulling because they needed a lot of fire hose in there. So I was part of the hose line with a lot of other firemen and all the rescue workers.[37]

Figure 290. (10/28/01) (a) This is the "basement" of WTC1. Where's the fire? (b) Why are the fumes coming out of wet soil?
(a)Original:http://www.photolibrary.fema.gov/photodata/original/5508.jpg,
(b)Original:http://www.photolibrary.fema.gov/photodata/original/5509.jpg

Deputy Chief Robert Browne, (Just after WTC2 Hole)

And I reported to the command post at that point.

Q. Where was the command post?

A. The command post was located on West Street, basically almost right in between Tower 1 and Tower 2 on the west side of the street over by the Winter Garden. That's where I—when I got into the command post, *Chief Gombo*, and *Chief Kowalczyk*—and *Chief Kowalczyk* appeared that he had just pulled up there himself, because he was just putting his coat.

I reported in and I asked them what they needed of me, what assignment they had for me, and they told me they wanted me to go to *Liberty* and *West Street* and run the operation at the corner of *Liberty* and *West Street.*

So with that I grabbed my aide, *Jason*, and we started heading back towards

> *Liberty* and *West Street.* We were walking back. We just left the command post. *There was a lot of debris coming down off the building, and I turned around to Jason and I told Jason, "Make sure you have your chin strap on you your helmet. Don't just have it sitting on your head. Make sure you secure it to your head," and with, that a large piece of debris was coming down,* sailing *off the building.* I remember looking up watching, because I was afraid that we were going to get hit with something, and it had to be almost the size of a *Volkswagen car,* a sheet of metal almost the size of a Volkswagen car, and it was—it was burned. It was glowing red, and it just landed in the street in front of us, maybe 20 feet in front of us. [38]

Figure 291a shows the approximate path of Deputy Chief Robert Browne on his way to the corner of West and Liberty Streets, shown in Figure 291b. WTC2 was still standing at the time Robert Browne crossed West Street and saw a piece of material land in front of him "the size of a *Volkswagen car*" that was glowing red. He describes it as "sailing off the building," but does not specify which building. Perhaps this was when the wheatchex stabbed into West Street. But why would it be glowing bright enough to be visible in bright sunlight?

(a) F *in the map in Figure 303.* *(b)* G *in the map in Figure 303.*

Figure 291. Photographs taken from (a) looking south, near the corner of West and Vesey Streets, showing the path of Deputy Chief Robert Browne with a glowing object superimposed and (b) looking northeast near the corner of West and Liberty Streets, where Robert Browne was going, before the 'collapse' of WTC2.
Figure 5-14. p. 101 (page 155 of 294 of pdf), http://wtc.nist.gov/NCSTAR1/PDF/NCSTAR%201-8.pdf,

Now let us consider at some length the testimony of emergency medical technician Patricia Ondrovic, who was there just as WTC2 went "poof".

Emergency Medical Technician, Patricia Ondrovic

> I saw a police captain that I knew, and he came out to me. He looked absolutely terrified, he was shaking, he was pale, he was sweating. I looked at him, I said what's wrong? He said there's another plane headed our way, and they just blew up the Pentagon. I said, another plane? What are you talking about? I hadn't realized that planes had hit this, I thought they just set bombs off. I didn't realize when I got there that planes hit it. I said, what do you mean another plane? He said two planes hit the World Trade Center. So I'm thinking a little Cessna. How can a little Cessna do all that damage? He said no, 757s. I said big things? See I was there for about 25 minutes before I knew that planes had crashed into this. We just got assigned to do stand-by. We didn't know what the stand-by was. I mean, who thinks something like that? You just think they hit it again. So I said,

what do you mean there's another one headed this way? He said, it's on the TV, there's a TV in there and it said that the Pentagon has been hit. Then we all went outside cause they had on the police radio that there was another plane headed in our direction, we all went outside and started looking up in the sky. Then the EMS captain said everyone grab your equipment, get to your vehicles and stay with your vehicles. My partner and I grabbed our stretcher, went to put it in the back of our vehicle, and at that time, I think it was the lobby of the building behind us blew out.

Everybody started running, I didn't see him again that day.

He got thrown one way, I got thrown the other way. I started running towards the West Side Highway, and there was another building on the corner, I guess it was a federal building, cause it was all the green and gray uniforms with the Smokey the Bear hats, the cops in there. I went to run in the lobby cause all of a sudden you couldn't see anything. *There was smoke, there was debris, there was everything flying around.* I ran into the lobby cause I had no idea what had happened and the cops that were in there were telling everybody get out, get out, get out. Where are you gonna go? Stuff's blowing up. So I ran back out and I started running west again. At that point, there was a car on the corner of I think I was here at that point, on the West Side Highway.

Q: West Side Highway and Vesey?

A: And Vesey, yeah. I was still on Vesey, cause the building that blew up on me was on Vesey, it was on the corner next to the West Side Highway. Cause I know I was running west, I didn't run that way. Thank God, I would have been dead had I run the other way. But I ran towards the West Side Highway, and I kept running up Vesey. As I was running up Vesey, the first car blew up on me on the corner of Vessey[sic] and the West Side Highway. *That set my turnout coat on fire, that set my hair on fire, and that set my feet on fire. I kept running. I got news for you, those turn out coats need to be called burn out coats, cause this thing caught up in flames.* They cut two inches off my hair in less that two minutes, my coat was completely engulfed, and *that was the only way I could see where I was running at that point, because I had a glow from my coat.* [39]

EMT Patricia Ondrovic Continues

There's hundreds of cops all running up there, and I ended up running through this park, and I couldn't even see where I was running anymore. I kept running North.

Q: Through North Park?

A: I guess that's North Park. It's a big green, grassy area, and there's nothing there. As I was running up here, two or three more cars exploded on me. They weren't near any buildings at that point, they were just parked on the street. The traffic guys hadn't gotten a chance to tow anything yet, cause this was all during the first hour I guess of this thing happening. So there were still cars parked on the street that were completely independent of that. Three cars blew up on me, stuff was being thrown. I went home all bruised that day. Thank God it was only bruises. I just ran into this park along with a bunch of other people, and stuff was still blowing up, I don't think I looked back, but you couldn't see anything, everything was just black. I was running and I was falling over people, cause people were crawling on the ground cause they couldn't see anymore. I just kept on running north. I could smell water, so I just kept on running towards the water, cause I knew that my coat was on fire, and I figured well, if I can see a boat over the water, I'm just gonna jump onto the boat and take that thing to Jersey, cause no one wants to blow up Jersey. Stuff is still blowing up behind me, as I'm running. I can hear stuff exploding. I could hear rumbling, the street under me was moving like I was in an earthquake. I've been in those, so I know what they feel like. It felt like an earthquake. [40]

Now let's look at the following map that tracks Ondrovic's path.

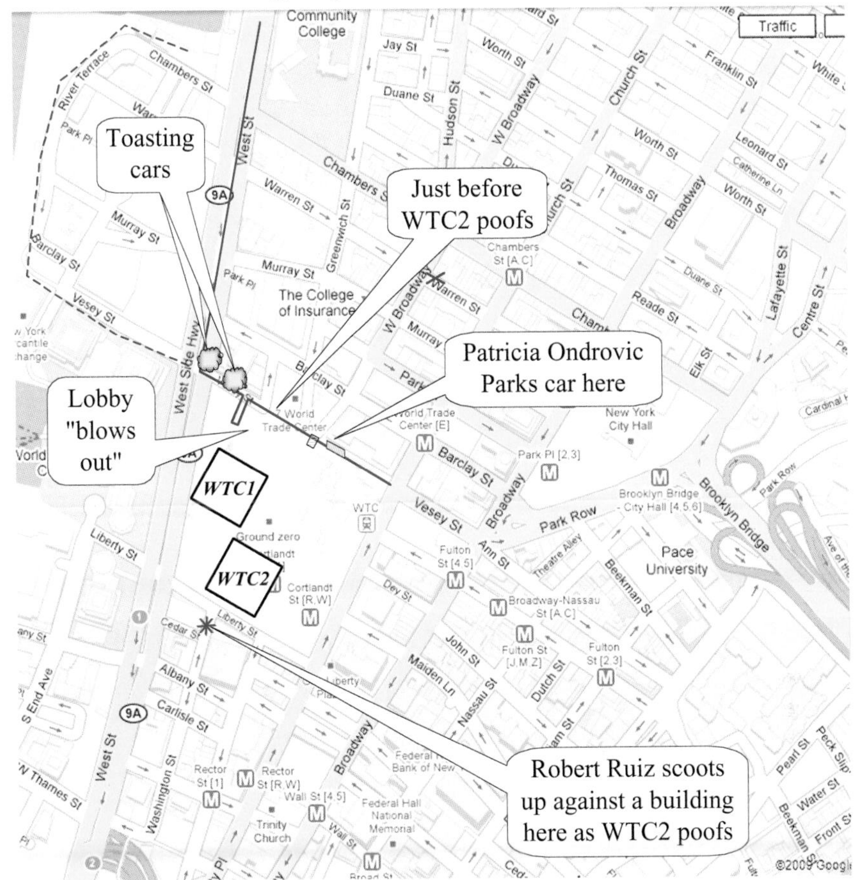

Figure 292. Path of Patricia Ondrovic. (maps.google.com)[41]

Ondrovic's testimony continues.

Q: Through North Park?

A: There was no where safe to go. As I was running north in this park, and then I could start seeing again a little bit, and I just kept looking in the sky. Cause the captain was saying there's another plane heading in our direction, I was looking for another plane. I saw something in the sky, it was a plane, but it was way out. It looked like it was over Jersey or something, then it wasn't there anymore. I saw a small fireball, and it was gone. I saw two other planes. One came in one way, and the other came in the other way, and there was a plane in the middle that was way far off in the distance. Then the plane in the middle just disappeared into a little fire ball. It looked like the size of a golf ball from where I could see it. And the other two planes veered off into opposite directions. I just kept on running north. [42]

About fifteen blocks later, I had no idea that that was just the first tower that had come down. I had no idea at that time that that's what that was, and the other buildings were being affected, of course, by that building falling. I found another ambulance, I believe it was the 08 Adam, because it was Valdivia and Jose Perez. Joseph Valdivia used to be my partner on tour 3, and when I finally caught up to them, I told them what was happening. I told them whatever you do, don't go back that way cause they just blew up triage. I thought that they blew up our triage sector, cause that's where the command was and everything. That was the only thing that I had to go by, everybody that was there was gone. Cause a couple of the police officers that are now missing are guys that I had known, and that's where they were. The paramedic from Cabrini, that's where he was. *I was just talking to him 20 minutes before everything blew up.* I don't know where he ran, I don't know if he ran the wrong way, cause I know I ran the right way. If you ran the opposite way, you were dead. As I got like 15, 20 blocks away, now I'm on the West Side Highway cause I came out of this park and I found the other ambulance. I saw my ex-partner, and I said get in this thing and drive it to Westchester. I told him get the hell out of the city. Get everyone was [sic] can get in this, I said shit's still blowing up down there. Whatever you do, don't go in that direction, start driving north.[42]

From Patricia Ondrovic's testimony, the question arises immediately: How could impacting airplanes, burning airplane fuel, or even thermite or controlled demolitions manage to catch Ondrovic's own clothing and hair on fire when she is clearly on the ground and several blocks away from the towers?

Again, in this connection, we must consider more anomalous vehicle fires.

Figure 293. (9/11/01) (a) A minivan is consumed by flames. The fire rages inside the vehicle, contained by the remaining windows. The fire appears more intense at the front where the engine is. (b) Image lightened to reveal the van in front, with fire on the right side.

(a)http://i24.photobucket.com/albums/c49/IgnoranceIsntbliss/911/BurnedCars/5772.jpg, (b)http://i24.photobucket.com/albums/c49/IgnoranceIsntbliss/911/BurnedCars/5772.jpg

Note the dust covering the paper. Flour is often used to smother kitchen grease fires, so why didn't the heavy dust coverage that blanketed all of southern Manhattan smother these fires? And, at the same time, why doesn't the *paper* burn?

Figure 294. View from location C *in the map in Figure 303.*
http://www.history.com/genericContent.do?id=60326#/trinity-church/
Trinity Church, Evan Fairbanks (just after WTC2 hole)(5:59)

Chief Louis Garcia struggled to explain the cause of these unusual car fires. He speculates that the cars were hit by falling debris. Falling debris would hit the *outside* of the car, first. The vehicle shown in Figure 293 (a and b), does not appear to have a hole in it showing where debris might have fallen inside the vehicle. Not only that, but the van in front of it has fire on the *side* of the vehicle, a strange place for a fire to start. There simply isn't evidence of large flaming debris having clobbered these vehicles.

Chief Louis Garcia

Q. Can you describe what you saw as you looked down West Street?

A. Down West Street, well, what I saw was two buildings down, smoke coming from two buildings. The cars that were in the parking lot, there was *a fire in the parking* lot in front of that bagel shop I was in. It was like a big, open, empty lot with cars in it. There was a fire that started there and it was spreading to other cars and once in awhile you'd have some explosions there and the fire actually after a period of time got pretty big because it was multiple cars. *Several cars were going at once. So some debris from the building was burning and must have landed there and started these cars on fire.* So you had this black smoke coming from these cars that were on fire, which no one paid attention to. We all said so what; let them burn themselves out. And you also heard what I thought were gunshots.[43]

There were anomalous burns and other damage not only to cars but to people as well. Firefighter James Curran, before the WTC2 was "impacted," had this to say:

Firefighter James Curran

We all jumped in the rig, geared up, drove down. We had a straight shot. All the cars were pulled off to the side and people were looking at the building. We went down, I believe we went down Lispenard to the West Side Highway. Made a U turn and we couldn't pull in any closer. We tried to get into the taxi indent, but there were people *on the ground burnt on the West Side Highway* to where you would have had to run them over to get any closer. So we got out of the rig, went in the lobby. Engine 7 pulled up right behind us and Ladder 1 was behind them I believe.

We went in through the revolving doors. There was a mini lobby. There was like brown haze, smoke in the lobby. *A lot of the* marble *slabs were falling off the wall, cracked. There were two people in like the little section of this lobby. One guy was burnt pretty much to a crisp and his jacket was the only thing left on him.* Put that out with a can and then there was a lady off to the right of us that was alive but she was screaming that she couldn't breathe. So I hit her with the can and cooled her down.

Q. Where did these people come from?

A. I don't know if they were in the elevator or what not, but *they were the only two people I saw in the lobby and they were right in the entranceway.* Like I said she *was still smoking* when we got in there. The other guy was dead and she was just screaming that she couldn't breathe.

After I used I guess about a half a can on her, we went through the lobby. All the elevator banks were kind of blown out at probably 70 degree angles, 60 degree angles, and there is all rubble and *spot fires* in the lobby.

Q. *Where did that come from?*

A. *I think they said that the fuel went all the way down the elevator shaft and when it finally hit rock bottom it blew out all the elevators.* [44]

If this is the "explanation," then it must have been the same airplane fuel that came down the elevator shafts that also managed to make its way over to the FDR Drive where it ignited and burnt the front ends of cars while not burning the back ends, melted door handles while leaving the upholstery, burned the insides of ambulances while leaving the external paint intact, and so on. Note carefully also that *the same airplane fuel apparently burnt one man to a crisp but, according to Curran, left his jacket more or less intact.* On the other hand, there *is* a mechanism that can explain all these examples of anomalous damage and burning. It is familiar to all of us who have used a microwave oven to cook food—that is, directed energy.

(a) (10/03/01) *(b) (9/21/01)*[45]

Figure 295. The "Cheeto" location.

(a)http://www.parrhesia.com/wtc/wtc070.jpg, (b)Photo credited to Frank Silecchia.[46],[47]

E. Fuming Dirt

Fumes from an unknown process emerged from the remains of the WTC for weeks and months and even years following 9/11, as will be discussed in Chapter 16. Toxic spills are often cleaned up with soil, peat, and/or clay.[48],[49] Dirt is not used to cover smoke from a fire.

(a) (10/03/01) *(b) (10/28/01)*

Figure 296. The fumes continued to emerge from the water-saturated soil.
(a)http://www.parrhesia.com/wtc/wtc070.jpg, (b)http://www.photolibrary.fema.gov/photodata/original/5508.jpg

Conventional "smoke" does not continue emerging from saturated dirt for months.

Figure 297. (10/31/01) Dirt was kept saturated with water, yet continued to fume.
A view out from the Century21 building (top of Figure 185a) is shown in Figure 405 on page 383.
http://www.w3.org/People/Jacobs/2001/10/wtc/pdrm1941.jpg

(a) (9/13/01) *(b) (9/20/01)*

Figure 298. WTC7 did not spill across the street, but turned to dirt?

(a)*http://www.hybrideb.com/source/eyewitness/complex/080.jpg, (b)http://forums.therandirhodesshow.com/index.php?act=Attach&type=post&id=19355*

(a) mid October 2001 *(b) (9/20/01)*

Figure 299. Con Edison transformer #7 (or #5) from debris pile of WTC7.[50]

(a)Figure A9, page 642 (pdf 304 of 382), http://wtc.nist.gov/media/NIST_NCSTAR_1-9_vol2_for_public_comment.pdf, (b)http://forums.therandirhodesshow.com/index.php?act=Attach&type=post&id=19357

(a) (9/11/01) *(b) (9/16/01)* 5 days *(c) (10/05/01)* 24 days

Figure 300. During the first few weeks of the "clean up" operation the amount of dirt appears to have grown.

(a)http://memory.loc.gov/service/pnp/ppmsca/02100/02110v.jpg, (b)http://www.photolibrary.fema.gov/photodata/original/3925.jpg, (c)http://www.photolibrary.fema.gov/photodata/original/5707.jpg

(a) (9/27/01) Dirt in *(b) (10/13/01) Dirt out*

Figure 301. Dirt was trucked in and out of the site for months and even years.
(a)http://www.photolibrary.fema.gov/photodata/original/5644.jpg, (b)http://www.photolibrary.fema.gov/photodata/original/5450.jpg

(a) (9/27/01) *(b) (10/13/01)*

Figure 302. Closer views of the images in Figure 301.
(a)http://www.photolibrary.fema.gov/photodata/original/5644.jpg, (b)http://www.photolibrary.fema.gov/photodata/original/5450.jpg

F. West and Liberty

Consider the comments of firefighter Richard Carletti.

Firefighter Richard Carletti

> At that point we stopped for a second and we heard impacts, which I guess was jumpers hitting the pavement. To our right, there was a parking lot right on West and Liberty. There were about seven cars on fire.[51]

The map below shows the location where Carletti saw the burning cars. Once again, the airplane fuel (or thermite, or demolitions explosives, or mini-nukes) is showing a remarkable ability to jump to various places around Ground Zero causing remarkable damage. Again, however, such regions or "zones" of damage are explicable as regions where directed energy interference could have been, or would have been, created.

Assistant Chief Fire Marshal Richard McCahey also attests to the anomalous nature of the "fires" that day. It is significant that, unlike Ondrovic, whose testimony we reviewed above, and O'Carroll, who reported the sensation of very high heat as far away as the Manhattan Bridge, McCahey, who was at the WTC complex itself, recorded something completely different.

Assistant Chief Fire Marshal Richard McCahey

> At that point, that's when I started to realize my mouth was filling up with like a sand ball. All of a sudden I realized when whoever said that, now I'm starting to pay attention to my surroundings. *I realized there was no heat, you could breathe. Stuff was going in your mouth but it was like a cool air coming* in when you breathe it, so I said maybe he was right. You couldn't see, it was gritty.[52]

View	Figure	Page	View	Figure	Page	View	Figure	Page
A	*305a*	292	**D**	*306*	293	**G**	*291b*	281
B	*305b*	292	**E**	*304*	291			
C	*294*	285	**F**	*291a*	281			

Legend 5. Legend of Figure views for the map in Figure 303.

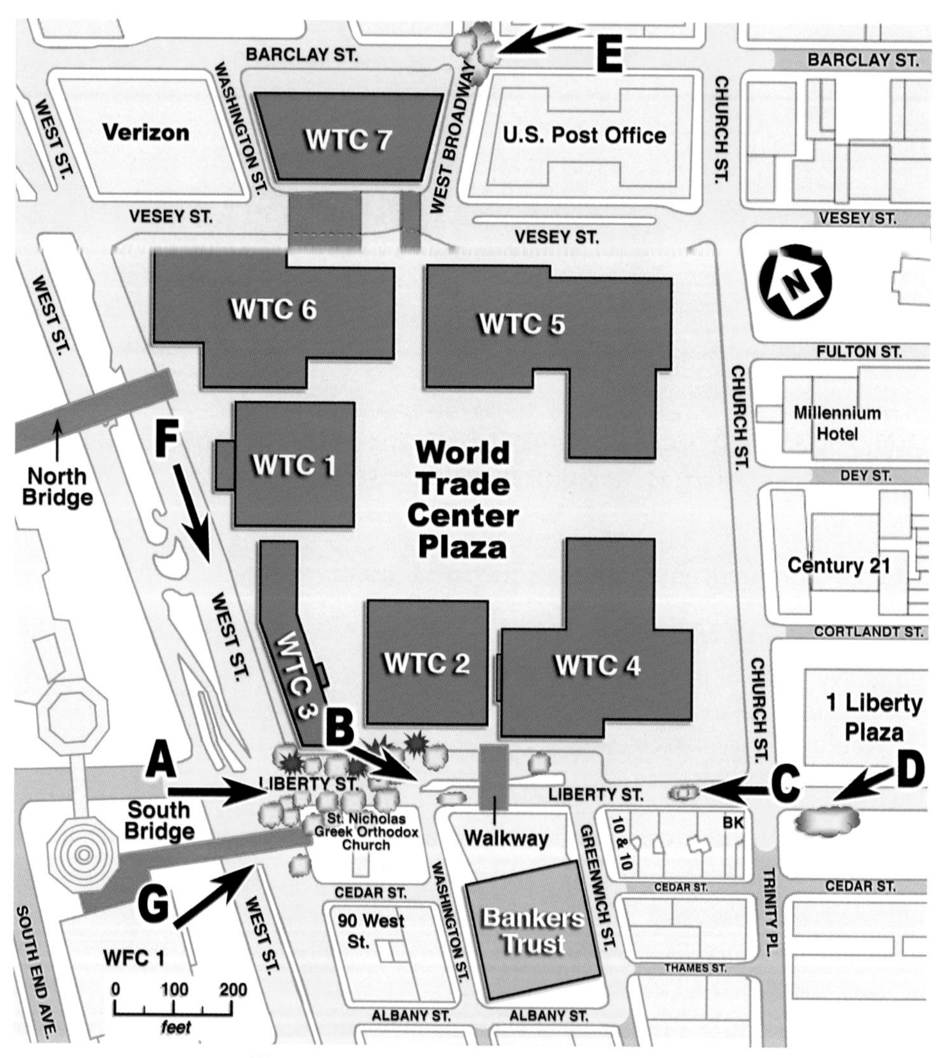

Figure 303. Image locations of the weird fires.
Map redrawn from NIST Report.[53]

What do we have thus far? We have, according to one person on site, a man "burnt to a crisp" yet with his jacket more or less intact, another person who states clearly that there was "no heat," and yet one more, on the Manhattan Bridge, who reports feeling high heat over a half a mile away from the WTC complex.

Clearly, airplane fuel, controlled demolitions explosives, mini-nukes and *thermite* do not explain these anomalies and disparities. But such things as regions of low and high heat, and as damage both to vehicles and people of the anomalous kinds we have seen—these phenomena *are* commensurate with interfered directed energy.

Firefighter Patrick Connolly

Firefighter Patrick Connolly, recording his observations just after WTC2 went "poof," notes that

> cars were—*cars with tires and cars were popping* and *they were just starting to light up spontaneously* and there was near zero visibility at this stage. It was better though than it was right after the collapse. And then we walked up two blocks. We walked up, there was a hot dog stand there. We broke the window in the hot dog stand, took bottles of water and we were washing our eyes, because our eyes were burning.[54]

Note, Connolly does *not* mention seeing any falling or burning debris striking the cars. They were simply "starting to light up spontaneously." An example of this is shown below in Figure 304. Connolly describes walking through this intersection as cars *"were popping* and *they were just starting to light up spontaneously..."*[54] [emphasis added]

Figure 304. View from location E on the map in Figure 303 (B in Figure 238 page 232). This is a frame taken from a video clip filmed near the corner of West Broadway and Barclay Street showing flames on a number of vehicles, one of the intersections Patrick Connolly walked through.
Figure 5–108. Page 194 (pdf page 238 of 404), http://wtc.nist.gov/media/NIST_NCSTAR_1-9_Vol1_for_public_comment.pdf

Armando Reno, another firefighter, makes the following observation.

Firefighter Armando Reno

> I was working by the south bridge. There were numerous car fires there. I was located by the south bridge and the chauffeur from 1 Engine was with me. There were two lengths of a 2-1/2 inch line stretched off the hydrant there on the south side of Liberty Street. *We were putting out the car fires, or attempting to, and there was no—the water had no effect on the car fires at the time.* I started thinking about getting the foam off the rig, and I also noticed there were numerous bodies by Cedar Street, and I was thinking of getting the EMS equipment off the rig, putting gloves on and starting to get the bodies, putting them in bags. Well, body pieces. [55]

Water failing to quench these car fires is a strong indicator that they are not ordinary fires. They are the result of another type of energy field, with effects that can be directed, or *directed energy.*

(a) 8:50 AM, location A. *(b) 8:51 AM, location B.*

Figure 305. Locations A and B, Figure 303 (page 290).
(a) The parking lot on West and Liberty, looking east on Liberty. (b) Liberty Street, facing Bankers Trust. (Deutsche Bank Building)
(a)http://www.swulinski.com/Images/USA/NY/NYC-Attack/NYC-Attack05-big.jpg, (b)http://wtc7lies.googlepages.com/Attack07.jpg/Attack07-full.jpg, http://www.swulinski.com/Images/USA/NY/NYC-Attack/NYC-Attack07-big.jpg

Consider the images in Figure 306, taken near the corner of Liberty Street and Church Street (which becomes Trinity south of this intersection). Aluminum melts at 660°C (1,220°F), and aluminum alloys melt at lower temperatures. The cladding of the WTC was made of an aluminum alloy. Not only is that cladding not melted, but the paper in the surrounding area is not burning.

(a) Plasma globe. *(b) (9/11/01) Location D of map in Figure 303.*

Figure 306. (a) Plasma globe.[56] (b) Near Church and Liberty Streets.[56]

(a)http://uw.physics.wisc.edu/~wonders/PlasmaB.jpg, (b)http://images30.fotki.com/v478/photos/1/115/34570/DSC_0194-vi.jpg,

G. Conclusions

Finally, what we have are several anomalies that the standard models—whether the official story of burning airplane fuel[57] or other theories of controlled demolitions by thermite[58] or even mini-nukes—fail to explain or account for. Bodies that burn to a crisp yet leave their owners' jackets visible, cars with front ends burnt but not the back ends, door handles disintegrated but upholstery intact, interiors of vehicles burnt but the exterior paint intact, vehicles burning near large amounts of loose paper that is not burning—all these phenomena are more in line with a force of some kind that is "cooking" things from the inside out. All these phenomena are in line with the deployment of an exotic directed energy technology, and a very sophisticated one at that.

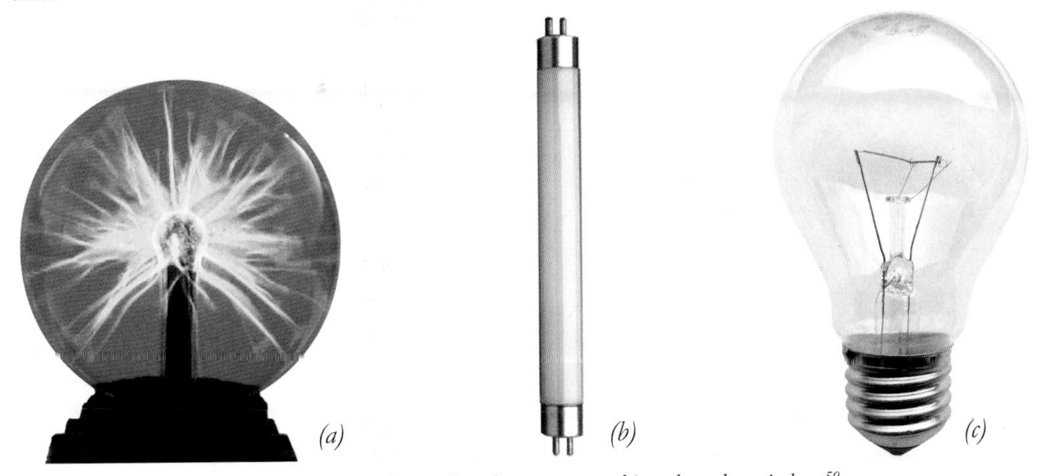

Figure 307. Hot things glow, but not everything that glows is hot.[59]

(a) http://uw.physics.wisc.edu/~wonders/PlasmaB.jpg, (b)redrawn with photo, (c)http://upload.wikimedia.org/wikipedia/commons/b/b4/Gluehlampe_01_KMJ.png,

---

[1] File No. 9110100, "Chief Jerry Gombo" October 17, 2001, p 17, *http://graphics8.nytimes.com/packages/pdf/nyregion/20050812_WTC_GRAPHIC/9110100.PDF*, emphasis added.

[2] Undisclosed Source (1), personal contact

[3] *http://graphics.nytimes.com/packages/audio/nyregion/20060410_MELISSADOI.mp3*

[4] *http://www.answers.com/topic/9-11-dispatcher-transcript*

[5] Brendan Casey, "Hydraulic Equipment Reliability: Beyond Contamination Control". Machinery Lubrication Magazine. July 2005, *http://www.machinerylubrication.com/article_detail.asp?articleid=772* (accessed 6/16/07)

[6] Although not conclusive

[7] EVOs and the Hutchison Effect, Nuclear Transmutation from Low-Voltage Electrical Discharge, Paper Presented at the MIT Cold Fusion Conference, May 21, 2005 by Ken Shoulders, *http://www.svn.net/krscfs/EVOs%20and%20Hutchison%20Effect.pdf*

[8] Quote by Richard Sparks, Scientific and Technical Intelligence/SBIR, Ottawa. *(1996)* in *The Hutchison File, (page 67 of 87)*, *http://drjudywood.com/pdf/HutchisonEffectReport_txt.pdf,*

[9] Quote by Richard Sparks, Scientific and Technical Intelligence/SBIR, Ottawa. *(1996)* in *The Hutchison File, (page 67 of 87)*, *http://drjudywood.com/pdf/HutchisonEffectReport_txt.pdf,*

[10] *http://doujibar.ganriki.net/english/e-05c-moltenmetal.html*

[11] The original image is no longer available: (*http://web.archive.org/web/20060116114119/www.wtcgodshouse.com/images/steel.jpg%22*) Alternate sites: *http://img201.imageshack.us/img201/4509/hotslagil3.jpg, http://911myths.com/assets/images/molten_steel.jpg,*

[12] *http://www.911myths.com/html/wtc_molten_steel.html*

[13] Brendan Casey, "Hydraulic Equipment Reliability: Beyond Contamination Control". Machinery Lubrication Magazine. July 2005, *http://www.machinerylubrication.com/article_detail.asp?articleid=772* (accessed 6/16/07)

[14] *Ibid.*

[15] Griffin, David Ray, Debunking 9/11 Debunking: An Answer to Popular Mechanics and Other Defenders of the Official Conspiracy Theory, Northampton MA: Olive Branch Press, 2007 (page 314).

[16] *Ibid*, (page 315).

[17] Jim Dwyer, U.S. moves to debunk 'alternative theories' on Sept. 11 attacks, SEPTEMBER 2, 2006, *http://www.iht.com/articles/2006/09/01/news/conspiracy.php?page=1, http://www.iht.com/articles/2006/09/01/news/conspiracy.php?page=2*

[18] *Ibid.*

[19] *Ibid.* (pages 181-182).

[20] Brendan Casey, "Hydraulic Equipment Reliability: Beyond Contamination Control". Machinery Lubrication Magazine. July 2005, *http://machinerylubrication.com/article_detail.asp?articleid=772* (accessed 6/16/07)

[21] *http://forums.randi.org/showpost.php?p=4521000&postcount=1751, http://groups.google.com/group/alt.a...e?dmode=source*

[22] *http://www.youtube.com/watch?v=MDuBi8KyOhw*

[23] *http://drjudywood.com/articles/short/misinfo.html*, Elizabeth F. Loftus, Hunter G. Hoffman, "Misinformation and Memory, The Creation of New Memories," *Journal of Experimental Psychology: General* 118(1):100-104 (March 1989). *http://faculty.washington.edu/eloftus/Articles/hoff.htm*

[24] Elizabeth F. Loftus, Hunter G. Hoffman, "Misinformation and Memory, The Creation of New Memories," *Journal of Experimental Psychology: General* 118(1):100-104 (March 1989): "Misleading information presented after an event can lead people to erroneous reports of that misinformation. Different process histories can be responsible for the same erroneous report in different people. We argue that the relative proportion of times that the different process histories are responsible for erroneous reporting will depend on the conditions of acquisition, retention, and retrieval of information. Given the conditions typical of most misinformation experiments, it appears that misinformation acceptance plays a major role, memory impairment plays some role, and pure guessing plays little or no role. Moreover, we argue that misinformation acceptance has not received the appreciation that it deserves as a phenomenon worthy of our sustained investigation. It may not tell us anything about impairment of memories, but it does tell us something about the creation of new memories. The collection of experiments seem to be teaching us an important lesson: When people do not have an original memory, they can and do accept misinformation

and adopt it as their own memory. However, it also appears that misinformation can sometimes impair an otherwise accessible original memory. But this conclusion leaves us with many unanswered questions. How much impairment occurs? What does it mean to say that memory has been impaired? Is it the memory traces themselves that are impaired, or is it our ability to reach those memories?" *http://faculty.washington.edu/eloftus/Articles/hoff.htm*

[25] "Planting misinformation in the human mind: A 30-year investigation of the malleability of memory," *LEARNING & MEMORY* 12:361-366, ISSN 1072-0502/05, Elizabeth F. Loftus. (July 18, 2005), *http://www.learnmem.org/cgi/content/full/12/4/361?*

[26] Max Weisbuch, *Why do first impressions stick with us so much?,* September 9, 2009, http://tuftsjournal.tufts.edu/2009/09_1/professor/01/: "Psychologists have known for some time that expectations influence visual perception. For example, whether you see a '13' or a 'B' in response to seeing the set of characters '13' depends on your expectations. Still, even for those of us who are psychologists, it can be very difficult to disbelieve something we see with our own two eyes. Hence, rejecting a first impression can require a very difficult act–to literally 'unsee' what seems like ongoing evidence in support of that first impression.... Beyond behavior interpretation, memory is surprisingly fragile and subject to interpretation. Even those memories that seem most real ("flashbulb memories") are vulnerable to bias. For example, well-intentioned people can honestly but falsely remember being sexually abused as a child. It is hardly surprising that more mundane memories of others' behavior can be biased."

[27] *http://pubs.usgs.gov/of/2001/ofr-01-0429/thermal.r09.html*

[28] Derived from *http://pubs.usgs.gov/of/2001/ofr-01-0429/thermal.r09.html*

[29] *http://pubs.usgs.gov/of/2001/ofr-01-0429/thermal.r09.html*

[30] *http://www.ae911truth.org/ Architects & Engineers for 9/11 Truth!,* /(check list of CD features on the far right) *http://www.ae911truth.org*: "13. Tons of molten Metal found by FDNY under all 3 high-rises (no other possible source other than an incendiary cutting charge such as Thermate)."

[31] File No. 9110116, "EMT Renae O'Carroll," October 18, 2001, p. 3, *http://graphics8.nytimes.com/packages/pdf/nyregion/20050812_WTC_GRAPHIC/9110116.pdf*

[32] Supplemented map of original from maps.google.com

[33] Supplemented map of original from maps.google.com

[34] File No. 9110116, "EMT Renae O'Carroll," October 18, 2001, pp. 7-8, *http://graphics8.nytimes.com/packages/pdf/nyregion/20050812_WTC_GRAPHIC/9110116.PDF*. emphasis added.

[35] File No. 9110333, "Paramedic Robert Ruiz," December 14, 2001, pp. 15-18, http://graphics8.nytimes.com/packages/pdf/nyregion/20050812_WTC_GRAPHIC/9110333.pdf

[36] File No. 9110333, "Paramedic Robert Ruiz," December 14, 2001, pp. 17-18, http://graphics8.nytimes.com/packages/pdf/nyregion/20050812_WTC_GRAPHIC/9110333.pdf

[37] File No. 9110845, "Paramedic Joel Daniel Pierce," January 23, 2002, pp. 15-17, http://graphics8.nytimes.com/packages/pdf/nyregion/20050812_WTC_GRAPHIC/9110845.pdf, emphasis added.

[38] File No. 9110155, "Robert Browne," October 24, 2001, pp. 5-6, *http://graphics8.nytimes.com/packages/pdf/nyregion/20050812_WTC_GRAPHIC/9110155.pdf*

[39] File No. 9110048, "EMT Patricia Ondrovic," October 11, 2001, p. 4-5, *http://graphics8.nytimes.com/packages/pdf/nyregion/20050812_WTC_GRAPHIC/9110048.pdf*

[40] File No. 9110048, "EMT Patricia Ondrovic," October 11, 2001, p. 5-6, *http://graphics8.nytimes.com/packages/pdf/nyregion/20050812_WTC_GRAPHIC/9110048.pdf*

[41] Supplemented map of original from maps.google.com

[42] File No. 9110048, "EMT Patricia Ondrovic," October 11, 2001, p. 6-8, *http://graphics8.nytimes.com/packages/pdf/nyregion/20050812_WTC_GRAPHIC/9110048.pdf*

[43] File No. 9110002, "Chief Fire Marshal Louis Garcia," October 2, 2001, pp. 23-24, *http://graphics8.nytimes.com/packages/pdf/nyregion/20050812_WTC_GRAPHIC/9110002.pdf*

[44] File No. 9110412, "Firefighter James Curran," December 30, 2001, pp. 3-4, *http://graphics8.nytimes.com/packages/pdf/nyregion/20050812_WTC_GRAPHIC/9110412.pdf*

[45] *http://doujibar.ganriki.net/english/e-05c-moltenmetal.html*

[46] The original image is no longer available: (*http://web.archive.org/web/20060116114119/www.wtcgodshouse.com/images/steel.jpg%22*) Alternate sites: *http://img201.imageshack.us/img201/4509/hotslagil3.jpg, http://www.physics.byu.edu/research/energy/hotSlag.jpg, http://911myths.com/assets/images/molten_steel.jpg,*

[47] *http://www.911myths.com/html/wtc_molten_steel.html*

[48] *http://www.peatsorb.com/*

[49] University Of Florida (2005, April 1). Right Blend Of Microbes And Plants Can Clean Up Toxic Spills. *ScienceDaily*. Retrieved March 30, 2010, *from http://www.sciencedaily.com-/releases/2005/03/050329134209.htm*

[50] Source for NIST report: G&S Technologies, reproduced with permission

[51] File No. 9110419, "Firefighter Richard Carletti," January 2, 2002, p. 5,*http://graphics8.nytimes.com/packages/pdf/nyregion/20050812_WTC_GRAPHIC/9110419.pdf*

[52] File No. 9110191, "Assistant Chief Fire Marshal Richard McCahey," November 2, 2001, pp. 14-15, *http://graphics8.nytimes.com/packages/pdf/nyregion/20050812_WTC_GRAPHIC/9110191.pdf*, emphasis added.

[53] Map redrawn from p. 3 (pdf p. 53 of 298), *http://wtc.nist.gov/NISTNCSTAR1CollapseofTowers.pdf*

[54] File No. 9110453, "Firefighter Patrick Connolly," January 13, 2002, pp. 10-11, *http://graphics8.nytimes.com/packages/pdf/nyregion/20050812_WTC_GRAPHIC/9110453.pdf*

[55] File No. 9110448, "Firefighter Armando Reno," January 13, 2002, p. 3, *http://graphics8.nytimes.com/packages/pdf/nyregion/20050812_WTC_GRAPHIC/9110448.pdf,* emphasis added.

[56] (a) Color adjusted, (b) (brightness and contrast adjusted)

[57] NIST NCSTAR 1 – Final Report on the Collapse of the World Trade Center Towers, *http://wtc.nist.gov/reports_october05.htm*

[58] thermite, super-thermite, thermate, nano-enhanced thermite, spray-on thermite, nano-thermite

[59] (a) Color and intensity adjusted

14.
Dust Cloud Rollout

It is the framework which changes with each new technology and not just the picture within the frame. —Marshall McLuhan

We live in a Newtonian world of Einsteinian physics ruled by Frankenstein logic. —David Russell

A. Introduction

Figure 308. The towers turn to dust.
http://911research.wtc7.net/wtc/evidence/photos/docs/nt_dust_aerial2.jpg[1]

We remember the huge volume of dust that rolled through New York City on 9/11. How could we forget? It left dust and paper everywhere. It rolled out and it rose up on a path that went up into the atmosphere and down the eastern seaboard. Dust covered southern Manhattan as if there had been a snowstorm. But often overlooked is the amount of dust that went not down but up, perhaps even into the upper atmosphere…for months. As soon as the air began to clear, we could see that very little of the material of the buildings had been left behind. News reporters began to ask, "Where did the Towers go?"

ABC news, Peter Jennings and George Stephanopolis wondered where the Towers had gone, noting the *lack* of visible debris.[2] ABC reporter Robert Krulwich described the solid material of the building as having turned to dust.

Reporter Robert Krulwich

> Engineers, at the firm that built the buildings, best guess to account for the missing 1200 feet of material from each tower, is t*hat large portions simply vaporized into the dust that rained down on New York immediately after the collapse.* It was that powerful. We're talking here about *43,600 windows, 600,000 sq. ft. of glass, 200,000 tons of structural steel, 5 million sq. ft. of gypsum, 6 acres of marble, 425,000 cubic yards of concrete,* turned, in good part, into a cloud, says Environmental Medical Doctor, Dr. Stephen Levin [from Mt. Sinai Hospital].[3],[4]

Dr. Stephen Levin then speaks almost in disbelief,

> I was astonished at the degree to which solid materials were turned into pulverized dust as a consequence of that building collapse! I think it was striking!

Robert Krulwich concludes his report, showing a dust cloud rolling up to, but not continuing past, the pedestrian bridge by Stuyvesant High School, north of the WTC.

Reporter Robert Krulwich's conclusion

> But most interesting, in the mix, they are looking, they think, at specks of steel that used to be beams and elevators, marble from the lobby floor and facings. So what were once the strongest architectural elements in the two towers were pulverized, large portions turned into clouds, like this one.
>
> *Still there is this mystery, if some of the hardest materials were vaporized, how to account for the presence everywhere of paper, fully intact letters, business, business forms, stationary.* Paper is so fragile, and combustible, yet somehow, maybe because we have so much of it, it was everywhere.[3,4,5]

Indeed, how *do* we account for the presence of so much paper when nearly everything else was turned to dust? Studying the process that resulted in the WTC turning to dust will reveal the mechanism of destruction. The question to be answered—and already hinted at—is "What can turn a 110 story building into powder in mid-air?"

The photograph in Figure 309 appeared in newspapers and magazines, capturing the horror of that day. Yes, it was horrible, and those of us who saw this photo on the front page of a newspaper the next day will likely have a conditioned response to the image. Conditioning of that kind is human nature and a part of our survival instinct. The sooner we recognize danger, the sooner we can run from it. On the other hand, as the disaster drew away into the past, it became easier to think of the image simply as a photograph *of* an historical event, itself posing no danger. Realizing this allowed me to look beyond the conditioned response to see what is really there. So, let us look beyond the foreground and into the background of this picture.

We see a huge wall of a dust cloud chasing people eastward on Fulton Street. The dust cloud is a uniform color and it stretches from the ground to the height of the picture. A pristine-looking WTC7 stands in the background.

Figure 309. (9/11/01) A view west along Fulton Street.
http://digitaljournalist.org/issue0110/original_images/Associated%20Press/apgsamoilova_WORLD_T_004JM.JPG, http://the.honoluluadvertiser.com/dailypix/2001/Sep/11/running_b.jpg

B. Cool Dust Cloud

The dust cloud overtook these people in Figure 309. They weren't able to outrun it, and, if this dust cloud had been very hot, we would have seen piles of burned bodies littering the streets. But there are no reports of burned bodies from the dust cloud. After the cloud passed, people stumbled out of their hiding places or crawled out from under cars, and they were alive. They were covered with dust; they were not burned (see Figure 310). We didn't see or hear of any streets filled with burned bodies. The grass, shrubs, and leaves on the trees remained green even after being exposed to the dust cloud. So there is no evidence that the cloud was hot.

In fact, there were reports that the dust cloud, initially, felt cooler than ambient temperature.[6] There were also reports that the dust on the skin was highly irritating and gave the sensation of being hot. It is likely that the dust was corrosive, but it could not have been hot. It didn't burn paper, even though it gave a burning sensation when it came into contact with skin.[7] It was also described by some as itchy.[8]

Figure 310. (a) Covered in dust. (b) Woman caught on the street as the dust cloud rolled by. (c) Just dust.
(a)http://hereisnewyork.org//jpegs/photos/6213.jpg, (b)Photo by Stan Honda/AFP, http://digitaljournalist.org/issue0110/original_images/AFP/10%20yellow%20woman.jpg, (c)http://digitaljournalist.org/issue0110/original_images/Reuters/rtr10.jpg

"A woman covered in dust takes refuge in an office building after one of the World Trade Center towers 'collapse'. The woman was caught outside on the street as the cloud of 'smoke' and dust enveloped the area."[9]

EMT Michael Ober did not note whether the dust was hot or cool, but only that it did not contain huge chunks, but "just dust." Neither did he recall hearing the sound of the building hitting the ground.[10]. As he experienced the event, it was as if the entire building had turned to dust.

C. Cloud Chase

Figure 309 shows the dust rollout from the destruction of WTC2. The photograph in Figure 311 was taken just after the one in Figure 309 and from nearly the same location on Fulton Street. The dust cloud is expanding so energetically that it appears to have "turned the corner" toward the east, even though there was nothing obstructing its straight-northward motion. The dust cloud was expanding laterally down Fulton Street, chasing pedestrians as they tried unsuccessfully to outrun it.

The cloud-chase photographs shown in Figures 309 and 311 were taken from the approximate location shown on the map in Figure 312. The letters on the map correspond to the figure-number of the photo from that location.

Figure 311. Pedestrians run from the scene as one of the World Trade Center Towers collapses. Photo by Doug Kanter/AFP.
http://digitaljournalist.org/issue0110/original_images/AFP/7%20tower%20collapse.jpg

View	Figure	Page	View	Figure	Page	View	Figure	Page
A	*309*	299	**D**	*314*	304	**G**	*316*	306
B	*311*	301	**E**	*315a*	305	**H**	*134*	140
C	*313*	303	**F**	*315b*	305	**H***	*353*	336

Legend 6. Legend of Figure views for the map in Figure 312.
H The image shown in Figure 353 (page 336) was taken from just to the right of the image shown.*

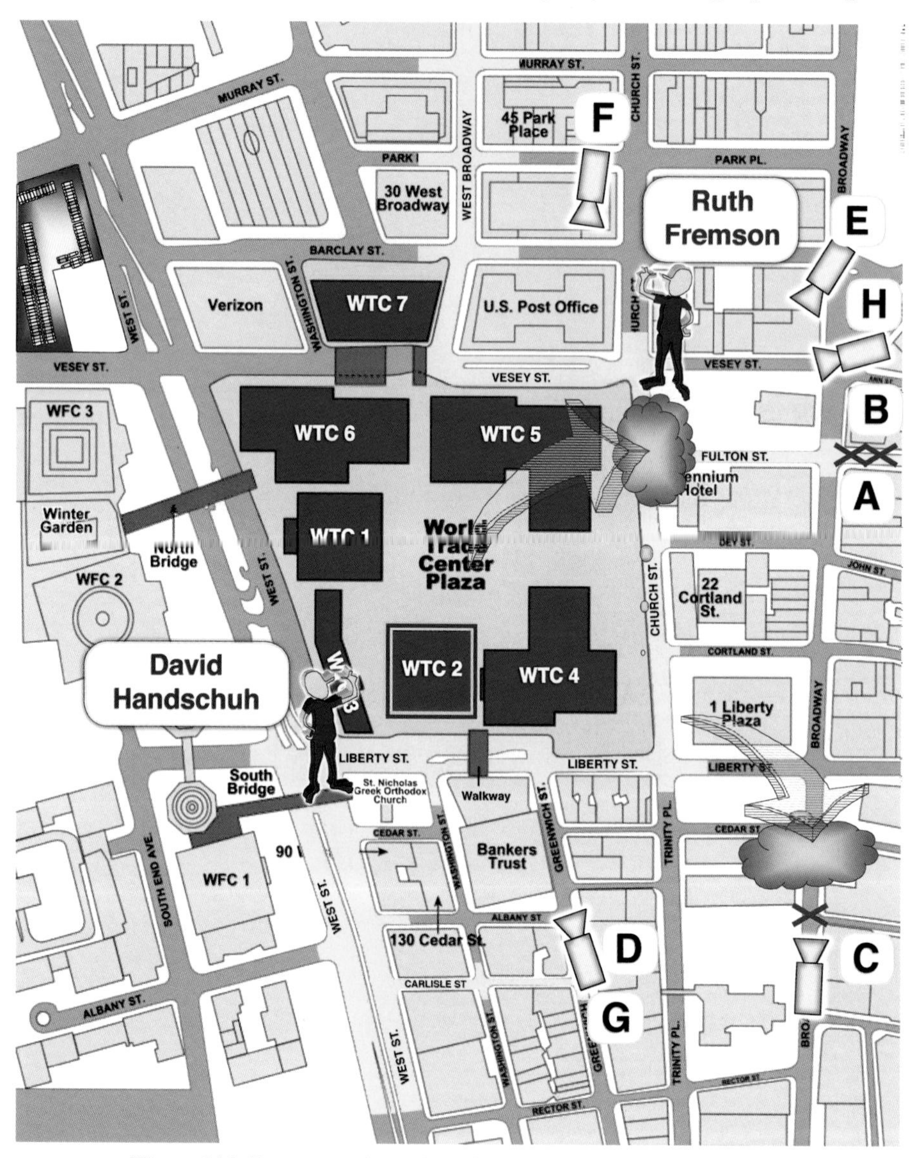

Figure 312. Locations of people and images referenced in this chapter.
Edit from p. 3 (pdf p. 53 of 298), http://wtc.nist.gov/NISTNCSTAR1CollapseofTowers.pdf

Figure 313 shows the vertical wall of dust smashing against buildings and turning south onto Broadway. That the dust-wall is high and vertical is very significant. If the building had in fact collapsed and smashed to the ground, dust would have been shot out at ground level first. A "collapsing" building is not a rigid structure, but an unknit and falling one. Here, however, we have a wall of dust at least ten stories high, and, further, the cloud is at the same distance away from the camera at its ten-story height as it is at street level. In a real "collapse," dust is generated at ground level and then billows *up*.

Figure 313. (9/11/01) A wall of dust moves like a plunger down Broadway toward Pine Street.
http://digitaljournalist.org/issue0110/original_images/Reuters/rtr08.jpg

Figure 313 looks north, up Broadway and across Pine Street toward the site of WTC2. The phenomenon of an enormous, towering dust cloud is not consistent with a building that has collapsed to the ground. It is consistent, however, with a 1368-foot building that is being turned to dust from the top down.

D. WTC2

How did the building's destruction begin? Exactly what happened in the moments before the huge wall of dust cloud blossomed?

The photo in Figure 314 reveals several very strange anomalies. The view is northward, toward the south face of WTC2. If the picture included the lower floors, we would see Bankers Trust in the foreground. On the south face is an extraordinarily odd formation that looks like a giant donut, covering the entire 200-foot width of the building. The middle of this weird formation looks like a black hole, a very dark and deep-looking region in the middle. Tubular protrusions coming out of the center look like fat worms and are lumpy.

The *southeast* corner of WTC2 at this moment is disjointed, with two kinks in the vertical structure of the building. A few stories below the very top of the building, there is a kink in the vertical line of the building, as if the very top floors are beginning to tip southward, independent of whatever is going on below them. There can't have been any sudden additional load on these floors, nor should there have been any prior weakening of them, since they are well above the floors that were initially damaged. So why would the building be fracturing there? This upper kink (marked 1st kink) is at about the 105th floor, while the initial damage occurred between floors 77 and 85. Also, what logical reason could there be for the 105th floor and the 78th floor, where the lower kink is (marked 2nd kink), to fail at exactly the same moment?

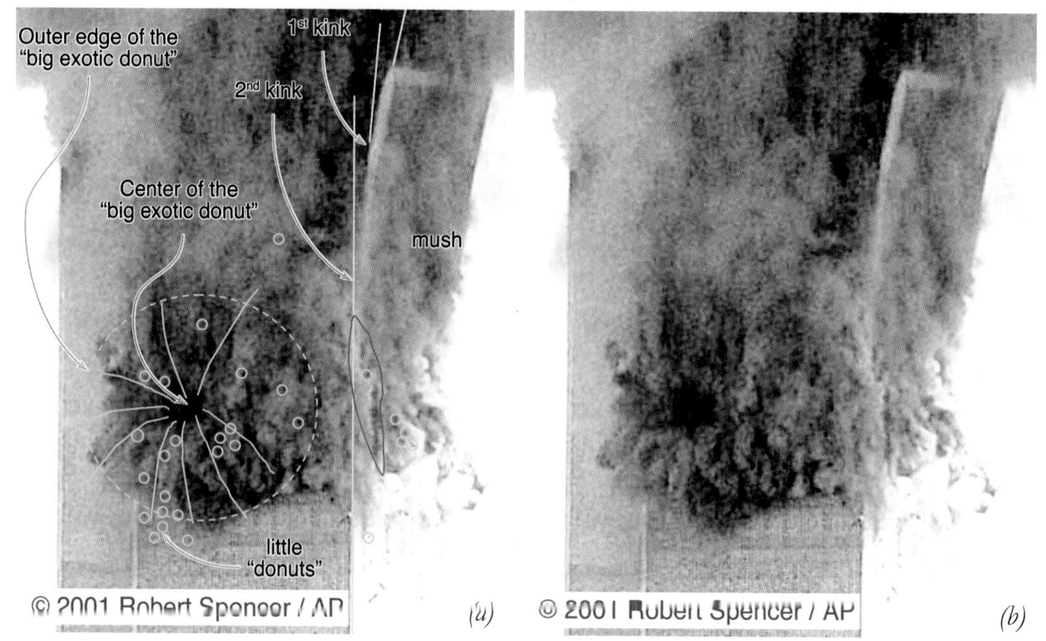

Figure 314. The WTC2 "exotic donut" viewed from the south.
NCSTAR1-6, page 183 (pdf 265), http://wtc.nist.gov/NISTNCSTAR1-6.pdf

The east face of WTC2, the right side of the building as shown in Figure 314, appears to be dissolving or turning to mush. The whole top third of the building appears to be tipping to the east, while the top few floors are tipping more to the south and east. Distinct parts of the top section are therefore moving in different directions, implying that the building is coming apart. That is, the top section is no longer part of a rigid body and is not even itself tipping as a unit.

The leaning top of the building appears to hinge at the 78th floor, along the east face, just above the 75th and 76th floors, the mechanical floors.

Photographer David Handschuh

David Handschuh, a photographer who was standing on Liberty Street at the time, in view of the *southwest* corner of WTC2, described "*the whole building…just disintegrating.*" He first described when the South Tower got its hole:

> *And then, this noise filled the air that sounded like a high-pressure gas line had been ruptured. And it seemed to come from all over, not from any one direction. And everybody's looking around like "what's that?" And then all of a sudden the second tower explodes.* That was the second aircraft plowing into the South Tower.[11]

Then shortly after that, David Handschuh describes the building as *"just disintegrating."*

> *I was just walking by myself and heard...another loud noise that kind of echoed the first noise... Looked up and [thought] OK, it's another aircraft...and the whole building was just disintegrating.* The building was starting to come down. My initial reaction was to grab my camera and to hold it up... My initial reaction was to start taking pictures. But somewhere in the back of my mind came a voice that said, "Run. Run, run, run." And in doing this for more than twenty years I have never run from anything. But there's no doubt in my mind that if I had not listened to the voice in my head that said "run," if I would have stopped and taken the pictures, I would not have been here, today.[12]

Photographer Ruth Fremson

Ruth Fremson, another photographer, was on the opposite side of WTC2 (near Church and Vesey), and in view of the northeast corner of WTC2. Her words:

> And I must have been there, ...it couldn't have been more than five minutes before the first building started to come down. I think it was probably less. I think it was two or three minutes of shooting and I heard this rumbling, and I thought it was a third plane coming to hit the trade centers again because it sounded like a plane. *So, I was looking up through the viewfinder trying to wait for the plane to come into view and instead I saw the building starting to come apart.* [13]

(a) WTC2 from the northeast *(b) WTC2 from the north*

Figure 315. The tipping top of WTC2 disintegrates.

(a)NCSTAR1-6, page 183 (pdf p. 265), http://wtc.nist.gov/NISTNCSTAR1-6.pdf,
(b)http://sf.indymedia.org/uploads/const_in_foreground.jpg,

Figure 314, as said, shows the top as it begins to tip. Figure 315a, taken from the northeast, very clearly shows the north face as it has tipped farther, hinging approximately at the 78th floor. It's of vital importance to note that there are approximately 32 floors *between* the hinge point and the tip of the northeast corner of the building.

In Figure 315b the tipping portion has diminished to many fewer floors, yet the rest of the building *below* the 78th floor appears still to be intact. In 315a we see the top rotating, and in 315b we see that the rotating portion has greatly diminished, as though it were dissolving *from the bottom up*. In Figure 315, the northeast corner of the 110th floor has moved closer to the northeast corner of the 78th floor. The bulges of dust protrude and blossom.

Between Figures 315 and 316, the protruding bulges have coalesced into one large bulge surrounding the building almost like a giant *snowball.* In Figure 316, this large bulge has expanded and moved down a floor or two. The top edge of the building is no longer visible in either Figure. At this point, it appears that the top two thirds of the building have turned into a huge ball of dust, but the lower third of the building is still intact.

(a) WTC2 from the south *(b) WTC2 from the south*

Figure 316. The snowball-shaped dust cloud forms and expands.

(a)http://digitaljournalist.org/issue0110/original_images/NYTimes/moment%20of%20explosion, (b)http://hereisnewyork.org//jpegs/photos/5245.jpg,

Again, between Figures 317a and 317b, the top of the structure has turned into a wall of dust while the lower part of the building appears to have remained intact, as shown also in the diagram in Figure 318.

The transition between Figures 318b and 318c *appears* to violate the laws of motion of conservation of momentum for rigid bodies. This therefore implies that we are not dealing here with a rigid body. The upper part of the building has turned to dust. Once the top begins to rotate, it develops angular momentum. That angular momentum predicts that the top will keep rotating until it falls to the ground. Figure 318c, however, shows that in this case the top did not continue to rotate but decreased in size.

(a) WTC2 from the south *(b) WTC2 from the south*

Figure 317. The tipping top becomes the snowball-shaped dust cloud.

(a)NCSTAR1-6, page 183 (265), http://wtc.nist.gov/NISTNCSTAR1-6.pdf, (b)http://hereisnewyork.org//jpegs/photos/5245.jpg,

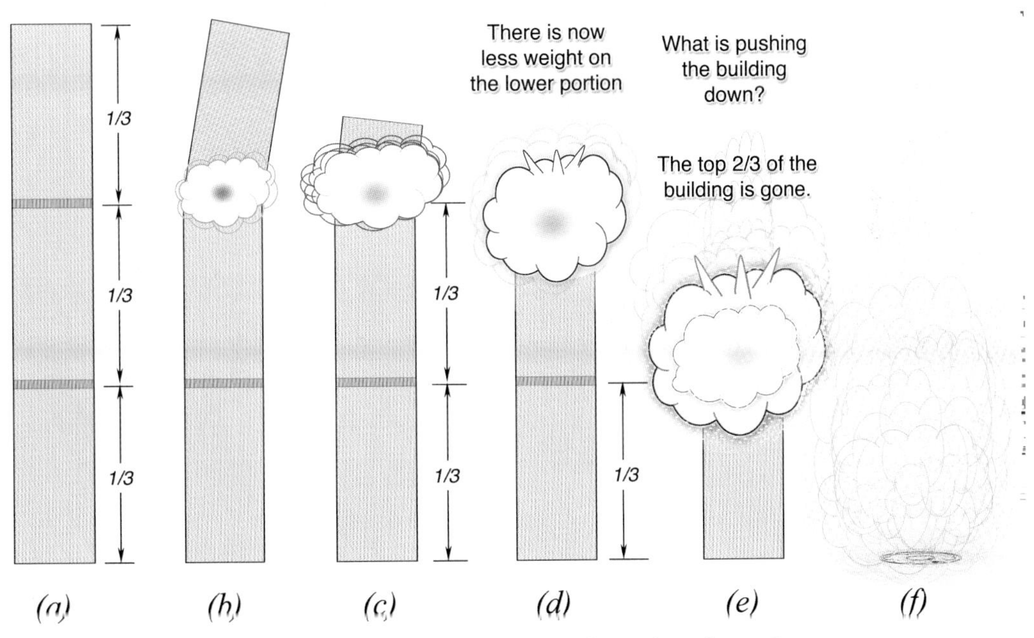

Figure 318. Diagram of WTC2 demise from the south.

The only way this phenomenon could occur is if the rotating top were in fact not to remain a rigid body but instead break down into individual particles. Only in that way would the laws of momentum not be violated, since each individual particle can continue to rotate, momentum being conserved on a piece by piece basis. The fact that the rotating top did not continue to tip over means, simply, that it did not fall as one rigid body. Its disintegration also explains why the size of the *snowball* increased. Again, in Figure 317a, we can see that the building is already turning into powder when the top of WTC2 has barely begun tipping over.

Figure 319. The building appears to be dissolving into powder.
We don't we see any solid parts of a falling building here.
http://hereisnewyork.org//jpegs/photos/5711.jpg,

Figure 320. WTC2 disintegrates.
(a)http://hereisnewyork.org//jpegs/photos/5712.jpg, (b)http://hereisnewyork.org//jpegs/photos/5711.jpg, (c)http://hereisnewyork.org//jpegs/photos/5710.jpg,

E. After the WTC2 Cloud Rollout

Figure 321 indicates that WTC5, 6, and 7 are still healthy. After WTC2 is gone, we can see here also that Vesey Street, between the main WTC complex and WTC7, has no major debris and is covered only in dust. Neither WTC5 or WTC6 appears yet to have gotten holes from damage. Considering the right half of the picture, if we were down at street level, it would be pretty dark there. In fact, if we were standing on Liberty Street in front of Bankers Trust, we might wonder if the sun had set, so little of its light is getting through. Not even the roof of Bankers Trust is receiving much sunlight. But what is blocking it? There's nothing above that roof except the fumes from WTC1. What this suggests is that the density of WTC1's fumes is greater than the density of the dust from WTC2. If you look across to the East River, there is plenty of sunlight, in spite of the presence of dust from WTC2.

The very abrupt shadows—look off-shore on the Hudson River—make it clear that something is blocking the sun to make things darker than just on an overcast day. It's more like an overcast day during a total solar eclipse. The cloud is denser than any "normal" cloud, and this density must be coming from WTC1. Conclusion? The fumes from WTC1 are much denser than the dust from the disappeared WTC2.

The photograph in Figure 322 was taken just after WTC2 went "poof" and the dust rolled out. The dust rolled up to Warren Street, about a block short of the pedestrian bridge, then abruptly stopped. See Figure 323.

Figure 321. Darkness covers southern Manhattan.
http://911wtc.freehostia.com/gallery/originalimages/GJS-WTC21.jpg

Now, if you were to spill vacuum-cleaner dust on the floor, the dust would spread out over an area with the pile deepest where the dust was spilled and thinner the farther it spread out. It would taper off with a distribution similar to the profile of a Gaussian curve, as shown in Figure 323. You would not expect to see an abrupt, cliff-like edge to the pile. And yet that kind of edge is effectively what we do see with the dust of WTC2.

Figure 322 and 324 show the dust cloud from WTC2 abruptly stopping a block before the pedestrian bridge.

Figure 322. After the dust rollout of WTC2.
http://911wtc.freehostia.com/gallery/originalimages/GJS-WTC17.jpg

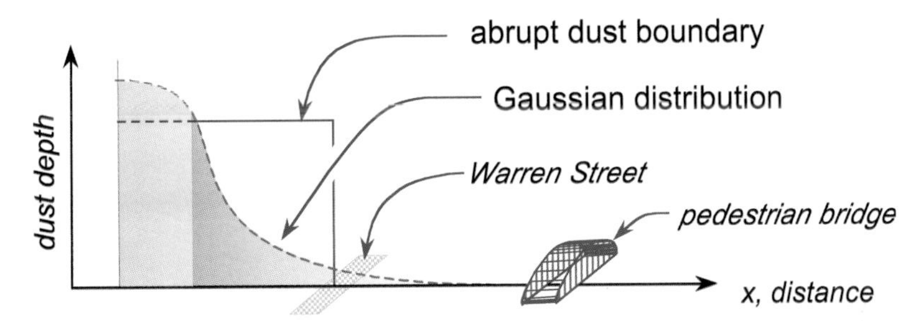

Figure 323. Diagram of abrupt dust boundary compared to a Gaussian distribution.

F. WTC1 Cloud Rollout

Figure 324 shows the rollout of dust from WTC1. The dust already lying on the pavement in front of the cloud is from WTC2, showing that that earlier dust cloud stopped about one block before the pedestrian bridge. The dust cloud from WTC1 stopped right *at* the pedestrian bridge. This additional distance is about the same as the distance between the two WTC towers. So it appears that the dust-cloud rollout from each tower traveled about the same distance before its abrupt halt. [14]

WTC2 went away at 09:59:04 AM and WTC1 went away at 10:28:31 AM, [15] a difference of 29:27, or about thirty minutes. Aside from the roll-out cloud itself, the air in Figures 324 and 325 appears to be quite clear, free of dust from the first roll-out. Fine dust cannot settle out of the air in thirty minutes. There being, in addition, only a light breeze that day, the dust must have been fairly coarse to settle out of the air that quickly. (The dust cloud from the controlled demolition of the Seattle Kingdome settled in less that 20 minutes.) [16]

Figure 324. The dust rollout from WTC1 approaching the settled dust of WTC2.
http://911wtc.freehostia.com/gallery/originalimages/GJS-WTC32.jpg

Figure 325. The WTC1 dust rolls past Warren Street.
http://911wtc.freehostia.com/gallery/originalimages/GJS-WTC35.jpg

Figure 326. The dust rolled out a particular distance then went up.
http://911wtc.freehostia.com/gallery/originalimages/GJS-WTC40.jpg

What appears to happen is that the dust clouds roll out for a certain distance and then rise upward, as can be seen in the video by Bob and Bri,[17] who filmed the rollout from their apartment in Battery Park City (see Figure 326), in the tall apartment building on the left. Their video (Figure 327) shows the dust cloud approaching, then, just before reaching their window, rising up and never quite reaching them. The wind speed, recorded at Newark, LaGuardia, and JFK airports, was averaging 9 mph from the north-northwest.[18]

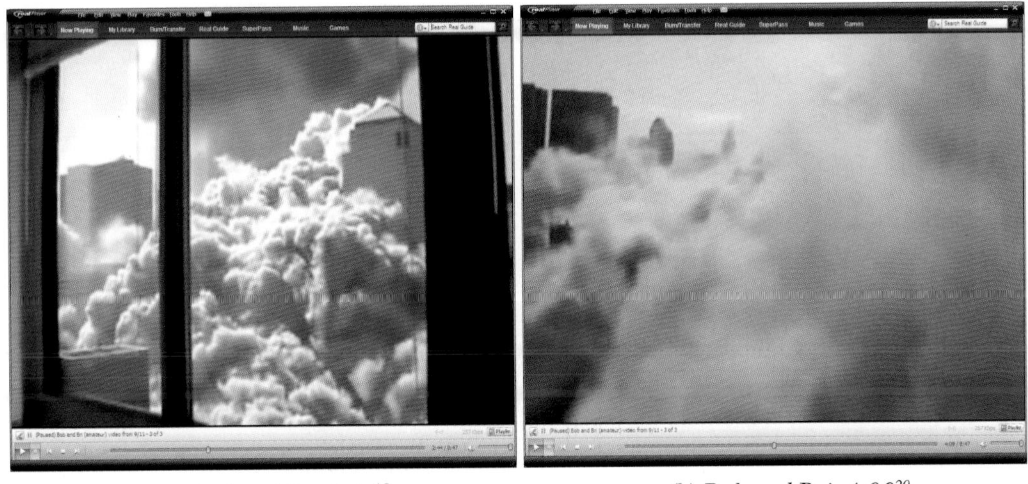

(a) Bob and Bri, 2:44[19] *(b) Bob and Bri, 4:09*[20]
Figure 327. Dust stops short of Bob and Bri's apartment, then goes up.

In the 85 seconds between Figure 327a and Figure 327b, from the video filmed by Bob and Bri, the dust cloud does not move any closer to the window but instead rises up like rapidly-rising yeast bread. This rise in the dust cloud can be seen also in Figure 328.

Bob and Bri's apartment is in the first tall building to the right of the dust cloud. When the dust cloud rolls out and hits a building, it appears to stick. Then when the wind blows it back, the dust seems to stream off the building. But nothing streams off of Bob and Bri's tower, so we know that the dust never reached it. You can see the dust wafting off every structure in Battery Park City except for Bob and Bri's.

Figure 328. A view from the northeast.
http://911wtc.freehostia.com/gallery/originalimages/GJS-WTC43.jpg

What we considered in regard to the conservation of angular momentum is true also for linear momentum in cases like the dust roll-out. When the dust cloud rolled up West Street, it slowed to a stop on the horizontal plane and began to rise upward on the vertical plane. As it rolled through the Vesey Street intersection, it may have been moving at 35-40 miles per hour. Perhaps a race horse could have outrun the dust cloud, but no human could do so. When the cloud rolled up West Street, it slowed dramatically and came to a complete stop before reaching the pedestrian bridge. Then it went up. The law of conservation of momentum tells us that the momentum of this dust cloud has to go somewhere. In this case, it came to a stop before it went *up*, indicating it encountered a lot of resisting force by the air.

Figure 329. Dust that rolled north is now floating up as it is blown southward.
http://911wtc.freehostia.com/gallery/originalimages/GJS-WTC46.jpg

It's true that, normally, the forward momentum of a dust cloud like this is slowed by wind resistance. But this resistance depends on the surface area of the particles. The finer the particles, the more surface area it has. More surface area means more wind resistance and greater resisting force. Very fine particles, or nano-particles, cannot travel far before the wind resistance slows them down. Only fairly *coarse* dust can carry enough momentum to travel a good distance, since it has enough linear momentum to buck the wind without encountering as much wind resistance.

A rock will go a lot further than a pebble if both are shot out of a cannon at the same speed, and, similarly, the distance a cloud spreads out is determined by the particle size. Interestingly in this case, instead of settling down after the rollout, the WTC dust cloud rose up like yeast bread. Dust kicked up by a car on a dirt road settles back to the ground; it does not continue to rise like yeast bread. There was a 5-8 mile/hour headwind, yes, but that should only slow down the dust cloud in the horizontal direction. A headwind shouldn't have caused it to *rise*.

G. Two-toned clouds

Now we must note something else about the dust cloud, and that is its two-toned coloration. See Figures 330 through 332. The contrasting coloration makes this something far different from an ordinary dust cloud.

Figure 330. (9/11/01) Two-toned fumes.
http://hereisnewyork.org//jpegs/photos/5704.jpg

Figure 331. (9/11/01) Two-toned fumes at the base of the tipping top.
http://www.studyof911.com/gallery/albums/userpics/10002/site1101.jpg

Figure 332. The WTC1 dust rollout, viewed across the Hudson.
Fumes pour out of the free-standing columns.
http://www.wtc911.us/911_photos/wtc46.jpg, http://www.wtc911.us/wtc_911_photos.html

H. Wafting away

A satellite image captured on 9/12/01 (Figure 333) clearly shows the dust cloud wafting for miles over New Jersey and Staten Island. It appears that the cloud emerging from the WTC site stays together until it reaches a particular altitude, where it begins to disperse. At this point, there appears to be a kink in the cloud trail.

Figure 333. (9/12/01) Space imaging. (A close-up image is shown in Figure 101b, page 108.)
http://archive.spaceimaging.com/ikonos/2/kpms/2001/09//browse.108668.crss_sat.0.0.jpg

As we know, the controlled demolition of the Seattle Kingdome left a dust cloud that rolled out and settled down in less than 20 minutes.[21] Yet dust from the WTC buildings went up and continued going up for at least 99 days.[22]

I. Conclusions

It is obvious to us now that there is a clear difference between the concrete chunks found in a typical controlled demolition and the micron-sized dust particles that were created on 9/11. The sheer amount of dust and the unusual behavior of that dust attest to the fact that no model of controlled demolition, thermite, or burning airplane fuel was at work in the destruction of the WTC towers.

Figure 334. (9/23/01) NOAA[23]: image from above the WTC complex.
http://www.noaanews.noaa.gov/wtc/images/wtc-photo.jpg

---

[1] This appears to be a GJS photo, from NYPD helicopter.

[2] Peter Jennings and George Stephanopolis, ABC news, September 12, 2001, 12:44 pm, *Where's all the rubble gone? ABC 12:44 pm, 9/12*, 2001 (12:44 pm), *http://youtube.com/watch?v=Bg4pEynzmTE*

[3] Robert Krulwich, ABC news, September 13, 2001, *World Trade Center (Steel turned into dust), 9/13, http://youtube.com/watch?v=Lixyp5ZE1XI* (0:40-1:22),(2:06-2:48)

[4] Robert Krulwich, ABC news, September 13, 2001, *WTC Dust Contains Steel: ABC 19:35, 9/13, http://youtube.com/watch?v=VIlwg-38bg0* (0:40-1:22),(2:06-2:48).

[5] Robert Krulwich, ABC news, September 13, 2001, *ABC Admits That WTC Steel Turned To Dust (Evidence of Exotic Weaponry?), http://youtube.com/watch?v=QNRn3WnQFlk* (0:03-0:45).

[6] File No. 9110191, "Assistant Chief Fire Marshal Richard McCahey," November 2, 2001, pp. 14-15, *http://graphics8.nytimes.com/packages/pdf/nyregion/20050812_WTC_GRAPHIC/9110191.pdf*

[7] File No. 9110039, "Paramedic Gary Smiley," October 10, 2001, pp. 12-13, *http://graphics8.nytimes.com/*

packages/pdf/nyregion/20050812_WTC_GRAPHIC/9110039.pdf

[8] File No. 9110146, "EMT Immaculada Gattas," October 17, 2001, p. 12, *http://graphics8.nytimes.com/packages/pdf/nyregion/20050812_WTC_GRAPHIC/9110136.pdf*

[9] *http://digitaljournalist.org/issue0110/images/m13.jpg, photo by Stan Honda/AFP*

[10] File No. 9110093, "EMT Michael Ober," October 16, 2001, p. 6, *http://graphics8.nytimes.com/packages/pdf/nyregion/20050812_WTC_GRAPHIC/9110093.pdf*

[11] *http://digitaljournalist.org/issue0110/video/dh03.mov,* (video 3, 0:0 - 3:38)

[12] *http://digitaljournalist.org/issue0110/video/dh04.mov,* (video 4, 0:0 - 2:44)

[13] *http://digitaljournalist.org/issue0110/video/rf02.mov*, (video 2, 0:00-3:13)

[14] *http://archives.seattletimes.nwsource.com/cgi-bin/texis.cgi/web/vortex/display? slug=4012219&date=20000327*

[15] *http://www.ldeo.columbia.edu/LCSN/Eq/20010911_WTC/fact_sheet.htm*

[16] *http://archives.seattletimes.nwsource.com/cgi-bin/texis.cgi/web/vortex/display? slug=4012219&date=20000327*

[17] Bob and Bri (home) video from 9/11 - 3 of 3, *http://www.youtube.com/ watch ?v=aMHT07 W3TM4*

[18] *http://www.almanac.com/weatherhistory/index.php?day=11&month=9 &year =2001*

[19] *Bob and Bri (home) video from 9/11 - 3 of 3*, (2:44), *http://www. youtube.com/watch?v=aMHT07W3TM4*

[20] Bob and Bri (home) video from 9/11 - 3 of 3, (4:09), *http://www.youtube.com/watch?v=aMHT07W3TM4*

[21] *http://archives.seattletimes.nwsource.com/cgi-bin/texis.cgi/web/vortex /display ?slug=4012219&date=20000327*

[22] "Sept. 11: For the record, USA TODAY takes a look at some of the facts and figures related to the attacks and aftermath of last Sept. 11." The fires at Ground Zero burned for 99 days, until Dec. 19. Sources: USATODAY research by April Umminger, Joan Murphy, Lori Joseph, William Risser, Darryl Haralson, Mary Cadden, *http://www.usatoday.com/news/sept11/2002-09-10-for-the-record_x.htm*

[23]NOAA's Aerial Photo of World Trade Center, Image taken by NOAA's Cessna Citation Jet on Sept. 23, 2001 from an altitude of 3,300 feet using a Leica/LH systems RC30 camera., *Image released Oct. 2, 2001 at 6:15 p.m. EDT Please credit "NOAA" http://www.noaanews.noaa.gov/stories/s798b.htm*

15.
FUZZBALLS

To acquire knowledge, one must study;
but to acquire wisdom, one must observe.
—Marilyn Vos Savant, Writer

A. Introduction

I call the following photo "the fuzzball intersection." For me, it was one of the most powerful images from 9/11.

Figure 335. (9/11/01) Coarse dust quickly settles to the ground. But fine dust can be seen around the feet, indicating that the dust continued to break down.
http://memory.loc.gov/service/pnp/ppmsca/02100/02121/0076v.jpg

The photo was taken looking west across the intersection of Murray & West Streets, a full block north of the WTC complex. The sign in the center of the photo can be seen in the view from above, shown in Figure 225 (page 224). The time was approximately 15 to 20 minutes after WTC1 "turned into a cloud of dust" or "went poof." We see here a deep blue and fairly clear sky. There is little haze in the air, which means that whatever "dust" had been blown this way had already settled out of the air. But if you look closely, you will notice a fuzzy haze around the feet of several policemen. I call these "fuzzballs." They look like a very fine dust that is kicked up and then seems to waft upward on its own. But dust that fine could not have settled from the air so quickly, if it could have settled at all. Also, I note that fuzzballs do not appear everywhere that people are walking, at least not yet.

Figure 336 shows people emerging from their hiding places immediately after the event. The fuzzballs and fuming haze seen here could not have settled out of the air in 15 to 20 minutes, if, again, they could have settled at all. Later photos reveal that these fuzzballs begin rising on their own, even without being stirred up by feet. I call this later stage, when the haze wafts upward on its own, "fuming." Fuming from the remains of the WTC continued for months after 9/11 and was referred to as "smoke from fires." But the evidence shows it not to have been a matter of "fire" at all. The sidewalk in Figure 336 was not on fire, nor were the fuming streets on fire. Thus, whatever process was at work, it was not airplane fuel that continued to burn, or any form of controlled demolition that caused continued burning. The breakdown of particulate matter was occurring because some other mechanism was in play.

Figure 336. Soon, finer and finer dust began rising from the ground. Dust this fine could not have settled from the air so quickly. It must be that coarse dust settled to the ground but then continued to break down.
(a)http://digitaljournalist.org/issue0110/images/m15a.jpg, (b)http://www.september11news.com/911WTCPeopleDW.jpg

Now consider another curious fact. The satellite image in Figure 337 shows whitish fumes emerging from the area of the WTC complex while blackish fumes emerge from the area's parking lots, where vehicles appeared to be burning. Amazingly, *these fumes appear to move in two different directions with entirely different flow behavior.* The black fumes appear to drift up and perhaps west, but they dissipate very quickly. In contrast, the white fumes flow south and upward and do not begin to dissipate until reaching a much higher elevation. This is astounding. The black fumes dissipate near ground level while the white fumes travel to the upper atmosphere before dissipating. Figure 338 shows extensive "fuming" still occurring four days after 9/11. Meanwhile, Figure 339b shows the relative iron content of various sites close to this fuming.

Now let us compare, in Figure 339a, the locations of "toasted cars," body parts, and "fuming" streets in relation, first, to the location of the WTC complex itself, and, second, to the locations where samples were collected for evaluating iron content from Figure 339b. Samples were collected from the locations shown in white in Figure 340b but curiously were not reported. Note the number of body parts found *blocks away* from the WTC and in directions where they are unlikely to have been sent as a result of shear and ejection from the "airplane impacts," not to mention being too far away to have landed because of having "jumped" there. Note also that some of the "fuming" wafts over the streets, up to the East River, and adjacent to the East River. Once again, we are clearly looking at another method of destruction than burning airplane fuel, controlled demolition, or thermite.

Figure 337. (9/12/01) (A close-up image is shown in Figure 101b, page 108.)
http://archive.spaceimaging.com/ikonos/2/kpms/2001/09//browse.108668.crss_sat.0.0.jpg

Figure 338. (9/15/01) Extensive fuming four days after 9/11.
http://www.911ea.org/images/manhattan_9_15_01.jpg

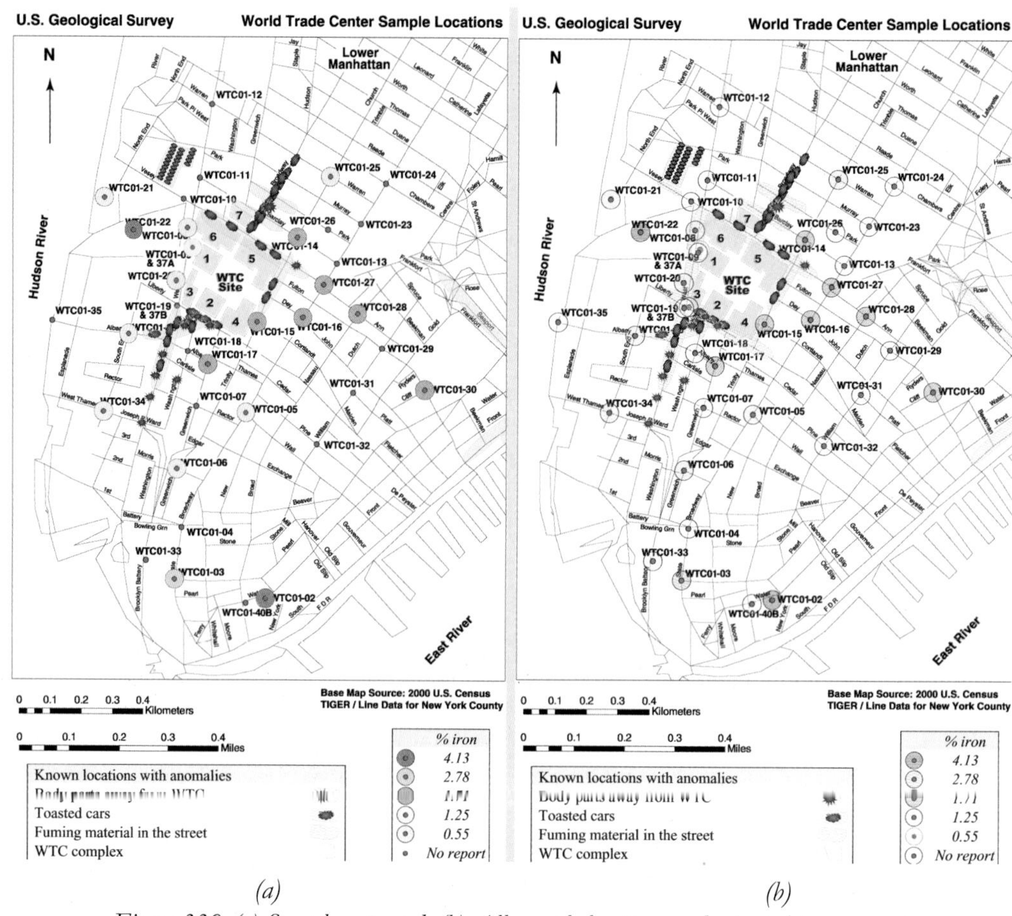

Figure 339. (a) Samples reported. (b) All sample locations and reported iron content.
http://pubs.usgs.gov/of/2001/ofr-01-0429/locmap10.29.01.large.gif

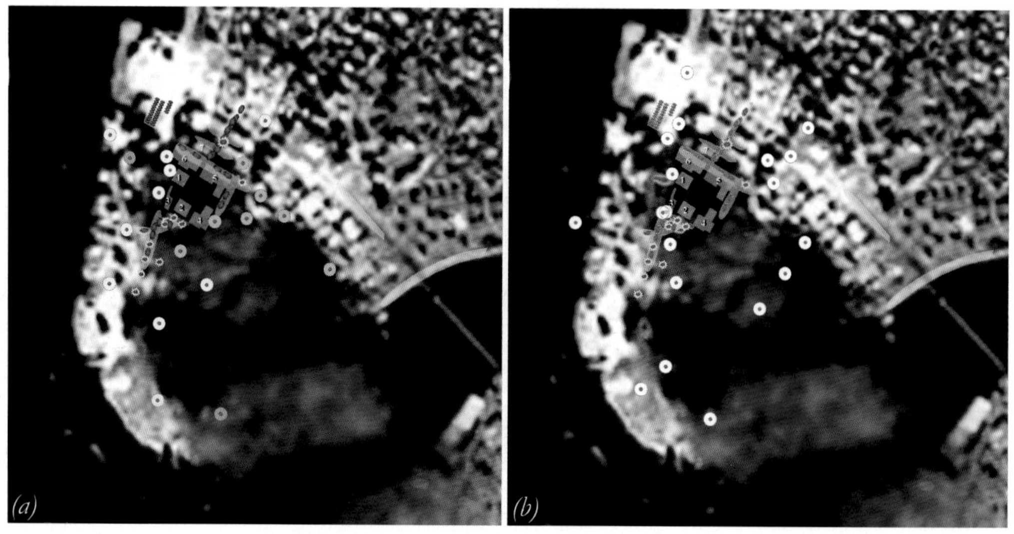

Figure 340. (a) Sample locations where iron content was reported. (b) Sample locations where samples were collected but iron content was not reported.
Adapted from: http://hybrideb.com/images/newyork/manhattan_temp.jpg, and http://pubs.usgs.gov/of/2001/ofr-01-0429/locmap10.29.01.large.gif

B. Dust Size

Figure 341 shows a truckload of fuming dirt being sprayed down with water as it is dumped, March 15, 2002. In spite of the watering, the fumes can be seen rising from the debris. On the other hand, water is hitting the pavement and there is no appearance of "fumes" from that.

Figure 341. (3/15/02) A truckload of fuming dirt is sprinkled with water while being dumped. The yellow platform then transferred the contents to a waiting barge on the Hudson River (at right). Enlargements of the two boxed-in regions are shown in Figure 342.
http://www.photolibrary.fema.gov/photodata/original/6032.jpg

Figure 342. (3/15/02) Enlargement of the regions shown in Figure 341.
http://www.photolibrary.fema.gov/photodata/original/6032.jpg

The surfaces the fumes are emerging from are not glowing. Also, there is no apparent steam (or condensation of water vapor) emerging from the surfaces hit by the water. The debris and truck bed appear simply to be getting wet. If water was used to keep the dust down, the fumes shown in Figure 342 would not be rising from wet surfaces or wet dirt (mud) in the debris. Another indication of there being no high heat is that the truck bed is being raised up by hydraulics. As discussed in Chapter 13, *Weird Fires*, hydraulic fluid temperatures above 82°C will permanently damage hydraulic equipment. (See pages 265 and 266.) These fumes, lacking heat, must be composed of a gas and/or very fine particulate that is not hot.

The fine dust and fumes pose an obvious problem for any conventional model of the WTC destruction. To see how it does so, and why, it is worth considering scientific studies. Lioy et al,[3] a group of researchers from five institutions, studied the dust samples:

> To begin assessing the exposure to dust and smoke among the residential and commuter population during the first few days, samples of particles that initially settled in downtown NYC were taken from three un-disturbed protected locations to the east of the WTC site. Two samples were taken on day 5 (16 September 2001) and the third sample was taken on day 6 (17 September 2001) after the terrorist attack. The purposes for collecting the samples were a) to determine the chemical and physical characteristics of the material that was present in the dust and smoke that settled from the initial plume, and b) to determine the absence or presence of contaminants that could affect acute or long-term human health by inhalation or ingestion. It was anticipated that the actual compounds and materials present in the plume would be similar to those found in building fires or implosion of collapsed buildings. The primary differences would be the simultaneous occurrence of each type of event, the intense fire (> 1,000°C), the extremely large mass of material (> 10 x 10^6 tons) reduced to dust and smoke, and the previously unseen degree of pulverization of the building materials. A summary of the potentially present types of carcinogenic and noncarcinogenic materials was reported in EHP in November 2001 (1).[3]

10,000,000? The WTC towers were about 500,000 tons each, so perhaps the authors actually meant greater than 1 x 10^6 tons. Either way, the figures indicate that the writers are assuming the majority of each tower to have been turned to dust.

The DELTA Group at the University of California at Davis and the Lawrence Berkeley Laboratory also undertook studies of the particles' size.[4,] The DELTA Group (Detection and Evaluation of Long-range Transport of Aerosols)[5] is a collaborative association of aerosol scientists at several universities and national laboratories that has made detailed studies of aerosols from the 1991 Gulf War oil fires, volcanic eruptions and global dust storms.

Samples were collected October 2-30, 2001, from a rooftop at 201 Varick Street, which is 1.8 km (1.1 mi.) north-northeast of the WTC site and about three blocks directly east of the location photographed in Figure 341. This site was upwind of the WTC on 9/11 and about a block north of the north edge of the dust rollout. (See Figure 326 page 313.)

> In the trade-center air samples, [group leader] Cahill identified four classes of particles that have been named by the EPA as likely to harm human health:
>
> - Fine and very fine transition metals, which interfere with lung chemistry.
> - Acids, in this case sulfuric acid, which attack cilia and lung cells directly.

- Very fine, un-dissolvable (insoluble) particles, in this case glass, which travel through the lungs to the bloodstream and heart.
- High-temperature organic matter, many components of which are known to be carcinogens.

> "For each of these four classes of pollutant, we recorded the highest levels we have ever seen in over 7,000 measurements we have made of very fine air pollution throughout the world, including Kuwait and China," Cahill said[6]

In their slide presentation,[7] the group also noted that the four categories above, "reached unprecedented ambient levels in the *very fine* [my emphasis] aerosol plumes from the WTC collapse piles," and added that "On most days, the plumes lofted above NYC so that only those on or near the WTC site breathed these aerosols." This study defined *very fine* as having a particle size in the range of 0.09 μm to 0.26 μm. A red blood cell is 6 μm to 8 μm in diameter,[8] up to 90 times the size of these particles. The thickness of a red blood cell is about 2 μm, or approximately 10 to 20 times the size of the particles collected near the WTC site. These particles are approximately the size of DNA.[9,10]

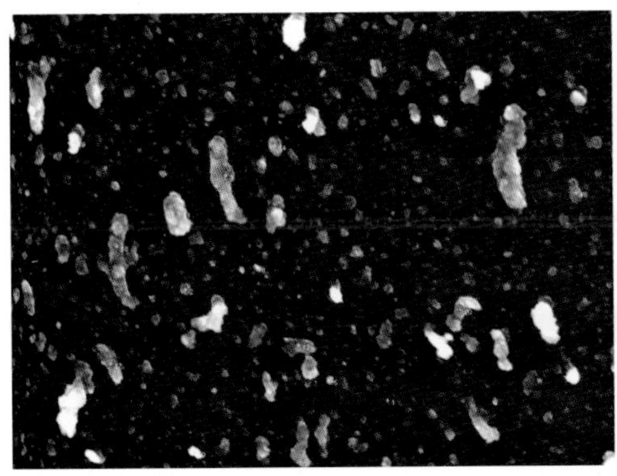

(a) Particles 0.09 to 0.26 μm near the WTC.

(b) Red blood cells, 6 to 8 μm diameter, 2 μm thick.

Figure 343. Aerosol particles near the WTC compared to red blood cells.
(a)DELTA,[11] (b)http://upload.wikimedia.org/wikipedia/en/4/4e/Red_Blood_Cells.jpg

Needless to say, these *very fine* particles were being inhaled by first responders and all others working in that area on 9/11 and for months afterward. Figure 341 (page 325) provides visual evidence of this ongoing fuming as of March 15, 2002. Particles this small can enter the bloodstream through the lungs, exactly like an intravenous drip of toxic chemicals.

While determining the source of the very fine sizes of particles they found, the investigating group noted that organic material made up part of the very fine particles. Although they recognized that organic material could not survive hot temperatures, they explained the very fine particle size as being the result of such hot temperatures. The following is from one of their slides:

> *Proposed explanation of very fine aerosols size and composition*
>
> **Problems:**
>
> - We see very fine aerosols typical of combustion temperatures far higher than the WTC collapse piles
> - We see some elements abundantly and others hardly at all, despite similar

abundances in the collapse dust

- We see organic species in the very fine mode that would not survive high temperatures

Explanation:

- The hot collapse piles are converting some species to gasses that can escape to the surface of the piles and then form aerosols, a process that yields very fine particles[12]

The DELTA Group determined that the ongoing source of this fuming was the WTC remains. However, it appears that they tried to fit their conclusions to the rumors that "fires burned for 99 days" and to the rumors that there was "molten metal" in the remains. It appears that they did not openly consider unknown mechanisms that might have caused the results they found. Still, to their credit, they recognized that hot fires for 99 days could not explain what they found. In a summary slide of their slide presentation were the following statements:

- The surface and near sub-surface debris pile was hot enough to melt aluminum, make steel red hot, and burned until Dec. 19.
- But this is still much cooler than typical sources of very fine particle metals such as power plants, smelters, and diesels.[13]

Some of the most interesting evidence presented by Cahill et al is shown in Figure 344. The color and color scale have been added for convenience. The melting temperatures and boiling temperatures for these elements are provided in Table 26 on page 196. The chart in Figure 344 shows almost as much zinc in the sample as iron, even though iron has a much higher boiling temperature than zinc (2,861°C vs. 907°C)(5,182°F vs. 1,665°F).

		ASTM A36 *(low-carbon)* Mild steel	1018 *(low-carbon)* Mild Steel	1144 *(Stressproof)* steel	A366/1008 *(rolled sheets)*	A8620 *(chrome-nickel-moly)* Alloy Steel
Ultimate Strength (ksi)		58.0-79.8	63.8	115.	43.9-51.9	97.0
Yield Strength (ksi)		36.3	53.7	100.	26.1-34.8	57.0
Elongation		20.0%	15.0%	8.0%	42-48%	25%
Iron	Fe_{26}	99%	98.81-99.26%	97.54-98.01%	99%	96.22-98.02%[14]
Manganese	Mn_{25}	0.75% (0.8-1.2%)[15]	0.6-0.9%	1.35-1.65%	0.06% max	0.7-0.9%
Carbon	C_6	0.26%	0.18%	0.40-0.44%	0.08%	0.18-0.23%
Copper	Cu_{29}	0.2%	-	-	0.2% min	-
Sulfur	S_{16}	0.05% max	0.05% max	0.24-0.33%	0.04%	0.4% Max
Phosphorus	P_{15}	0.04% max	0.04% max	0.04% max	0.035% max	0.35% Max
Nickel	Ni_{28}	-	-	-	-	0.4-0.7%
Chromium	Cr_{24}	-	-	-	-	0.4-0.6%
Silicon	Si_{14}	(0.15-0.30)[15]	-	-	-	0.15-0.35%
Molybdenum	Mo_{42}	-	-	-	-	0.15-0.25%

Table 14. Composition of common steel with Yield Strength and Ultimate Tensile Strength.[16]

Each WTC tower had about 100,000 tons of steel, a material that is mostly iron. The composition of various common types of steel are shown in Table 14. The yield strength and elongation are also provided as a general guide. The report by the

DELTA group noted that there were two very different types of air pollution from the 9/11/01 event.

> The collapse of the World Trade Center structures (South Tower, North Tower, and WTC 7) presented two very different types of air pollution events:
>
> 1. Initial fires and collapse-derived "dust storm"
> 2. Continuing emissions from the debris piles
>
> Both cases shared the unusual aspect of a massive ground level source of particulate matter in a highly populated area with potential health impacts. [17]

There have been no studies of "continuing emissions" resulting from a normal building collapse or from a normal controlled demolition. Ongoing fires can emit pollutants into the air, but according to Dr. Thomas Cahill, the DELTA group had never recorded such high levels as this in "over 7,000 measurements [studies] [they] have made of very fine air pollution throughout the world...".[18] The image in Figure 341 shows visible fuming from the truck bed in March 2002. Some fuming was still visible when I, along with others[19] visited the site in October 2007 and in January 2008 and photographed the fuming. Some of these photographs are available on my website. *http://drjudywood.com/towers*

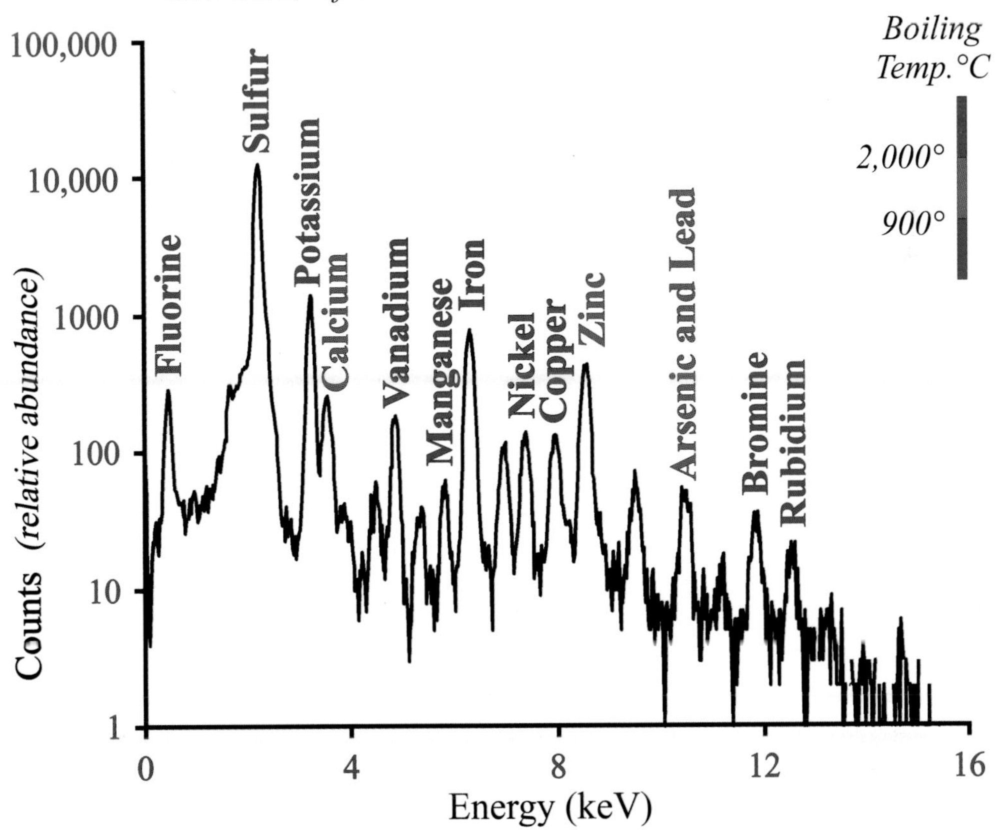

Figure 344. Relative quantity of various elements in a typical S-XRF sample by Cahill, et al.[20] *[Colors, including background, have been changed and "(relative abundance)" added for clarity.]*

If the very fine aerosols measured were from "hot fires vaporizing materials," one would expect to see a pattern of elements in the samples. All of the samples came from 201 Varick Street, about a mile north-northeast of the WTC site. Elements with a similar boiling temperature might be expected in a sample an abundance appropriate to their source. If steel were being vaporized, one would expect to see components of steel in similar proportions, absent other sources. So if steel were being vaporized, we might expect that for every 962 to 993 counts of iron, there might be 6 to 16 counts of manganese and 0.5 to 3.3 counts of sulfur. But these are not the relative proportions appearing in the sample by the DELTA group.

The relative distribution of elements in a typical sample by Cahill et al, the DELTA group, is shown in Figure 344. Vanadium, which is used in paint, is much more abundant than manganese, nickel, or copper. The DELTA group noted in their slide (quoted on page 327) that the samples showed "fine aerosols typical of combustion temperatures far higher than the WTC collapse piles."[21] Vanadium has the highest melting temperature and boiling point of the elements shown, at 1,910°C and 3,407°C, respectively. Fluorine, which they found was only slightly more abundant than Vanadium, has the lowest melting temperature and boiling point of the elements shown, at -220°C and -188°C, respectively. The quantity of fluorine, vanadium, nickel, copper, and calcium are similar, which is puzzling. The DELTA group were puzzled by this as well, as shown on page 328. But the most troublesome evidence for concluding that the results could be explained by high heat was also stated in their slide quoted on page 328. They found very fine organic material that would not survive high temperatures. That is, not heat, but some other mechanism must be causing the ongoing breaking-down of material into a very fine aerosol.

It was concluded by Thomas Cahill and the DELTA group that these very fine aerosol particles were not from the "collapse" of the buildings, but were being emitted by some ongoing reaction in the rubble. In their conclusion slide, they state the following:

Conclusions – WTC Aerosols

There were heavy and continuing emissions of aerosols *in narrow plumes of unusual size* and composition from the WTC collapse site that on 9 to 15 occasions impacted 201 Varick St, 1.8 km NNE.[emphasis added]

Coarse particles were similar to the initial collapse aerosols (cement, dry wall, glass, ...) but had chemicals and soot from the ongoing combustion. *Little asbestos was expected or observed.* [emphasis added]

The presence of unprecedented (vis. Beijing, Kuwait) levels of very fine ($0.26 > D_p > 0.09$ µm) particles by mass and number in *narrow plumes* was more typical of an industrial source, specifically a chlorine rich municipal incinerator, than any normal ambient air situation. Upwind sources were a very minor contribution. [emphasis added]

The very fine silicon and sulfur and many of the coarse metals like vanadium decreased steadily during October. Very fine particles near the WTC site in May, 2002, were generally < 10% of the October, 2001 plume impact days at Varick Street. (except S, Ni)[22]

Let us now turn to what the DELTA group referred to as the *unusual size* of the dust particles in the *narrow plumes* of material emerging from the remains. The graphs are shown in Figure 345.[23] Figure 345a shows a typical profile and Figure 345b shows the WTC profile of concentration versus particle size.

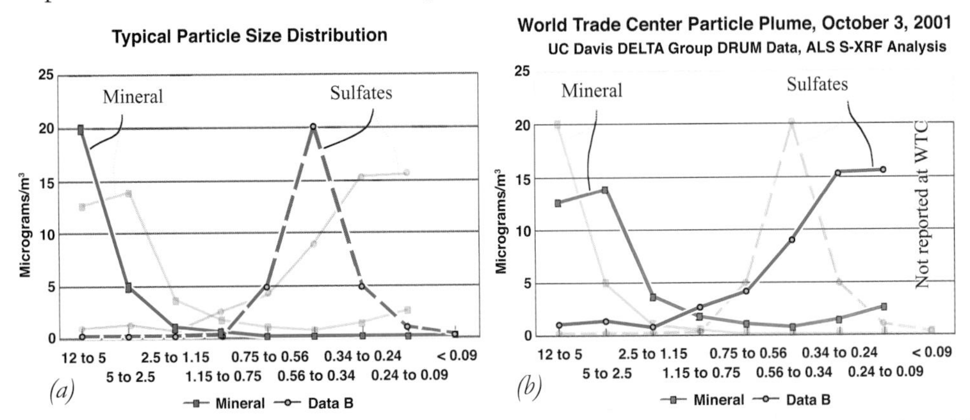

Figure 345. Comparison Between WTC and Typical Aerosol Compositional Size Distribution from 201 Varick St., NYC, Oct 3, 2001, by Lawrence Berkeley Lab.[24] (Size in micrometers.)
(Curves shown in the foreground of Figure 345a appear in the background of Figure 345b and vice versa.)
Adjusted from: http://delta.ucdavis.edu/SizeDist.jpg[25]

The difference between Figure 345a and Figure 345b is the addition of an increasing volume of finer material with a decrease in the volume of course material. Figure 346 provides an explanation of the UC Davis study data. In particular, note the one entitled "Molecular Dissociation."

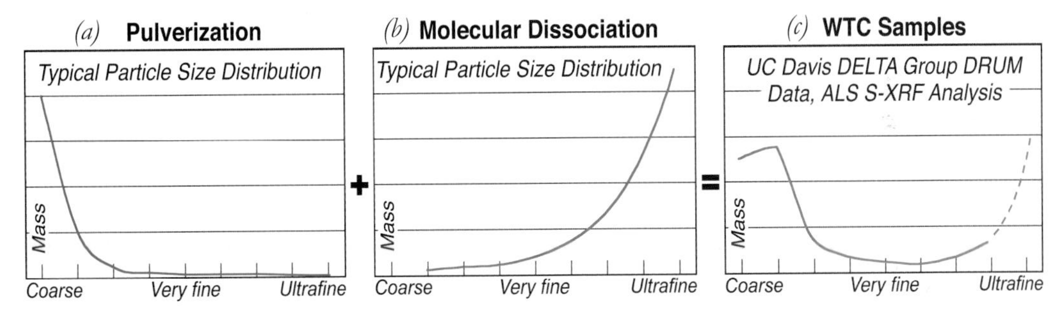

Figure 346. Particle-size distributions from (a) pulverization, (b) molecular dissociation, (c) WTC samples.

As we have seen, dust from conventional controlled demolition settles out of the air in 15 to 20 minutes and doesn't rise much higher than the highest point in the pre-existing structure, since the dust is fairly coarse. The Seattle Kingdome is an example of such a collapse, imploded by conventional demolition. In that instance, the dust settled out in about 20 minutes.[26]

C. Microscope Analysis and Conclusions

Before we proceed to conclusions, carefully examine the following photos of the 9/11 particles.

1. Iron Spheres at the WTC Site

Figure 347. (10μm - scale) Iron-rich sphere found in dust from the WTC. Using the scale on the image, the diameter of the iron-rich sphere is about 25 μm.
http://pubs.usgs.gov/of/2005/1165/graphics/IRON-04-IMAGE.jpg[27]

2. Iron Spheres at Tunguska Site (Not WTC)

The Tunguska Event, an unexplained energy phenomena, occurred on June 30, 1908 not far from the Tunguska River in what is now Krasnoyarsk Krai in Russia. Although various theories about the cause have been proposed (including Tesla's Wardenclyffe Tower, a meteorite, an antimatter weapons test),[28] none have been conclusive. As scientific methods developed, new samples from the area were obtained and analyzed.

> *Expeditions sent to the area in the 1950s and 1960s found microscopic silicate and magnetite spheres in siftings of the soil. Similar spheres were predicted to exist in the felled trees, although they could not be detected by contemporary means. Later expeditions did identify such spheres in the resin of the trees…*[29]

3. Iron Spheres in Crop Circle (Not WTC)

Iron-rich magnetic spheres have also been found in coatings on plants and inside of plants where crop circles have appeared. The magnetic spheres in Figure 348 (indicated by the arrow) are of a similar size to the one found at the WTC site shown in Figure 347. The plants did not appear burned, indicating a process that did not involve high heat.

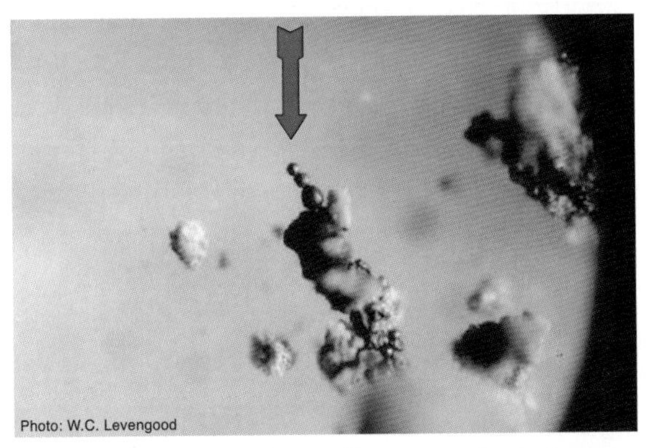

Figure 348. Photomicrograph of 10-40 micron diameter, spherical, magnetic particles of the type regularly found in crop circle soils. EDS reveals these spheres to be pure iron; their being magnetized reveals they were formed in a magnetic field.[30,31,32]
(red arrow enhanced) http://www.bltresearch.com/images/magnetic/magmat2.jpg

Now, back to the WTC.

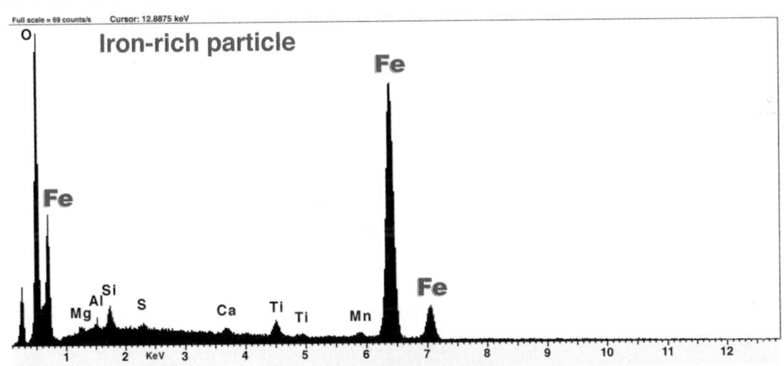

Figure 349. Spectrograph of the sample shown in Figure 347.
enhanced from: http://pubs.usgs.gov/of/2005/1165/graphics/IRON-04.jpg[33]

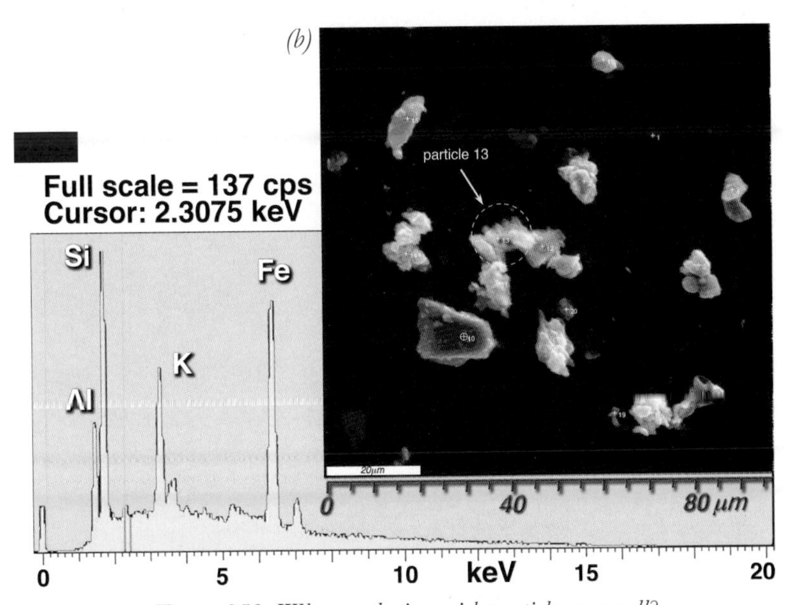

Figure 350. Why are the iron-rich particles so small?
enhanced from: (a)http://delta.ucdavis.edu/Particle13.jpg,[34] *(b)http://delta.ucdavis.edu/WTCdust.jpg*[35]

Why are the iron-rich particles so small? If the iron were hot enough to have melted, it would have ended up stuck together in larger blobs of material. Concrete is not made of iron aggregates, nor would iron aggregates be pulverized to the degree seen here. Now compare the image of particle 13 in Figure 350b (and in Figure 351b) with the image of the chrysotile bundle (the most common type of asbestos[36]) in Figure 351a.

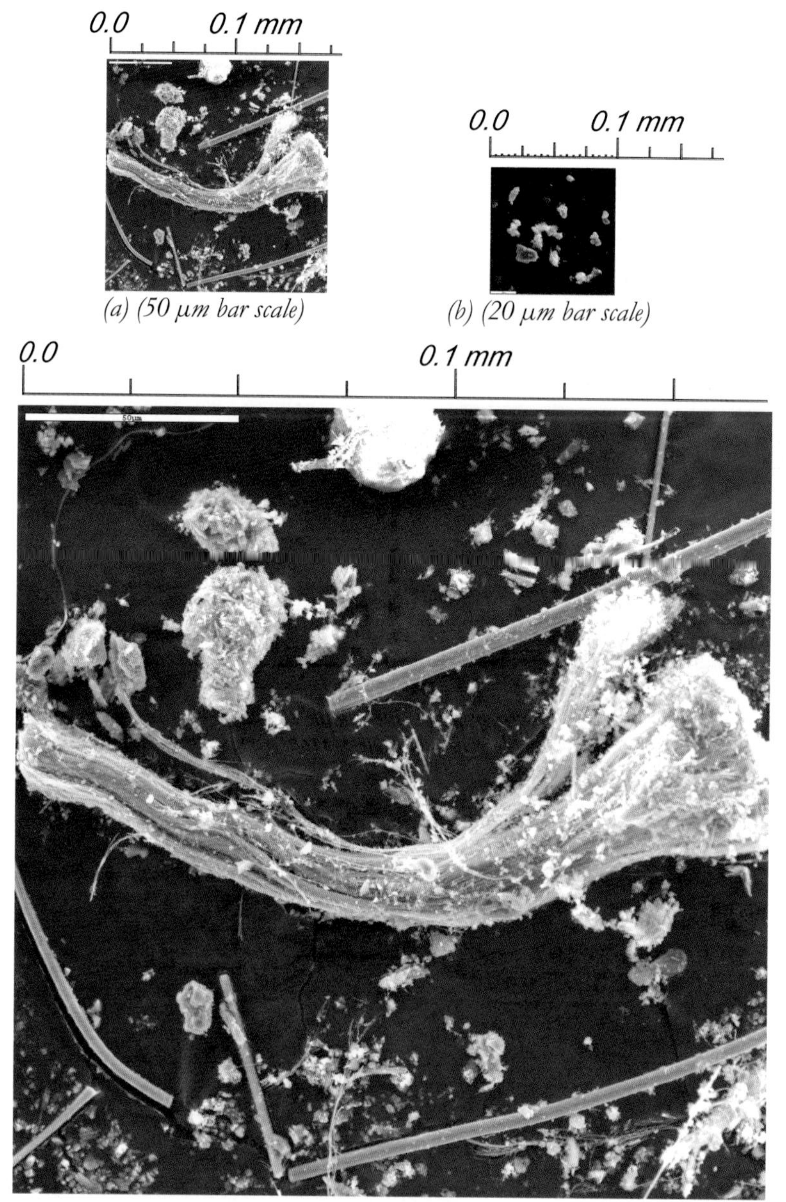

(a) (50 μm bar scale) *(b) (20 μm bar scale)*

Figure 351. (50μm) SEM image of a chrysotile (asbestos) bundle and glass fibers. Figure 351a and Figure 351b are shown at approximately the same scale for comparison. Figure 351b is the same image as 350b.
(b) http://delta.ucdavis.edu/WTCdust.jpg, (a)(c)http://pubs.usgs.gov/of/2001/ofr-01-0429/sem1/wtc01-08.sem.im3.gif[37]

Now note the following graph in Figure 352, and the commentary following it:

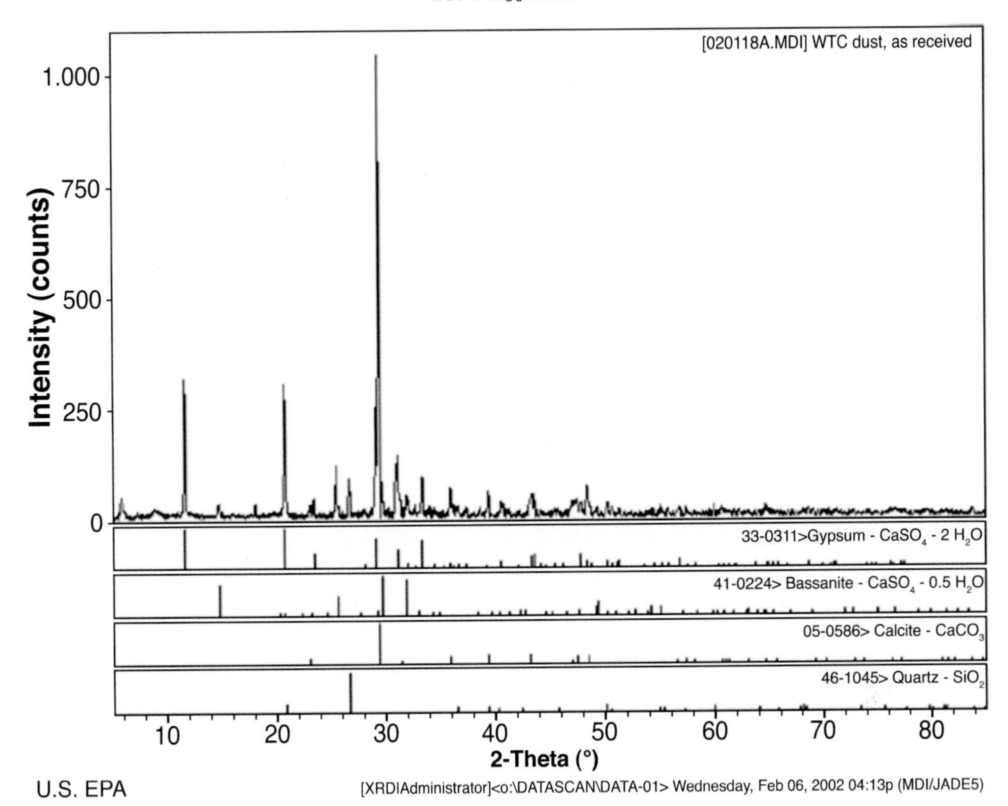

Figure 352. (2/6/02) Low levels of carbon (C) are inconsistent with hydrocarbon fires.
enhanced from: http://ehp.niehs.nih.gov/members/2003/5930/fig5.jpg

> Levels of carbon were relatively low, *suggesting that combustion-derived particles did not form a significant fraction of these samples recovered in the immediate aftermath of the destruction of the towers.*[38] [emphasis added]

One would expect higher carbon if there had in actuality been a lot of fires.

In any case, what we are looking at in these pictures, graphs, and analyses is evidence of the almost total pulverization, right down to "molecular dissociation," the coming apart of molecules. This molecularly-dissociated dust has to have been the result of another mechanism of destruction than one caused by burning fuel or controlled demolition, for those mechanisms cannot bring about *molecular dissociation* and the almost complete *dustification* of a building. Molecular dissociation occurs with ionization[39,40] and is seen in cold fusion. A group of researchers also determined that electric field fluctuations in liquid water cause molecular dissociation. [41]

The photo in Figure 353 shows fine dust around the feet of a survivor shortly after the destruction of the first tower, WTC2. The photo appears to have been taken near the intersection of Park Row and Broadway, where Vesey Street intersects, just to the right of position H in Figure 312, page 302. The clock in the distance, in the upper-right portion of the photo (Figure 353), shows a time of 10:14. WTC2 went away at 9:59 and WTC1 went away at 10:29.[42] So this gives a good indication of how long the dust from WTC2, alone, has been falling out of the air. At 10:14, WTC1 was still standing. There is a lot of fine dust around the man's feet and behind him, yet the air looks much clearer above the level of his knees.

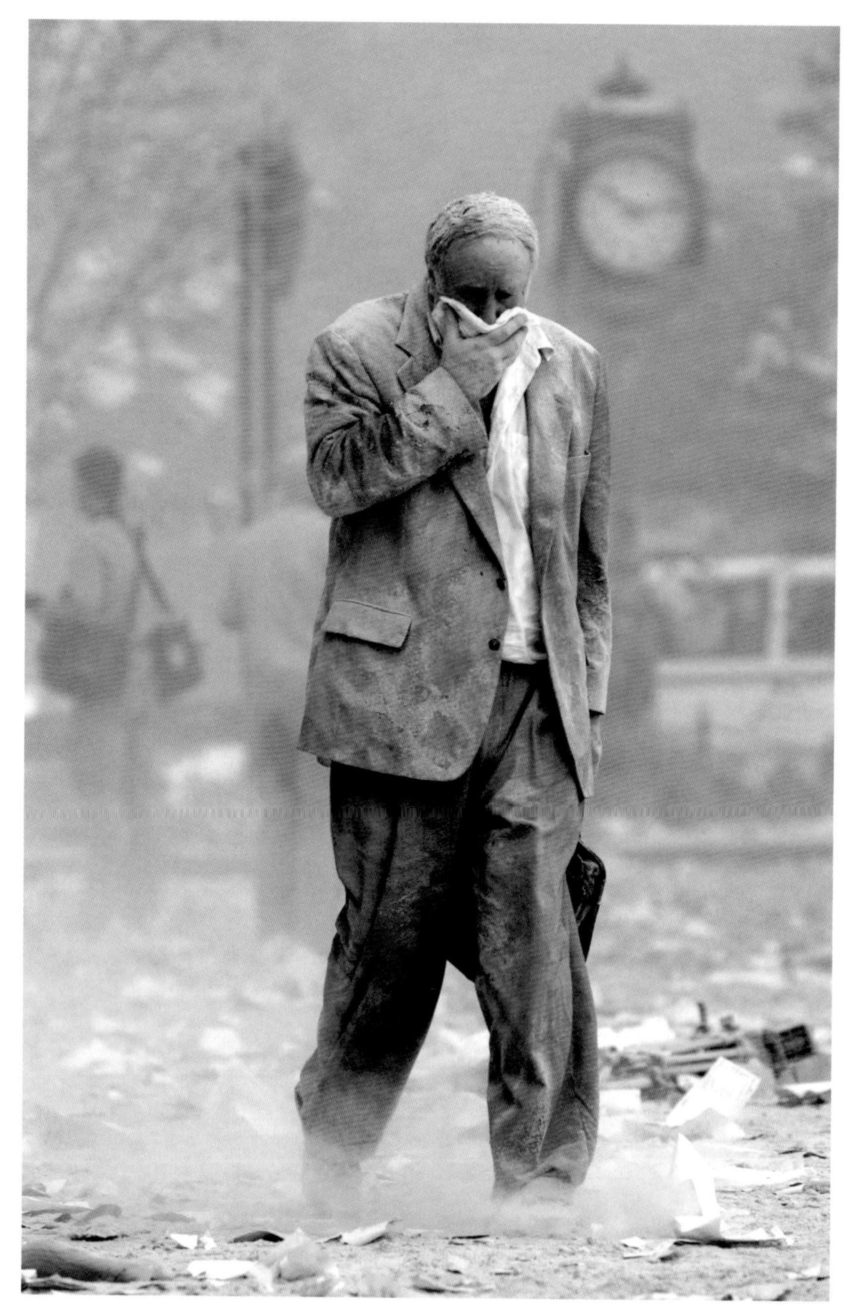

Figure 353. (9/11/01, 10:14 AM) Fine dust rises from the ground as a man walks home.[43]
http://digitaljournalist.org/issue0110/original_images/AFP/6%20man%20smoke.jpg

With briefcase in hand, a man heads home from a rough day—a very rough day—at the office. He was one of the survivors who got to go home that day. I hope he's still ok.

---

[1] http://pubs.usgs.gov/of/2001/ofr-01-0429/chem1/WTCchemistrytable.html

[2] http://pubs.usgs.gov/of/2001/ofr-01-0429/chem1/WTCchemistrytable.html

[3] Paul J. Lioy, Clifford P. Weisel, James R. Millette, Steven Eisenreich, Daniel Vallero, John Offenberg, Brian Buckley, Barbara Turpin, Mianhua Zhong, Mitchell D. Cohen, Colette Prophete, Ill Yang, Robert

Stiles, Glen Chee, Willie Johnson, Robert Porcja, Shahnaz Alimokhtari, Robert C. Hale, Charles Weschler, and Lung Chi Chen, "Characterization of the Dust/Smoke Aerosol that Settled East of the World Trade Center (WTC) in Lower Manhattan after the Collapse of the WTC 11 September 2001," *http://www.ehponline.org/members/2002/110p703-714lioy/lioy-full.html*

[4] Cahill, T.A., Cliff, S.S., Perry, K.D., Jimenez-Cruz, M., Bench, G., Grant, P.G., Ueda, D., Shackelford, J.F., Dunlap, M., Meier, M., Kelly, P.B., Riddle, S., Selco, J. and Leifer, R., "Analysis of Aerosols from the World Trade Center Collapse Site, New York, October 2 to October 30, 2001," Aerosol Science and Technology, 38: 165-183, 2004, American Chemical Society Meeting 2003

[5] *http://delta.ucdavis.edu/WTC.htm*

[6] The Delta Group, "Trade Center Debris Pile Was a Chemical Factory, Says New Study" September 10, 2003, *http://delta.ucdavis.edu/WTC.htm*

[7] Thomas A. Cahill, Steven S. Cliff, Kevin D. Perry (U. Utah) , James Shackelford, Michael Meier, Michael Dunlap, Graham Bench, (LLNL), and Robert Leifer (DOE EML), *"Very fine aerosols from the World Trade Center collapse piles: Anaerobic Incineration?"* American Chemical Society Meeting, Sept. 7-11, 2003, DELTA* (δ) Group, University of California, Davis (* Detection and Evaluation in Long-range Transport of Aerosols), slide 17, *http://delta.ucdavis.edu/WTC%20aersols%20ACS%202003.ppt*

[8] *http://en.wikipedia.org/wiki/Red_blood_cell*

[9] Near field laser 'tweezers' reveal new vista for medical diagnosis, January 28, 2009, Swinburne University, *http://www.sciencealert.com.au/features/20092801-18725.html*

[10] Ogris M, Steinlein P, Kursa M, Mechtler K, Kircheis R, Wagner E., Institute of Biochemistry, University of Vienna, Austria, "The size of DNA/transferrin-PEI complexes is an important factor for gene expression in cultured cells," *Journal of Gene Therapy*, 1998 Oct., 5(10), pp. 1425-33, *http://www.ncbi.nlm.nih.gov/pubmed/9930349*

[11] Thomas A. Cahill, Steven S. Cliff, Kevin D. Perry (U. Utah) , James Shackelford, Michael Meier, Michael Dunlap, Graham Bench, (LLNL), and Robert Leifer (DOE EML), *"Very fine aerosols from the World Trade Center collapse piles: Anaerobic Incineration?"* American Chemical Society Meeting, Sept. 7-11, 2003, DELTA* (δ) Group, University of California, Davis (* Detection and Evaluation in Long-range Transport of Aerosols), slide 53, *http://delta.ucdavis.edu/WTC%20aersols%20ACS%202003.ppt*

[12] *Ibid.*, slide 26

[13] *Ibid*, slide 18

[14] Calculated from maximum and minimum values of the other ingredients.

[15] J.R. Barnett, R.R. Biederman, and R.D. Sisson, Jr., "An Initial Microstructural Analysis of A36 Steel from WTC Building 7," The Minererals, Metals, and Materials Society, *Journal of Materials,* 53 (12) (2001), p. 18., *http://www.wpi.edu/Academics/Research/MPI/News/aninit697.html, http://www.tms.org/pubs/journals/JOM/0112/Biederman/Biederman-0112.html*

[16] Carbon Steel Grades, *http://www.eaglesteel.com/download/techdocs/Carbon_Steel_Grades.pdf*

[17] Thomas A. Cahill, Steven S. Cliff, Kevin D. Perry (U. Utah) , James Shackelford, Michael Meier, Michael Dunlap, Graham Bench, (LLNL), and Robert Leifer (DOE EML), *"Very fine aerosols from the World Trade Center collapse piles: Anaerobic Incineration?"* American Chemical Society Meeting, Sept. 7-11, 2003, DELTA* (δ) Group, University of California, Davis (* Detection and Evaluation in Long-range Transport of Aerosols), slide 2, *http://delta.ucdavis.edu/WTC%20aersols%20ACS%202003.ppt*

[18] The Delta Group, "Trade Center Debris Pile Was a Chemical Factory, Says New Study" September 10, 2003, *http://delta.ucdavis.edu/WTC.htm*

[19] names withheld for protection.

[20] Thomas A. Cahill, Steven S. Cliff, Kevin D. Perry (U. Utah) , James Shackelford, Michael Meier, Michael Dunlap, Graham Bench, (LLNL), and Robert Leifer (DOE EML), *"Very fine aerosols from the World Trade Center collapse piles: Anaerobic Incineration?"* American Chemical Society Meeting, Sept. 7-11, 2003, DELTA* (δ) Group, University of California, Davis (* Detection and Evaluation in Long-range Transport of Aerosols), slide 18, *http://delta.ucdavis.edu/WTC%20aersols%20ACS%202003.ppt*

[21] *Ibid.*, slide 26

[22] *Ibid.*, slide 33

[23] Comparison Between WTC and Typical Aerosol Compositional Size Distribution from NYC at 201 Varick St., Oct 3, 2001 by Synchrotron X-Ray Fluorescence (XRF) at The Advanced Light Source, Lawrence Berkeley Laboratory, *http://delta.ucdavis.edu/WTC.htm*

[24] "Comparison Between WTC and Typical Aerosol Compositional Size Distribution from NYC at

201 Varick St., Oct 3, 2001 by Synchrotron X-Ray Fluorescence (XRF) at The Advanced Light Source, Lawrence Berkeley Laboratory," *http://delta.ucdavis.edu/WTC.htm, http://delta.ucdavis.edu/BG_WTC_TradeCtr_01Oct.xls*

[25] *http://delta.ucdavis.edu/WTC.htm*

[26] Dome's Final Roar, by Jeff Hodson, Eric Sorensen, Alex Fryer, Beth Kaiman, Dionne Searcey, Sara Jean Green, John Zebrowski, Phil Loubere, Seattle Times staff reporters, Monday, March 27, 2000, *http://archives.seattletimes.nwsource.com/cgi-in/texis.cgi/web/vortex/display?slug=4012219&date=20000327* (Last visited 4/10/09)

[27] *http://pubs.usgs.gov/of/2005/1165/table_1.html*

[28] Tunguska event, *http://en.wikipedia.org/wiki/Tunguska_event*

[29] Tunguska event, *http://en.wikipedia.org/wiki/Tunguska_event*

[30] "Magnetic Materials in Soils," *http://www.bltresearch.com/magnetic.php*

[31] W.C. Levengood and John Burke, "Semi-Molten Meteoric Iron Associated with a Crop Formation," *Journal of Scientific Exploration, http://www.jse.com/*, Vol 9, No. 2, pp. 191-199, 1995, 0892-3310/95, *http://www.bltresearch.com/published/semi-molten.php*

[32] From the research of Andrew Johnson, *http://www.checktheevidence.co.uk/cms/index.php?option=com_content&task=view&id=273&Itemid=60*

[33] *http://pubs.usgs.gov/of/2005/1165/table_1.html*

[34] *http://delta.ucdavis.edu/WTC.htm*

[35] *http://delta.ucdavis.edu/WTC.htm*

[36] *http://ntp.niehs.nih.gov/ntp/roc/eleventh/profiles/s016asbe.pdf*

[37] *http://pubs.usgs.gov/of/2001/ofr-01-0429/sem1/wtc01-08.sem.im3.html*

[38] Chemical Analysis of World Trade Center Fine Particulate Matter for Use in Toxicologic Assessment, by John K. McGee, Lung Chi Chen, Mitchell D. Cohen, Glen R. Chee, Colette M. Prophete, Najwa Haykal-Coates, Shirley J. Wasson, Teri L. Conner, Daniel L. Costa, and Stephen H. Gavett, *Environmental Health Perspectives* Volume 111, Number 7, June 2003, *http://ehp.niehs.nih.gov/members/2003/5930/5930.html*

[39] I Ben-Itzhak *et al*, "Ionization and dissociation of molecular ion beams caused by ultrashort intense laser pulses," *Journal of Physics: Conference* Series 88 (2007) 012046, *http://iopscience.iop.org/1742-6596/88/1/012046/media, media: http://iopscience.iop.org/1742-6596/88/1/012046/media/MDI-FieldFree.avi, http://iopscience.iop.org/1742-6596/88/1/012046/media/MDI-EfieldLong.avi*

[40] H. O. Folkerts, R. Hoekstra, and R. Morgenstern, K.V.I. Atomic Physics, "Velocity and Charge State Dependences of Molecular Dissociation Induced by Slow Multicharged Ions" Physics Review Letters 77, 3339-3342 (1996), *http://link.aps.org/doi/10.1103/PhysRevLett.77.3339*

[41] Geissler, P. L., Dellago, C., Chandler, D., Hutter, J., Parrinello, M. (2001) "Autoionization in liquid water," *Science*, 16 March, 2001, **291** (5511): 2121–2124. doi: 10.1126/science.1056991. *http://www.ncbi.nlm.nih.gov/pubmed/11251111, http://www.sciencemag.org/cgi/content/full/291/5511/2121*

[42] If the destruction of a tower stopped the clock, it would likely show a time of 10:14 or a time of 10:29, not 10:14.

[43] The Caption from Stan Honda's photo, *http://digitaljournalist.org/issue0110/images/m14.jpg, identifies tha man as George Sleigh:* "George Sleigh, who was pulled from the debris by coworkers on the 91st floor of the World Trade Center's north tower, covers his mouth as he walks through debris after the collapse of one of the World Trade Center Towers. Photo by Stan Honda/AFP." However, other photos identify a man wearing different clothes as George Sleigh. (See *http://tamaraheater.wordpress.com/2009/09/11/remembering-911-through-george-sleigh/*, and *http://www.parksidechurch.com/ministries/life-stages/singles/latest-info/2009/9/remembering-911-george-sleigh/*)

16.
Lather

Crime is naught but misdirected energy. —Emma Goldman

A. Introduction

Careful analysis of photographs shows evidence of what I call "lathering up" of buildings in the WTC just prior to their going "poof." WTC1 lathered up just as WTC2 went poof and continued lathering after the destruction of WTC2.

Figure 354. (9/11/01) Immediately after WTC2 was destroyed, WTC1 lathered up.
http://911wtc.freehostia.com/gallery/originalimages/GJS-WTC10.jpg

Note that just prior to its own "collapse," WTC1 "lathered up" *along the entire face of one side of the building and that this is not a signature of a typical controlled demolition.* It is a distinctive phenomenon and it occurred prior to the "initiation of collapse" of WTC1. In Figure 167, we see the process continuing as WTC1 disappears.

Figure 355. (9/11/01) "Shaving cream"/ "Alkaseltzer" WTC1 disintegrated while falling.
Photo by Shannon Stapleton, Reuters[1]

Assistant Chief Fire Marshal Richard McCahey's testimony is significant for its indication that some other process is at work here than mere fire:

Assistant Chief Fire Marshal Richard McCahey

> Somebody, which actually helped me, I don't know who it was, after somebody said I can't breathe, somebody, I don't know which direction it came from, screamed out don't panic or relax, relax.. It's not smoke. *It's just dust.* Just relax. [2] [emphasis added]

In Figure 356, WTC1 and WTC7 are "lathering-up" simultaneously. The photographer stood at Church and Duane, a few blocks north of the Brooklyn Bridge, looking southward. It appears from the haze between WTC7 and the camera location that WTC2 has just been destroyed. The street here is not coated with powder, but the dust cloud from WTC2 may not have gone as far north as Duane St. Later photos of this area, after the disappearance of WTC1, show a fine whitish dust coating the street. At this moment, WTC1 is still standing, so the familiar argument that "debris from the collapse of WTC1" is what explains the "smoke" pouring out of WTC7 is proven incorrect.

Figure 356. (9/11/01) WTC1 and WTC7 lather-up together.
http://www.magnumphotos.com/CoreXDoc/MAG/Media/TR3/F/P/J/F/NYC20193.jpg[3]

B. WTC7 Lathers-Up

The "lathering-up" of WTC7 appears much darker in late afternoon (See Figure 357a), while the foreground air is clear of dust or fumes over the space where WTC1 stood.

Figure 357. (9/11/01) WTC7 Lathering up.
(a)http://www.magnumphotos.com/CoreXDoc/MAG/Media/TR3/F/W/L/Y/NYC14148.jpg[4]
(b)http://www.jnani.org/mrking/writings/911/king911_files/image001.jpg

WTC7 also showed the extraordinary phenomenon of lather emanating from just *one* face of the building.

Figure 358. Fumes from the east face of WTC7, 3:53 p.m. - 4:02 p.m.[5]
http://wtc.nist.gov/media/NIST_NCSTAR_1-9_Vol1_for_public_comment.pdf, p. 227 (pdf p. 271 of 404), Fig. 5–141

Figure 358 is a frame from a video shot from near the corner of West Broadway and Barclay Street. It shows the eastern side of the north face of WTC7 and was shot between 3:53 p.m. and 4:02 p.m.

(a) WTC7 10.0 s ± 0.2 s *(b) WTC7 11.0 s ± 0.2 s*
Figure 359. The remains of WTC7 erupt into lather in all directions for its final demise.
(a)http://wtc.nist.gov/media/NIST_NCSTAR_1-9_Vol1_for_public_comment.pdf, p. 283 (pdf p. 327 of 404), Fig. 5–209,
(b)http://wtc.nist.gov/media/NIST_NCSTAR_1-9_Vol1_for_public_comment.pdf, p. 284 (pdf p. 328 of 404), Fig. 5–212,

Figure 359a is a frame from the Camera 6 video clip showing the north face of WTC7 10.0 s ± 0.2 s after the east penthouse began to descend,[6] while Figure 359b is a frame from the same video clip, showing the north face of WTC7 11.0 s ± 0.2 s after the east penthouse began to move downward.[7]

As we have seen also in the cases of WTC2 and WTC1, if the top portion of WTC7 had "fallen" at near free-fall speed, it would have to have encountered no more resistance from the lower portions than it would have from air. But in fact the building disintegrated while falling, as if it encountered very *high* resistance. Here we have conditions that contradict one another, a contradiction that The National Institute of

Standards and Technology (NIST) fails to address, much less explain. The observed conditions, paradoxical within conventional parameters of thinking, are at the same time logically consistent with unusual energy effects that clearly both deserve and require explanation.

C. Smoke Rings

In some photos, the towers appear to be blowing "smoke rings."

Figure 360. (9/11/01) WTC1 is "blowing smoke rings."
http://911wtc.freehostia.com/gallery/originalimages/GJS-WTC19.jpg

D. Eye Witnesses

Firefighter Robert Byrne was evacuating WTC1 shortly before its final moments. When reaching the third or second floor, he came upon what he at first mistook as smoke, then later realized that that's not what it was:

Firefighter Robert Byrne

> I remember later on we went up to—I don't know if it was still on the 35th floor and that's when we all dove into the staircase because basically the whole tower shook and we heard the noise of something going on. We didn't know what it was.
>
> What it was was the south tower collapsing. We didn't know. Finally we got some sort of transmission on the radio saying there was a collapse on the 60th floor. Meanwhile the south tower happened to come down.

> We were still on a rest period. We started going back. We were supposed to meet up with another unit; I don't remember who it was. We made it as far as, I believe it was the 37th floor, and I believe it was a chief from the 11th Battalion that popped up on the staircase. His exact words were "*Drop everything and get out.*" We looked to Lieutenant Hanson, and he said, "Drop everything and get out." That's when we ...basically evacuated.
>
> I remember going up the stairs took us over the [sic] hour. Getting down the stairs took maybe ten minutes, not even. By that time the staircase was empty. The same staircase we took up was empty on the way down.
>
> We got as far as I believe the 10th floor, 10th or 15th—I'm not a hundred percent sure—and we knew something was bad at that time anyway.
>
> There was a radio transmission for—they needed help. Lieutenant Hanson told me to get out because my—when the chief told me to drop everything, because I'm a proby, I followed orders to the T, I guess, and I dropped everything, except for my bunker gear, of course. *But I dropped my Scott tank and everything.* When I got down to that floor, he said, "All right, Byrne, you don't have your face piece. Just get out of the building."
>
> Basically, I got as far as the third floor, where I ran into—it looked like there was a collapse down there. *It was pretty bad. It was all smoky and dusty. I thought it was smoke, and I got a little nervous. I was at the point where I was going to go up and get another Scott tank, but I realized it wasn't smoke.*
>
> That's when I saw it was a collapse. It looked like a collapse; either that or the collapse and [sic] just closed up the staircase, I think it was the second floor, third or second floor, whatever it was. That's where I ran into a Port Authority cop, and he directed me out.
>
> *It was a good thing I had my flashlight on still, because it was pitch-black.* I followed a pitch-black hallway, and that's where I ran into a group of civilians. When we got to the point, I think it was the lobby...[8] [emphasis added]

It seems probable that what Robert Byrne found on the second or third floor that blocked out all light was what I call *lather.* He described leaving the building with the group of civilians he had met and being about 150 feet away from the building saying "We were about 150 feet away when the building came down."[9] The width of the WTC tower was 200 feet, indicating that he was still very close to the building when it "came down." He described what I am calling *lather* shortly before he left the building, indicating that in all likelihood he encountered the *lather* within the last few minutes that the building was still standing. It is noteworthy that Byrne was about 150 feet from a 1368-foot tall building and not only walked away unharmed but helped carry someone else along with him to safety.

Firefighter Hugh Mettham records another oddity.

Firefighter Hugh Mettham

> Suddenly the north tower starts to rumble and shake violently. We all head to stairwell B and huddle near the door while *the floor we are on shakes and rumbles for 30 seconds for [sic] more.* Lights go out, and we are thrown into total darkness. Stairways and hallways fill with smoke and dust as the rumble and roar subsides.
>
> All of us start to speculate on the cause of the rumble. It could be another

plane crash or localized collapse. Someone in the darkness mentions a bomb. There is little time now to ponder what just happened. What could shake the north tower so violently?

Lieutenant Borega tells us to start a lighting relay with our flashlights to assist the civilians that are leaving the staircases and converging with us on the fifth floor. Apparently smoke and dust had filled it up or down the staircases and elevator shafts, causing civilians to look for another way down.

Communication abruptly stops on our department radios. Only a few mayday and urgent messages are heard and then abruptly end.[10]

Note first that the shaking and rumbling inside the building took place, according to Mettham, for about half a minute "or more," which is far beyond the actual time it took the buildings to go away. They are coming apart, in other words, prior *to* their disappearance, which is a phenomenon consistent with the interference of energy causing a resonance within the structure. And while many things can cut off radio communications, one of them is high electromagnetic energy, which could, conceivably, jam transmissions.

We pick up with Mettham's statement just after he has reached the outdoors and the North Tower begins to "collapse." Mettham also notes other oddities.

Hugh Mettham continues

...near the corner of Vesey and West Street, we hear a terrible roar behind and above us. *The upper section of the north tower begins to collapse and push out a wave of smoke, ash and debris.*

We run in fear of our lives. The entire north tower collapses, and we are not going to outrun the tremendous cloud of ash and debris that is toppling down and bulging outwards.

Kevin and I dive under a fire truck while Lieutenant Borega, Harry and Steve take refuge behind or under other parked apparatus. I hug the ground in a fetal position, protecting my eyes and face from pellets of concrete that are bouncing off my fire helmet and body.

A dark cloud of dust and ash engulfs the undercarriage of the fire truck. I gasp for air and pull my hood over my mouth and nose. The noise from the collapse is intense.

Shall I continue?

Q. Sure.

A. Unexpectedly, the loud crashing sound subsides. It is strangely quiet.

[...]

The dark brown cloud begins to diminish. I make out outlines of vehicles ahead. Kevin moves next to me. Squinting and peering through the haze, we try to focus our eyes while stumbling and moving north on West Street.

We leave the terrible fire and destruction behind us and start to look for Lieutenant Borega and the rest of Ladder 18. A few minutes later emergency workers and firefighters can be seen moving in all directions. One stops and tells Kevin and me to sit down on the curb. He pours water on our heads, and we try to clean away some of the grit that is lodged around our eyes.

With our vision improved, Kevin and I continue to look for Ladder 18.

> Lieutenant Borega, Charlie, Harry and Steve spot us about ten minutes later. It was a tremendous relief we see all of them, especially Charlie Maloney who had left us earlier inside the tower.
>
> We also find Bobby Newman from Battalion 4 who tells us how he narrowly escaped from the collapse of both towers. *He surely looks that way with both of his shoes blown off his feet.*[11] [emphasis added]

We now have a remarkable list of things: (1) interrupted radio communications, (2) shaking and rumbling and *lathering* of the building for a full half a minute *prior to* its "collapse," and (3) a firefighter who somehow manages to lose his boots in the process. While there are many possibilities that could explain the first, controlled demolitions would not really explain the second (for why was there no seismic signature from that half-minute?). With the *lathering and rumbling*, it must be that we are in the presence of another mechanism, one that was *not* recorded on local seismographs but that the fireman clearly felt. Again, this is consistent with the types of effects one obtains when interfering fields of electromagnetic energy within a zone or region. Such a thing might also account for the abrupt cessation of radio communications. And, as they themselves attest, survivors of lightning strikes might be blown clean out of their shoes. But it seems that whatever caused the building to disintegrate may also have caused Bobby Newman's shoes to disintegrate.

Is there an example of such directed energy technology that we might consider as a possible source for the anomalous types of damage we have discussed? There is, and we'll look at it now.

---

[1] *http://upload.wikimedia.org/wikipedia/en/c/c5/DustifiedWTC2.jpg,* (now removed), *http://hereisnewyork.org//jpegs/photos/1539.jpg,*

[2] File No. 9110191, "Assistant Chief Fire Marshal Richard McCahey," November 2, 2001, pp. 14, *http://graphics8.nytimes.com/packages/pdf/nyregion/20050812_WTC_GRAPHIC/9110191.pdf,* emphasis added.

[3] *http://www.magnumphotos.com/Archive/C.aspx?VP=Mod_ViewBox.ViewBoxZoom_VPage&VBID=2K1HZOQIW7CLO&IT=ImageZoom01&PN=985&STM=T&DTTM=Image&SP=Search&IID=2K7O3RBZ4JP6&SAKL=T&SGBT=T&DT=Image*

[4] *http://www.magnumphotos.com/Archive/C.aspx?VP=Mod_ViewBox.ViewBoxZoom_VPage&VBID=2K1HZOQIW7CLO&IT=ImageZoom01&PN=985&STM=T&DTTM=Image&SP=Search&IID=2K7O3RBZ4JP6&SAKL=T&SGBT=T&DT=Image*

[5] The intensities have been adjusted, column and floor numbers have been added, as stated in the NIST report.

[6] The intensity levels were adjusted, as stated in the NIST report.

[7] *http://wtc.nist.gov/media/NIST_NCSTAR_1-9_Vol1_for_public_comment.pdf,* p. 283 (pdf p. 327 of 404),

[8] File No. 9110266, "Firefighter Robert Byrne," December 7, 2001, pp. 6-9, *http://graphics8.nytimes.com/packages/pdf/nyregion/20050812_WTC_GRAPHIC/9110266.pdf,* emphasis added.

[9] File No. 9110266, "Firefighter Robert Byrne," December 7, 2001, p. 11, *http://graphics8.nytimes.com/packages/pdf/nyregion/20050812_WTC_GRAPHIC/9110266.pdf,* emphasis added.

[10] File No. 9110441, "Firefighter Hugh Mettham," January 10, 2002, pp. 7-8, *http://graphics8.nytimes.com/packages/pdf/nyregion/20050812_WTC_GRAPHIC/9110441.pdf*

[11] File No. 9110441, "Firefighter Hugh Mettham," January 10, 2002, pp. 7-14, *http://graphics8.nytimes.com/packages/pdf/nyregion/20050812_WTC_ GRAPHIC/9110441.pdf,* emphasis added.

16.5
Perceptual Conformity

Believe none of what you hear and half of what you see.
—Benjamin Franklin

Any sufficiently advanced technology is indistinguishable from magic.
—Arthur C. Clarke

...[This is] the story of the first military jet aircraft to fly in the United States—an aircraft that apparently no one could see. The date was 1942; the location was Muroc Army Air Field (today Edwards Air Force Base). Whenever it was on the ground, the P-59 was fitted with a fake propeller for the sake of secrecy. Unfortunately for secrecy, at the local watering hole, test pilots mixed with P-38 pilots stationed nearby. After slugging down a few drinks, the test pilots bragged about flying a propellerless aircraft and were immediately labeled as liars by the P-38 crowd—fighting words for sure. Subsequently, test-pilot Jack Woolams decided to put them in their place, not with his fists but with something far more effective.[1]

(a) *Secret-Bell P-59*

(b) *Secret-Bell P-59, incognito*

Figure 361. Secret-Bell-P-59, incognito.
(a)http://sbiii.com/avpix/p59jaf03.jpg, (b)http://fredsboringpictures.com/images/Chino%2006/images/4-0006-Secret-Bell-P-59-of-.jpg

He rented a gorilla suit and took off wearing it along with a big cigar protruding from his mouth and a derby hat on his head. Once airborne, he found a lone P-38 pilot, pulled alongside, giving the P-38 pilot a clear view of the jet and gorilla suit, then waved, much to the shock of his intended target. The next day when queried at the local watering hole, not a single P-38 pilot had seen an "escaped gorilla" or knew anything about it. The explanation: why of course, it must be that P-38 pilots could only see what they believed was possible. Yeah, right. Apparently, the P-38 pilots never again questioned the possibility of propellerless aircraft, let alone the honesty of test pilots.

(a) Bell P-59 Airacomet[2] *with Bell P-63 Kingcobra (background)*

(b) Artist's image of P-59 pilot

Figure 362. Bell P-59 and Bell P-63 Kingcobra, side-by-side, and P-59 pilot.
(a)http://www.military.cz/usa/air/post_war/p59/p59&p63king.jpg, (b)http://ny-image0.etsy.com/il_155x125.37892272.jpg

Although the events are not even a century old, already there are more than one version of the Jack Woolams tale. All are slightly different. One version relates that there were multiple sightings of the gorilla-piloted jet and that the base psychiatrist talked several P-38 pilots out of believing what they saw.[3,4] Who knows? The fact is, that even if someone sees and believes a phenomenon, it doesn't mean they will honestly talk about it. And if they do, it doesn't mean that the details will be perfectly remembered in the historical record—especially if there isn't one.[5]

---

[1] Gorilla Story, *http://yarchive.net/mil/jet_gorilla.html, 2005*

[2] American postwar aircraft, *http://www.military.cz/usa/air/post_war/p59/p59_en.htm*

[3] A Flyin' Gorilla?, by William B. Scott, *http://www.vintage-aircraft-7.org/newsletters/VAA7_NL_0702.pdf*

[4] *A FLYIN' GORILLA?,* by William B. Scott, submitted by Dave Petri, PIKE'S PEAK FLYER (newsletter), The Voice of EAA Chapter 72, Meadow Lake Airport - Colorado Springs, Colorado, August 2002, pp.2-3, *http://www.eaa72.org/news/2002/aug02news.pdf*

[5] *http://www.intuitor.com/moviephysics/bleep.html*

17.
The Tesla-Hutchison Effect

I never let schooling interfere with my education. —Mark Twain

If we all did the things we are capable of doing, we would literally astound ourselves.
—Thomas Alva Edison

It's not difficult really - the secret is in knowing how. —Edward Leedskalnin

A. Introduction

Many people have criticized my research into the destruction of the WTC complex because I have not named the exact technology that was used in the destruction, including its make, model, and serial number. But it is erroneous to blindly discard evidence that does not conveniently describe a known weapon or blindly discard evidence that contradicts one's pet theory. Remember, *empirical evidence is the truth that theory must mimic,*[1] not the other way around. The pages of this book include a very great amount of evidence–evidence that must be explained. This evidence unarguably rules out kerosene-fueled fire, conventional controlled demolition, thermite (and its variations) and mini nukes as being the cause of the WTC destruction or even making a significant contribution to it. At the same time, the evidence in this book also strongly implicates a particular class of technology. This class of technology produces effects on various materials that are similar to the effects produced on various materials by whatever exact technology was used in the WTC destruction.

Jellification	Reduced mass of material
Bent Beams	Rounded Holes in Glass
Slow Bending of Metals	Lather
Shredded Metal Structures	Fuming
Fractured Metal Structures	Crumbling
Peeling appearance	Transmutation
Fusion of Dissimilar Materials	Weird Fires
Thinning and Rapid Aging	Melting Without Heat
Lift or Disruption	Metal Luminance Without Heat
Toasted-Looking Metal	EVO[2] Strikes Abounding in Sample
Circular holes in material	Propulsion—-Both Slow and Impulsive

Table 15. Characteristics of the Hutchison Effect and the WTC remains.

Table 15 shows a correlation between phenomena observed in the WTC destruction and phenomena consistent with what has come to be known as "*The Hutchison Effect.*" A Canadian inventor and experimental scientist, John Hutchison has been working with "field effects" for over thirty years. Notably, almost all of these effects are the result of interferometry, that is, the result of interfering several beams or fields of varying frequencies of electromagnetic energy in a target zone. A beautifully-descriptive analogy was recently given to me by a former student who now designs audio systems.[3] He noted that if you locate speakers a certain way, you may end up with dead zones where the sound is canceled or other zones where glass and china could be damaged. This is a type of destructive and constructive interference, where the interfering sound waves cancel or intensify the effects, respectively. Also, anyone familiar with the old rabbit-ears TV reception can probably remember the interference from jet airplanes.[4] Jet airplanes interfered with the TV signal.

The result of the interference created using the Hutchison Effect sets up a kind of "grid" or template within a target zone that causes materials in that zone to undergo extreme modification and, in some instances, disruption from within. Think of photographic emulsion as an analogy. Light does not burn an image into a piece of paper, but it triggers a chemical reaction in the photographic emulsion. Light controls where the reaction will take place. When you develop the film, you get the picture. Controlling how much light goes where on the photographic emulsion is what produces the image in a photograph. So think of the field effect, or the general field, as being the photographic emulsion. And then another beam tells where the reaction will occur.

John Hutchison

In the words of John Hutchison:

> I use a different combination of radio waves, along with threshold high voltage, and electrostatic operators. And I only use weak magnetic fields to steer the electrostatic [field] around...and to conform to certain patterns. Now mind you, this is a fairly accidental discovery I made. And I had a real problem trying to replicate it in the early years. And I didn't think it was important, but it leaked out into the scientific community and they started coming down with other scientists, and they started asking for demonstrations, which I did give.[5]

The Hutchison Effect is not just a singular effect but a combination of many. The effects it produces can include the following:

1) Levitation of heavy objects,
2) Fusion of dissimilar materials, such as metal and wood (as portrayed in the movie, The Philadelphia Experiment),
3) Anomalous "melting" of metals without burning adjacent material,
4) Spontaneous fracturing of metals (sliding sideways),
5) Both temporary and permanent changing in crystalline structure and physical properties

What Hutchison demonstrates is not a new technology, but an old one that he has rediscovered and developed. The Hutchison Effect is in fact a collection of phenomena that John Hutchison discovered in 1979 when he was making attempts to study the longitudinal waves of Nikola Tesla as well as the work of others, such as Thomas Townsend Brown, George Piggott, Francis E. Nipper, and Martin L. Perl. Here is a brief introduction to some of these scientists. More information is available at the end of this chapter.
Also see: *http://drjudywood.com/towers*

B. Some History of this Technology

1. Nikola Tesla (1856-1943)[6]

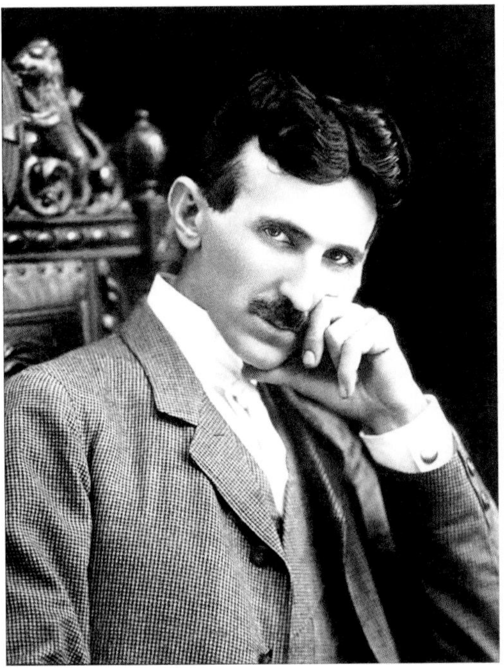

(a) circa 1896 *(b)*

Figure 363. Nikola Tesla, electrical engineer and inventor. (July 10, 1856 - January 7, 1943)
(a)http://upload.wikimedia.org/wikipedia/commons/d/d4/N.Tesla.JPG,
(b)http://www.serbianunity.net/people/tesla/images/tesla-38-big.jpg,

Nikola Tesla was an inventor in the fields of mechanical and electrical engineering. He is best known for his many revolutionary developments in the field of electromagnetism in the late 19th and early 20th centuries, especially alternating current (AC). This includes the polyphase system of electrical distribution and the AC motor, which helped usher in the Second Industrial Revolution. Our world runs on AC power and wireless communication; few recognize that these very significant contributions were made by Nikola Tesla.[7] Tesla first demonstrated wireless communication (radio) in 1894.[8] The Serbian Unity Congress provides the following:

> Nikola Tesla, American scientist of Serbian origin, gave his greatest contribution to science and technological progress of the world as the inventor of the rotating magnetic field and of the complete system of production and distribution of electrical energy (motors, generators) based on the use of alternate currents. His name was given to the SI unit for magnetic induction ("tesla"). Tesla also constructed the generators of high-frequency alternate currents and [the] high-voltage coreless transformer known today as "Tesla Coil".
>
> Nikola Tesla was one of the greatest electrical inventors who ever lived, and he is probably the first cross-over scientist. He is known and respected in scientific and engineering circles, but he also appeals to a youthful and general audience with no formal background in science. *He was a contemporary of Thomas Edison and Guglielmo Marconi, and these two men are frequently credited for Tesla's invention*

of AC power transmission and radio. His technological achievements transformed America from a nation of isolated communities to a country connected by power grids where information was available upon demand. In the 20th century, it was Tesla's technology that united the United States and eventually the world. [emphasis added]

Tesla was also a visionary thinker, and in his papers and interviews he anticipated the development of radio and television broadcasting, robotics, computers, faxes, and even the Strategic Defense Initiative. Tesla's great dream was to find the means to broadcast electrical power without wires in between.

[. . .]

Tesla was so far ahead of his time that many of his ideas are only appearing today. His legacy can been[sic] seen in everything from microwave ovens to MX missiles. But more than this, Tesla's life inspires us to believe that anything we can imagine can be accomplished—especially with electricity.[9]

2. George Piggott—Overcoming Gravitation (late 1800s-1900s)

George Piggott applied for a patent for his "Space Telegraphy" machine June 19, 1903, and it was granted October 24, 1911. (Pat. No. 1006786) This machine transmitted "radiant electric energy."

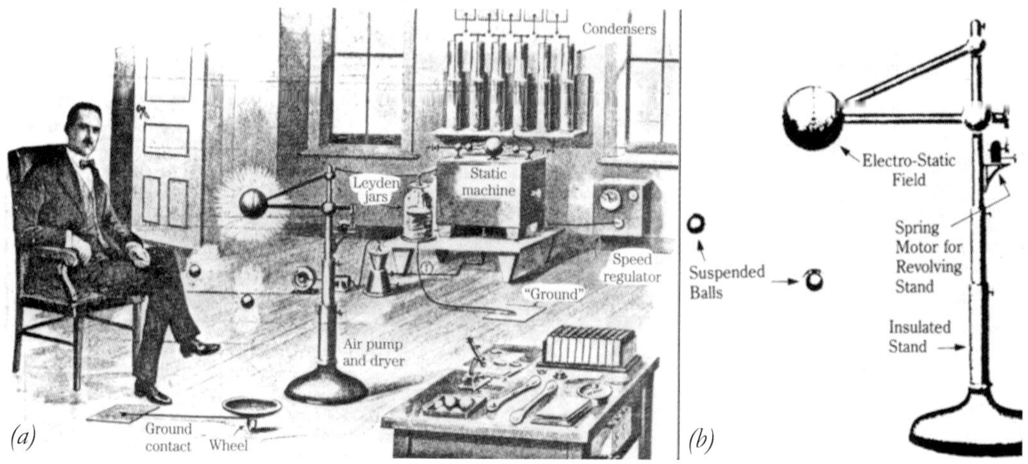

Figure 364. (Illustration: 1903-1920) (a) Mr. George Piggott and his lab. He was able to suspend silver balls, each with a diameter of 11 mm and weighing 1.3 grams. (b) A close-up diagram of the charged metal sphere he used, mounted on a revolving stand with a spring-driven motor, which he used to levitate the two silver balls.[10]

(a)http://rexresearch.com/piggott/fig1.jpg, (b)http://rexresearch.com/piggott/fig3.jpg,[11]

3. Edward Leedskalnin (1887-1951)[12]

Edward Leedskalnin,[13] a Latvian emigrant, is known for his unusual understanding of the interaction between magnetism and gravity. He single-handedly built the home he called Coral Castle, in Florida City, cutting and moving limestone pieces weighing up to 35 tons[14] using simple tools and a chain hoist that could not in "real" terms support such a load. He later moved his Coral Castle about ten miles to Homestead, Florida.[15] As given in Wikipedia,

> Working alone at night, Leedskalnin, who weighed less than 100 lb (45 kg), eventually quarried and sculpted over 1,100 short tons (1,000 t) of coral into a monument that would later be known as the Coral Castle…
>
> [. . .]
>
> Leedskalnin's other four pamphlets addressed his theories on magnetism, detailing his theories on the interaction of electricity, magnetism and the body; Leedskalnin also included a number of simple experiments to validate his theories.[16]

The above refers to 1,100 short tons or 1,000 [metric] tons. So, it is assumed the following is referring to metric tons for Ed Leedskalnin's rocks:

> It's estimated that 1,000 tons of coral rock were used in construction of the walls and towers, and an additional 100 tons of it were carved into furniture and art objects:
>
> - An obelisk he raised weighs *28 tons.*
> - The wall surrounding Coral Castle stands 8 ft. tall and consists of large blocks each weighing several tons.
> - Large stone crescents are perched atop 20-ft.-high walls.
> - A 9-ton swinging gate that moves at the touch of a finger guards the eastern wall.
> - The largest rock on the property weighs an estimated *35 tons.*
> - Some stones are *twice the weight of the largest blocks* in the Great Pyramid at Giza.[17] [emphasis added]

(a) Coral Castle *(b) John Hutchison at the Great Pyramid*

Figure 365. (a) Ed Leedskalnin's Coral Castle. (b) John Hutchison, at the Great Pyramid.
(a)http://image22.webshots.com/22/0/90/85/239009085SNJrWQ_fs.jpg, (b)John Hutchison photo

Let us compare the Coral Castle with the Great Pyramid of Khufu, the largest single building ever constructed.[18]

> Researchers estimate that 2.3 million blocks were used to build the Great Pyramid[18], with an *average weight of about 2.5 metric tons* per block.[18] *The largest block weighs as much as 15 metric tons.*[18,19,20] [emphasis added]

Ed Leedskalnin indeed appears to have moved and lifted rocks weighing more than *twice* the weight of the largest block in the Great Pyramid.

Figure 365a is a photo of the living quarters Ed Leedskalnin built for his Coral Castle. Note the height of the blocks relative to the doorway. In Figure 365b, John Hutchison stands in front of the Great Pyramid, while a person at the base is shown for scale. Ed Leedskalnin, with only a fourth-grade formal education, cut and lifted blocks for his Coral Castle that were about the same size as those used to build the Great Pyramid and is thus said to have discovered the secrets of Stonehenge and the great pyramids.[21] The Great Pyramid of Khufu was built for the pharaoh Khufu (Cheops to the Greeks), the second king of the Fourth Dynasty (c. 2575 to c. 2465 BC).[22]

(a) Thomas Townsend Brown[23] *(b) Edward Leedskalnin*[24]

Figure 366. Thomas Townsend Brown and Edward Leedskalnin.

(a)http://www.reallycoolscience.com/images/brown-01.jpg, (b)http://www.intalek.com/Index/Projects/CoralCastle/Photos/CoralCastle01.jpg, http://www.intalek.com/Index/Projects/CoralCastle/CoralCastle.htm

In Figure 366b, Edward Leedskalnin demonstrates the electrical generator he designed and invented. It was his only source of power for his radio and his experiments involving levitation and stone cutting.[25] According to the official Coral Castle website, "two-time Grammy nominated rock artist Billy Idol wrote and recorded his song *Sweet Sixteen* as a tribute to Ed and the Coral Castle. The video was also shot on location at the Castle."[26]

4. Thomas Townsend Brown (1905-1985)

Brown was blessed with the unique ability to *"see what others have seen and think what nobody has thought."* As a teenager in the 1920s, working in a well-equipped laboratory in the basement of his prominent Ohio family's opulent home, Brown noticed an unusual effect when high voltage was applied to a Coolidge X-Ray tube. With that observation, he came to believe he had discovered a link between electricity and gravity and a way to lift and propel flying vehicles by purely electrical means.[27]

5. Project Winterhaven (1953-)

Project Winterhaven was an important and ambitious proposal calling for the continuation of the study of field effects, an area of science that T.T. Brown himself (1905-1985) had already explored to a remarkable and advanced extent.

In 1938, Brown formed The Townsend Brown Foundation as a non-profit corporation in the state of Ohio, and it was under the auspices of this Foundation that in 1953 he proposed Project Winterhaven. The program of research Brown was calling for (his appeal was titled "A Proposal for Joint Services Research and Development Contract") included the use of researchers and facilities at The Franklin Institute in Philadelphia, the Institute for Advance Studies at Princeton, the University of Chicago, the Stanford Research Institute, and others. Above all, the project had to do with "Research on the Control of Gravitation."

> For the last several years, accumulating evidence along both theoretical and experimental lines has tended to confirm the suspicion that a fundamental interlocking relationship exists between the electrodynamic field and the gravitational field.
>
> It is the purpose of Project WINTERHAVEN to compile and study this evidence and to perform certain critical or definitive experiments which will serve to confirm or deny the relationship. If the results confirm the evidence, it is the further purpose of Project WINTERHAVEN to examine the physical nature of the basic "electro-gravitic couple" and to foresee and develop possible long-range practical applications.[28]

In exploring the "electro-gravitic couple," Brown had already brought about the levitation of materials in his own experimentation, but he was convinced there was much more to be learned about the process he had begun to control:

> In further confirmation of the existing hypothesis [namely, that "a gravitational field can be effectively controlled by manipulating the space-energy relationships of the ambient electrostatic field"], experimental demonstrations actually completed in July 1950, together with subsequent confirmations with improved materials, tend to indicate that a new motive force, useful as a prime mover, has in reality been discovered. While the first experiments with new dielectric materials of higher K indicated the presence of a noteworthy force, the tests were mainly qualitative and imperfect because of other factors, and the ultimate potential in terms of thrust still remains highly theoretical. The behavior of the new motive force nevertheless does appear to be in agreement with the

> hypothesis that *there is an interaction between the electrical field and the gravitational field and that this interaction may be electrically controlled.*[29] [emphasis added]

Brown felt certain that with further research and development, applied control of this "interaction" would make possible enormous advances not only in communication but also and more notably in propulsion:

> It is believed by the sponsors of Project WINTERHAVEN that the technical development of the electrogravitic reaction would usher in a new age of speed and power and of revolutionary new methods of transportation and communication. Theoretical considerations would predict that, because of the privilege of sustained acceleration, top limits of speed may be raised far beyond those of jet propulsion or rocket drive, with possibilities eventually of approaching the speed of light in "free space." The motor which may be forthcoming will be essentially soundless, vibrationless and heatless. As a means of propulsion in flight, its potentialities already appear to have been demonstrated in model disc-shaped airfoils, a form to which it is ideally adapted. These model airfoils develop a linear thrust like a rocket and may be headed in any direction.[30]

Of the "model disc-shaped airfoils" that he had already created and demonstrated, Brown wrote that

> The discs contain no moving parts and do not necessarily rotate while in flight. In atmospheric air they emit a bluish-red electric coronal glow and a faint hissing sound.[31]

What advances actually took place in the study of Brown's concept of "the basic 'electro-gravitic couple'" is something not fully known. Wikipedia gives some sense of the uncertainty surrounding that question:

> Top U.S. aerospace companies had also become involved in such research (see United States gravity control propulsion research (1955 - 1974) which may have become a classified subject by 1957. Others contend Brown's research simply reached a dead end and lost support. Though the effect he discovered has been proven to exist by many others, Brown's work was controversial because others and even he himself believed that this effect could explain the existence and operation of unidentified flying objects (UFOs).[32]

For continued reading, go to *http://www.ttownsendbrown.com/*

6. John Hutchison (1945-)[33]

My looking at the characteristics of the Hutchison Effect must not be taken as synonymous with my arguing that it is *the exact device* responsible for the mechanism of the WTC destruction. I use the Hutchison effect as an *illustrative example or model* of what kinds of known extraordinary effects can result from electromagnetic interference. I use Hutchison's work also because a great deal of information is available about it, especially on the internet. A selection of links to information is included in these pages.[34]

All of the effects and events that were observed at or in the vicinity of the World Trade Center during and after its destruction are comparable with observed results of the *Hutchison Effect* on various materials. The Hutchison Effect has similarities to other known technologies (shown in Table 16) as well as to naturally-occurring events of the kind to be discussed in Chapters 18-20.

Hutchison Effect	Slow Bending of Metals Shredded Metal Structures Fractured Metal Structures Propulsion—-Both Slow and Impulsive Melting Without Heat Metal Luminance Without Heat EVO[35] Strikes Abounding in Sample
Cold Fusion	Thermal Cycling Method Has EVO Evidence Gas Discharge Method Has EVO Evidence Electrolytic Method Has EVO Evidence Sonic Method Has EVO Evidence
The energetics of these technologies all have a common basis in electron clustering	Plasma focus Hutchison Effect Adamenko Work EVO energy production & transmutation Cold fusion (LENR)

Table 16. The Hutchison Effect, Cold Fusion, and EVOs.[36]

C. Weird Fires

As we have seen, fires near the toasted cars failed to ignite close-lying paper. Some photos show firemen walking very close to or even through these flames. Are they "cold" fires?

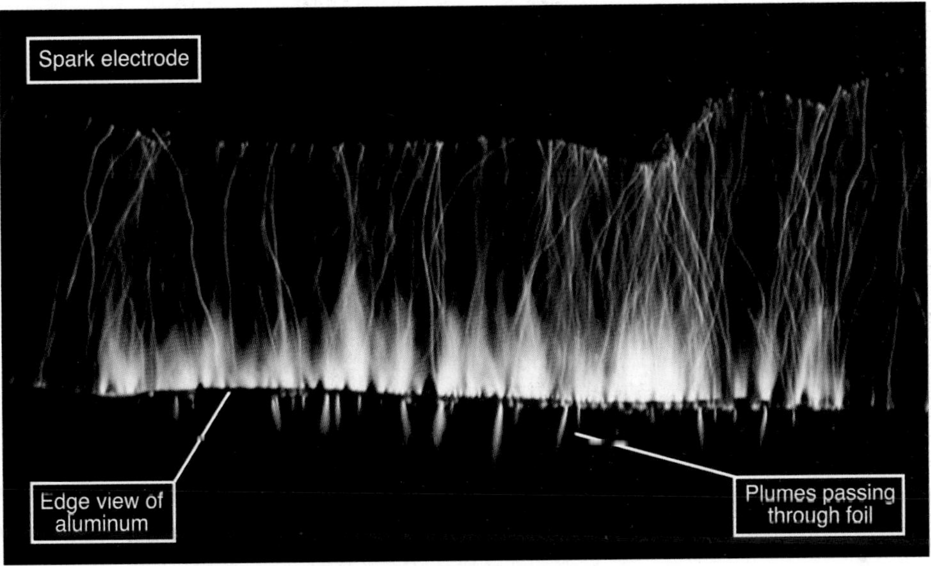

Figure 367. Side view of aluminium foil, coated with silicon carbide, being subjected to EVOs.[37]
The small jets seen coming from under the foil have penetrated through it.
Image: Ken Shoulders[37] *http://blog.hasslberger.com/img/EVO.jpg*

Richard Sparks described several of the phenomena seen with the Hutchison Effect, including luminance.

> A less frequent phenomenon is the apparent heating to incandescence of iron and steel specimens having high length to width ratios. This event is not accompanied by the [apparent] heating, [apparent] charring or the burning of combustible materials in contact with the specimen throughout the duration of the event of about two minutes. The fracturing of certain iron and steel geometries accompanied *by an anomalous residual magnetic field, permanent in nature, is not uncommon.*[38] [emphasis added]

Let us remember that hot things glow, though not everything that glows is hot. Now let us consider some anomalies at the WTC.

Figure 368. (9/11/01) Weird fires near the corner of Church and Liberty Streets.
(a)http://www.pixelpress.org/september11/sept11_pix/index_pix/alan2.jpg, (b)http://images6.fotki.com/v1/photos/1/115/34570/DSC_0195-vi.jpg

D. Bent Beams

Samples that John Hutchison has produced show highly unusual effects on metal. Sometimes severe bending occurs:

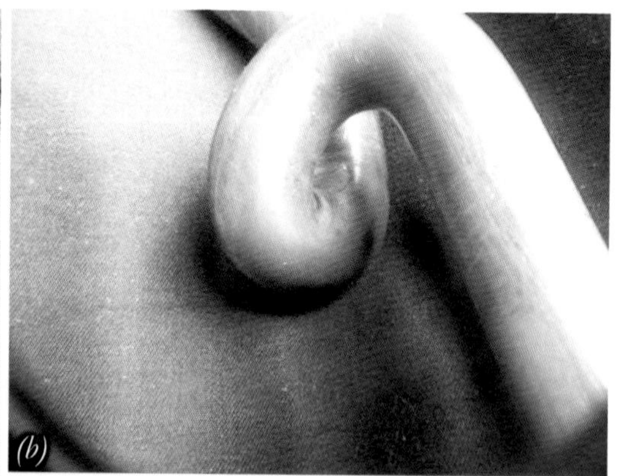

Figure 369. Solid circular bars, 2.5-inch or 3-inch diameter, bent from the Hutchison Effect. (a) Solid copper bar.[39] *(b) Solid molybdenum bar, now in the possession of Col. John Alexander.*[40]
Images courtesy of John Hutchison.

Now consider the WTC:

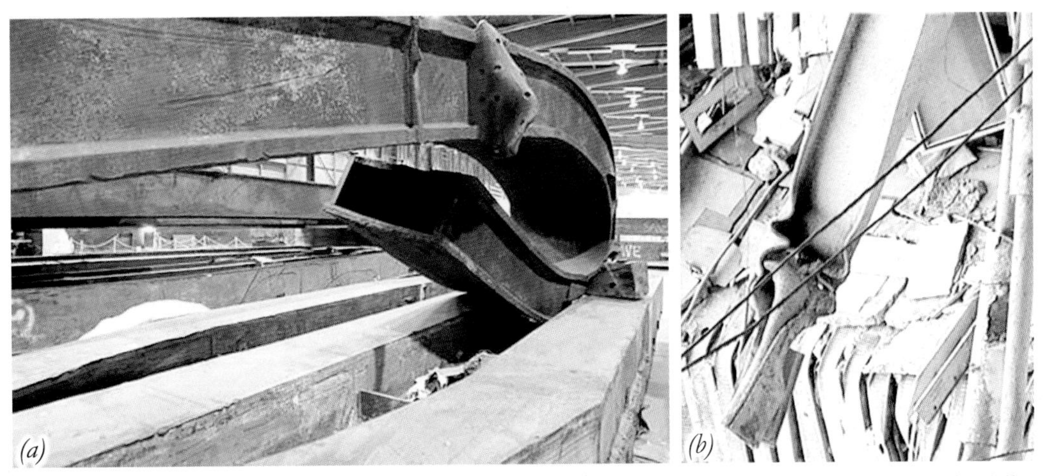

Figure 370. (a) WTC core column curled, not buckled. A gravity-driven "collapse" would not do this. The beam above has smooth curves, without kinks. (b) This beam shriveled up and has see-through holes in it, hanging in the gash/opening of Bankers Trust which had no fires.[41]
(a)http://911research.wtc7.net/wtc/evidence/photos/docs/hanger17/core1.jpg, (b)http://www.fema.gov/pdf/library/fema403_ch6.pdf, Fig6-10, page 6-9.

E. Jellification

In Hutchison's experiments, the metal sometimes "jellifies." Other effects are also seen.

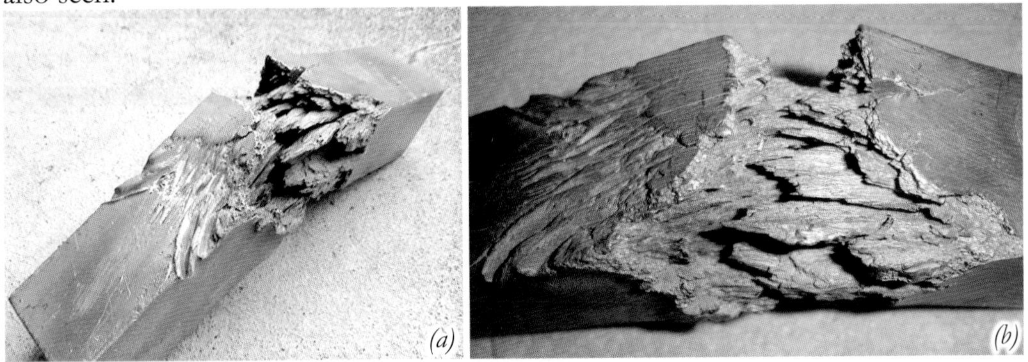

Figure 371. Split-open aluminum bar resulting from the Hutchison Effect.
(a)http://americanantigravity.com/graphics/Hutchison-Photos-Sept05.zip, (b)personal photo[42]

And now consider this beam from the WTC:

Figure 372. Steel columns from the WTC that look like paper maché.
http://wtc.nist.gov/images/WTC-003_hires.jpg[43]

F. Peeling appearance

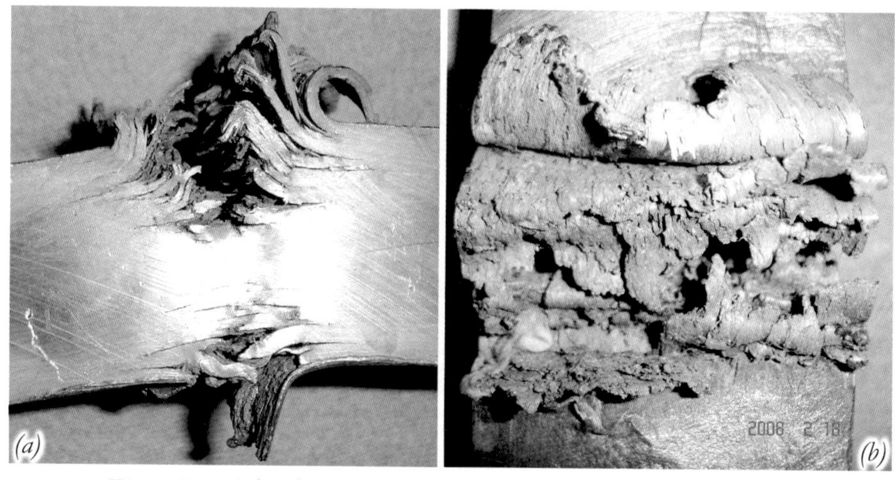

Figure 373. (2/19/08) Hutchison Effect sample of peeled aluminum.[42]

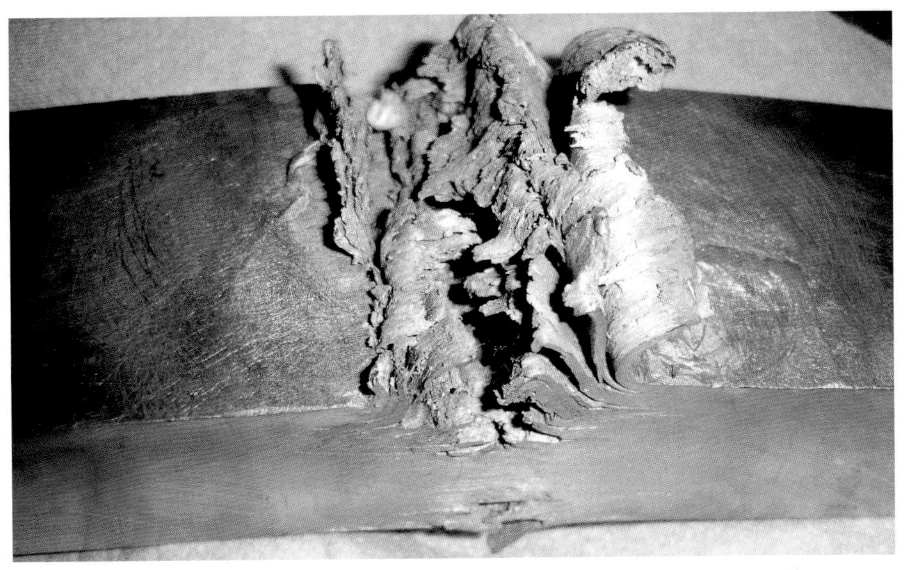

Figure 374. (2/19/08) Hutchison Effect sample of peeled aluminum.[42]

Figure 375. (a) The door of a truck appears peeled down at a split-thickness as if it delaminated. But these doors are not laminated. (b) (9/13/01) A car near FDR Drive with this peeling appearance.
(a)http://hereisnewyork.org//jpegs/photos/1805.jpg, (b)http://nyartlab.com/bombing/09-13/DSC08111.jpg

G. Fusion of Dissimilar Materials

Figure 376. Wood in aluminum, Hutchison-Effect sample.
(a)http://www.americanantigravity.com/graphics/Hutchison-Photos-Sept05.zip, (b)http://www.americanantigravity.com/graphics/Hutchison-Photos-Sept05.zip

Another result obtained in Hutchison's experiments is the fusion of entirely dissimilar materials, such as wood and aluminum, as seen in Figure 376.

And now the WTC: Figure 377a shows the only file cabinet found in the WTC remains. Figure 377b shows coins fused together.

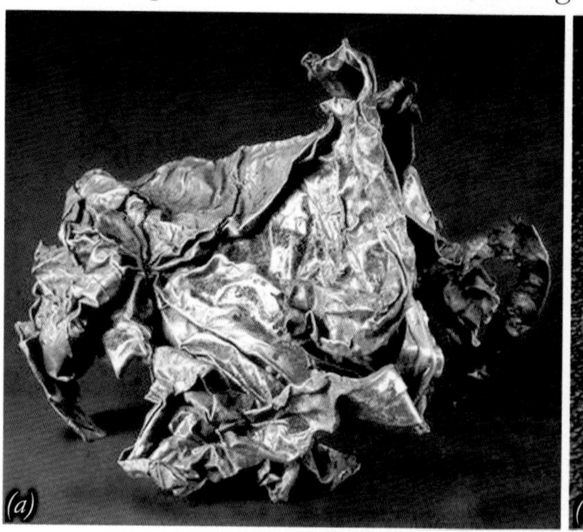

Figure 377. (a) File cabinet from Ben & Jerry's. (b) Fused coins.
(a)http://americanhistory.si.edu/september11/images/large/40_92.jpg,[44] (b)http://americanhistory.si.edu/september11/images/large/70_131.jpg,[45]

The U.S. *penny*[46] is made of zinc (97.5%) with a copper plating (2.5%).
The U.S. *nickel*[47] is made of copper (75%) with a nickel outside (25%).
The U.S. *dime*[48], *quarter*[48], *half-dollar*[49] is copper (91.67%) with a nickel outside (8.33%).

Table 17 on page 368, which cites the melting and boiling temperatures of various elements, shows us that a penny would have exploded if the temperature had been above 907°C, the boiling point of zinc. Even more importantly, all of the coins would have melted before the temperature rose high enough to melt iron. (See Table 26 on page 496 for these and other relevant elements.)

David Shayt September 11 Collecting Curator, Museum Specialist (Division of Cultural History), describes below how he got the file cabinet. The cabinet came from the Ben & Jerry's ice cream shop, which was located between WTC1 and WTC2, adjacent to the northeast corner of WTC3. It is shown on the map in Figure 183 on page 190.

David Shayt, Museum Curator

> It is a file cabinet, but it doesn't look like a file cabinet. It's a two-drawer file cabinet from the Ben and Jerry's ice cream shop. I saw it first when it was a ball of metal about the size of a basketball, delivered to the Port Authority Police compound because there was money coming out of it--little edges of $20 bills were seen in this compressed ball of metal, indicating it was something other than a ball of metal.
>
> [...]
>
> They logged in the money and were about to pitch the carcass of the file cabinet in the trash when I realized, we've been looking for a file cabinet from the

> World Trade Center—typical office equipment. So I asked if we could have that. And they said, "Sure, catch." So I came away with that.
>
> In fact they washed it, they cleaned it of debris, and I drove back with it that night. *However, just in the space of two or three hours it began to rust.* I didn't want it rusty. It was still messy and shiny when I saw it, although completely crumpled. So I visited a food store in East Rutherford, New Jersey, and bought a bottle of, a can of oil, and sprayed it to stop the rusting from occurring before I could get back to the museum.[50]

If it had been exposed to high heat, wouldn't the $20 bills have burned? Why was a steel file cabinet rusting in two or three hours? Another view of the same file cabinet is shown in Figure 378a.

Figure 378. (a) The paper retained its color, indicating this file cabinet did not shrivel due to conventional heat. (Intensity and lettering adjusted for visibility.) (b) The so-called "meteorite" found in the remains showed a stack of paper embedded in other materials.

(a)http://thewebfairy.com/911/h-effect/filingcabinet.htm[51], (b)http://www.amny.com/media/photo/2006-08/24912492.jpg, (no longer available)

H. Thinning and Rapid Aging

At other times, materials subjected to the Hutchison Effect seem to change at an "elemental" level—could this be the explanation for the rapid rusting we observed earlier, as steel is turned into iron?

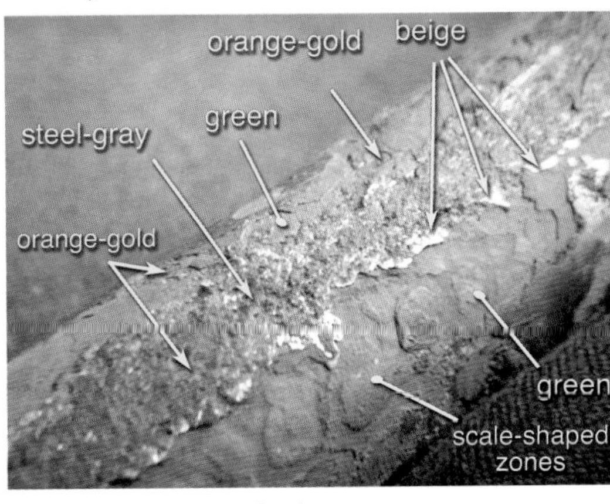

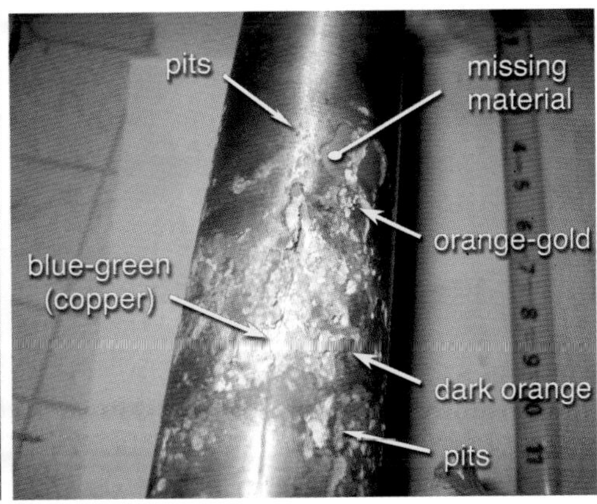

(a) Steel *(b) Stainless steel*

Figure 379. Rapid degradation of steel as a result of the Hutchison Effect.

(a)John Hutchison's blog[52], (b)John Hutchison's blog[53]

And again, the WTC:

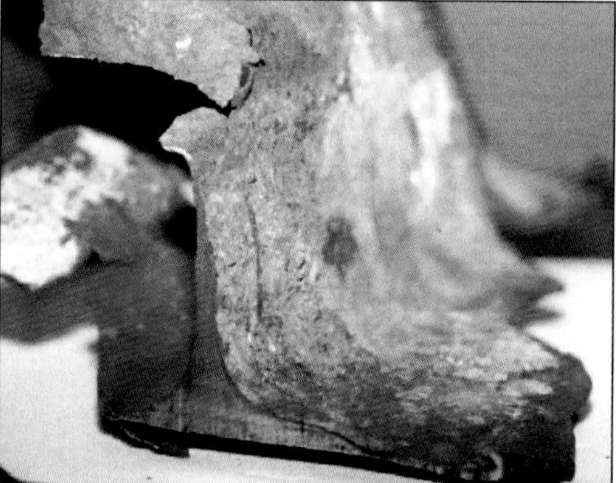

(a) (after 9/11/01) approx. 9/18/01 *(b) 2001*

Figure 380. "Rustification" and rapid aging of steel from the WTC.

(a)http://BocaDigital.smugmug.com/photos/10697258-O.jpg, http://BocaDigital.smugmug.com/photos/10697258_aaPUV-X2-1.jpg, (b)http://www.fema.gov/pdf/library/fema403_apc.pdf

I. Apparent Transmutation

Dr. Eugene Mallove, in his Foreword to Dr. Tadahiko Mizuno's book, *Nuclear Transmutation: The Reality of Cold Fusion,*[54] gave an excellent description of the emergence of the field of "cold fusion."

> When a scientific discovery seems to break all the rules, when it appears to violate cherished theories held for decades or hundreds of years, it breaks a fundamental paradigm and there is hell to pay. Such is the case in the scientific revolution that began on March 23, 1989, when electrochemists Drs. Martin Fleischmann and Stanley Pons announced "cold fusion" at a press conference at the University of Utah.
>
> In the history of science there will be few peaks higher or stranger than the discovery of cold fusion. From that moment, a long-held notion was to be smashed forever: that atoms could not change their nuclear identities in near-room temperature reactions–reactions that were presumed to be chemical, not nuclear. Following the Fleischmann-Pons announcement, intense scientific investigations in electrochemistry uncovered a whole new class of low-temperature nuclear reactions. The astounding claims of Fleischmann and Pons had involved primarily large excess energy production, *but also* tritium formation *and the appearance of low levels of neutrons.* Later, investigators began to observe heavier elements and strange isotopes that were not present when their experiments began…
>
> […]
>
> It is now clear that Fleischmann and Pons discovered the mere tip of an iceberg within physics and chemistry. This new realm may eventually be called *electro-nuclear reactions,** so encompassing has it become. It was not merely a new "island" of physics that had come into view, but a whole new continent. Other names have been put forward for these alchemy-like reactions: "chemically assisted nuclear reactions" or LENRs (low energy nuclear reactions). Whatever the name, it seems that twentieth century physics took a wrong turn long ago by denying that such reactions could occur. There may be an error in the foundations

of physics. [emphasis added]

* From what I have discovered in preparing this book, an even more accurate term may be *magnetic-electrogravitic-nuclear reactions.*

Transmutation, or nuclear transmutation, is the conversion of one chemical element or isotope into another, which occurs through nuclear reactions.[55,56] Elements such as iron ($_{26}Fe$), nickel ($_{28}Ni$), Copper ($_{29}Cu$), and zinc ($_{30}Zn$), are defined by the number of protons they possess. Changing the element requires changing the atomic (proton) number. Until the era of "cold fusion,"[57] it has been thought that such changing could occur only in the presence of enormous heat, such as that of a star or supernova.

> In nature, new elements are created by adding protons and neutrons to *hydrogen* atoms within the nuclear reactor of a star, producing increasingly heavier elements, up to *iron* (atomic number 26). This process is called nucleosynthesis. Elements heavier than iron are formed in the stellar explosion of a supernova. [58]

The Coulomb barrier is very high for such reactions, so it has been assumed that significant energy is required. However, this process discovered by Drs. Stanley Pons and Martin Fleischmann implies that the Coulomb barrier, no matter how large, can be reduced to insignificant values[59] by a process already available within a solid material, as if unlocking a door. So, with the discovery of "cold fusion" has come the discovery also that *transmutation* can actually occur at or near room temperature. This may in fact be the same mechanism that causes the Hutchison Effect.

As discussed earlier (e.g. smoke vs. fumes), words can bias our observations, especially when used derisively. Perhaps the inappropriate name of "cold fusion" helped to shape public opinion into believing it was nonsense instead of a newly discovered phenomenon that could provide clean energy at a fraction of the cost of other methods. As Dr. Edmund Storms points out, the term was an especially poor description.

> In 1989, Pons and Fleischmann [1] (P-F) caused a media storm by claiming to cause fusion to take place in an ordinary electrolytic cell containing D_2O. This process was first named "Cold Fusion" by Steven Jones—an especially poor description. The names "Chemically Assisted Nuclear Reactions" (CANR) and "Low Energy Nuclear Reactions" (LENR) more correctly describe the phenomenon. …
>
> The phenomenon is claimed to produce fusion as well as a complex mixture of transmutation reactions. Twelve different methods, listed in Table 1, have been reported to produce anomalous energy (AE) and/or nuclear products (NP), some with good reproducibility and some with difficulty.… Regardless of the explanation du jour, *more energy appears to be produced than is being applied*, thereby violating basic thermodynamic expectations. The important question is, *"Why"*? [59] [emphasis added]

Similar to the Hutchison Effect, there appears to be a process here analogous to unlocking the front door rather than breaking in with a bulldozer. Dr. Edmund Storms adds the following:

> Nevertheless, nuclear behavior is quite different from that experienced when high energy is used to initiate such reactions. *Especially novel is the absence of radioactivity and energetic radiation, including neutron and gamma emissions...*[emphasis added][59]

Turning lead ($_{82}Pb$) into gold ($_{79}Au$) has been referred to as Alchemy, but has actually been documented. Glenn Seaborg, 1951 Nobel Laureate in Chemistry, succeeded in transmuting lead into gold in 1980 (possibly en route from bismuth), demonstrating this phenomenon.[60] Those replicating the work of Pons and Fleischmann have also found evidence of transmutation.[61] One of these researchers is Dr. Tadahiko Mizuno, who wrote the book, *Nuclear Transmutation: The Reality of Cold Fusion,*[62] documenting this phenomenon.

In addition to transmutation, the CANR process generates magnetic precipitate, meaning that a solid has formed (in tiny particles) in the electrolytic cell containing the reaction and that this solid is magnetic. This implies an interaction between electricity and magnetism is involved. As explained by Dr. Edmund Storms,

> *Transmutation* reactions have been reported to occur in all environments to which the CANR process has been applied. The easiest method involves creating a plasma under water. This can be done by applying sufficient voltage (up to 150 V) to form an arc between two carbon rods immersed in an electrolyte containing various salts dissolved in water [122-124].[63] *The method is reported to generate a* magnetic precipitate *in addition to various elements and is easy to duplicate.* ... The most complete study was undertaken by Prof. Miley [130-132] using electrolytic current applied to a nickel cathode in H_2O-based electrolyte.[64],[65]

A few years ago, while examining some of John Hutchison's steel samples, I experimented with a magnet and discovered that the Hutchison-Effect-affected (HE-affected) area was magnetic while the unaffected regions were not. As quoted above, Richard Sparks noted that the HE-affected samples were "accompanied *by an anomalous residual magnetic field, permanent in nature...*"[66]

1. Metallurgical Analysis of Hutchison-Effect Samples

Metallurgical Analysis of several Hutchison-Effect (HE) samples were done by Dr. George Hathaway[67] using a Hitachi Scanning Electron Microscope (SEM) with an XRF detector. The results are shown in Figures 381, 382, and 383, for steel, aluminum, and brass, respectively. For each of these three figures, (a) shows the unaffected (virgin) sample and (b) shows the Hutchison-Effect-affected (HE-affected) area on that same sample.

> In general, the analyses were done using an area scan, as opposed to a single spot on the specimen. The elemental analysis given by XRF is only a surface (or a few microns depth) analysis. The vertical axis of each plot represents specifically the number of x-ray counts at a particular energy, but can be roughly taken to represent the relative amounts of each element in the sample area. The horizontal axis is the X-Ray energy in kilo-electron volts (keV). Each element has several specific energies (called "lines") which are used to identify it. The first and most important of these are the so-called K-alpha and k-beta lines, the alpha line being

more prominent than the beta line. For instance, the K-alpha line for nickel is 7.4 keV and the K-beta line for nickel is 8.2 keV.[68]

(a) Hutchison Steel Sample

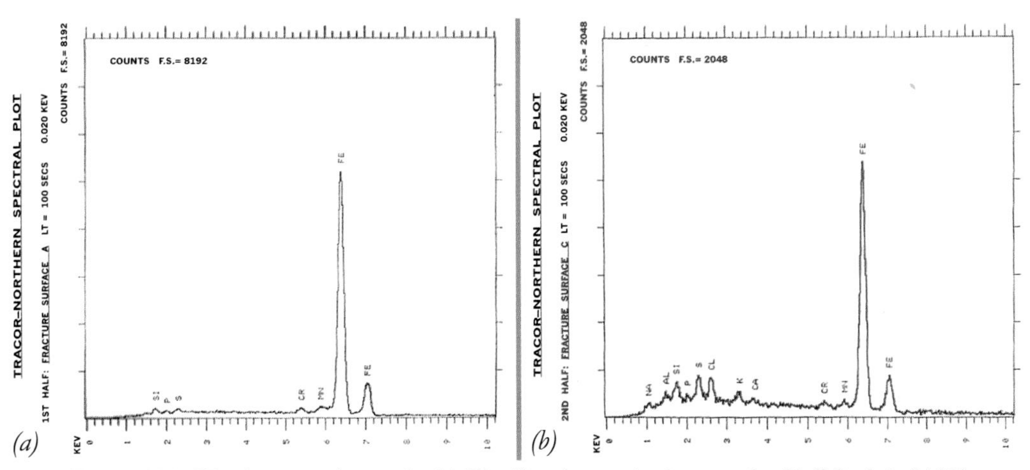

Figure 381. Hutchison steel *sample. (a) Unaffected area A of a sample. (Full Scale is 8192) (b) HE-affected area C of the same sample. (Full Scale is 2048)*[69]

In Figure 381, the iron ($_{26}$Fe), manganese ($_{25}$Mn), and chromium ($_{24}$Cr) peaks are about the same height in both (a) and (b), although on a different scale. But other elements ($_{13}$Al, $_{14}$Si, $_{15}$P, $_{16}$S, $_{17}$Cl, $_{11}$Na, $_{19}$K, $_{20}$Ca) have risen in proportion. Except for Silicon (Si), all of these have *lower* boiling and melting temperatures than iron, manganese, and chromium. (See Table 17 on page 368 and Table 27 on page 496.)

(b) Hutchison Aluminum Sample

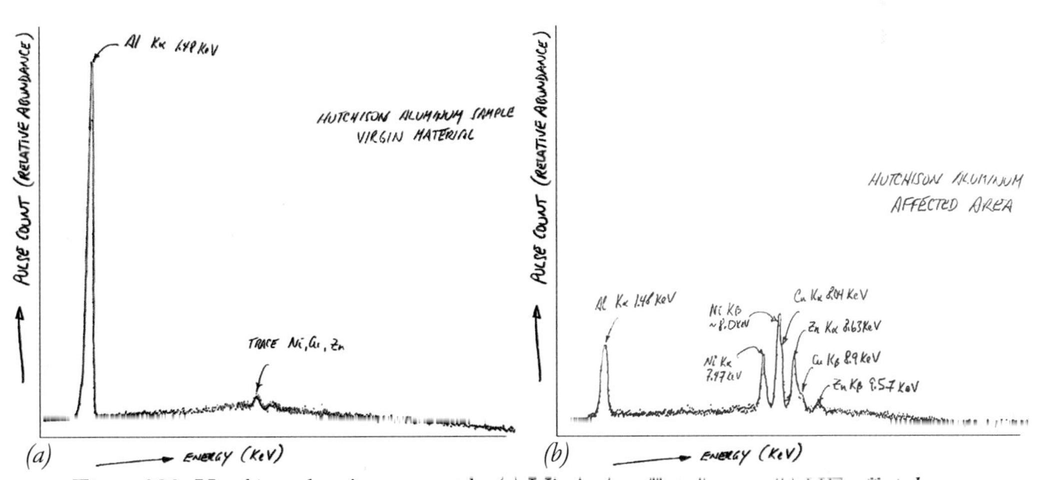

Figure 382. Hutchison aluminum *sample. (a) Virgin (unaffected) area. (b) HE-affected area. The vertical scales are not known for either chart.*[70]

The aluminum analysis, shown in Figures 382, are from a long two-inch diameter solid bar that had a Hutchison-Effect (HE)-affected zone in the center.[71] The virgin material shows it as almost entirely aluminum, as expected, with trace amounts of copper, nickel and zinc. The HE-affected area shows that the relative abundance of the trace elements has risen spectacularly relative to the host material, aluminum ($_{13}$Al). Here we see that nickel, copper and zinc ($_{28}$Ni, $_{29}$Cu, $_{30}$Zn) are present in more-or-less equal amounts to the aluminum.

As said, one might be tempted to argue that the rise in abundance of the alloying or trace elements is due to heating a specific portion of the sample to a temperature hot enough to drive off some of the host elements ($_{29}$Cu) by local boiling (e.g. using an acetylene torch).

element	Symbol	Atomic number	melting point °C[72]	boiling point °C[73]	boiling point °F[73]
Nickel	Ni	28	1,453	2,913	5,275
Copper	Cu	29	1,084.6	2,562	4,643
Aluminum	Al	13	660.25	2,519	4,566
Zinc	Zn	30	419.73	907	1,665

Table 17. Melting and boiling temperatures for selected elements.

So, if the heat were the agent, one would expect that the zinc ($_{30}$Zn) would be completely driven off the surface. But it is still there, in significant amounts.

(c) Hutchison Brass Sample

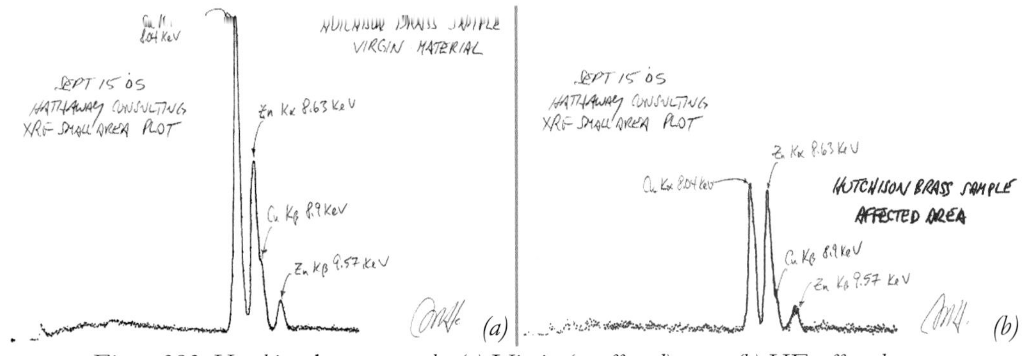

Figure 383. Hutchison brass *sample. (a) Virgin (unaffected) area. (b) HE-affected area. The vertical scales are not known or if they are the same for both charts.*[74]

The brass analysis, in Figure 383, is from a one-inch [diameter] bar which was bent severely and broken off at an end where the HE is clearly in evidence.[75] The plot of the virgin material, 383a, is typical for a commercial brass rod (more copper than zinc). The copper ($_{29}$Cu) peak is about twice the magnitude of the zinc ($_{30}$Zn) peak. In Figures 383b, the HE-affected area shows that the zinc has risen in relative abundance compared to the copper, becoming substantially equal to it. That is, the copper has decreased and/or the zinc has increased.

One might be tempted to argue that the rise in abundance of the alloying or trace elements is due to heating a specific portion of the sample to a temperature hot enough to drive off some of the host elements ($_{29}$Cu) by local boiling (e.g. using an acetylene torch). However, the boiling temperatures of copper and zinc are 2,562°C and 907°C, respectively and the melting temperatures of copper and zinc are 1,085°C and 420°C, respectively. (See Table 26 on page 496 for other values.) So the difference cannot be explained by vaporizing or melting the copper away because the zinc would vaporize or melt before the copper would.

Let us remember, again, that elements, such as iron ($_{26}Fe$), copper ($_{29}Cu$), and zinc ($_{30}Zn$) are defined by the number of protons they possess. Changing the element requires changing the atomic (proton) number. One explanation for the shift from copper to zinc is that copper (29 protons) becomes zinc (30 protons) with the addition of one proton.

2. Analysis of WTC Samples

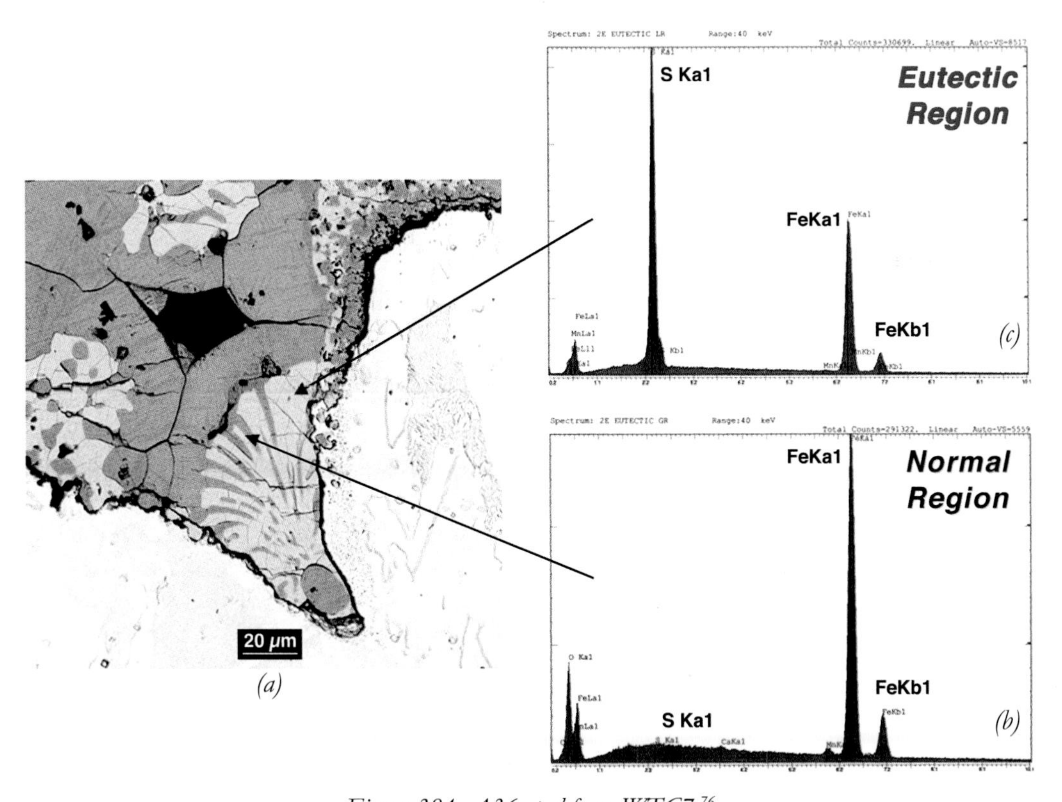

Figure 384. A36 steel from WTC7.[76]
(a) Eutectic formation (iron oxide-iron sulfide), etched 4% natal. (b) EDX Analysis of normal region, darker zone. (c) EDX Analysis of eutectic region, lighter zone.
http://www.tms.org/pubs/journals/JOM/0112/Biederman/fig4.gif

(a) Limited Metallurgical Examination

Limited Metallurgical Examination-FEMA Report
(Appendix C, FEMA Report)[77]
Jonathan Barnett, *Ronald R. Biederman, R. D. Sisson, Jr.*

C.2 Sample 1 (From WTC 7)

Several regions in the section of the beam shown in Figures C-1 and C-2 were examined to determine microstructural changes that occurred in the A36 structural steel as a result of the events of September 11, 2001, and the subsequent fires. Although the exact location of this beam in the building was not known, the severe erosion found in several beams warranted further consideration. In this preliminary study, optical and scanning electron metallography techniques were used to examine the most severely eroded regions as exemplified in the metallurgical mount shown in Figure C-3. Evidence of a severe high temperature corrosion attack on the steel, including oxidation and sulfidation with subsequent intergranular melting, was readily visible in the near-surface microstructure. A liquid eutectic mixture containing primarily iron, oxygen, and sulfur formed during this hot corrosion attack on the steel. This sulfur-rich liquid penetrated preferentially down grain boundaries of the steel, severely weakening the beam and making it susceptible to erosion. The eutectic temperature for this mixture strongly suggests that the temperatures in this region of the steel beam approached 1,000 °C (1,800 °F), which is substantially lower than would be expected for melting this steel.

Figure C-8 illustrates the results of this analysis.

How does something melt only *between* grains boundaries? Figure C-8 is very similar to Figure 384, discussed by the same authors in a another publication.[78]

An Initial Microstructural Analysis of A36 Steel from WTC Building 7

J.R. Barnett, R.R. Biederman, and R.D. Sisson, Jr.,[79]
Journal of Materials, December 2001

A section of an A36 wide flange beam retrieved from the collapsed World Trade Center Building 7 was examined to determine changes in the steel microstructure as a result of the terrorist attack on September 11, 2001. This building was not one of the original buildings attacked but it indirectly suffered severe damage and eventually collapsed. While the exact location of this beam could not be determined, *the unexpected erosion of the steel found in this beam warranted a study of microstructural changes that occurred in this steel.* …

ANALYSIS

Rapid deterioration of the steel was a result of heating with oxidation in combination with intergranular melting due to the presence of sulfur. The formation of the eutectic mixture of iron oxide and iron sulfide lowers the temperature at which liquid can form in this steel. This strongly suggests that the temperatures in this region of the steel beam approached ~1,000°C, forming the eutectic liquid by a process similar to making a "blacksmith's weld" in a hand forge.

The "Deep Mystery" of Melted Steel[80]

There is no indication that any of the fires in the World Trade Center buildings were hot enough to melt the steel framework. Jonathan Barnett, professor of fire protection engineering, has repeatedly reminded the public that steel—which has a melting point of 2,800 degrees Fahrenheit—may weaken and bend, but does not melt during an ordinary office fire. Yet metallurgical studies on WTC steel brought back to WPI reveal that a novel phenomenon—called a eutectic reaction—occurred at the surface, causing intergranular melting capable of turning a solid steel girder into Swiss cheese.

[. . .]

"The important questions," says Biederman, "are how much sulfur do you need, and where did it come from? *The answer could be as simple—and this is scary—as acid rain."* [emphasis added]

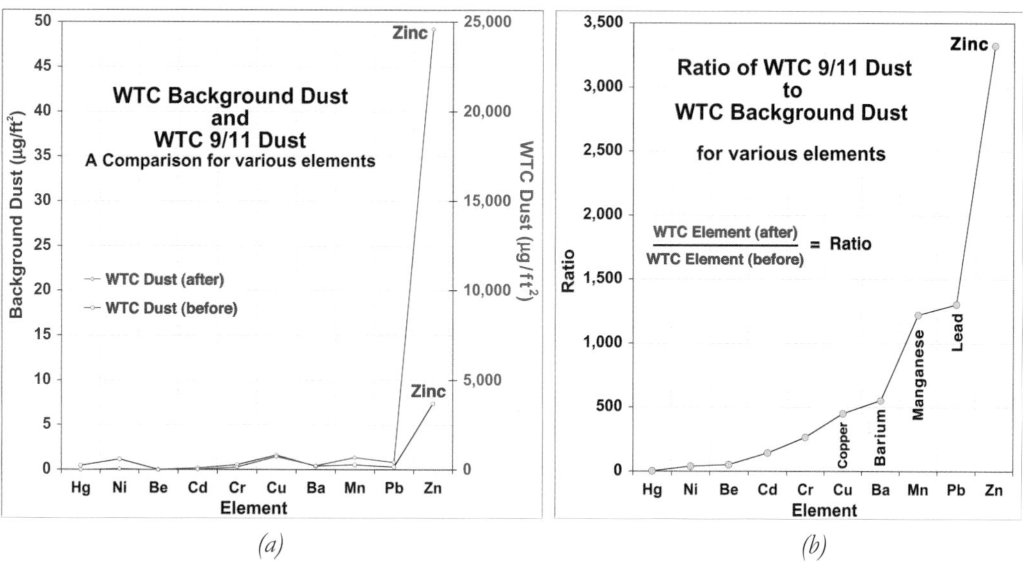

(a) *(b)*

Figure 385. Comparison of concentrations of elements in WTC dust to background dust.
Data from report by RJ Lee Group, Inc.[81]

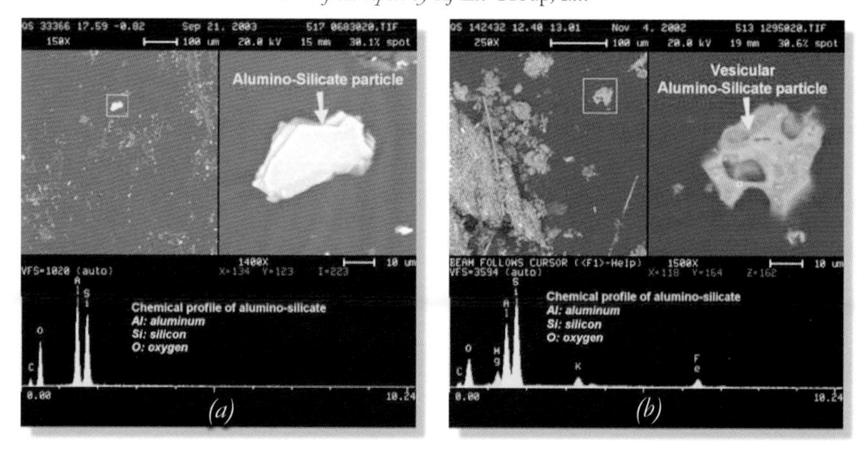

(a) *(b)*

Figure 386. (a) SEM image and EDS of alumino-silicate in Background Building.
(b) SEM image and EDS of vesicular alumino-silicate.
Photos by the RJ Lee Group, Inc.[82]

3. A Comment About Cold Fusion (LENR, CANR)

There exists a process which (1) produces excess energy, (2) causes transmutation of elements, (3) forms tritium, (4) generates a magnetic precipitate, (5) occurs at room temperature, and (6) does all this without producing radioactivity. When first presented in 1989, cold fusion was quickly dismissed as junk science and the careers of those who discovered it were destroyed. But on March 23, 2009, the 20th anniversary of the announcement by Pons and Fleischmann, the two were vindicated on network television. The CBS show, *60 minutes,* aired a show called "Cold Fusion Is Hot Again."[83] The archive of this show, presented by Scott Pelley, "More Than Junk Science," is also available.[84] An excellent survey article by Dr. Edmund Storms gives references to at least 34 studies[87] with positive results using the method of Pons and Fleishmann.[85,86,87]

J. Tritium at the WTC

Tritium was identified in samples taken from a WTC storm sewer and from the basement of WTC6[88,89] three days and ten days, respectively, after the 9/11 events. Tritium is a radioactive form of hydrogen that is used in research,[90] fusion reactors,[91] and neutron generators.[92] The radioactive decay product of tritium is a low energy beta that cannot penetrate the outer dead layer of human skin. Therefore, the main hazard associated with tritium is internal exposure from inhalation or ingestion.[92] Tritium is also used in watch faces and exit signs with chemicals (such as phosphor) that emit light in the presence of radiation. Rifle sites have about 12mCi of tritium and exit signs contain "several curies of tritium."[92]

The curie is a unit measure of an amount of radioactivity. A curie (Ci) is the amount of a radioactive substance that has 3.7 x 10^{10} decays per second, or 1 Becquerel (Bq).[93]

The WTC contained no exit signs with tritium, according to the group studying the tritium samples found at the WTC.[94,95] They concluded that the tritium must have come from exit signs on the alleged two planes. However, as the Idaho State University Tritium Information page states, *"Signs often have several curies of tritium in them. If the exit signs were severely damaged, HT gas might escape into the local area, but it should be dispersed by ventilation or wind quickly."*[96] So it does not seem plausible that all of the tritium in the four exit signs on the alleged planes made it into the groundwater of WTC6, especially when you consider that rain and fire hoses would have diluted it. Yet there are researchers who have suggested that this tritium is a sign that the WTC was destroyed by "mini nukes."[97]

In any case, let us consider this tritium data presented in the studies by Parekh et al[98] and Semkow et al[99] and its relative concentration. Table 18 provides values of tritium reported at the WTC. The values are provided in both units of *Becquerel per liter* (Bq/L) and *Curie per liter* (nCi/L) for convenience. The samples were collected in the basement of WTC6, the building that had the middle portion missing.

C. WTC Tritium data		Bq/L	nCi/L
WTC storm sewer	(9/14/01) (day 4)	6.07±2.74	0.164±.074
WTC 6, basement B5	(9/21/01) (day 11)	130.6±6.3	3.53±0.17
WTC 6, basement B5	(9/21/01) (day 11)	104.7±5.6	2.83±0.15
Background for WTC		4.1-4.8	0.11-0.13

Table 18. Tritium values reported by Parekh, et al[100] *and Semkow, et al.*.[101]

A more comprehensive table of values is given in Table 27 (page 497) along with the acceptable limits set by various authorities for drinking water. Again, these values are provided in both units for convenience.

The values of measured tritium at the WTC are plotted on a logarithmic scale in Figure 387 along with representative values from a variety of known causes, providing a visual comparison.

In the late 1950s and early 1960s, atmospheric nuclear bomb testing produced a considerable amount of atmospheric tritium that entered lakes and rivers through precipitation, producing high levels of tritium in the Great Lakes. That is, nuclear explosions in the atmosphere contaminated the entire planet.

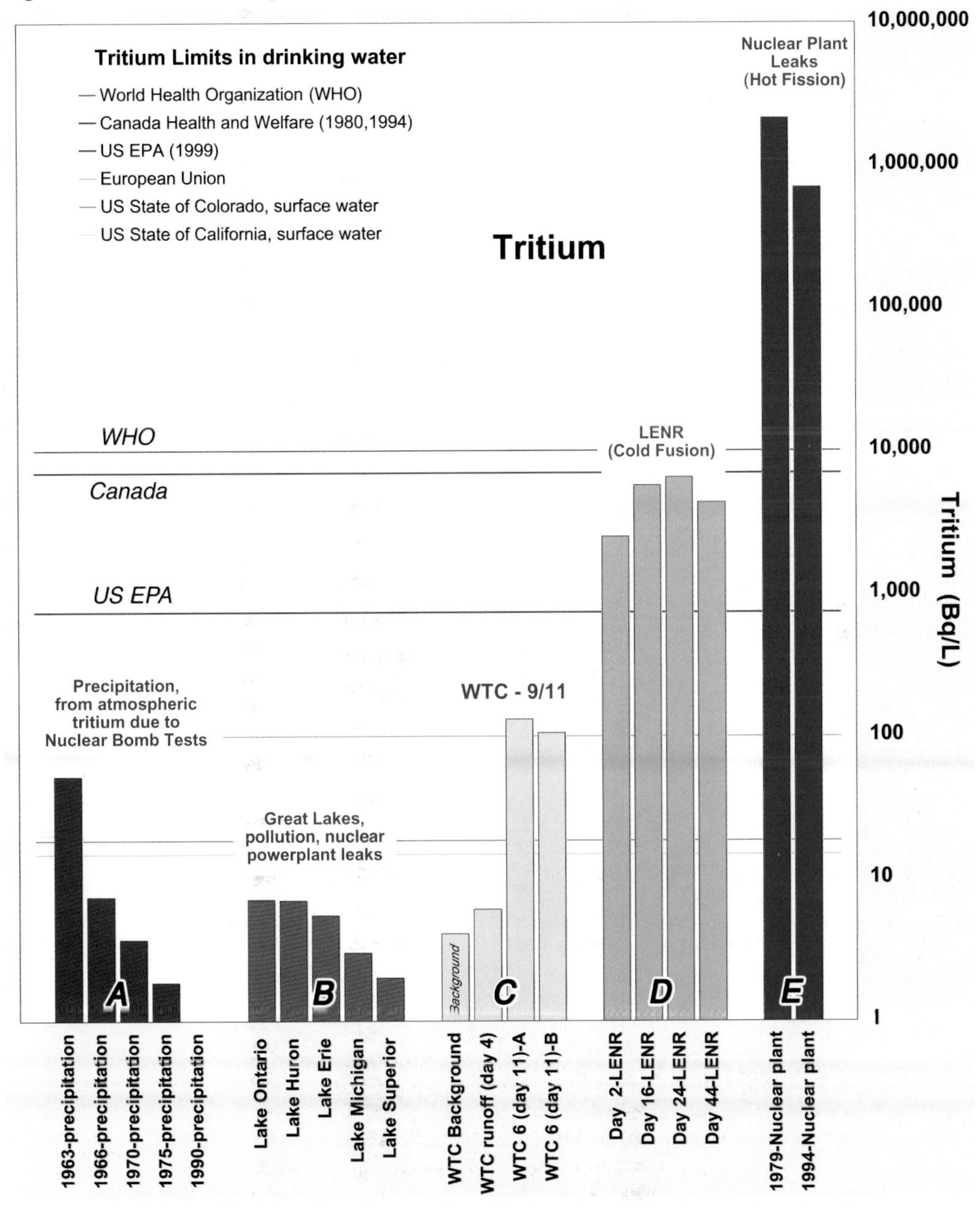

Figure 387. Tritium values shown on logarithmic scale resulting from various situations. Precipitation,[102,103] Great Lakes,[104] WTC - 9/11,[100,101] LENR,[105] Power-plant leaks.[106]

The first group (brown bars) on the left side of Figure 387 reflects the decrease in atmospheric nuclear detonations. According to the USGS, Tritium decays spontaneously to helium-3 (^3He) through ejection of a beta particle (essentially a high-energy electron). The half-life of tritium is about 12.32 years.[107,108]

The second group (green bars) shows the most recent tritium levels in the Great Lakes. There are no nuclear power plants on Lake Superior, so it is a good indicator for "background levels" of tritium. The middle group (yellow bars) are of tritium found at the WTC by Parekh, et al[109] and Semkow, et al.[110] These values measured at the WTC following the 9/11 event are about *50 times greater* than background values. Is this significant? Let us compare these values with two other categories.

The fourth group (blue bars) are values of tritium measured *in the cell* of a Low Energy Nuclear Reaction (LENR) experiment (also referred to as "cold fusion"). Tritium in the LENR cell is about *50 times greater* than what was seen at the WTC. The fifth and final group (red bars) are values measured in groundwater following a leak from a nuclear power plant ("hot fission"), which is the type of reaction in a "nuke" or "mini-nuke". These values are 360 times greater than LENR values, or about 18,000 times greater than the WTC values.

Values for contaminated drinking water, which are diluted by large bodies of water, are shown in category E of Table 28 (page 497). A leak at the Chalk River Nuclear Laboratories caused tritium levels in Ottawa drinking water, 200 km (124 miles) down-river from the site, of 150 Bq/L

Do you think they could have kept the <u>Chernobyl</u> <u>disaster</u> a secret?

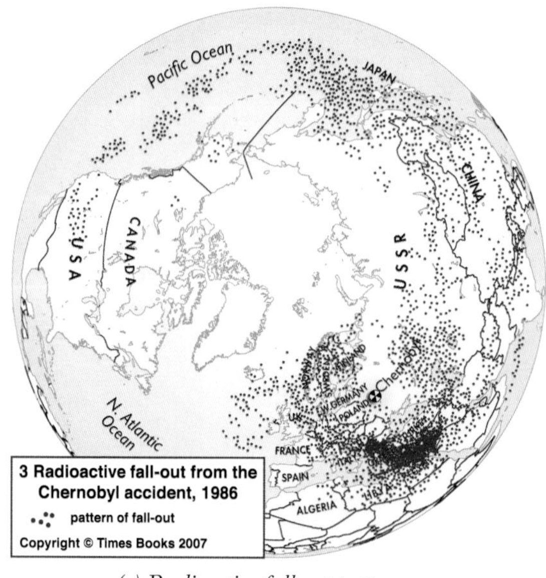

(a) Radioactive fall-out pattern

(b) Fuming reactor 4 core

Figure 388. (4/26/86) Effects of the Chernobyl nuclear disaster spread around the world.[111]

(a)Enhanced from: https://qed.princeton.edu/getfile.php?f=Radioactive_fall-out_from_the_Chernobyl_accident.jpg, http://planetliberty.wikidot.com/nuclear-power-incidents, (b)http://upload.wikimedia.org/wikipedia/en/f/f6/Chernobyl_burning-aerial_view_of_core.jpg

A nuclear hot-fusion nuclear event (nukes, suitcase nukes, pocket nukes, mini-nukes, nano-nukes…) if even realistic, would have produced a noticeable seismic signal (see Table 4, page 79), but it would also have produced tremendous heat and radiation. Although the tritium levels at the WTC were significantly above background levels, they were not high enough to be consistent with "a nuke."

Consider what happened at Chernobyl in the early hours of April 26, 1986. Figure 388b shows the remains of reactor 4 after it was destroyed by a steam explosion. (Water and molten metal don't mix well.) Although the damage was not caused by a nuclear explosion, radioactive fall-out from this accident was measured around the world, as shown in Figure 388a. If a nuclear bomb had been used to destroy the WTC, radiation from it would have been detected around the world, and there also would have been a seismic signature.

A closer look at the fall-out pattern from the Chernobyl disaster

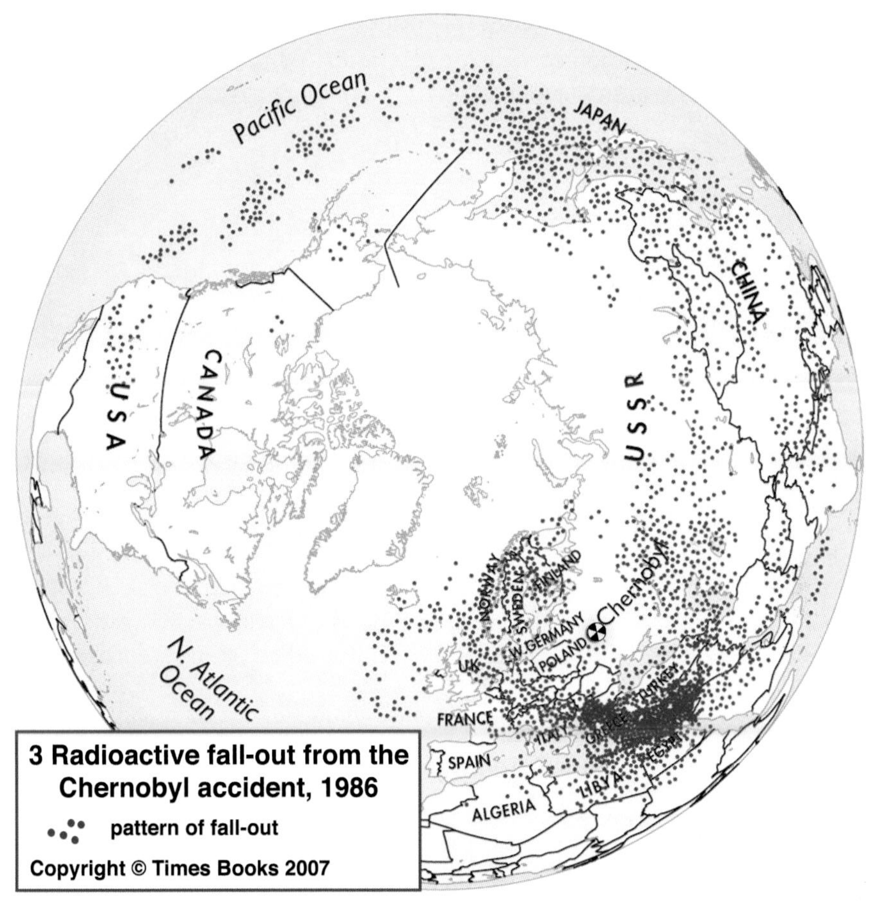

Figure 389. (4/26/86) Effects of the Chernobyl nuclear disaster spread around the world.[112]
Enhanced from: https://qed.princeton.edu/getfile.php?f=Radioactive_fall-out_from_the_Chernobyl_accident.jpg,

Group	Group	Tritium (Bq/L)	Relative to WTC
A	*Background (Great Lakes)*	*2.5*	*.02 (1/50)*
C	WTC	120	1
D	LENR	6,000	50
E	Nuclear-event Leak	2,150,000.	360*50 = 18,000

Table 19. Approximate magnitudes of tritium, relative to the WTC.
Great Lakes,[104] *WTC - 9/11,*[100,101] *LENR,*[105] *Power-plant leaks.*[106]

And, again, the order of magnitudes of tritium resulting from various processes suggest the same conclusion. As shown in Table 19, the amount of tritium found in a sample from WTC6 is approximately 50 times the background level. The tritium measured in a cold fusion cell is 50 times greater than that. The amount of tritium found in groundwater from a leaky nuclear power plant is 360 times that in a cold-fusion cell, or 18,000 times what was measured in a sample from the WTC.

K. Lift or Disintegration

As we have seen, some WTC pictures show cars upside down. One of the key effects John Hutchison has repeatedly reproduced is a "levitation" or "anti-gravity" effect, where objects are sometimes seen to levitate—or to disintegrate.

Figure 390. The pliers lift off the table, gaining distance from their shadow.[113]

Mark A. Solis wrote the following in an article about Hutchison:

> The levitation of heavy objects by the Hutchison Effect is not—repeat not—the result of simple electrostatic or electromagnetic levitation. Claims that these forces alone can explain the phenomenon are patently ridiculous… Challengers should note that their apparatus must be limited to the use of 75 Watts of power from a 120 Volt AC outlet**,** as that is all that is used by Hutchison's apparatus to levitate a 60-pound cannon ball.[114]

Figure 391. The Hutchison Effect. (a) A wrench is moving upward. (b) Water is levitating.[115]

(a)(Google video[116]), (b)http://bp1.blogger.com/_SkyVfBO47ts/R13bc9gnzLI/AAAAAAAAAHZQ/79r5gtldeiw/s1600-h/Picture+642.jpg

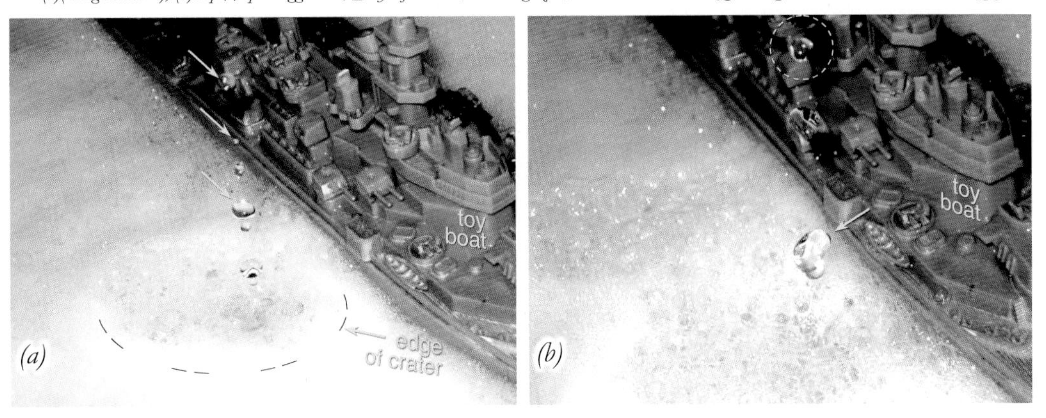

Figure 392. Water levitating from the Hutchison Effect.

(a)http://bp2.blogger.com/_SkyVfBO47ts/R18hhtgn1QI/AAAAAAAAAHp4/_2mfORLMwKg/s1600/Picture%2B752.jpg, (b)http://bp1.blogger.com/_SkyVfBO47ts/R18fidgn1MI/AAAAAAAAAHpY/f2XSfbcnPkM/s1600/Picture%2B756.jpg,

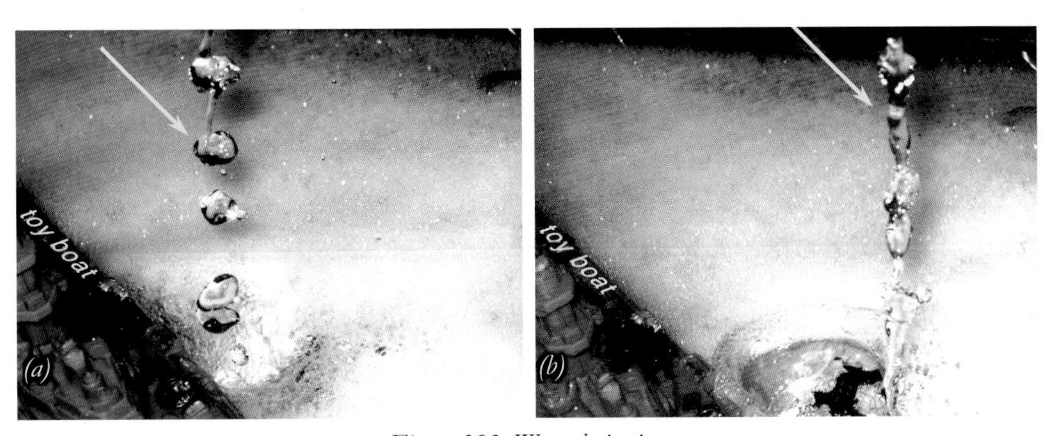

Figure 393. Water levitating.

(a)http://bp2.blogger.com/_SkyVfBO47ts/R18djtgn1GI/AAAAAAAAAHoo/7jB5gJ9E6zQ/s1600/Picture%2B762.jpg, (b)http://bp2.blogger.com/_SkyVfBO47ts/R18djtgn1HI/AAAAAAAAAHow/1eLqsu3kjTg/s1600/Picture%2B761.jpg

Now consider these flipped cars from the WTC:

(a) (9/12/01) *(b) (9/11/01)*

Figure 394. Flipped cars.

("Cortlandt St." and "Church St." highlighted for clarity.)

(a)http://911pictures.com/images/previews_lg/911-1392.jpg, (b)http://memory.loc.gov/service/pnp/ppmsca/02100/02102v.jpg[117]

Figure 395. Flipped cars appear less damaged than adjacent cars.
http://i24.photobucket.com/albums/c49/IgnoranceIsntbliss/911/NonBurnedCars/a3.jpg

The result of "lift" or "disruption" can be explained by field effects, with a demonstration where a soda pop can is either crushed or launched by a magnetic field.[118]

L. Toasted Metal and Similar Effects

A number of metal effects have been observed in samples from the WTC that show similar features to samples made by John Hutchison. For example, consider the Red Bull drink can (like a soda pop can) from a Hutchison experiment.

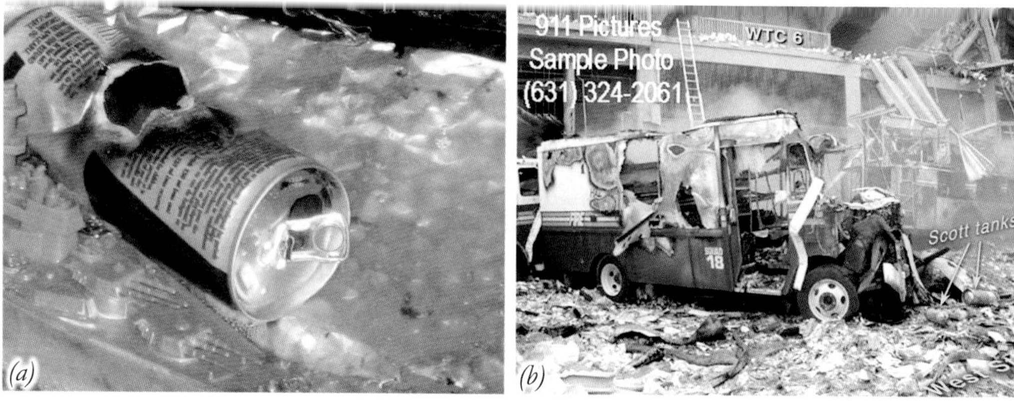

Figure 396. (a) Hutchison Effect affected Red Bull Can.[119] *(b) (9/11/01) A FDNY Hazmat truck with toasted holes in front of WTC6 on West Street.*
The remaining upper part of the truck has been peeled and evaporated in areas. The upper part of the cab is gone and the engine block seems to have disappeared.
(a)http://bp1.blogger.com/_SkyVfBO47ts/R13We9gny6I/AAAAAAAAHXI/lolE_aOZ0P4/s1600-h/Picture+625.jpg,
(b)http://911pictures.com/images/previews_lg/911-1328.jpg

M. Holes

The Hutchison Effect also causes cylindrical holes in materials. In the example shown in Figure 397, the holes are in one side of the once-solid aluminum bar but do not extend out the other.

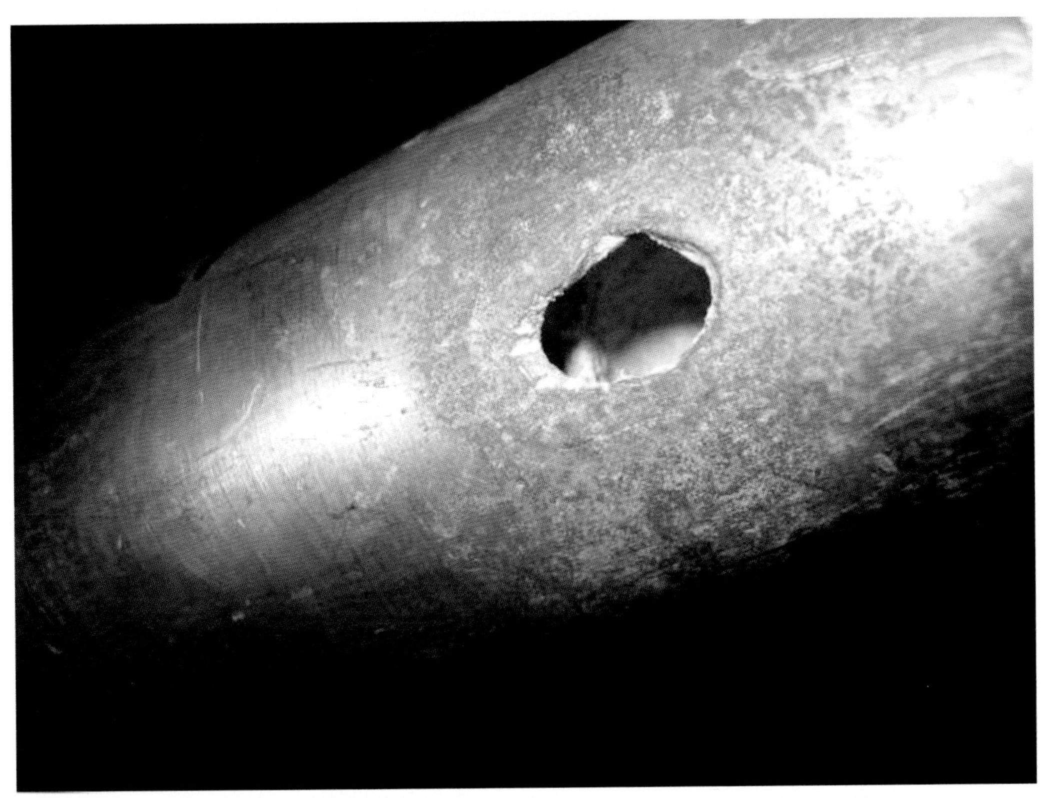

Figure 397. A hole in a solid-aluminum bar caused by the Hutchison Effect.
http://www.americanantigravity.com/graphics/Hutchison-Photos-Sept05.zip

A closer look reveals that the hole leads to a horizontal tunnel approximately parallel to the surface. This tunnel in Figures 397 and 398 extends only to the left, not to the right.

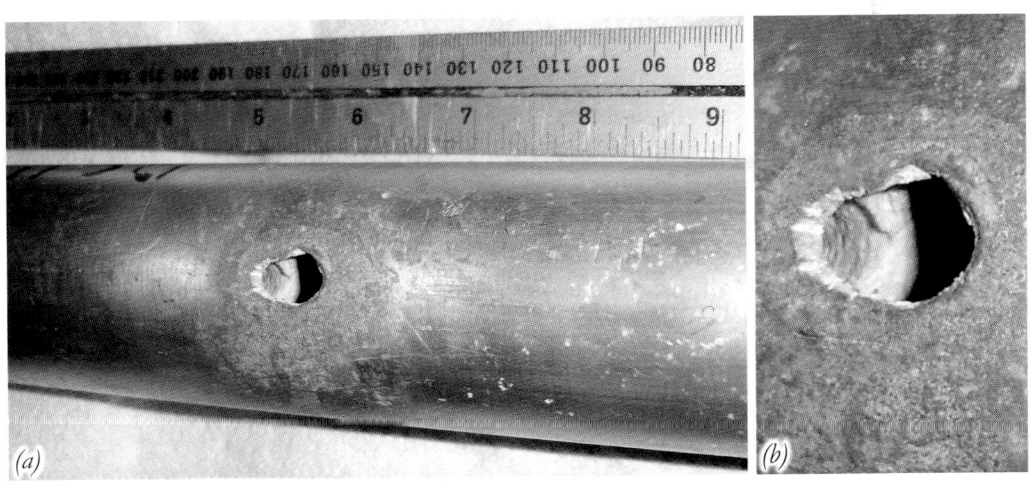

Figure 398. More views of the hole.
Photos (a) and (b) by the author

And now consider these cylindrical voids of material at the WTC:

Figure 399. (9/23/01) Holes (voids)
http://www.noaanews.noaa.gov/wtc/images/wtc-photo.jpg

N. Fuming

The WTC site fumed for months and even years. An example of this fuming is seen in the haze looming over WTC5 in Figure 399. Fuming cars on 9/11 are shown in Figure 233 (page 228), where fuming takes place *before* the WTC1 dust cloud arrives. Fuming cars are also seen in Figure 234 (page 229), where Figure 234b shows fuming cars on Vesey Street immediately after the event.

The Hutchison Effect also produces fuming from the samples, albeit on a smaller scale. This fuming was caught on camera in a recent demonstration. Fuming can be seen in this video from about 23 seconds to 28 seconds, but especially at 25 seconds (Figure 400) and at 26 seconds (Figure 401). The solid iron block, approximately 2in. x 2in. x 7in., appears to shrivel and bend. The fumes become visible as the block appears to begin disintegrating.

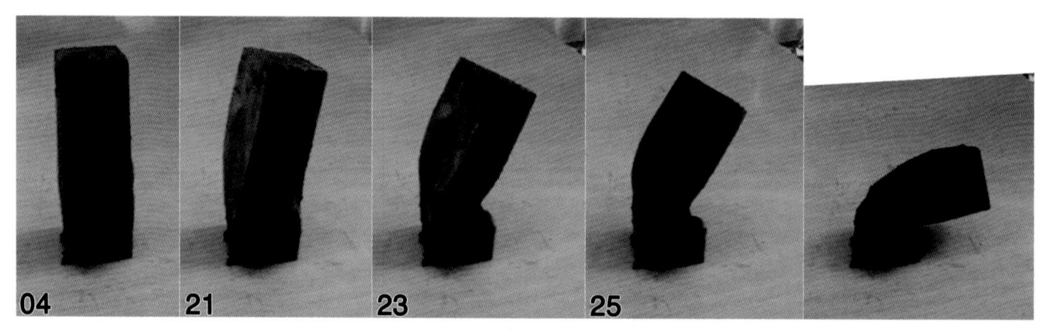

Figure 400. Frames from the video at 4, 21, 25, 25,seconds and one from the next clip. Fuming can be seen in frame 25, shown, and in Figure 401.
Source: http://www.youtube.com/watch?v=BjWE0K8zPGY, http://www.youtube.com/watch?v=tnBdhsXl088

A video released by NIST shows a similar fuming from a door handle of a car on West Broadway, before the demise of WTC7.[120] Later photos taken along this street show many vehicles with missing door handles.

To view this video, see: http://drjudywood.com/towers

Figure 401. Originally a solid block of iron, approximately 2 x 2 x 7in.
Source: http://www.youtube.com/watch?v=BjWE0K8zPGY, http://www.youtube.com/watch?v=tnBdhsXl088

To view this video, see: http://drjudywood.com/towers

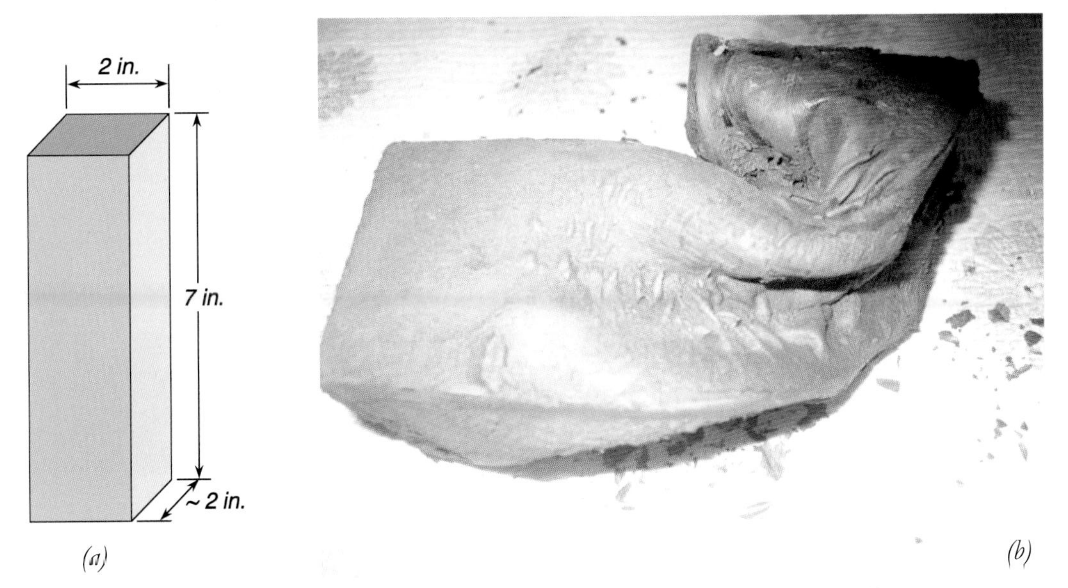

Figure 402. Solid block of iron, partly disintegrated by the Hutchison Effect.
Source: http://www.youtube.com/watch?v=BjWE0K8zPGY, http://www.youtube.com/watch?v=tnBdhsXl088

O. Lather

Figure 403. (9/11/01) Steel being pulled apart.
(a)http://hereisnewyork.org//jpegs/photos/1539.jpg, (b)Photo by Shannon Stapleton, Reuters[121]

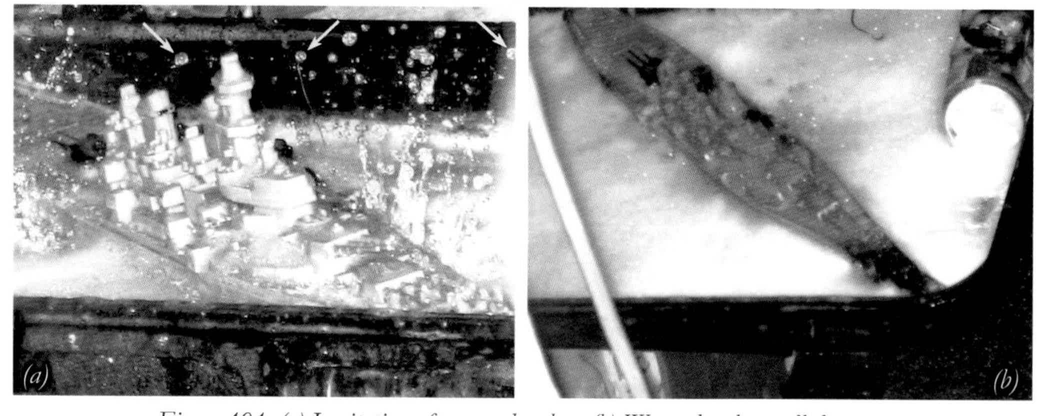

Figure 404. (a) Levitation of water droplets. (b) Water droplets pulled apart.
(a)http://bp0.blogger.com/_SkyVfBO47ts/R13bctgnzKI/AAAAAAAAHZI/vkeBbjbcqY0/s1600/Picture%2B642.jpg,
(b)http://bp3.blogger.com/_SkyVfBO47ts/R13f6dgnzUI/AAAAAAAAHaY/D118u8fvx60/s1600/Picture%2B651.jpg

P. Rounded Holes in Glass

Consider these round holes through glass. The view is from the Century-21 building, looking over the remains of WTC4 and WTC5, with WFC2 in the distance (through the far-left window).[122]

Certain buildings adjacent to the WTC complex were fitted with double-paned windows. In some of these cases, the round holes and breakage occurred only in the outer pane of the glass but not the inner. An example of this is shown in Figures 406 and 409.

Figure 405. Circular holes in windows of the Century-21 building.
An outside view of these windows can be seen in Figure 185a on page 192 and in Figure 297 on page 287.
http://bocadigital.smugmug.com/photos/10746124-D.jpg

Figure 406. Circular holes in outer pane of windows. FEMA photo.[123]

These round holes were in windows across the street from WTC4, between the Burger King and the 10&10 Firehouse. (See Figures 179 and 212 on pages 188 and 212, respectively.)

Hutchison also describes the creation of round holes in glass during his experiments. The physical process that we are considering here can be compared to the dimensional changes that thermal expansion can bring about, although in the Tesla-Hutchison case the dimensional change is not necessarily heat-induced. But consider how water expands when it freezes, explaining why a full and sealed beverage can will rupture or explode if left in the freezer.

Figure 407. Illustration suggesting how longitudinal waves may cause circular holes in windows.
(a)http://open.salon.com/files/ripple1253522867.jpg, (b)adapted from: http://www.bnl.gov/rhic/images/black_hole.jpg

Or use the comparison of ice freezing on a pond. If the water freezes slowly, it will freeze uniformly with relatively smooth ice good for skating. But if the pond freezes from a dramatic and very rapid drop in the temperature, the surface will develop cracks, buckles, and ruptures that will provide nothing like a smooth skating surface. The ruptures and buckles come about because the freezing water needed to expand but was confined by adjacent ice (or structures). These constraints on the ice produce a compressive force. If the compressive force is strong and confining enough, the ice explodes upward. Or consider holding the flame of a propane torch to a patch of concrete floor. The heat causes rapid expansion in a small region while the adjacent concrete constrains that expansion. If you're not careful, divots of concrete will suddenly explode out from the constrained concrete floor, leaving little craters.

For a similar illustration, consider what happens when you drop a stone into a pond on a day without wind, when the surface of the water is smooth. When the stone plops into the water, it displaces water by the volume of space the stone's mass occupies, including any air pockets it might have inside it. In the fluid pond, this displaced water sends a ripple outward. In this case, we have a vertical drop of the stone causing a horizontal wave propagating outward from the disturbance.

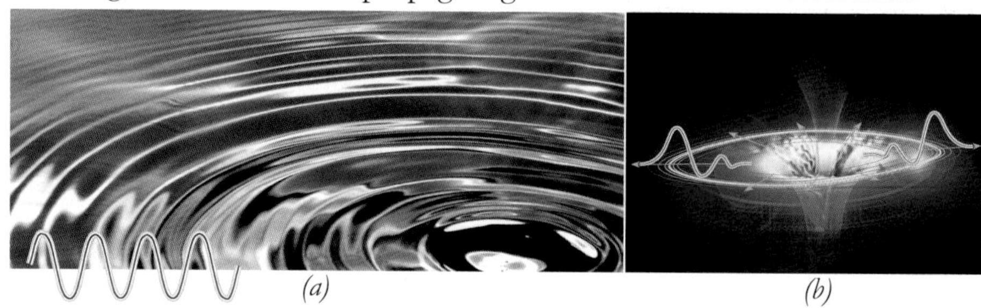

Figure 408. (a) Ripples on water. (b) Resulting transverse waves.
(a)http://i.pbase.com/o6/21/538321/1/72841823.OJV3XioI.ripple.jpg,
(b)adapted from: http://www.bnl.gov/rhic/images/black_hole.jpg

Suppose we imagine a vertical pane of glass at the WTC complex as being the equivalent of the horizontal surface of a perfectly calm pond. And suppose we consider longitudinal electromagnetic waves striking its surface as being equivalent to a stone being dropped into the still pond. Now, in the case of the glass pane at the

WTC complex, this dropping of the "stone" causes an expanding ripple of energy analogous to that caused by the stone in the pond.

Figure 409. Windows with circular holes in the outer pane.
Figure 179 on page 188 shows the location of these windows on the south side of Liberty Street, directly across from WTC4. FEMA photo.[122]

In a conventional situation, a rock or baseball thrown at a window causes a large bend in the pane, causing the surface opposite where contact is made to be loaded in tension. When brittle, glass will have low tensile strength and will therefore break. A wave of energy in the plane of the glass causes a ripple of compression and tension in the plane of the glass, traveling radially outward. The tension part of the ripple will be in an approximately circular pattern, perhaps causing the glass to fail along this circular zone of greatest tension.

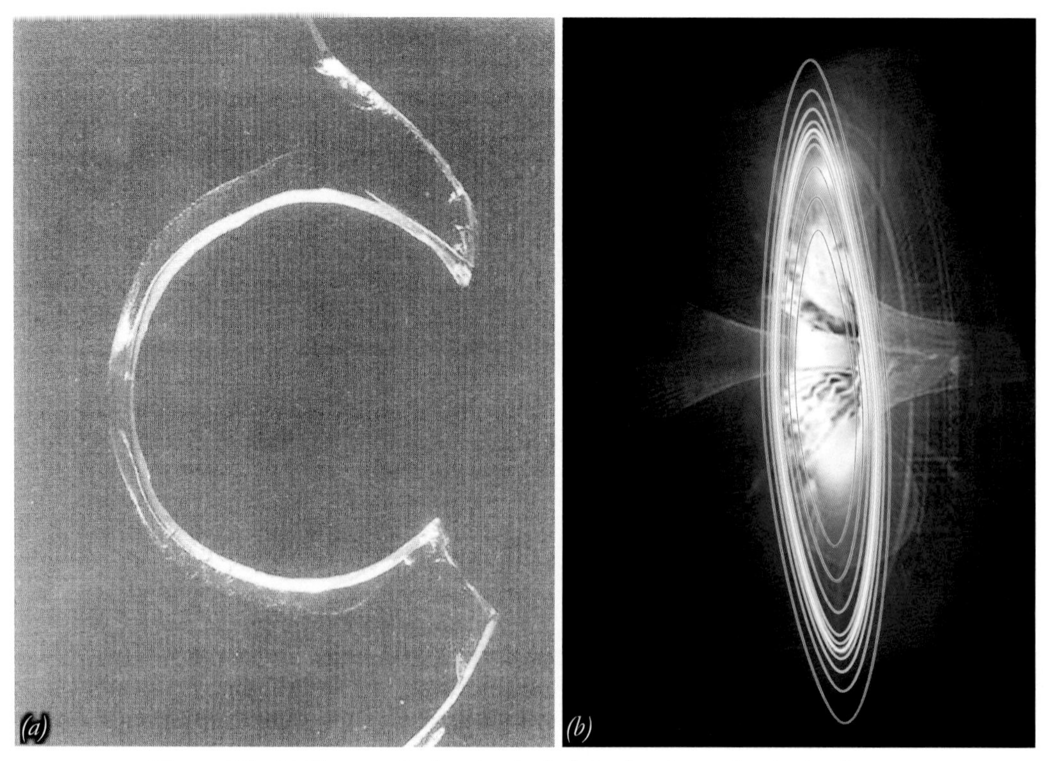

Figure 410. (a) Photograph of a circular hole in glass from a longitudinal wave, (b) Illustration of how a longitudinal wave could cause a circular hole.
(a)Photo by D.H.McIntosh, Electric Ufos: Fireballs, Electromagnetics and Abnormal States, by Albert Budden, photo section, page iv, (b)adapted and redrawn from: http://www.bnl.gov/rhic/images/black_hole.jpg

Q. Conclusions

The more a person studies the essential aspects and characteristics of experimental effects obtained by Hutchison, the more reasonable it becomes to conclude that a type of technology *similar to* the Hutchison Effect was employed in the destruction of the WTC complex. Such technology, whoever or whatever entity possesses it, is undoubtedly very, very highly classified and kept most strictly away from both public view and public knowledge.

By using radio frequency and electrostatic generators, Hutchison obtains his own effects in a volume of space where fields intersect and interfere. The results, once more, include

1. Levitation of heavy objects;
2. Fusion of dissimilar materials such as metal and wood;
3. Anomalous melting (without heating) of metals without burning adjacent material;
4. Spontaneous fracturing of metals (which separate by sliding in a sideways fashion); and,
5. Both temporary and permanent changes in the crystalline structure and physical properties of metal samples.

Hutchison has reproduced these experiments a great many times. His metal samples, further, have been repeatedly tested scientifically, notably at the Max Planck

Institute in Germany, an institution that has confirmed the validity—by merit of the repeatability—of the Hutchison Effects.

Metal samples produced by the Hutchison Effect show highly unusual results. Sometimes the metal "jellifies," turning soft and losing form, leading in turn to severe bending or fracturing of the sample. Sometimes samples erupt from the center. And sometimes *they turn to dust.*

Of an enormous *potential* importance, Hutchison's samples sometimes exhibit a type of ongoing reaction, a continuation of material change even after the energy field is removed, which has been referred to as a "non-self-quenching" reaction. Richard Sparks warns of this issue.

> 15. Serious considerations should be given to the idea that exceeding a certain critical mass of any relatively pure material may result in a reaction that is not self-quenching.[124]

This "non-self-quenching" reaction appears to take place at the nuclear level of the material. The same "non-self-quenching" reaction appears to me to be taking place at Ground Zero (GZ). I have been able to determine that the WTC site is apparently still undergoing "decontamination," with trucks moving dirt into and out of the site, while "hosing down operations" continue. Andrew Johnson and I photographed and recorded these operations on video in January 2008.

Similarly, materials subjected to the Hutchison Effect sometimes appear to transmute at the molecular or even atomic level. Reactions of this "deep" kind could be the explanation for the apparently very rapid rusting at GZ, where steel is seen rusting as if it were iron. Also, photographs show effects on the WTC towers' aluminum cladding similar to the effects produced on Hutchison's aluminum samples.

Finally, Hutchison has observed "spontaneous combustion" in some of his experiments, where "fires appeared out of nowhere." Hutchison has confirmed that in 1983 Col. John Alexander and others from the U.S. military visited him with a team from Los Alamos National Laboratories (LANL) and filmed his experiments.[125] Hutchison was visited again in 1986 by Canadian MP Chuck Cook and Dr. Lorne A. Kuehne of the Canadian Security Intelligence Service (CSIS), who told him his work was "a matter of National Security."[126] Hutchison says he has been told that defense contractor S.A.I.C. has similar technology and has been developing it.[127]

Whatever one makes of John Hutchison's work, one thing remains crystal clear. Many, if not most, of the anomalous examples of damage at the WTC on 9/11 are explicitly paralleled in his experiments concerning the effects of directed and interfered electromagnetic energy. No other known technology points to a mechanism of the kind seen in the destruction at the World Trade Center.

While the Hutchison Effect adequately models and parallels some of the anomalies seen on September 11, 2001, there is one more very significant piece of information we must look at, one that very few people consider in conjunction with the destruction of the World Trade Center, or that very few people, for that matter, even know about.

[1] A powerful statement by someone who has taught me well.

[2] EVO refers to Exotic Vacuum Objects, a phenomenon discovered and named by Ken Shoulders

[3] Thank you Dan!

[4] The old rabbit-ears reception refers to the analog signal used prior to 2009 that was received directly by an anenna on the TV set.

[5] John Hutchison gave this information on *The Ralph Winterrowd Show*, April 4, 2010.

[6] *http://en.wikipedia.org/wiki/Nikola_Tesla*

[7] Alternating current, an invention of Nikola Tesla, is sold by companies named after Thomas Edison, such as Southern California Edison, Con Edison, Ohio Edison, and Toledo Edison.

[8] *http://en.wikipedia.org/wiki/Nikola_Tesla*

[9] "Nikola Tesla - electrical engineer and inventor," Serbian Unity Congress. *http://www.serbianunity.net/people/tesla/index.html.* [Retrieved 2010-04-10]

[10] George Piggott, Overcoming Gravitation, *Electrical Experimenter* (July 1920), (Reprinted in Richard A. Ford: *Homemade Lightning -- Creative Experiments in Electricity*; McGraw-Hill; ISBN 0-07-137323-3), *http://rexresearch.com/piggott/piggott.htm*

[11] *http://rexresearch.com/piggott/piggott.htm*

[12] *http://en.wikipedia.org/wiki/Edward_Leedskalnin*

[13] Ed Leedskalnin, Magnetic Current (Illustrated), (1945), *http://www.scribd.com/doc/242432/Ed-Leedskalnin-Magnetic-Current-Illustrated?autodown=pdf*

[14] *http://paranormal.about.com/od/moremadscience/a/coral-castle-secrets.htm*

[15] *http://www.intalek.com/Index/Projects/CoralCastle/CoralCastle.htm, http://www.gizapower.com/Coral%20Castle/coralcastle.html*

[16] *http://en.wikipedia.org/wiki/Edward_Leedskalnin*

[17] *http://paranormal.about.com/od/moremadscience/a/coral-castle-secrets.htm*

[18] *http://www.sacred-destinations.com/egypt/giza-pyramids*

[19] *http://en.wikipedia.org/wiki/Giza_pyramid_complex*

[20] 15 metric tons = 33,069.3393 lbs. = 16.53467 (US) short tons, 2.5 metric tons = 5,511.55655 lbs. = 2.76 (US) short tons

[21] *http://coralcastle.com/whos-ed/*

[22] *http://www.sacred-destinations.com/egypt/giza-pyramids*

[23] *http://www.reallycoolscience.com/ttbrown-01.html*

[24] *http://www.intalek.com/Index/Projects/CoralCastle/CoralCastle.htm*

[25] *http://www.intalek.com/Index/Projects/CoralCastle/CoralCastle.htm*

[26] The Coral Castle website: http://coralcastle.com/faqs/, Billy Idol videos: *http://www.youtube.com/watch?v=ICP7vevaHcg, http://www.youtube.com/watch?v=FGxwaYyjfUU,*

[27] *http://en.wikipedia.org/wiki/T._Townsend_Brown*

[28] Project Winterhaven, page 1 (pdf p. 5 of 66), *http://www.americanantigravity.com/documents/military-research/Project-Winterhaven.pdf*

[29] Project Winterhaven, page 5 (pdf p. 11 of 66), *http://www.americanantigravity.com/documents/military-research/Project-Winterhaven.pdf*

[30] Project Winterhaven, page 6 (pdf p. 12 of 66), *http://www.americanantigravity.com/documents/military-research/Project-Winterhaven.pdf*

[31] Project Winterhaven, page 7 (pdf p. 13 of 66), *http://www.americanantigravity.com/documents/military-research/Project-Winterhaven.pdf*

[32] Thomas Townsend Brown, *http://en.wikipedia.org/wiki/Thomas_Townsend_Brown*

[33] *http://en.wikipedia.org/wiki/John_Hutchison*

[34] Links: (a) http://thehutchisoneffect.com, (b)John Hutchison's Affidavit for the federal Qui-Tam case I filed, *http://drjudywood.com/articles/JJ/JJ8.html, http://drjudywood.com/pdf/AffJHutchison4.pdf,* (c) More Links to Technical papers, Technical presentatons, videos, and other informaton on the Hutchison Effect, *http://drjudywood.com/towers*

[35] EVO refers to Exotic Vacuum Objects, a phenomenon discovered and named by Ken Shoulders

[36] EVOs and the Hutchison Effect, Nuclear Transmutation from Low-Voltage Electrical Discharge, Paper Presented at the MIT Cold Fusion Conference, May 21, 2005 by Ken Shoulders, *http://www.svn.net/krscfs/EVOs%20and%20Hutchison%20Effect.pdf*

[37] Ken Shoulders' EVOs - Exotic Vacuum Objects Challenge Particle Theory, by Sepp Hasslberger, *http://blog.hasslberger.com/2007/10/ken_shoulders_evos_exotic_vacu.html*

[38] Quote by Richard Sparks, Scientific and Technical Intelligence/SBIR, Ottawa. (1996) in The Hutchison File, (page 67 of 87), *http://drjudywood.com/pdf/HutchisonEffectReport_txt.pdf,*

[39] *http://hutchisoneffect2008.blogspot.com/2007/11/starting-few-metals-more-to-come.html*

40 *http://hutchisoneffect2008.blogspot.com/2007/12/blog-post_05.html, (a)http://bp0.blogger.com/_SkyVfBO47ts/R0zDfYhJliI/AAAAAAAAGu0/QDbXJ3u33_c/s1600-h/Picture+317.jpg, (b)http://bp1.blogger.com/_SkyVfBO47ts/RosgV_fdmxI/AAAAAAAAADWs/5K5FrRzjs74/s1600/y1pl90TVU-OcIwN4ecOXONeyFZgomAw9RZx2PHqelwTKFLrC0ynOHnyr5TI2t5vgeTY.jpg*

[41] *http://www.fema.gov/pdf/library/fema403_ch6.pdf, page 6-1.*

[42] personal photos

[43] *http://wtc.nist.gov/media/gallery.htm#metal*

[44] *http://americanhistory.si.edu/september11/collection/record.asp?ID=40*

[45] *http://americanhistory.si.edu/september11/collection/record.asp?ID=70*

[46] Since 1982, *http://en.wikipedia.org/wiki/Coins_of_the_United_States_dollar, http://www.coinresource.com/articles/frb_united_states_coins.htm*

[47] Since 1938, *http://en.wikipedia.org/wiki/Coins_of_the_United_States_dollar*

[48] Since 1965, *http://en.wikipedia.org/wiki/Coins_of_the_United_States_dollar*

[49] Since 1970, *http://en.wikipedia.org/wiki/Coins_of_the_United_States_dollar*

[50] *http://americanhistory.si.edu/september11/collection/record.asp?ID=40*

[51] *http://thewebfairy.com/911/h-effect/filingcabinet.htm*

[52] *http://bp0.blogger.com/_SkyVfBO47ts/RmRZQQJNkfI/AAAAAAAACDU/Tuht-Ky6rOE/s1600/y1pl90TVU-OcIwKX5Iy-uhazrus4Knwykeh6kb3OqqV5XnV7U57UWrDhQXPLnX559EL,*

[53] *http://bp2.blogger.com/_SkyVfBO47ts/R1H3rNgnuZI/AAAAAAAAGzI/K5BBL8RnGb4/s1600-R/Picture+348.jpg*

[54] Tadahiko Mizuno, *Nuclear Transmutation: The Reality of Cold Fusion*, (Translated by Jed Rothwell), Infinite Energy Press, 1998, Infinite Energy Press, Concord, New Hampshire, *http://www.books-by-isbn.com/1-892925/1892925001-Nuclear-Transmutation-The-Reality-of-Cold-Fusion-1-892925-00-1.html, http://www.amazon.com/exec/obidos/ASIN/1892925001/ref=nosim/schildnet0c,*

[55] Transmutation, *http://www.britannica.com/EBchecked/topic/602997/transmutation*

[56] Nuclear Transmutation, *http://en.wikipedia.org/wiki/Nuclear_transmutation*

[57] Eugene F. Mallove, *Fire from Ice: Searching for the Truth Behind the Cold Fusion Furor*, 1991, John Wiley & Sons.

[58] Anne Marie Helmenstine, Turning Lead into Gold: Is Alchemy Real?, *http://chemistry.about.com/cs/generalchemistry/a/aa050601a.htm*

[59] Edmund Storms, *COLD FUSION: An Objective Assessment*, Energy K. Systems, Santa Fe, NM 87501, (12/16/01) *http://pw1.netcom.com/~storms2/review8.html*

[60] Anne Marie Helmenstine, Turning Lead into Gold: Is Alchemy Real?, *http://chemistry.about.com/cs/generalchemistry/a/aa050601a.htm*

[61] Hideo Kozima and Kunihito Arai, Localized nuclear transmutation in PdH_x observed by Bockris and Minevski revealed a characteristic of CF phenomenon, International Journal of Hydrogen Energy, Volume 25, Issue 6, 1 June 2000, Pages 513-516

[62] Tadahiko Mizuno, *Nuclear Transmutation: The Reality of Cold Fusion*, (Translated by Jed Rothwell), Infinite Energy Press, 1998, Infinite Energy Press, Concord, New Hampshire, *http://www.books-by-isbn.com/1-892925/1892925001-Nuclear-Transmutation-The-Reality-of-Cold-Fusion-1-892925-00-1.html, http://www.amazon.com/exec/obidos/ASIN/1892925001/ref=nosim/schildnet0c,*

[63] References from Edmund Storms, *COLD FUSION: An Objective Assessment*, Energy K. Systems, Santa Fe, NM 87501, (12/16/01) *http://pw1.netcom.com/~storms2/review8.html*

130. Miley, G. "On the Reaction Product and Heat Correlation for LENRs," in *8th International Conference on Cold Fusion*. 2000. Lerici (La Spezia), Italy: Italian Physical Society, Bologna, Italy.

131. Miley, G.H. and J.A. Patterson, "Nuclear transmutations in thin-film nickel coatings undergoing electrolysis," *J. New Energy*, 1996. 1(3): p. 5.

132. Hora, H., J.C. Kelly, and G. Miley, "Energy gain and nuclear transmutation by low-energy p- or d-reaction in metal lattices," *Infinite Energy*, 1997. 2(12): p. 48.

[64] Edmund Storms, *COLD FUSION: An Objective Assessment*, Energy K. Systems, Santa Fe, NM 87501, (12/16/01) *http://pw1.netcom.com/~storms2/review8.html*

[65] *http://www.newenergytimes.com/v2/library/2001/2001StormsE-ObjectiveAssessment.pdf*

[66] Quote by Richard Sparks, Scientific and Technical Intelligence/SBIR, Ottawa. (1996) in The Hutchison File, (page 67 of 87), *http://drjudywood.com/pdf/HutchisonEffectReport_txt.pdf,*

[67] George D. Hathaway, P. Eng. Hathaway Consulting Services, Toronto, Ontario, Canada, September 15, 2005.

[68] George D. Hathaway, P. Eng. Hathaway Consulting Services, Toronto, Ontario, Canada, September 15, 2005.

[69] Contrast adjusted, color added, and descriptive text replaced or added, for clarity, Courtesy of John Hutchison and George Hathaway, P.E.

[70] Contrast adjusted, color added, and descriptive text moved, for clarity. Axis added for reference. Courtesy of John Hutchison and George Hathaway, P.E., *http://www.americanantigravity.com/galleries/hutchison-effect-spectroscopy/, (a)http://www.americanantigravity.com/images/gallery/6/233_original.jpg, (b)http://www.americanantigravity.com/images/gallery/6/239_original.jpg*

[71] From personal correspondance with John Hutchison and documentation he provided from Dr. George Hathaway.

[72] *http://en.wikipedia.org/wiki/List_of_elements_by_melting_point*

[73] *http://en.wikipedia.org/wiki/List_of_elements_by_boiling_point*

[74] Contrast adjusted, color added, and descriptive text moved, for clarity. Courtesy of John Hutchison and George Hathaway, P.E.

[75] From personal correspondance with John Hutchison and documentation he provided from Dr. George Hathaway.

[76] J.R. Barnett, R.R. Biederman, and R.D. Sisson, Jr., "An Initial Microstructural Analysis of A36 Steel from WTC Building 7," reposted from *the Journal of Materials*, JOM, 53 (12) (2001), pp. 18, *http://www.tms.org/pubs/journals/JOM/0112/Biederman/Biederman-0112.html,* also shown in Figure C-8, p. C-5, "Limited Metallurgical Examination," FEMA Report - Appendix C, 2002, *http://www.fema.gov/pdf/library/fema403_apc.pdf.*

[77] J.R. Barnett, R.R. Biederman, and R.D. Sisson, Jr., "Limited Metallurgical Examination," FEMA Report - Appendix C, 2002, *http://www.fema.gov/pdf/library/fema403_apc.pdf.*

[78] J.R. Barnett, R.R. Biederman, and R.D. Sisson, Jr., "An Initial Microstructural Analysis of A36 Steel from WTC Building 7," reposted from *the Journal of Materials*, JOM, 53 (12) (2001), pp. 18, *http://www.tms.org/pubs/journals/JOM/0112/Biederman/Biederman-0112.html,* also shown in Figure C-8, p. C-5, "Limited Metallurgical Examination," FEMA Report - Appendix C, 2002, *http://www.fema.gov/pdf/library/fema403_apc.pdf.*

[79] J.R. Barnett, R.R. Biederman, and R.D. Sisson, Jr., "An Initial Microstructural Analysis of A36 Steel from WTC Building 7," reposted from *the Journal of Materials*, JOM, 53 (12) (2001), pp. 18, *http://www.tms.org/pubs/journals/JOM/0112/Biederman/Biederman-0112.html.*

[80] Transformations, newsletter of Worcester Polytechnic Institute (WPI), *http://www.wpi.edu/News/Transformations/2002Spring/steel.html*

[81] RJ Lee*Group*, Inc., 350 Hochberg Road, Monroeville, PA, 15146, "Damage Assessment, 130 Liberty Street Property," WTC Dust Signature Report, *Composition and Morphology,* Summary Report, December 2003, Table 1. , page 7 (pdf p. 11 of 34), *http://www.docstoc.com/docs/11423437/httpwwwnyenvirolaworgWTC13020Liberty20StreetMike20Davis20LMDC2013020Liberty20DocumentsSignature20of20WTC20dustWTC20Dust20SignatureComposition20and20MorphologyFinalpdf*

[82] RJ Lee*Group*, Inc., 350 Hochberg Road, Monroeville, PA, 15146, "Damage Assessment, 130 Liberty Street Property," WTC Dust Signature Report, *Composition and Morphology,* Summary Report, December 2003, Figures 23-24, page 19 (pdf p. 23 of 34), *http://www.docstoc.com/docs/11423437/httpwwwnyenvirolaworgWTC13020Liberty20StreetMike20Davis20LMDC2013020Liberty20DocumentsSignature20of20WTC20dustWTC20Dust20SignatureComposition20and20MorphologyFinalpdf*

[83] Cold Fusion Is Hot Again, 60 Minutes: Once Considered Junk Science, Cold Fusion Gets A Second Look By Researchers, April 19, 2009, *http://www.cbsnews.com/stories/2009/04/17/60minutes/main4952167.shtml*

[84] More Than Junk Science, *http://www.cbsnews.com/video/watch/?id=4967330n*

[85] Edmund Storms, "FUSION: An Objective Assessment," Energy K. Systems, 12/16/01, *http://www.newenergytimes.com/v2/library/2001/2001StormsE-ObjectiveAssessment.pdf* (248 kb)

[86] References from Edmund Storms, *COLD FUSION: An Objective Assessment*, Energy K. Systems, Santa

Fe, NM 87501, (12/16/01) *http://pw1.netcom.com/~storms2/review8.html*

130. Miley, G. "On the Reaction Product and Heat Correlation for LENRs," in *8th International Conference on Cold Fusion*. 2000. Lerici (La Spezia), Italy: Italian Physical Society, Bologna, Italy.

131. Miley, G.H. and J.A. Patterson, "Nuclear transmutations in thin-film nickel coatings undergoing electrolysis," *J. New Energy*, 1996. 1(3): p. 5.

132. Hora, H., J.C. Kelly, and G. Miley, "Energy gain and nuclear transmutation by low-energy p- or d-reaction in metal lattices," *Infinite Energy*, 1997. 2(12): p. 48.

[87] Edmund Storms, "*FUSION: An Objective Assessment*," Energy K. Systems, 7/31/02, *http://www.lenr-canr.org/acrobat/StormsEcoldfusiond.pdf* (648 kb)

[88] Thomas M. Semkow, Ronald S. Hafner, Pravin P. Parekh, Gordon J. Wozniak, Douglas K. Haines, Liaquat Husain, Robert L. Rabun and Philip G. Williams, "Study of Traces of Tritium at the World Trade Center," Proceedings of the Symposium on Radioanalytical Methods at the Frontier of Interdisciplinary Science: Trends and Recent Achievements. 223rd American Chemical Society National Meeting, Orlando, FL, April 7-11, 2002, October 1, 2002 Preprint NYS DOH 02-116, *http://www.osti.gov/energycitations/servlets/purl/15002340-YM5IJp/native/15002340.PDF*

[89] Pravin P. Parekh, Thomas M. Semkow, Liaquat Husain, Douglas K. Haines, Gordon J. Wozniak, Philip G. Williams, Ronald S. Hafner, and Robert L. Rabun, "Tritium in the World Trade Center September llth, 2001 Terrorist Attack, It s Possible Sources and Fate," Proceedings of the Symposium on Radioanalytical Methods at the Frontier of Interdisciplinary Science: Trends and Recent Achievements. 223rd American Chemical Society National Meeting, Orlando, FL, April 7-11, 2002, May 3, 2002 Preprint NYS DOH 02-116, *http://www.llnl.gov/tid/lof/documents/pdf/240430.pdf*

[90] Tritium production reported in electrochemical cells (Cell 73) by Drs. Edmund Storms and Carol Talcott of the Los Alamos National Laboratory. (Courtesy of Drs. Edmund Storms and Carol Talcott) in Eugene F. Mallove, *Fire from Ice: Searching for the Truth Behind the Cold Fusion Furor,* 1991, John Wiley & Sons.

[91] Ian Fairlie, Tritium Hazard Report: Pollution and Radiation Risk from Canadian Nuclear Facilities, June 2007, (p. 21 0f 92) *http://www.greenpeace.org/raw/content/canada/en/documents-and-links/publications/tritium-hazard-report-pollu.pdf*

[92] Idaho State University, Radiation Information Network's "Tritium Information Section," *http://www.physics.isu.edu/radinf/tritium.htm*

[93] *http://nuclearweaponarchive.org/Nwfaq/Nfaq12.html*, Idaho State University, Radiation Information Network's "Tritium Information Section," *http://www.physics.isu.edu/radinf/tritium.htm*

[94] Thomas M. Semkow, Ronald S. Hafner, Pravin P. Parekh, Gordon J. Wozniak, Douglas K. Haines, Liaquat Husain, Robert L. Rabun and Philip G. Williams, "Study of Traces of Tritium at the World Trade Center," Proceedings of the Symposium on Radioanalytical Methods at the Frontier of Interdisciplinary Science: Trends and Recent Achievements. 223rd American Chemical Society National Meeting, Orlando, FL, April 7-11, 2002, Preprint UCRL-JC-150445, *U.S. Department of Energy*, Lawrence Livermore National Laboratory, October 1, 2002. *http://www.llnl.gov/tid/Library.html, http://www.osti.gov/energycitations/servlets/purl/15002340-YM5IJp/native/15002340.PDF*

[95] Pravin P. Parekh, Thomas M. Semkow, Liaquat Husain, Douglas K. Haines, Gordon J. Wozniak, Philip G. Williams, Ronald S. Hafner, and Robert L. Rabun, "Tritium in the World Trade Center September llth, 2001 Terrorist Attack, It s Possible Sources and Fate," Proceedings of the Symposium on Radioanalytical Methods at the Frontier of Interdisciplinary Science: Trends and Recent Achievements. 223rd American Chemical Society National Meeting, Orlando, FL, April 7-11, 2002, Preprint UCK-JC-148360, *U.S. Department of Energy*, Lawrence Livermore National Laboratory, May 3,2002. *http://www.llnl.gov/tid/Library.html, http://www.llnl.gov/tid/lof/documents/pdf/240430.pdf*

[96] Idaho State University, Radiation Information Network's "Tritium Information Section," *http://www.physics.isu.edu/radinf/tritium.htm*

[97] "The Anonymous Physicist", with a list of links to other articles proposing "mini-nukes" were used. *http://wtcdemolition.blogspot.com/*

[98] Thomas M. Semkow, Ronald S. Hafner, Pravin P. Parekh, Gordon J. Wozniak, Douglas K. Haines, Liaquat Husain, Robert L. Rabun and Philip G. Williams, "Study of Traces of Tritium at the World Trade Center," Proceedings of the Symposium on Radioanalytical Methods at the Frontier of Interdisciplinary Science: Trends and Recent Achievements. 223rd American Chemical Society National Meeting, Orlando, FL, April 7-11, 2002, October 1, 2002 Preprint NYS DOH 02-116, *http://www.osti.gov/energycitations/servlets/purl/15002340-YM5IJp/native/15002340.PDF*

[99] Pravin P. Parekh, Thomas M. Semkow, Liaquat Husain, Douglas K. Haines, Gordon J. Wozniak, Philip G. Williams, Ronald S. Hafner, and Robert L. Rabun, "Tritium in the World Trade Center September llth, 2001 Terrorist Attack, It s Possible Sources and Fate," Proceedings of the Symposium on Radioanalytical Methods at the Frontier of Interdisciplinary Science: Trends and Recent Achievements. 223rd American Chemical Society National Meeting, Orlando, FL, April 7-11, 2002, May 3, 2002 Preprint NYS DOH 02-116, *http://www.llnl.gov/tid/lof/documents/pdf/240430.pdf*

[100] Thomas M. Semkow, Ronald S. Hafner, Pravin P. Parekh, Gordon J. Wozniak, Douglas K. Haines, Liaquat Husain, Robert L. Rabun and Philip G. Williams, "Study of Traces of Tritium at the World Trade Center," Proceedings of the Symposium on Radioanalytical Methods at the Frontier of Interdisciplinary Science: Trends and Recent Achievements. 223rd American Chemical Society National Meeting, Orlando, FL, April 7-11, 2002, October 1, 2002 Preprint NYS DOH 02-116, *http://www.osti.gov/energycitations/servlets/purl/15002340-YM5IJp/native/15002340.PDF*

[101] Pravin P. Parekh, Thomas M. Semkow, Liaquat Husain, Douglas K. Haines, Gordon J. Wozniak, Philip G. Williams, Ronald S. Hafner, and Robert L. Rabun, "Tritium in the World Trade Center September llth, 2001 Terrorist Attack, It s Possible Sources and Fate," Proceedings of the Symposium on Radioanalytical Methods at the Frontier of Interdisciplinary Science: Trends and Recent Achievements. 223rd American Chemical Society National Meeting, Orlando, FL, April 7-11, 2002, May 3, 2002 Preprint NYS DOH 02-116, *http://www.llnl.gov/tid/lof/documents/pdf/240430.pdf*

[102] Cordy, G.E., Gellenbeck, D.J., Gebler, J.B., Anning, D.W., Coes, A.L., Edmonds, R.J., Rees, J.A.H., and Sanger, H.W., 2000, Water Quality in the Central Arizona Basins, Arizona, 1995–98: U.S. Geological Survey Circular 1213, 38 p., on-line at *http://pubs.water.usgs.gov/circ1213/*, Tritium in precipitation from 1950 to 1998, *http://pubs.usgs.gov/circ/circ1213/images/cab_fig18a.gif, http://pubs.usgs.gov/circ/circ1213/major_findings2.htm*

[103] Toxic Substances Hydrology Program, *http://toxics.usgs.gov/definitions/tritium.html*

[104] Ian Fairlie, "Tritium Hazard Report: Pollution and Radiation Risk from Canadian Nuclear Facilities," June 2007, (p. 20 0f 92) *http://www.greenpeace.org/raw/content/canada/en/documents-and-links/publications/tritium-hazard-report-pollu.pdf*

[105] Tritium production reported in electrochemical cells by Drs. Edmund Storms and Carol Talcott of the Los Alamos National Laboratory. (Courtesy of Drs. Edmund Storms and Carol Talcott) in Eugene F. Mallove, *Fire from Ice: Searching for the Truth Behind the Cold Fusion Furor,* 1991, John Wiley & Sons, (p. 227).

[106] "Ian Fairlie, Tritium Hazard Report: Pollution and Radiation Risk from Canadian Nuclear Facilities," June 2007, (p. 21 0f 92) *http://www.greenpeace.org/raw/content/canada/en/documents-and-links/publications/tritium-hazard-report-pollu.pdf*

[107] Toxic Substances Hydrology Program, *http://toxics.usgs.gov/definitions/tritium.html*

[108] Idaho State University, Radiation Information Network's "Tritium Information Section," *http://www.physics.isu.edu/radinf/tritium.htm*

[109] Thomas M. Semkow, Ronald S. Hafner, Pravin P. Parekh, Gordon J. Wozniak, Douglas K. Haines, Liaquat Husain, Robert L. Rabun and Philip G. Williams, "Study of Traces of Tritium at the World Trade Center," Proceedings of the Symposium on Radioanalytical Methods at the Frontier of Interdisciplinary Science: Trends and Recent Achievements. 223rd American Chemical Society National Meeting, Orlando, FL, April 7-11, 2002, October 1, 2002 Preprint NYS DOH 02-116, *http://www.osti.gov/energycitations/servlets/purl/15002340-YM5IJp/native/15002340.PDF*

[110] Pravin P. Parekh, Thomas M. Semkow, Liaquat Husain, Douglas K. Haines, Gordon J. Wozniak, Philip G. Williams, Ronald S. Hafner, and Robert L. Rabun, "Tritium in the World Trade Center September llth, 2001 Terrorist Attack, It s Possible Sources and Fate," Proceedings of the Symposium on Radioanalytical Methods at the Frontier of Interdisciplinary Science: Trends and Recent Achievements. 223rd American Chemical Society National Meeting, Orlando, FL, April 7-11, 2002, May 3, 2002 Preprint NYS DOH 02-116, *http://www.llnl.gov/tid/lof/documents/pdf/240430.pdf*

[111] Reduced ledgend size, highlighted Chernobyl location, *http://planetliberty.wikidot.com/nuclear-power-incidents.*

[112] Reduced ledgend size, highlighted Chernobyl location, *http://planetliberty.wikidot.com/nuclear-power-incidents.*

[113] *http://video.google.com/videoplay?docid=-1259460515501131928, http://video.google.com/videoplay?docid=-1259460515501131928#00h06m55s*

[114] Mark A. Solis, "The Hutchison Effect -- An Explanation," February 16, 1999, *http://rexresearch.com/*

hutchisn/hutchisn.htm#1, http://www.geocities.com/ResearchTriangle/Thinktank/8863/HEffect1.html

[115] *http://hutchisoneffect2008.blogspot.com/2007/12/dec-philly-testing-mini-warship.html*

[116] *http://video.google.com/videoplay?docid=-1259460515501131928, http://video.google.com/videoplay?docid=-1259460515501131928#00h06m55s*

[117] Photo trimmed.

[118] "Can you crush a can? Or launch it in the air?" A physics demonstration by *The Wonders of Physics* group from the University of Wisconsin Physics Department, *http://uw.physics.wisc.edu/~wonders/CanCrusher.html*

[119] *http://hutchisoneffect2008.blogspot.com/2007/12/dec-philly-testing-mini-warship.html*

[120] *http://www.youtube.com/watch?v=o6hi9pcpfQM*

[121] *http://upload.wikimedia.org/wikipedia/en/c/c5/DustifiedWTC2.jpg, (now removed), http://hereisnewyork.org//jpegs/photos/1539.jpg,*

[122] *Wilson10746124-D_windowhole.jpg.,* (after 9/11/01)

[123] I thank Kurt Sonnenfeld, FEMA photographer, for providing me with this photo.

[124] Richard Sparks, Scientific and Technical Intelligence, SBIR(Ottawa, pdf p. 69 of 87, *The Hutchison File, http://drjudywood.com/pdf/HutchisonEffectReport_txt.pdf*

[125] *http://www.weourselves.org/mp3/wpfw_011808_judy-andrew2.mp3*

[126] *http://www.drjudywood.com/articles/JJ/JJ8.html*

[127] *http://drjudywood.com/media/071212_JohnHutchison-TruthH.mp3*. See also *http://www.prlog.org/10048184-scientists-see-wtc-hutchison- effect-parallel.html*

18.
Hurricane Erin

I have no doubt that we will be successful in harnessing the sun's energy... If sunbeams were weapons of war, we would have had solar energy centuries ago.
—Sir George Porter

A. Introduction

It was a beautiful early-autumn morning in New York City. September 11, 2001, started out calm, with pleasant temperatures and crystalline blue skies. Some had taken time to do early morning errands.[1] But very few of those people, in fact very few among the entire population of New York City, knew that a massive hurricane was located at that very same time just off the shore of Long Island. That storm was Hurricane Erin, as seen in Figure 411:

Figure 411. Hurricane Erin on September 11, 2001.
http://svs.gsfc.nasa.gov/vis/a000000/a002500/a002521/wtc_terra1.tif

Hurricane Erin was born on September 1, 2001, the fifth Atlantic storm of the 2001 season to be given a name and the first to reach hurricane strength.[2] Merging with another storm system on September 17, Erin was also the longest-lived Atlantic storm of the 2001 season. Before that merger, however, starting back on September 7, Erin began a four-day march on a path taking her directly toward New York City. (See

Figure 420.) By September 10, Hurricane Erin had become a Category 3 hurricane with wind speeds of 120 mph[2] just as it passed Bermuda along its straight path toward New York City. And yet we heard nothing about this storm. The World Trade Center Towers were built to withstand wind loads of up to 140 mph (225 km/hr.)[3], the equivalent force of a Category 4 hurricane and *only 20 mph more than the wind speed of Hurricane Erin.* The danger to tall buildings aside, even if Hurricane Erin changed course and failed to make landfall, the threat of flooding from storm surges was very real. Hurricane Ike (2008) made landfall in Texas as a Category 2 hurricane and had peak storm surge values of 15-20 ft.[4] And yet, once again, we heard *nothing* about this storm.

People remember 9/11 as being one of the clearest days they had ever seen along the East Coast. The satellite that took the images shown in Figures 411 and 412 had a clear view of New York City and it also had a clear view of Hurricane Erin. The outer bands of the storm reached over Cape Cod and the end of Long Island.

Not even I knew there was a hurricane just outside of New York City on 9/11/01 until I discovered the fact while looking for weather-satellite images to study the rising dust plume from the destroyed WTC towers.

B. Hurricane Erin on 9/11/01

In fact, as I was to discover, Erin came closest to NYC, and also reached its largest size, on 9/11 itself.[5] Interestingly, the National Hurricane Center projected Erin to be of stronger force than it was to project in the case of Katrina four years later.[6] How curious it is, then, that this hurricane was not mentioned or shown by graphics on morning weather reports.[7] Were meteorologists absolutely certain that this hurricane would make a sharp right-hand turn away from New York and head back out to sea before there was a major threat of storm surges? No—at least not according to the National Weather Service. [8] Approximately 500 miles in diameter, Hurricane Erin was approximately the same size as the later Hurricane Katrina, and yet the public was not widely alerted to it, none of the major morning news shows mentioning the storm. Still, Erin was the subject of extended study. The forecast was for Erin[9] to be stronger than Katrina was later to be.[10] In fact, Hurricane Erin actually did have more cyclonic energy than Hurricane Katrina, as measured by each storm's Accumulated Cyclone Energy (ACE).

The storm was a tropical depression on the first of September, reached minimum pressure late on September 9 and maximum speed on September 10, and then it reached maximum diameter on September 11. *On the morning of September 11, the storm stopped at its closest approach to New York City, then in the afternoon it veered dramatically to the east.* Joey Moore, a flight attendant flying out of Boston's Logan Airport that morning, was unaware of this hurricane! He had only just learned of Hurricane Erin from my website (2010).

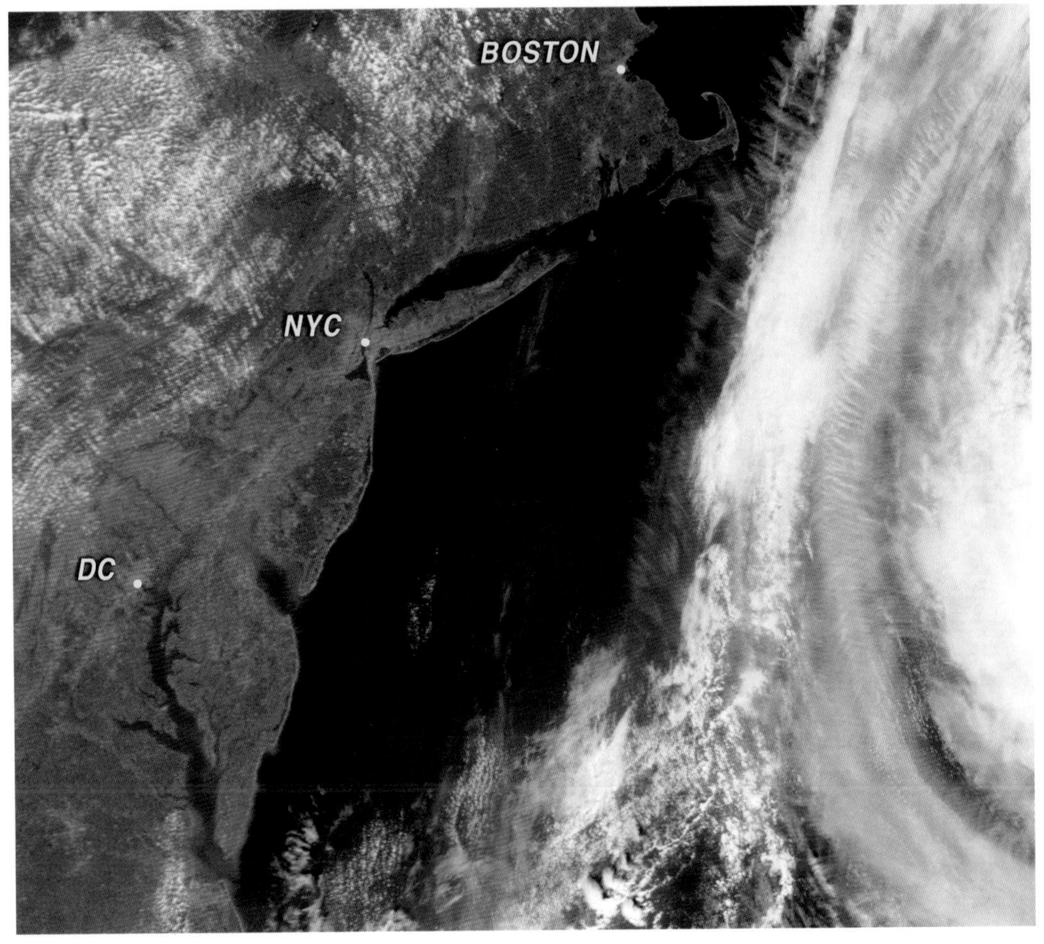

Figure 412. Hurricane Erin off the end of Long Island on 9/11/01.
http://svs.gsfc.nasa.gov/vis/a000000/a002500/a002521/wtc_terra1.tif

Joseph (Joey) Moore, United Airlines Flight Attendant

> I worked that very strange morning out of Boston's Logan on the 7AM nonstop to San Fran. [I saw] a handful of my co-worker flying partners in the office that morning, unknowingly moving toward their deaths on flt 175…[and it] could've easily been me. I probably worked 175 the entire month prior…I only remember it being a nice day and no mentions of the Hurricane.[11]

C. Controlled Environment?

This remarkable fact leads us to posit the question: Was Erin somehow *steered* away from New York City? Erin did not gradually turn northward, but actually reversed direction.

Figure 413 plots the wind speed, air pressure and relative distance from New York City in the few days preceding and following 9/11. The data shows that Erin slowed down as it approached New York, *and then remained almost stationary during the morning of 9/11.* Immediately after the WTC was attacked, Erin began to move away from New York City, as can be seen by the red and blue lines above. For the 24 hours surrounding the events of 9/11, Hurricane Erin maintained the same wind speed, the same pressure, and approximately the same distance from New York City.

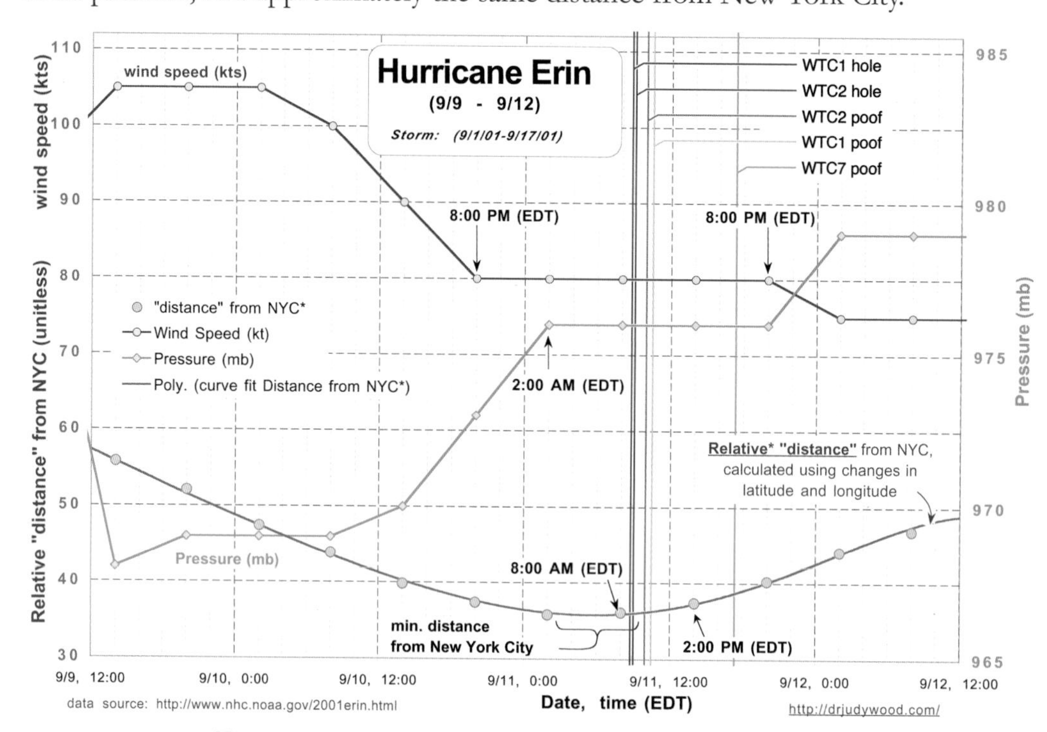

Figure 413. Wind speed, pressure, and relative location of Erin.[12]
Data source: http://www.nhc.noaa.gov/2001erin.html

D. Unprecedented Data Collection

As mentioned, hurricane Erin began on September 1 as a tropical depression and ended on September 17 after reverting to a tropical storm. With a life span extending from September 1 through September 17, 2001,[13] Hurricane Erin was studied more than any other hurricane had been studied before, and more was learned from it than had been learned from any hurricane before it, as is discussed in the following article:

> Hurricane Erin raced across the North Atlantic and along the eastern seaboard in September 2001. She was used as an experiment for a study to improve hurricane tracking and intensity predictions, *allowing meteorologists to provide more accurate and timely warnings to the public.* Studies show that temperatures measured at an extremely high altitude collected from a hurricane's center or eye can provide improved understanding of how hurricanes change intensity.
>
> Hurricane Erin was analyzed during the fourth Convection And Moisture Experiment (CAMEX-4), which took place *from August 16 through September 24, 2001. The mission originated from the Naval Air Station in Jacksonville, Fla. The mission united researchers from 10 universities, five NASA centers and the National Oceanic and Atmospheric Administration.* CAMEX-4 is a series of field research investigations to study tropical cyclones & storms commonly known as hurricanes.
>
> [...]

For the first time, researchers were able to reconstruct the structure of the eye in three dimensions from as high as 70,000 feet, down to the ocean surface, in great detail.[14]

But at least one network, CNN, was confident it would not strike New York City or the East Coast. The question is: Why?

September 10, 2001 Posted: 6:59 PM EDT

MIAMI, Florida (CNN) -- Hurricane Erin posed no immediate threat to land on Monday after grazing past Bermuda overnight and moving into the open Atlantic.

[...]

The worst part of the storm, with maximum sustained winds of 120 mph (195 km/h), passed to the northeast of Bermuda on Sunday, according to the National Weather Service.

At 5 p.m. EDT, the weather service placed the center of Erin, the first hurricane of the 2001 Atlantic season, 540 miles (875 kilometers) south of Yarmouth, Nova Scotia. The storm was moving north-northwest near 8 mph (13 km/h). [15]

The CNN article states that Hurricane Erin was *"moving north-northwest"* after *"grazing past Bermuda"* but was *"moving into the open Atlantic."* New York City is north-northwest of Bermuda, almost exactly in Erin's path toward "the open Atlantic." Erin was 540 miles directly south of Yarmouth, Nova Scotia, just about the same distance that it was also away from New York City, as shown in Figure. 416. Yet the CNN article described the storm as *"moving into the open Atlantic," as if* the hurricane were moving out to sea, eastward.

At the same time, there was a cold front moving from the Midwest towards New York City that would have slowed the hurricane and turned it northward, but how sure could meteorologists have been about the *timing* of the turn? How sure could they have been that the storm wouldn't pose a serious threat to Cape Cod? If Erin had stalled a little bit longer where it was, storm surges would have flooded JFK and LaGuardia airports as well as Cape Cod. Not only is New York City near sea level, but so is most of Long Island. Evacuation from these areas would be a mammoth undertaking and could not be organized at a moment's notice—and yet the public remained uninformed.

Figure 414 shows screen shots of weather maps and forecasts from four networks on the morning of September 11, 2001. These images were shown on network TV between 8:31:33 and 8:36:02 am, 10 to 15 minutes before WTC1 got its hole at 8:46:26 a.m.,[16] yet none of them showed an icon of a hurricane where Hurricane Erin was located offshore in the Atlantic—*even though* that area of each weather map was visible in the on-screen image as shown above as well as elsewhere. [17] Former TV meteorologist Scott Stevens[18] asked, "How much effort does it take to make one click of a mouse button to put the hurricane icon there?"

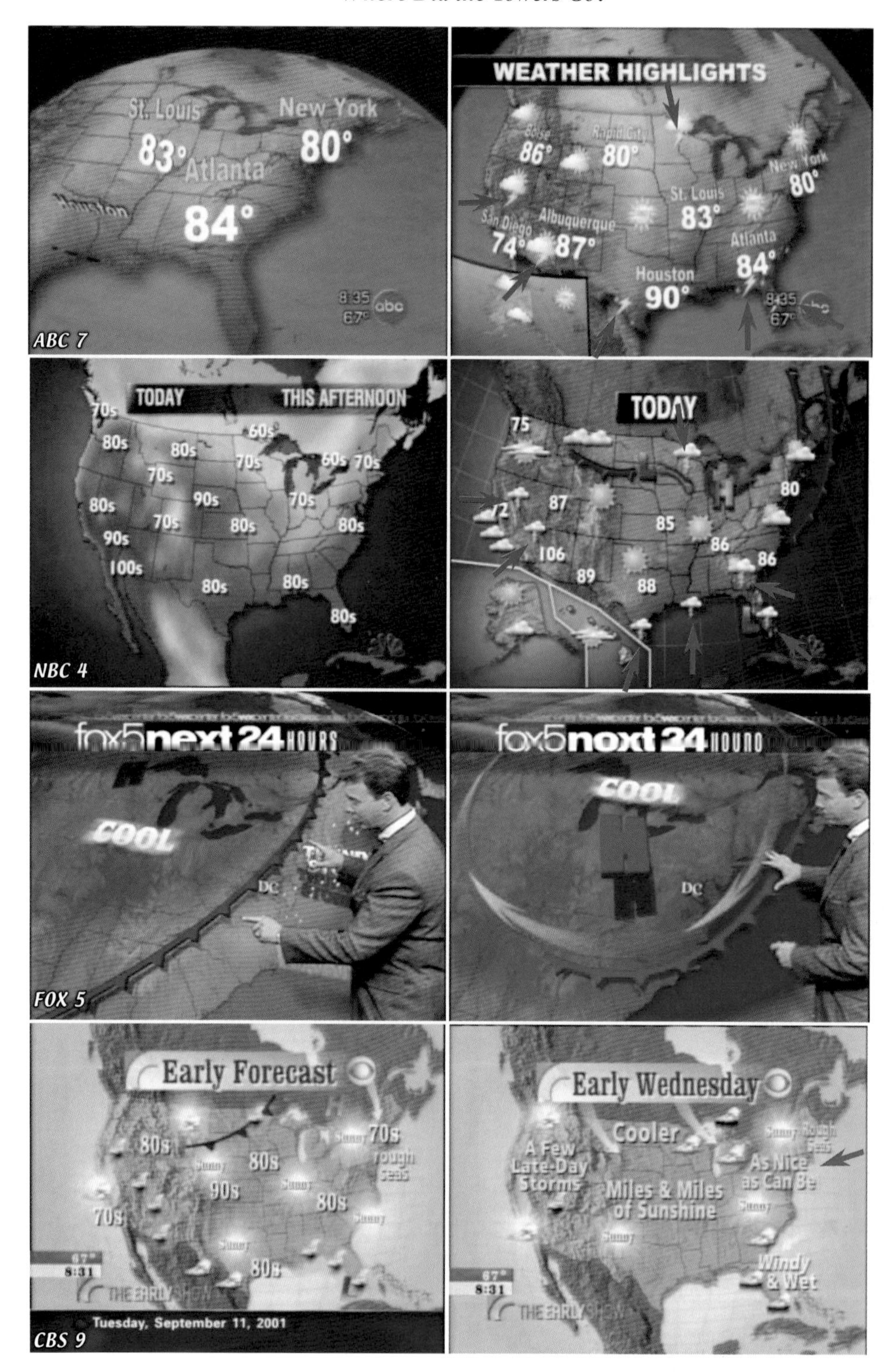

Figure 414. (9/11/01) "As Nice as Can Be"[22] *The red arrows have been added to the ABC and NBC images to highlight the thunderstorms shown.*
(a) ABC7[19]*, 8:35:02 - 8:36:02 am, (b)* NBC4[20]*, 8:32:29 - 8:33:33 am,*
(c) FOX5[21]*, 8:33:06 - 8:34:08 am, (d)* CBS9[22]*, 8:31:21 - 8:32:35 am.*

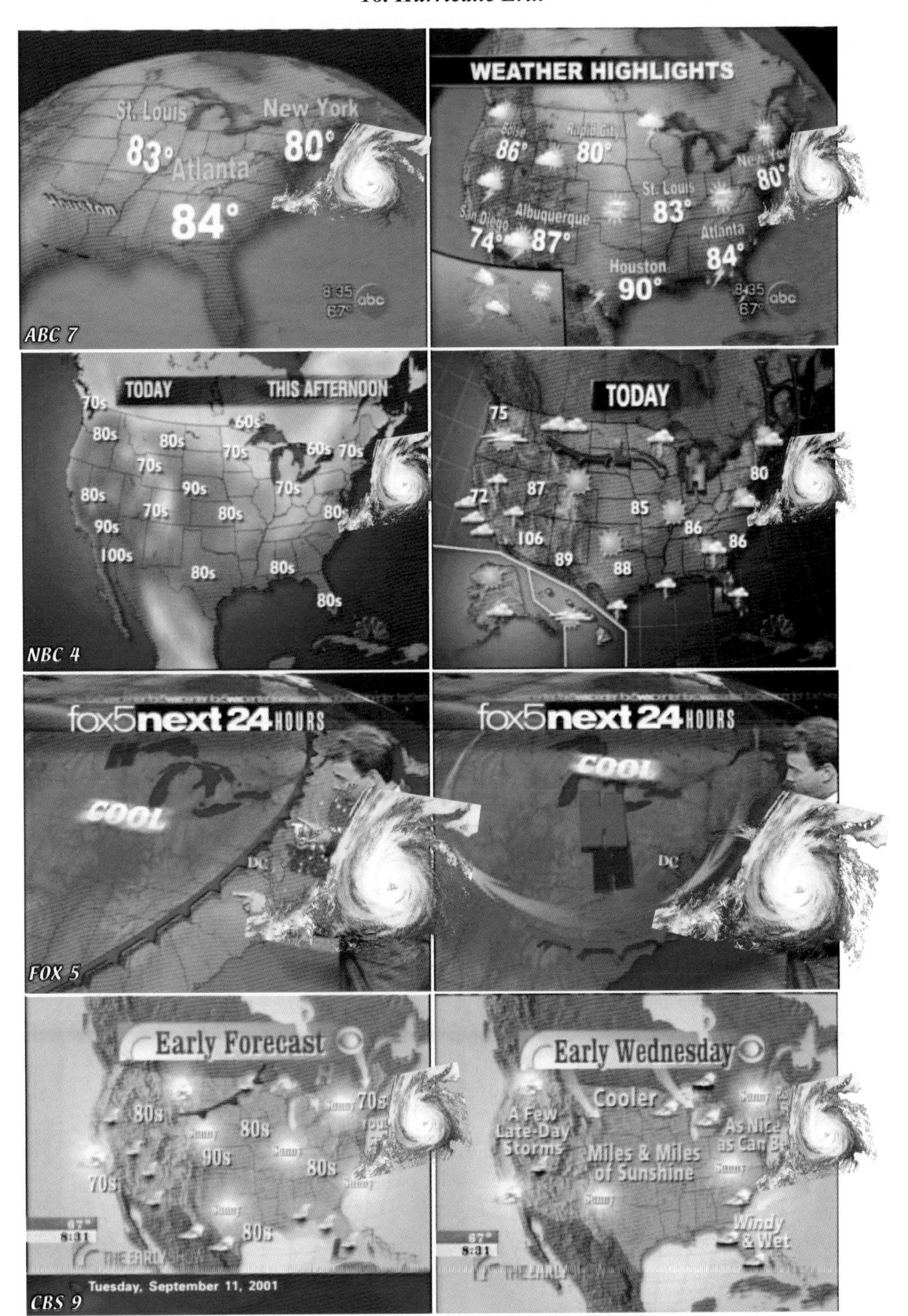

Figure 415. (9/11/01) The image shown in Figure 411 is superimposed to show the approximate size and location of Hurricane Erin at around 8:30 AM, at the time of those screen images. The curved arrows have been added to the FOX image to highlight the counter-rotating weather systems. (a) ABC7[19], *8:35:02 - 8:36:02 am, (b)* NBC4[20], *8:32:29 - 8:33:33 am, (c)* FOX5[21], *8:33:06 - 8:34:08 am, (d)* CBS9[22], *8:31:21 - 8:32:35 am.*

Network	Time of Weather Report	Time Before WTC1 Hole[16] *(min:sec)*
ABC7[19]	8:35:02 - 8:36:02 am	*10:24 - 11:24*
NBC4[20]	8:32:29 - 8:33:33 am	*12:53 - 13:57*
FOX5[21]	8:33:06 - 8:34:08 am	*12:18 - 13:20*
CBS9[22]	8:31:33 - 8:32:35 am	*13:51 - 14:53*

Table 20. (9/11/01) The times when the network weather reports aired.[23]

Figure 415 shows the same images as Figure 414, but with the image of Hurricane Erin (Figure 411) superimposed on each weather map, showing the approximate location of Hurricane Erin on the morning of 9/11/01. The shoreline was used to match the scale and orientation of the Figure 411 image. For example, the silhouette of Long Island can be seen superimposed on the right hand of the FOX weatherman and the Maine coastline on his left cheek. It should be noted that the location of Hurricane Erin was actually closer to Manhattan at 8:00 am than what is shown (see Figure 413.) The image in Figure 411 was taken later in the day, as can be seen in Figure 417, showing the fumes emerging from southern Manhattan.

Network	hurricane on map	Hurricane mentioned	Erin's name mentioned	cold front on map[24]	cold front discussed
ABC7[19]	no	no	no	no	no
NBC4[20]	no	no	no	*yes*	no
FOX5[21]	no	*yes*	*yes*	*yes*	*yes*
CBS9[22]	no	*yes*	no	no	no

Table 21. Network weather reports.

We have seen that the national weather map for each of these four networks did not show an icon for a hurricane where Hurricane Erin was located offshore in the Atlantic—*even though* that area of each weather map was visible in the on-screen image as shown above as well as elsewhere.[25] ABC did not show or mention a hurricane, nor did it show or mention that there was a significant cold front along the entire east coast from New England to Florida. Al Roker of NBC said there would be "up to seven inches of rain in parts of Florida" but made no mention of a hurricane and did not discuss the cold front, although two such fronts were shown on the national weather map, but only in the northeast. Mark McEwen of CBS noted that there was in "Florida, a little rain" and added that there were "rough seas ...from that hurricane that's going away." He later said there were "rough seas from the ...chop from that hurricane. Other than that, kinda quiet around the country and we like quiet."

The most informative weather report came from Tom Sater, the FOX5 weather man, who told the listening audience about the significant high-pressure cold front moving into the area. He said that the humidity was down to 73% from the 88% earlier that morning, and that it would continue to fall. "We're going to have just a beautiful day today. This is a real cold front, folks [indicates on the map]. That means along the boundary line there is very cool-dry air. And it is going to make its way well offshore and help to push Hurricane Erin up to the north. It's not going to affect us at all."

The significant high-pressure cold front shown in the on-screen image by FOX (Figure 414) covers the continental United States. Around the perimeter of this enormous weather system are thunderstorms, which appear on the ABC and NBC weather maps (Figure 414) and are highlighted by the red arrows. The thunder icons appear around the perimeter of the weather system except for the northeast, although thunder was recorded there (see Figures 418 and 419), which will be discussed later, beginning on page 434.

At the same time, however, there were quiet Erin Advisories for coastal North Carolina on Friday, September 7.

> Apparently the NWS is pretty concerned about this since they also spent the day calling all the local Ocean Rescue depts and warning them firsthand.
>
> SOUTHEAST SWELL WAVES THIS WEEKEND WILL PRODUCE LIFE THREATENING RIP CURRENTS AND DANGEROUS BOATING NEAR NORTH CAROLINA INLETS UNTIL 5:00AM EDT
>
> Marine Weather Statement National Weather Service Newport/Morehead City NC 400 PM EDT Fri Sep 7 2001
>
> ... Southeast Swell Waves This Weekend Will Produce Life Threatening Rip Currents And Dangerous Boating Near North Carolina Inlets...
>
> [. . .]
>
> Swell Waves Will Begin Approaching The Coast Late Tonight And Increase In Height Through The Weekend. For Saturday Night Through Monday... On Beaches From Cape Hatteras South To North Topsail Beach... Waves Are Expected To Reach A Height Of 5 To 6 Feet On The Open Ocean... *And 6 To 8 Feet On The Beaches.* For Areas From Cape Hatteras North... Swell Waves Will Reach 5 To 6 Feet On The Open Ocean... *With 6 To 8 Feet Possible On The Surf On Sunday And Monday.*[26]

And, in another report, Erin is kicking up heavy surf along the East Coast, 09/10/ 2001—Updated 06:28 PM ET,

> The first Atlantic hurricane of the season swirled northward with sustained wind of 115 mph, down from its peak of 120 mph during the weekend. But Erin was expected to turn away from the United States.
>
> "We haven't had the big seas they're talking about, but the rip currents are sucking pretty hard," said Robert Levy, chief of the Atlantic City Beach Patrol. "We'll keep our beaches open and play it by ear."[27]

The National Weather Service in Miami issued various reports from September 8th through 12th, warning of large swells along the East Coast, with a report at 5 AM, EDT, on 9/11/01, stating that "ERIN REMAINS A SIGNIFICANT HURRICANE.".[28]

According to this map (Figure 416) from the Canadian Hurricane Centre (CHC), Hurricane Erin entered the Canadian "Response Zone" on Sept. 14, meaning that the hurricane, being in this zone, should presumably trigger a "response."

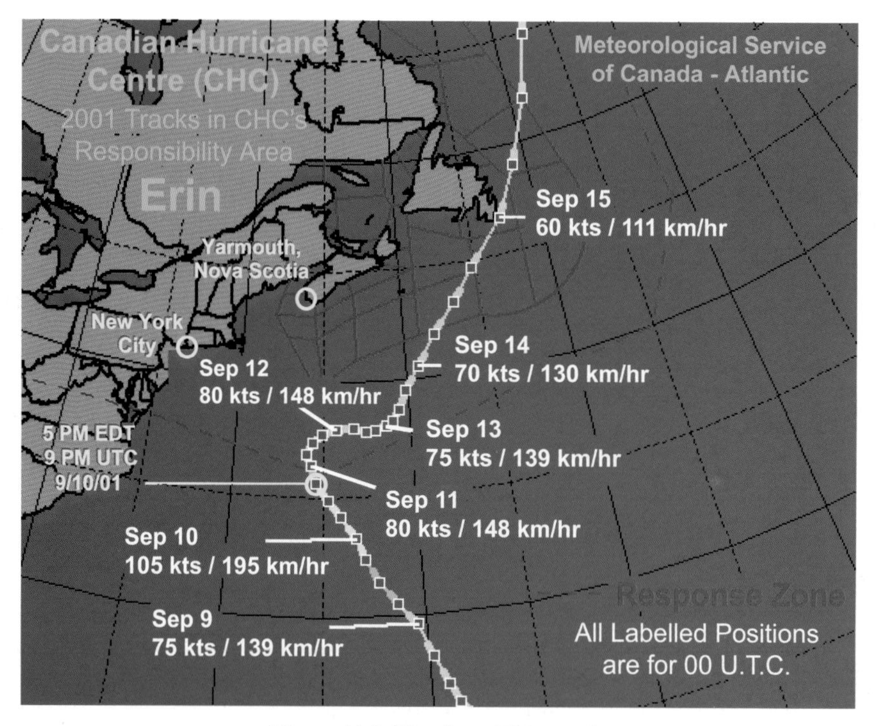

Figure 416. Hurricane Erin track.
http://www.atl.ec.gc.ca/weather/hurricane/images/2001/erin2001_e.gif

Yet, for four entire days, Hurricane Erin had been heading towards the American response zone on a direct track towards New York City. Just as it was entering the response zone, at about 11 p.m. on the evening of September 10, the National Weather Service issued this report:

NATIONAL WEATHER SERVICE MIAMI FL,
11 PM EDT MON SEP 10 2001

> THERE MAY BE SOME FLUCTUATIONS IN INTENSITY OVER THE NEXT 24-36 HOURS...AS THE INNER EYEWALL COLLAPSES AND THE OUTER ONE TAKES OVER. *GLOBAL MODELS SUGGEST THAT ERIN COULD BECOME A VERY VIGOROUS EXTRATROPICAL SYSTEM* AROUND THE END OF THE FORECAST PERIOD...AND THE OFFICIAL FORECAST DOES NOT SHOW AS MUCH WEAKENING AS INDICATED BY THE SHIPS GUIDANCE…
>
> *LARGE SWELLS GENERATED BY ERIN* WILL LIKELY BE A HAZARD ALONG THE NORTHEAST U.S. COAST DURING THE NEXT COUPLE OF DAYS.[29]

Figure 417 shows Hurricane Erin near its closest approach to New York City. The inset is a close up view of the fumes rising from the World Trade Center site. The dark fumes are moving west and dissipating, but the lighter fumes are heading due south. Unique cloud formations are also visible in the eye of the hurricane. The eye wall is in the shape of a pentagon and the interior of the eye looks like a maze that winds back and forth. How, and why, are these unique formations there?

Figure 417. Inset shows the rising fumes from the WTC site.
http://svs.gsfc.nasa.gov/vis/a000000/a002500/a002521/wtc_terra1.tif

E. Accumulated Cyclone Energy (ACE)

The Accumulated Cyclone Energy (ACE) Index is calculated by summing the squares of the estimated 6-hourly maximum sustained wind speed in knots (V_{max}^2) for all periods in which the tropical cyclone is a tropical storm or greater intensity storm (wind speed 34 knots or higher).[30] The numbers are divided by 10,000 to make them more manageable, so the unit of ACE is 10^4 kt^2. When the value is used as an index, the unit is assumed.[31]

$$\sum_{n=1}^{n_i} \frac{v_{max}^{\;2}}{10^4} \; (10^4 \; kt^2)$$

where v_{max} is maximum sustained wind speed in knots for each six-hour interval and n_i is the total number of intervals in which the wind speed was 34 knots or higher.[30]

Storm	ACE (10^4 kt^2)	n_i	Total Days	Dates
Hurricane Erin	22.40[32]	52[32]	17	Sept. 1-Sept. 17, 2001[32]
Hurricane Katrina	20.00[33]	24[33]	9	Aug. 23 - Aug. 31, 2005[33]

Table 22. Ace values for Hurricanes Erin and Hurricane Katrina.

F. Field Effects

It is often said that people can sense approaching storms,[34] an ability especially true of those suffering from arthritis.[35] This is because the ion balance of the atmosphere is affected by approaching storm fronts, and negatively-charged cloud bases induce positive charges on the earth's surface. Once a storm arrives, the customary or normal ion balance returns and people feel better.[36] There really is, then, something that can be sensed ahead of an approaching storm, an actual change in the atmosphere. Just ahead of a storm, also, there is often dry lightning accompanied by thunder, another indication of the field effects of a storm.

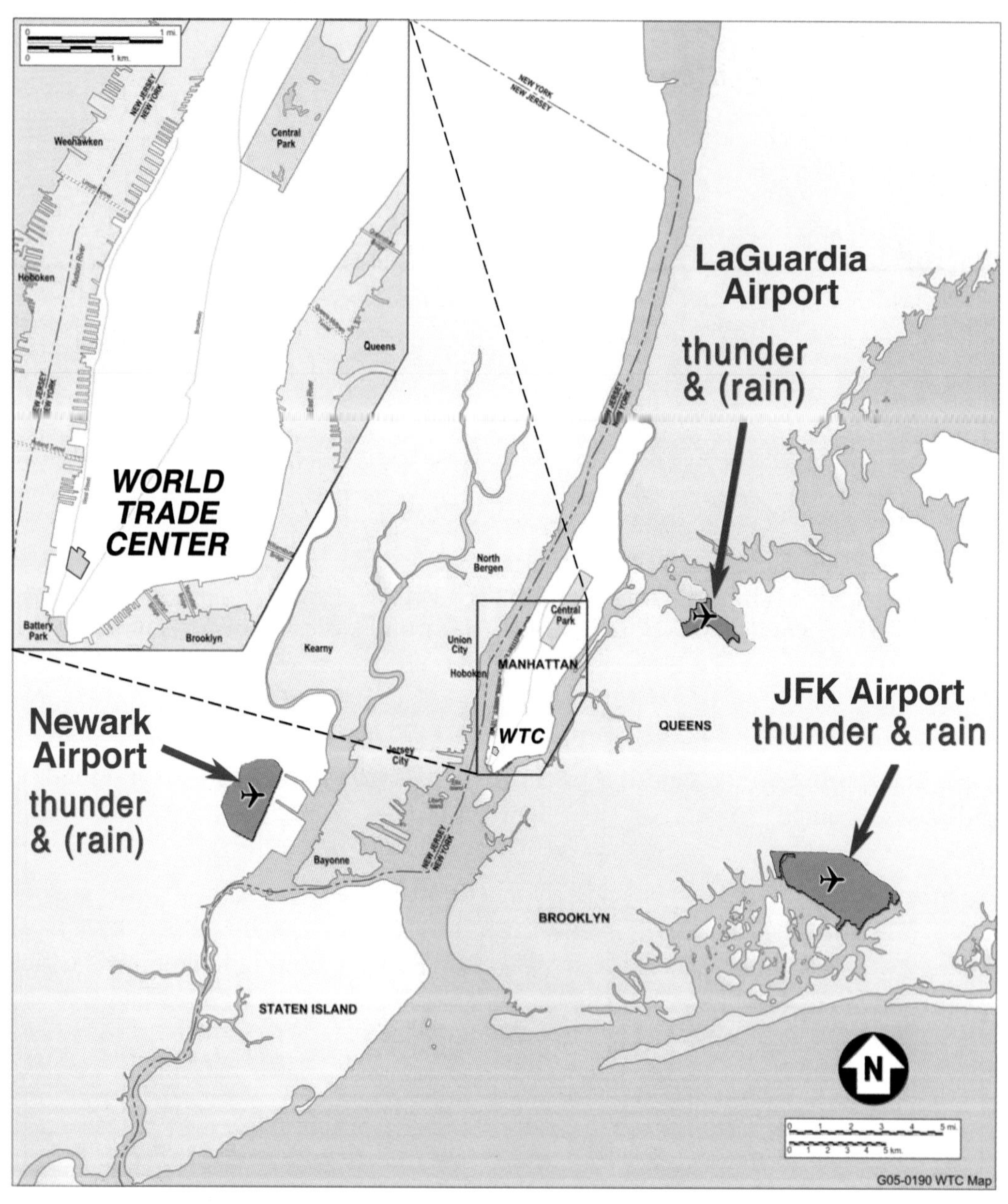

Figure 418. Map of lower Manhattan and the WTC relative to the local airports.[37]

On 9/11/2001, thunder was reported at nearby airports, including Newark Airport, JFK and LaGuardia. Rain was either reported or measured at all three of these airports, too.

As mentioned before, the outer bands of Hurricane Erin passed over Cape Cod and Long Island, but the *field effects* accompanying the storm spread out even farther. Weather-data shown in Figure 419 is evidence that the New York City area experienced the field effects of Hurricane Erin *on the morning of 9/11*. This fact can be seen also in Figure 420, where the approximate cloud cover of Hurricane Erin is shown in blue and the area covered by field effects in light yellow-orange.

Now let's look at the curious resemblance of a hurricane to a well known technological invention, the Tesla Coil:

	Tuesday, September 11, 2001		
JFK Airport (a)	**PRECIPITATION**		
	TOTAL PRECIPITATION **0.69**IN Rain and/or melted snow reported during the day.	VISIBILITY **9.9**MI Mean visibility for the day.	SNOW DEPTH ***NO DATA AVAILABLE*** Last report for the day if reported more than once.
	OCCURENCES		
	✓ Rain ✗ Snow ✗ Hail ✓ Thunder ✗ Tornado ✗ Fog		

	Tuesday, September 11, 2001		
LaGuardia Airport (b)	**PRECIPITATION**		
	TOTAL PRECIPITATION **0.02**IN Rain and/or melted snow reported during the day.	VISIBILITY **9.7**MI Mean visibility for the day.	SNOW DEPTH ***NO DATA AVAILABLE*** Last report for the day if reported more than once.
	OCCURENCES		
	✗ Rain ✗ Snow ✗ Hail ✓ Thunder ✗ Tornado ✓ Fog		

	Tuesday, September 11, 2001		
Newark Airport (c)	**PRECIPITATION**		
	TOTAL PRECIPITATION **0.00**IN Rain and/or melted snow reported during the day.	VISIBILITY **9.8**MI Mean visibility for the day.	SNOW DEPTH ***NO DATA AVAILABLE*** Last report for the day if reported more than once.
	OCCURENCES		
	✓ Rain ✗ Snow ✗ Hail ✓ Thunder ✗ Tornado ✗ Fog		

Figure 419. (9/11/01) Rain and thunder were recorded at (a) J.F. Kennedy International Airport, NY, (b) LaGuardia International Airport, New York, and (c) Newark International Airport, Newark, NJ.
http://www.almanac.com/weatherhistory/index.php?day=11&month=9&year=2001,

Hurricanes are born in the steamy late-summer environment of the tropics when rapidly evaporating ocean waters combine with strong wind currents to spawn them. Several hundred miles wide and packing winds of over 100 m.p.h., hurricanes cool the Earth by sucking heat from the Earth's surface and drawing it up into the atmosphere, frequently above 40,000 feet. But it is curious why some storms become hurricanes while others do not. Some have suggested a correlation between Coronal Mass Ejections (CME) and Cyclonic Earth Storms.[38] The paths hurricanes take can be influenced by various competing forces including high-pressure areas, prevailing winds, and even water temperatures and ocean currents.

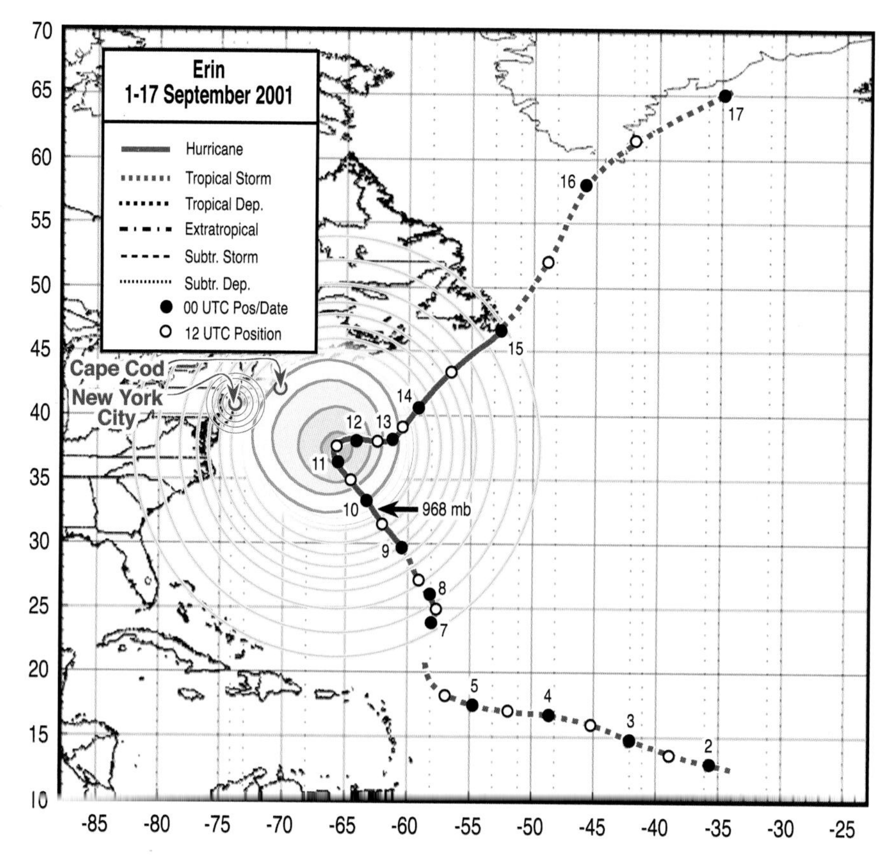

Figure 420. Eye of Hurricane Erin at approximately 8 AM(EDT), [12 UTC] on 9/11/01, and track of Hurricane Erin, September 1-17, 2001.
http://www.nhc.noaa.gov/prelims/2001erin1.gif,

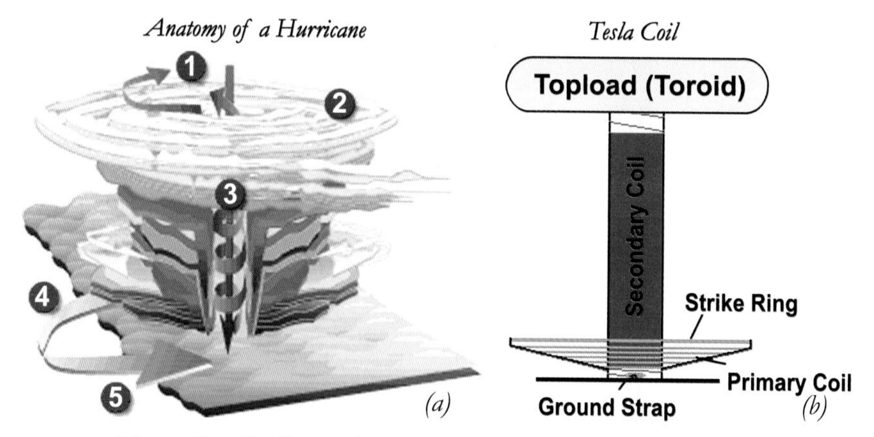

Figure 421. Similarities between (a) a hurricane and (b) a tesla coil.
(a)http://img.coxnewsweb.com/B/01/36/62/image_62361.gif, (b)redrawn from: http://www.powerlabs.org/images/tcdraw.gif

1. Exhaust: Hot air drawn into the atmosphere
2. Spiraling Storm Clouds
3. Eye: Cool air descends into the eye (~ 20 mi. wide) creating a small center of calm weather.
4. High Winds: In the lower few thousand feet of the hurricane, air flows in toward

the low pressure-center and it whirls upward. These spiraling winds gain speed as they approach the central eye, just as currents do in a whirlpool. The narrower the eye, the stronger the winds.

5. Spiraling winds spin counterclockwise in the Northern Hemisphere, clockwise in the Southern Hemisphere.

Now, though, consider the path of Erin in relation to the lines of magnetic declination of the earth's magnetic field: This image shows that "something" dramatic must have happened to get Erin, with such a dramatic turn, off of the dead track she had been following along the minus-15 line.

What was that "something"? Let's consider more information about Erin and the Earth's magnetism. In Figure 423, the light green is approximately when the magnetometer was excited.

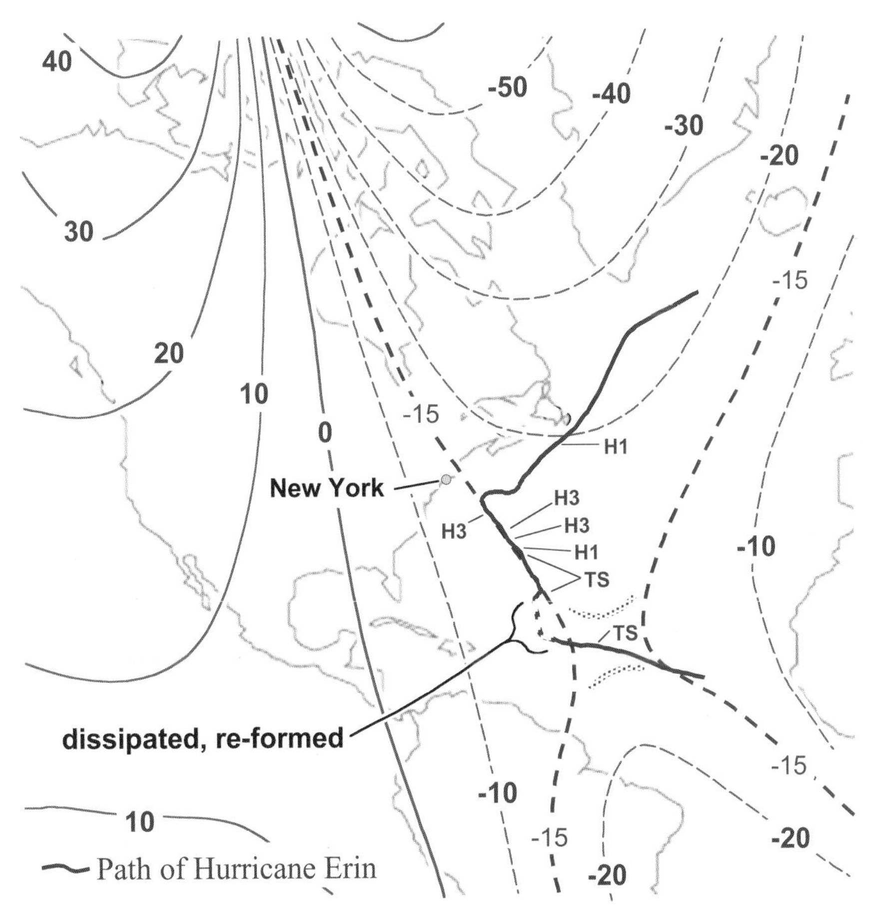

Figure 422. Magnetic declination and the path of Hurricane Erin.
Adjusted from: http://upload.wikimedia.org/wikipedia/commons/6/68/IGRF_2000_magnetic_declination.gif

The two green zones (B) and (C) are at the beginning and end of the storm's long straight path toward New York. These two zones have much more dramatic fluctuations than the earlier green zone (A) (September 2-5) and the later zone (D) (September 15). It thus appears that *the turns in the hurricane are closely related to the wider fluctuations in the magnetometer readings.* Unquestionably, more data is needed before a conclusion could be made that magnetic activity caused course-change, or that both course-change *and* magnetic activity are both the result of something else, or that it is all just a coincidence. On the other hand, the use of directed energy technology on the huge scale of its use during 9/11 might very well have had weather-altering effects. Such technology might have been able to draw upon the vast energies and field effects of the enormous Tesla Coil known as Hurricane Erin. Thoughts of this kind are justified, even necessitated, by the fact of Erin's having been treated as a carefully kept secret, much like a state secret.

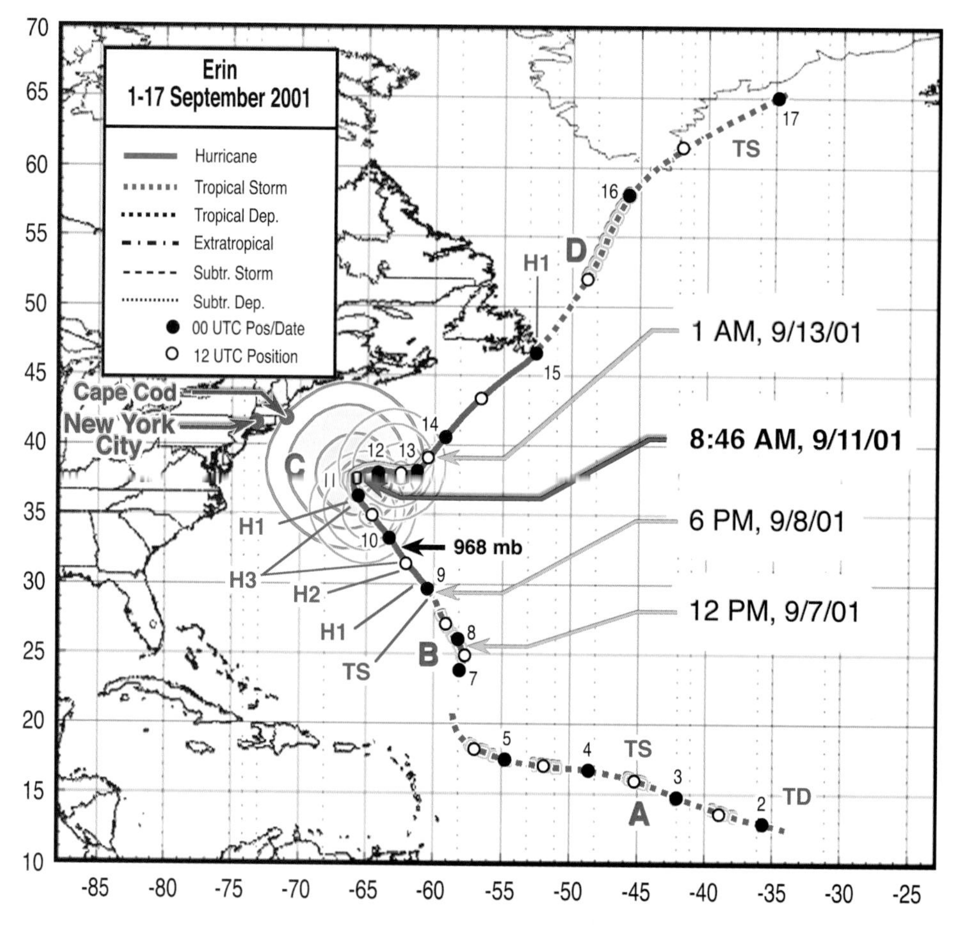

Figure 423. Best track of Hurricane Erin, September 1-17, 2001.
Adapted from: http://www.nhc.noaa.gov/prelims/2001erin1.gif

---

[1] page 51 of 115 of pdf (page 13 of report), *http://wtc.nist.gov/media/NIST_NCSTAR_1A_for_public_comment.pdf*

[2] *http://www.nhc.noaa.gov/2001erin.html*

[3] Thomas W. Eagar and Christopher Musso, "Why Did the World Trade Center Collapse? Science, Engineering, and Speculation," Journal of Materials, JOM, 53 (12) (2001), pp. 8-11., *http://www.tms.org/pubs/journals/JOM/jom.html, http://www.tms.org/pubs/journals/JOM/0112/Eagar/eagar-0112.html, http://doc.tms.org/JOM/contents-0112.html*

[4] *http://www.nhc.noaa.gov/aboutsshs.shtml*

[5] Largest size refers to the diameter of the storm system, not the wind speed. Hurricane Erin's greatest

wind speed was on 9/10/01, but its greatest size was on 9/11/01.

[6] *http://drjudywood.com/articles/erin/erin2.html, http://www.nhc.noaa.gov/ms-word/TCR-AL122005_Katrina.doc, http://www.nhc.noaa.gov/2001erin.html*

[7] *http://drjudywood.com/articles/erin/erin1.html, http://youtube.com/watch?v=wEoyN44oYh4, http://youtube.com/watch?v=tYEJUzuxFoM*

[8] ZCZC MIATCDAT1 ALL, TTAA00 KNHC DDHHMM, HURRICANE ERIN DISCUSSION NUMBER 35, NATIONAL WEATHER SERVICE, MIAMI FL, 11 PM EDT MON SEP 10 2001, *http://www.nhc.noaa.gov/archive/2001/dis/al062001.discus.035.html*

[9] Richard J. Pasch and Daniel P. Brown, "Tropical Cyclone Report, Hurricane Erin, 1 - 15 September 2001," National Hurricane Center, 20 November 2001, Revised: 25 January 2002, *http://www.nhc.noaa.gov/2001erin.html*

[10] *http://drjudywood.com/articles/erin/erin2.html, http://www.nhc.noaa.gov/2001erin.html, http://www.nhc.noaa.gov/2005atlan.shtml, http://www.nhc.noaa.gov/pdf/TCR-AL122005_Katrina.pdf*

[11] Personal communication from Joey Moore (4/5/10), flight attendant (BOS-SFO flt. 163), the morning of 9/11/01.

[12] *Relative "Distance"* = $\sqrt{(40.77 - LAT)^2 + (73.78 - LONG)^2}$, where LAT and LONG are the latitude and longitude of the eye of Hurricane Erin, respectively. The values of 40.77 North and 73.78 West are the latitude and longitude of New York City. The value of *Relative Distance* was used to indicate whether Hurricane Erin was moving toward or away from New York City, not as an absolute value of distance from New York City.

[13] *http://www.nhc.noaa.gov/2001erin.html*

[14] "NASA Makes A Heated 3-D Look Into Hurricane Erin's Eye," *http://www.sciencedaily.com/releases/2005/10/051007090048.htm* Adapted from materials provided by, *http://www.nasa.gov/goddard,* ScienceDaily (Oct. 11, 2005)

[15] CNN.com/Weather, "Hurricane Erin weakens, but still strong storm," September 10, 2001 Posted: 6:59 PM EDT (2259 GMT), *http://archives.cnn.com/2001/WEATHER/09/10/erin/*

[16] Seismology Group, Lamont-Doherty Earth Observatory, Columbia University. *http://www.ldeo.columbia.edu/LCSN/Eq/20010911_wtc.html*

[17] *http://drjudywood.com/articles/erin/erin1.html, http://youtube.com/watch?v=wEoyN44oYh4, http://youtube.com/watch?v=tYEJUzuxFoM*

[18] Scott Stevens, personal communication

[19] News from ABC 7, Washington, D.C., 8:35:02 am - 8:36:02 am (September 11, 2001), *http://www.archive.org/details/abc200109110831-0912*, recorded by the Television Archive, a non-profit archive, (03:45 - 04:45 in video segment) *http://www.archive.org/details/sept_11_tv_archive*

[20] News from NBC 4, Washington, D.C., 8:32:29 am - 8:33:33 am (September 11, 2001), *http://www.archive.org/details/nbc200109110831-0912*, recorded by the Television Archive, a non-profit archive, (01:20 - 02:24 in video segment) *http://www.archive.org/details/sept_11_tv_archive*

[21] News from Fox 5, Washington D.C., 8:33:06 am - 8:34:08 am (September 11, 2001), *http://www.archive.org/details/fox5200109110831-0912*, recorded by the Television Archive, a non-profit archive, (01:53 - 02:55 in video segment) *http://www.archive.org/details/sept_11_tv_archive*

[22] News from CBS 9, Washington, D.C., 8:31:21 am - 8:32:35 am (September 11, 2001), *http://www.archive.org/details/cbs200109110831-0912*, recorded by the Television Archive, a non-profit archive, (00:12 - 01:26 in video segment) *http://www.archive.org/details/sept_11_tv_archive*

[23] Seismology Group, Lamont-Doherty Earth Observatory, Columbia University. *http://www.ldeo.columbia.edu/LCSN/Eq/20010911_wtc.html*

[24] East coast cold front

[25] *http://drjudywood.com/articles/erin/erin1.html, http://youtube.com/watch?v=wEoyN44oYh4, http://youtube.com/watch?v=tYEJUzuxFoM*

[26] Erin Advisories for coastal NC, Fri Sep 07 2001 - 17:27:45 EDT, *http://www.weathermatrix.net/archive/stormreports/200108-200110/0027.html*

[27] Hurricane Erin kicks up heavy surf along East Coast, 09/10/2001 - Updated 06:28 PM ET, *http://www.usatoday.com/weather/hurricane/2001/atlantic/erin/2001-09-10-usa-heavy-swells.htm*

[28] ZCZC MIATCDAT1 ALL, TTAA00 KNHC DDHHMM, HURRICANE ERIN DISCUSSION NUMBER 36, NATIONAL WEATHER SERVICE MIAMI FL, 5 AM EDT TUE SEP 11 2001, *http://www.nhc.noaa.gov/archive/2001/dis/al062001.discus.036.html*

[29] ZCZC MIATCDAT1 ALL, TTAA00 KNHC DDHHMM, HURRICANE ERIN DISCUSSION NUMBER 35, NATIONAL WEATHER SERVICE, MIAMI FL, 11 PM EDT MON SEP 10 2001, *http://www.nhc.noaa.gov/archive/2001/dis/al062001.discus.035.html*

[30] *http://lwf.ncdc.noaa.gov/oa/climate/research/2004/oct/oct-2004-ace.html*, *The Accumulated Cyclone Energy (ACE) Index is calculated by summing the squares of the estimated 6-hourly maximum sustained wind speed in knots (V_{max}^2) for all periods in which the tropical cyclone is a tropical storm or greater intensity. For a complete description of the ACE Index see Bell et al. (2000), *Bulletin of the American Meteorological Society* (81) S1-S50.

[31] *http://www.aoml.noaa.gov/hrd/tcfaq/E11.html*

[32] *http://www.nhc.noaa.gov/2001erin.html*

[33] *http://www.nhc.noaa.gov/pdf/TCR-AL122005_Katrina.pdf*

[34] Dolores LaChapel, "D.H. Lawrence Future Primitive," University of North Texas Press (1996) p. 40.

[35] Miranda Hitti, "Arthritis Pain Increases With Cool Temps, Barometric Changes. Weather, Arthritis Pain Link Confirmed," (*http://www.webmd.com/miranda-hitti*), WebMD Health News (Oct. 18, 2004), *http://arthritis.webmd.com/news/20041018/weather-arthritis-pain-link-confirmed*

[36] Dolores LaChapel, "D.H. Lawrence Future Primitive," University of North Texas Press (1996) p. 40, *http://books.google.com/books?id=fZ7ZA4kHe3kC&pg=PA40&lpg=PA40&dq=ion+feel+%22approaching+storm%22&source=bl&ots=9w8TKQajE1&sig=WQVPYeL-_5ZwuTJu1bK7S7KjSAg&hl=en&ei=crhjSvX9C8L7tgey-aSyAg&sa=X&oi=book_result&ct=result&resnum=1*

[37] Adapted from: Figure 5–3, p. 94 (pdf p. 138 of 404), *http://wtc.nist.gov/media/NIST_NCSTAR_1-9_Vol1_for_public_comment.pdf*

[38] John Thomas Bryant, Jr., (Courtesy of Justin Shaw) Relationship of Coronal Mass Ejections are Cyclonic Earth Storms, *http://www.youtube.com/user/astrotometry, http://astrotometry.com/html/formative_issues.html, http://www.astrotometry.com/html/formative_issues.html, http://www.coreweather.com/*

19.
Earth's Magnetic Field on 9/11

There are some reports, for example, that some countries have been trying to construct something like an Ebola Virus, and that would be a very dangerous phenomenon, to say the least... Others are engaging even in an eco- type of terrorism whereby they can alter the climate, set off earthquakes, volcanoes remotely through the use of electromagnetic waves. —William S. Cohen

I think it's fair to say there has never been an accident that you can clearly say was caused by electromagnetic interference.—Granger Morgan

It's not difficult really - the secret is in knowing how. —Edward Leedskalnin

An electromagnetic fishing pole —Robert Clark

A. Introduction

We know there is a relationship between electricity and magnetism. As a child, I remember making a magnet by wrapping a coil of wire around a nail and connecting the ends to a nine-volt battery. Then, with things going the other way, I also learned that, through electromagnetic induction, voltage can be generated from a magnet. So we all know not only that electricity and magnetism are related, but we know that they *interact.* After conceptualizing and studying the similarities between a hurricane and a Tesla Coil, I became curious as to what effect a large hurricane might have on the Earth's magnetic field.

The data I found was recorded by the Geophysical Institute Magnetometer Array (GIMA) of the University of Alaska, an institution that operates magnetometer sites at locations across Alaska and western Canada. Six stations were active and recording data during the time around September 11, 2001. An example of their recordings is shown below.

The magnetometer readings from the six different research stations in northern Alaska reveal anomalous changes in the Earth's magnetic field at the exact moments that key events were taking place in New York City on 9/11. Figure 424 shows deviations from average values during a four-day period.

Until about twenty minutes before the first two events—those being the creating of the holes in WTC1 and in WTC2—values hovered close to the average. And *then* they revealed changes (see Figure 425).

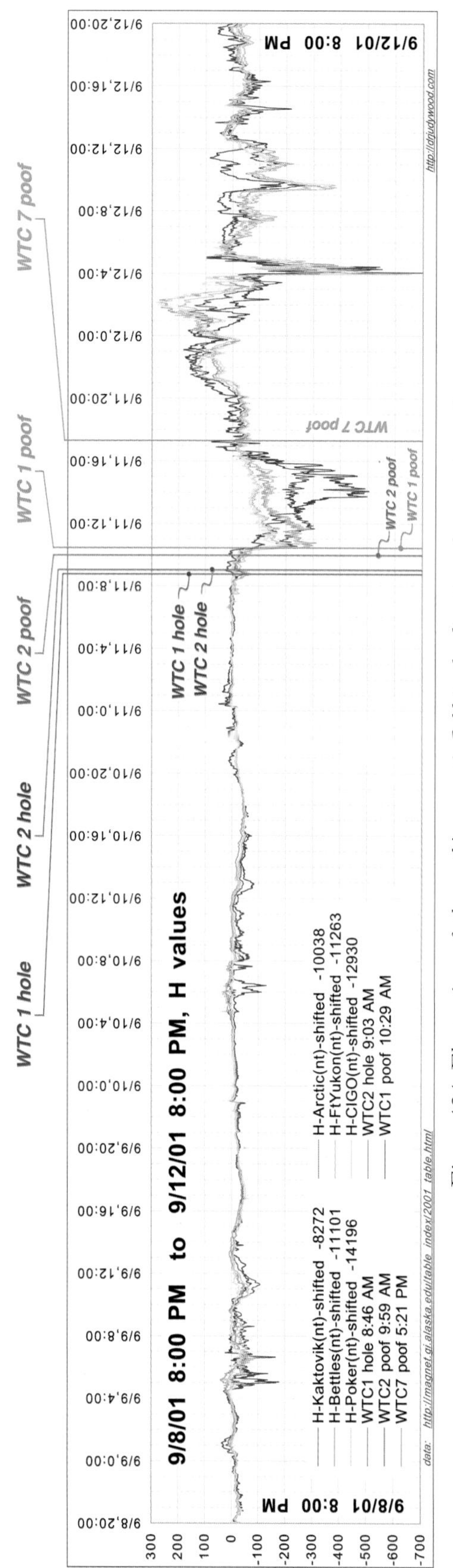

Figure 424. Fluctuations of the earth's magnetic field in the direction of magnetic north, (H: N/S), 8 PM 9/8/01 to 8 PM 9/12/01 (4 days).[1]

Immediately before the first event, at WTC1, the magnetometer readings began to fluctuate from the average. These fluctuations increased, and at each destructive event from then on, the values shifted dramatically either up or down. In all cases, the fluctuations of the magnetometer readings were dramatically different from normal, especially in their timing.

Though the data clearly seem related to the events of the day, the importance of these fluctuations has not yet been described fully, and they unquestionably deserve further research. By themselves, fluctuations in the magnetometer readings are not unusual. But the timing, magnitude, and relationships of these fluctuations *are* unusual. In fact, the timing of these anomalous fluctuations is downright uncanny.

A detailed view of the 14 hour period surrounding the events of 9/11 is shown in Figure 425.

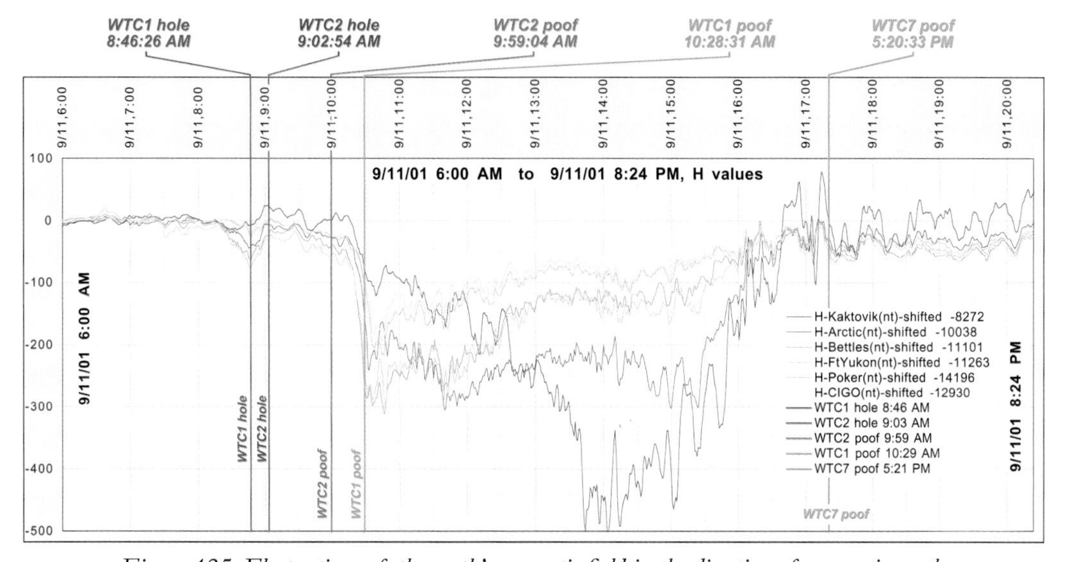

Figure 425. Fluctuations of the earth's magnetic field in the direction of magnetic north, (H: N/S), 6:00 AM to 8:25 PM 9/11/01 (14 hrs).[2]

Approximately twenty minutes before WTC1 got its hole, the magnetometer readings began to diverge from their average values. As shown in the Event Legend for Figure 425, the vertical blue line corresponds exactly with the time when the North Tower got its hole. The next event, at 9:03 a.m., shown in red, corresponds with the South Tower getting its hole. Then at 9:59, the South Tower goes poof at the moment indicated by the orange line. WTC1 poofed at the time shown by the light-blue line. Later in the day, at 5:21 p.m., WTC7 went poof at the time indicated by the green line.

The alert observer will notice that approximately twenty minutes before the North Tower got its hole, the values show a downward trend before turning abruptly upward. They continue upward until 9:03. Then they level off when the South Tower got its hole. (The latitude and Longitude of each recording station is given in Table 24.)

Event	EDT*	Magnitude (Richter)	Duration (seconds)
Downward trend begins	*8:20 AM*	-	-
WTC1 gets hole	8:46:26±1 AM[3] 8:46:30 AM[4] 8:46:40 AM[5]	**0.9[3]**	**12[3]**
WTC2 gets hole	9:02:54±2 AM[3] 9:02:59 AM[6] 9:03:11 AM[7]	**0.7[3]**	**6[3]**
Pentagon Event	*9:37:46 AM[8]*	-	-
WTC2 goes poof	9:59:04±1 AM[3]	**2.1[3]**	**10[3]**
WTC1 goes poof	10:28:31±1 AM[3]	**2.3[3]**	**8[3]**
WTC7 goes poof	5:20:33±1 PM[3] 5:20:52 p.m.[9]	**0.6[3]**	**18[3]**

Table 23. The time of each event.

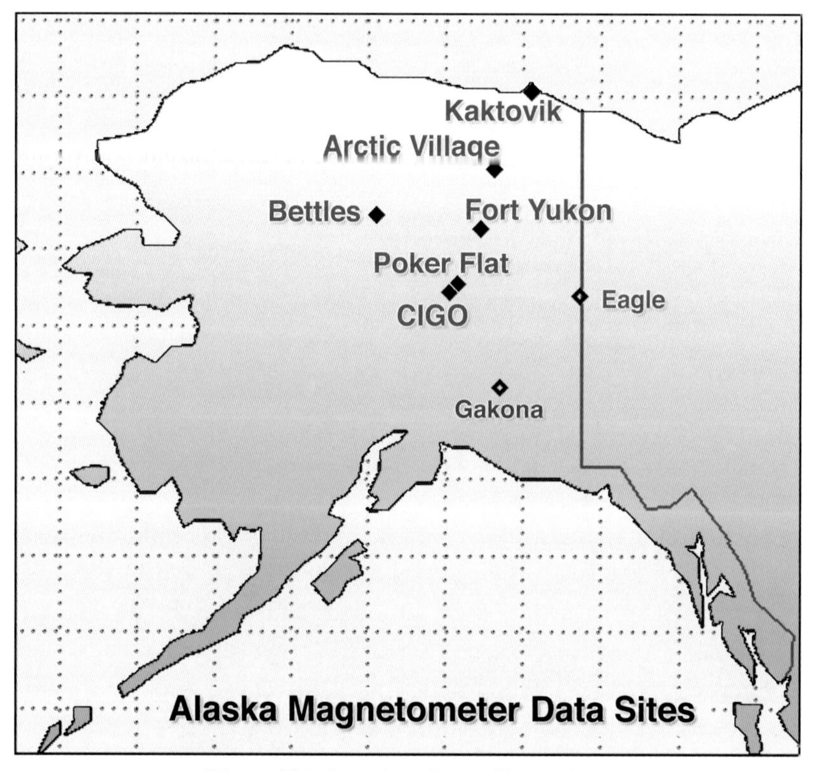

Figure 426. Location of recording stations.
Redrawn from: http://magnet.gi.alaska.edu/images/akmap.gif

All six magnetometer values declined and then reversed direction just at the moment when WTC1 got its hole. The magnetometer data available was recorded every 60 seconds. So, it is not possible to determine whether or not the values reverse at the precise moment of the WTC event. Also, the exact time at which WTC1 was hit[10] is not well defined. The times reported by three government sources are slightly different. The Lamont-Doherty Earth Observatory reported 8:46:26 AM[2], the National Institute of Standards and Technology (NIST) reported 8:46:30 AM[3], and The *9/11 Commission* reported 8:46:40 AM.[1]

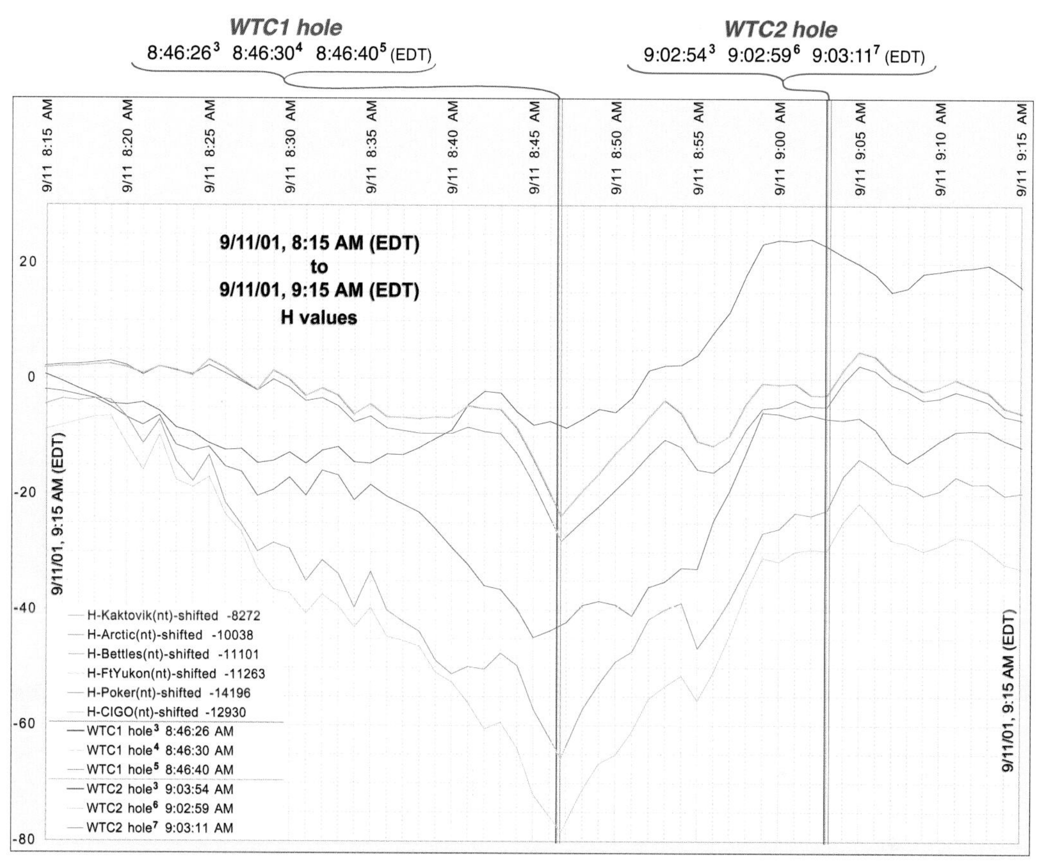

Figure 427. Fluctuations of the earth's magnetic field in the direction of magnetic north, (H: N/S), 9/11/01, 8:15 to 9:15 AM (1 hr).[11]

The magnetometer data from the six Alaska locations (Figure 427) are given in Figure 428 as three components: horizontal intensity (H), geomagnetic declination (D), and vertical component (Z). These plots cover a 14-hour period.

For all magnetometer charts shown here, the vertical grid is in nano-Tesla and the horizontal scale represents eastern daylight-savings time (EDT). The mean value (in nano-Tesla) that was subtracted from each data set is shown in the legend of each plot. This value is the average value for each component, each station. The magnetic deviation (D-component) displayed for each set was computed using the average value by GIMA. The H-component (deviations down) indicates that the local field has dipped southward. The D-component dips are magnetic west deviations as the magnetometer head is oriented in magnetic coordinates, not geographic. The Z component is the vertical component of the magnetic field.

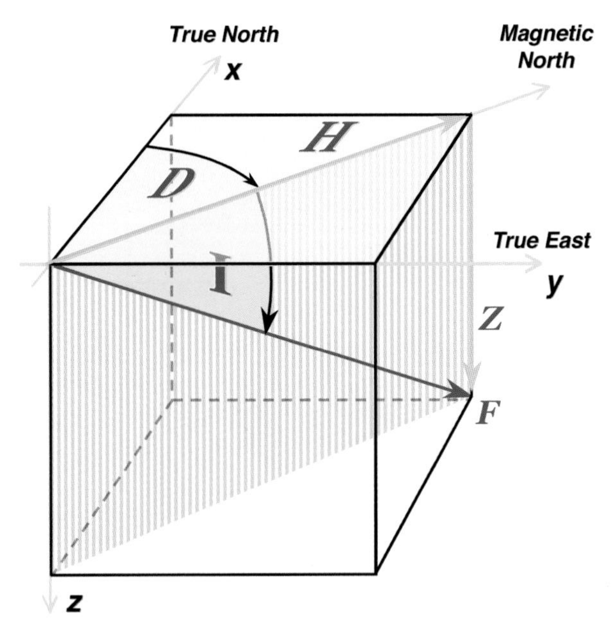

Figure 428. The Magnetic Elements.

The Magnetic Elements[12]

F - Total Intensity of the geomagnetic field
H - Horizontal Intensity of the geomagnetic field
X - North Component of the geomagnetic field
Y - East Component of the geomagnetic field
Z - Vertical Component of the geomagnetic field
I (DIP) - Geomagnetic Inclination
D - Geomagnetic Declination (Magnetic Variation, or the angle between magnetic north and geographic north.)

Between WTC1's going "poof" and WTC2's going "poof," the values, especially seen in the vertical direction (Z), in Figure 429c, began to wander way far from home. The precise moment that WTC1 went poof seemed to coincide with the initiation of a significant geomagnetic event, which gradually subsided at just about the same time WTC7 went poof. After this, all six values seemed to resonate together at the same frequency (a type of magnetic resonance), but with slightly different amplitudes. This behavior of the electromagnetic values is most apparent in Figure 429a.

In the time between WTC1 going "poof" and WTC7 going "poof," some *strange* magnetic events were taking place. In Figure 429b it can be seen that the declination (D) varied over a wide range. The geomagnetic declination is the angle between magnetic north and geographic north. In other words, if you were standing at one of those geomagnetic stations holding a compass, *you would have seen the needle swinging wildly.*

Most of us can remember holding a compass for the first time and experimenting with it. Holding the compass and moving it around will cause the dial to waver and fail to hold steady.

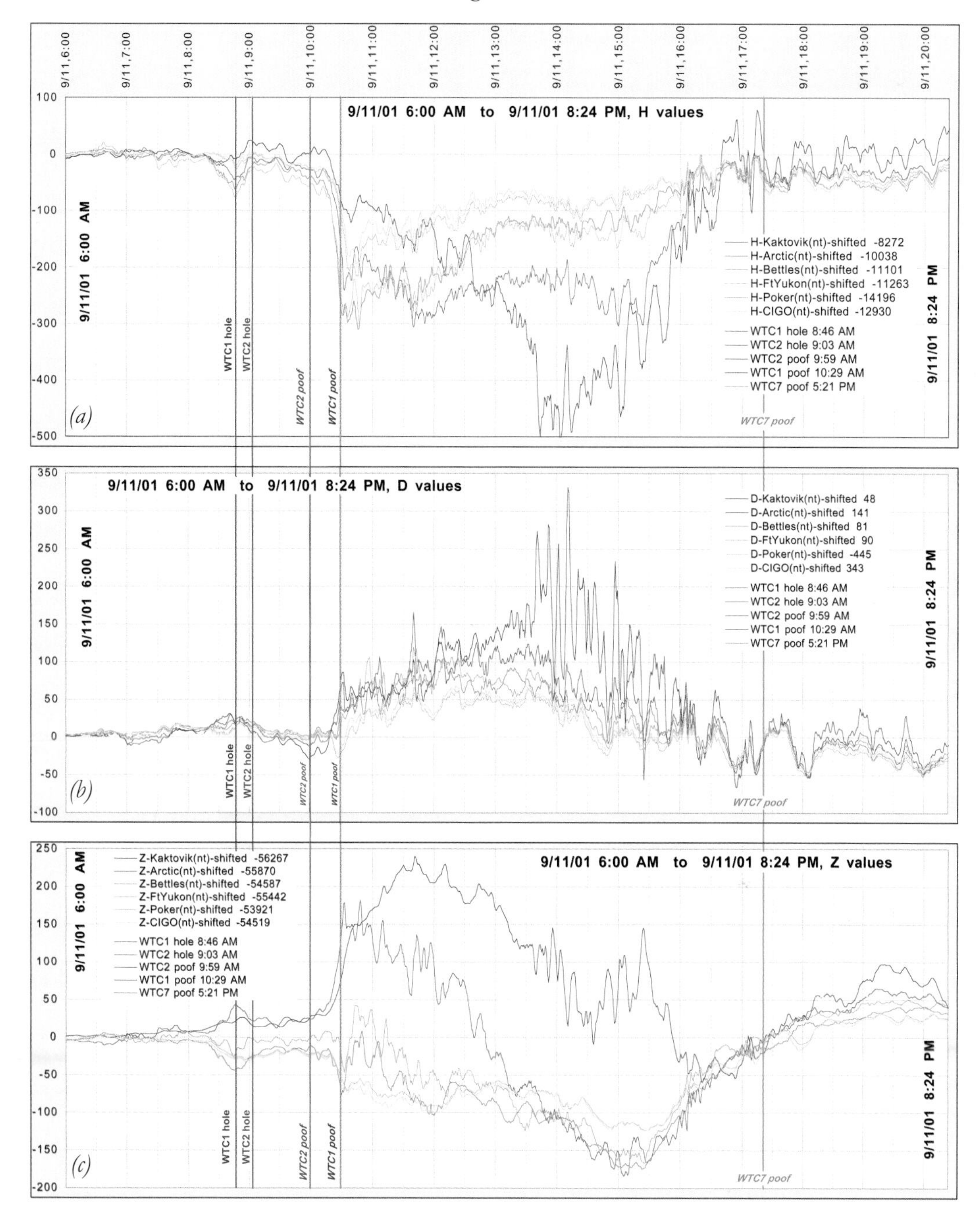

Figure 429. Fluctuations of the earth's magnetic field, 6:00 AM to 8:25, PM 9/11/01 (14 hrs).13 (a) Fluctuations of the earth's magnetic field in the direction of magnetic north, (H: N/S), 6:00 AM to 8: 25 PM 9/11/01 (14 hrs).14 (b) Fluctuations of the earth's magnetic field deviation from magnetic north, (D: E/W), 6:00 AM to 8:25 PM 9/11/01 (14 hrs).15 (c) Fluctuations of the earth's magnetic field in the vertical direction, (Z: up/down), 6:00 AM to 8:25 PM 9/11/01 (14 hrs).[16]

But if we set the compass down on a rock, we can expect to see the needle quickly slow down and hold one direction, nice and steady. The data in Figure 429b indicate that if we had had a compass placed on a rock in Kaktovik, AK, on 9/11/01, its needle would not have remained pointing in the same direction.

The geomagnetic declination is the angle between magnetic north and true north. The green data traces from Kaktovik show the most dramatic fluctuations of the six recording stations, while the orange traces, from Arctic Village, AK, show the second most dramatic fluctuations. The sensitivity of these stations appears to be related to their proximity to the magnetic north pole. Kaktovik is the closest to the magnetic north pole and Arctic Village the next closest. CIGO and Poker Flat are the farthest from the magnetic north pole, and they show the least dramatic excursions in the data.

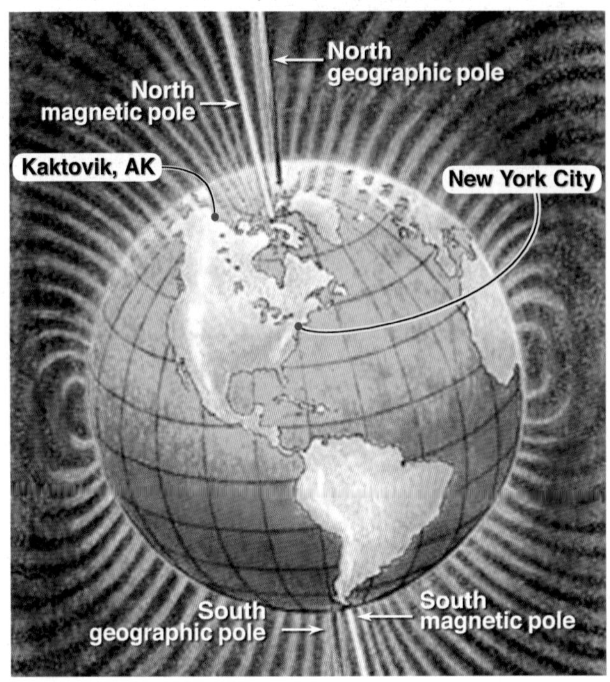

Figure 430. Earth's magnetic field
http://www.mnh.si.edu/earth/text/images/4_0_0_0/4_1_5_0_magnetic_02.jpg

Location Name	**Latitude (N)**	**Longitude (W)**
Kaktovik	70.13478	143.64706
Arctic Village	68.1333	145.5667
Bettles	66.90139	151.55
Fort Yukon	66.560	145.2183
Poker Flat	65.11908	147.43119
CIGO	64.87345	147.86045
Magnetic North (2001)[17]	81.3	110.8
Magnetic North (2005)[18]	82.7	114.4
New York City, NY[19]	40.77	73.98 W

Table 24. Magnetometer stations,[20] magnetic north, and New York.

Consider a bar magnet. The field lines of a bar magnet are closer together at the poles, so that's where there could be a greater sensitivity to disturbances of the magnetic field. Kaktovic is 11.2°N, 32.8°W of magnetic north. New York City is

40.6°N, 36.8°E of magnetic north. The red dots in Figure 431 suggest where each of those locations would be, if you were to think of the Earth as a bar magnet, would be placed—with Kaktovic closer to a pole of the bar magnet than New York.

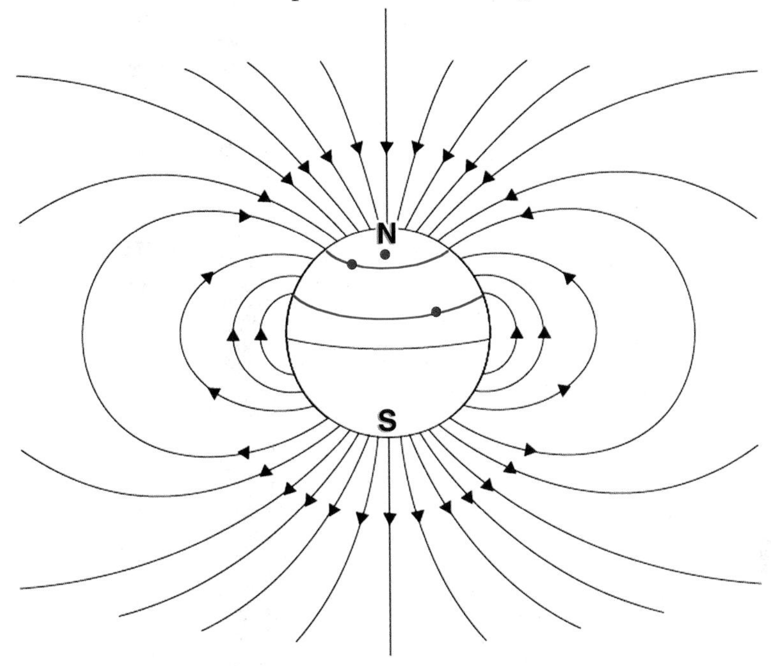

Figure 431. Earth is like a bar magnet.
altered from: http://upload.wikimedia.org/wikipedia/commons/9/94/Dipole_field.svg

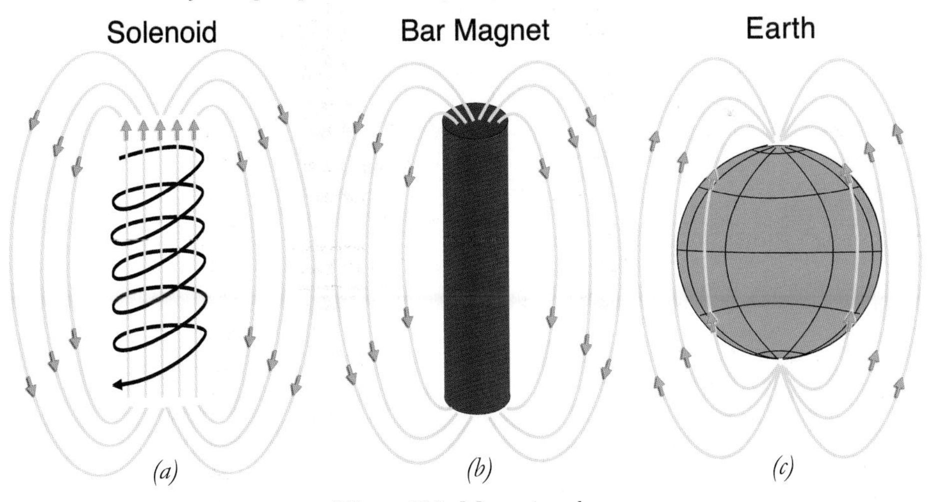

Figure 432. Magnetic poles.
Redrawn from: http://cache.eb.com/eb/image?id=63378&rendTypeId=4

In 1820, Hans Christian Oersted, a Danish scientist, discovered a connection between electricity and magnetism when he accidentally found that the magnetized needle of a compass would realign if brought near a current-carrying wire.[21] That is, Oersted discovered that electrical fields and magnetic fields interact.

The diagrams in Figure 432 show how a compass needle would point if it were placed at various positions in a plane perpendicular to a conductor that is carrying electrons.

Earth is surrounded by a magnetic field that was disturbed on 9/11 at the very moment that the destructive events in New York City took place. Field effects do not necessarily arise from a point source but instead they are directional. We do know Hurricane Erin created field effects near New York City, since all three local airports reported thunder.

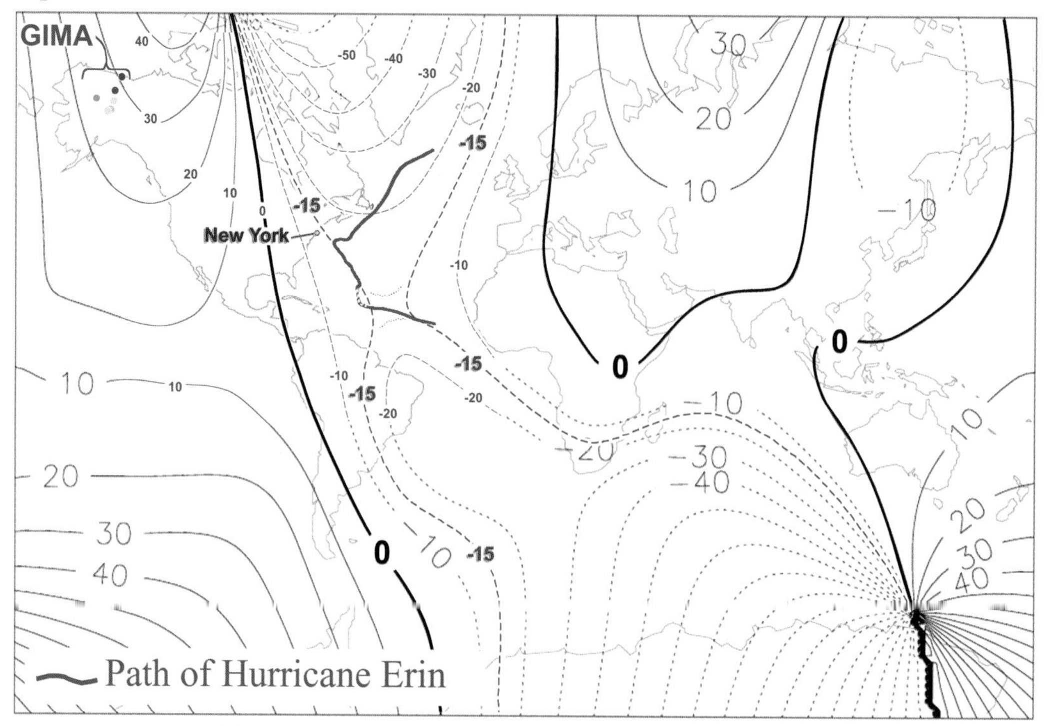

Figure 433. International Geomagnetic Reference Field (IGRF), Geophysical Institute Magnetometer Array (GIMA).
http://upload.wikimedia.org/wikipedia/commons/6/68/IGRF_2000_magnetic_declination.gif

The previous chapter discussed field effects. This chapter speaks about magnetometer data, and we clearly know that *field effects* and *magnetism* can interact. This leads to questions about the weather conditions in the New York City area at the time of the disasters. The questions have to do with air pressure (at sea level), dew point, and relative humidity. These are shown in Figure 435a on the same time scale as the magnetometer data in Figure 435b.

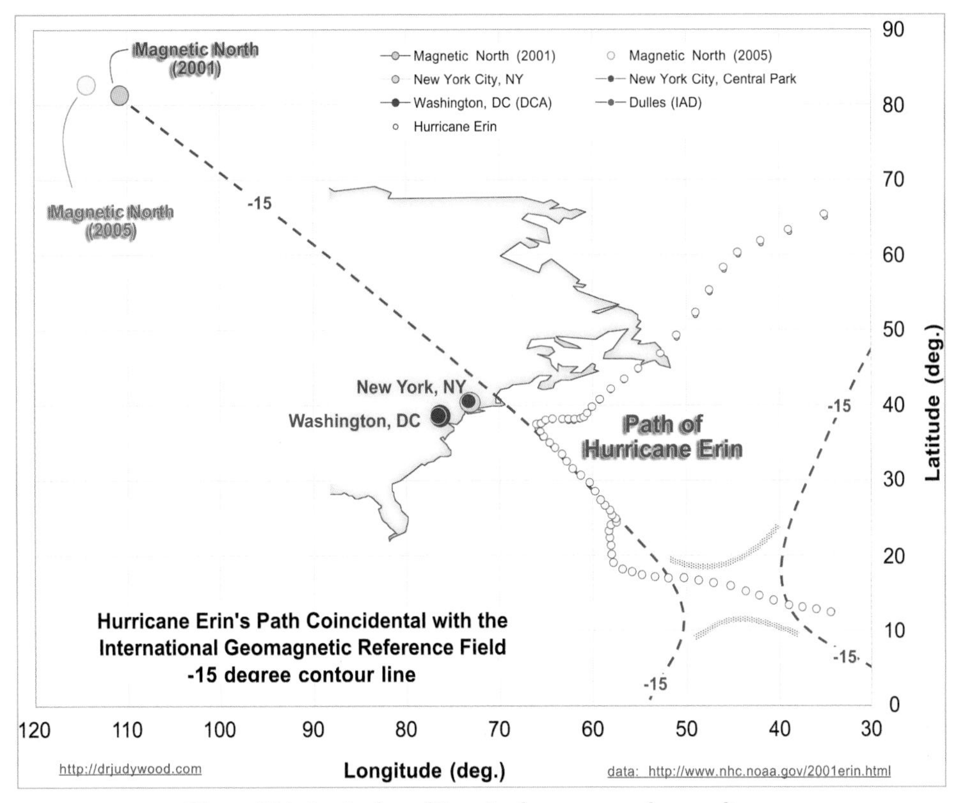

Figure 434. Latitude and Longitude as rectangular coordinates.
data source: http://www.nhc.noaa.gov/2001erin.html

B. Atmospheric Changes

As noted in the previous chapter, a cold front coming from the west was expected to turn Hurricane Erin north, although the timing of that turn was unknown. As it happened, the cold front arrived in New York City on the morning of September 11. Figure 435 shows the sea level air pressure, dew point, and relative humidity as recorded at JFK airport over the thirty-six hours beginning at 8:00 AM on 9/10/01.

During a typical day, we can expect the temperature to rise during the daylight hours and decrease during the nighttime, with fluctuations. If the moisture content in the air remains constant when the temperature rises, the relative humidity decreases, because warmer air can hold more moisture.

Curiously, however, just before 7:00 AM on 9/11/01, the relative humidity (pink) abruptly began a linear *decline* that ended at 11:30 AM. Rarely is there a change in the weather that is linear, with an abrupt beginning and end, but this time there was. The relative humidity (pink) moved from 80% to 40%, linearly, with a downward slope of about 9% (relative humidity) per hour. The dew point (dark blue) remained fairly constant until 8 p.m. on 9/10/01, when it began a gradual downward trend for 24 hours, falling from 73°F to 52°F.

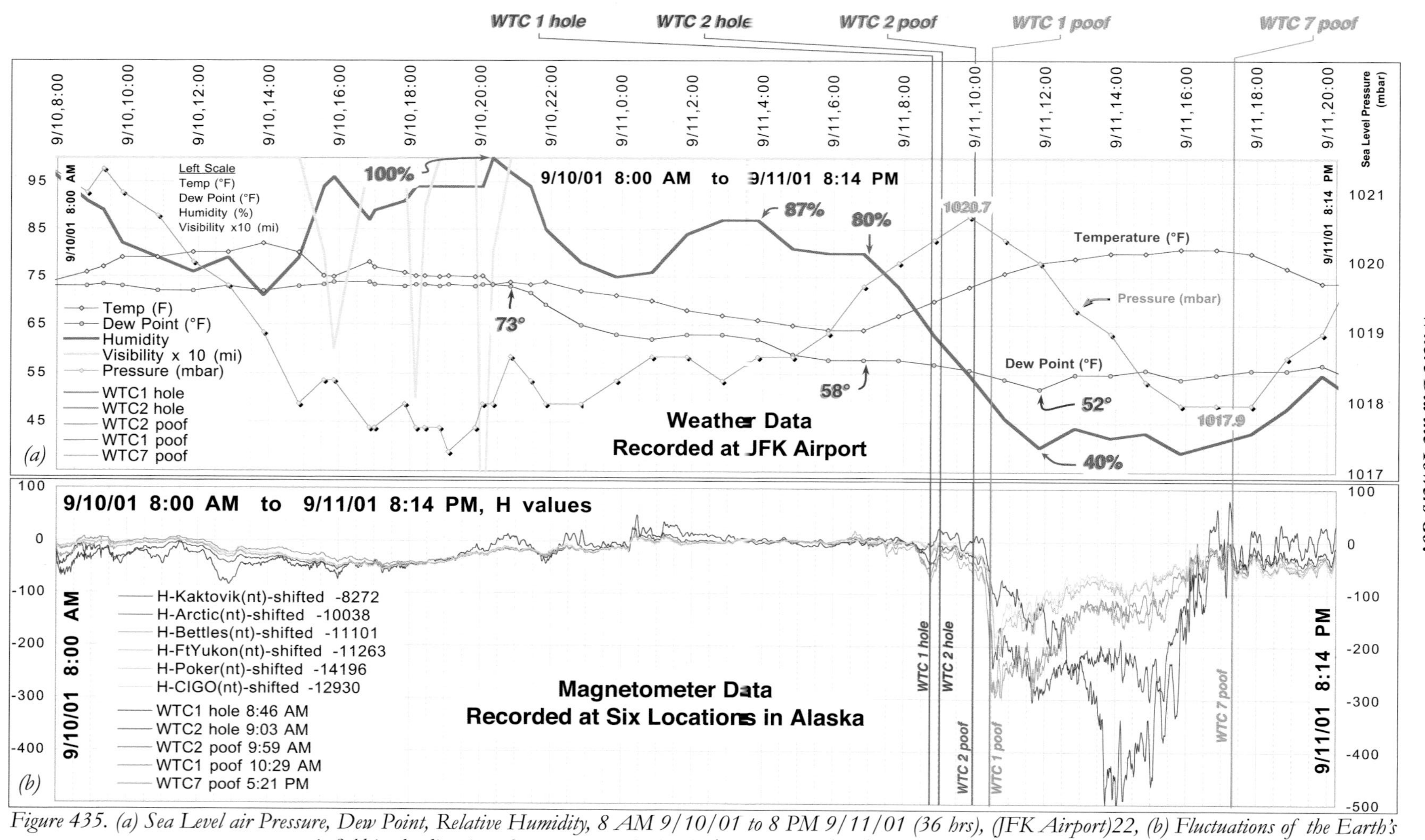

Figure 435. *(a) Sea Level air Pressure, Dew Point, Relative Humidity, 8 AM 9/10/01 to 8 PM 9/11/01 (36 hrs), (JFK Airport)22, (b) Fluctuations of the Earth's magnetic field in the direction of magnetic north, (H: N/S)_ 8 AM 9/10/01 to 8 PM 9/11/01 (36 hrs).*[23]

On the morning of 9/11/01, during the time when the relative humidity decreased linearly, the dew point decreased from 58°F to 52°F, indicating that a mass of cooler and drier air had moved into the area. Cold fronts are usually associated with high-pressure systems. Figure 435 shows the sea-level air pressure (light blue) increasing for about 12 hours with maximum pressure at 10:00 AM on 9/11/01, which was just when WTC2 went poof. So, as indicated by the atmospheric data, the high-pressure cold front from the west arrived at JFK airport *from the west* just as Hurricane Erin's arrived *from the east.*

As discussed on page 402, Tom Sater, the FOX5 weather reporter, told the viewing audience about the high-pressure cold front moving into the area. He emphasized the strength of this weather system by saying "This is a real cold front, folks," as he outlined this front on his weather map (see Figure 414). He also told the audience that the humidity was down to 73% from 88% earlier that morning. That is, by 8:33 to 8:34 AM, when the FOX5[24] weather report aired in the Washington, DC area, the humidity had dropped about the same as it had at JFK airport. (Assuming the same rate of decline, the 88% to 73% decline would have taken about an hour and 40 minutes.)

So by all indications, the high-pressure cold front had indeed moved into the area just as Hurricane Erin had also arrived there. From a careful study of Figure 435, we can see that the atmospheric pressure at JFK airport had been increasing steadily—over a period of fifteen or sixteen hours—to reach its peak at almost exactly 10:00 AM on 9/11. And exactly then, astonishingly, it began an immediate and sharp downward turn. If a high pressure system alone had passed through the area, that atmospheric pressure would have remained high for a substantial length of time. But in this case, immediately after 10:00 AM, the pressure began declining even faster than it had previously been increasing, in fact falling in *only six hours* the same amount that it had risen over the previous sixteen. Such a sharp decline reveals the presence of a powerful low-pressure system nearby. In meteorological terms, it suggests a kind of duel-to-the-death between the immense, continent-wide high pressure area coming from the west on the one hand, and, on the other, the remarkably nearby and extraordinarily low-pressure area of Hurricane Erin just to the east.

C. Space Weather

Another cause of fluctuations in the Earth's magnetic field can be solar storms. However, there were no solar storms or other significant space-weather events in the days leading up to 9/11/01.[25] The earth's magnetic field, as measured by the GEOS-10 satellite, is plotted in nano-Tesla units (nT) in Figure 436. The horizontal scale is in Universal Time (UT), but Eastern Standard Time (EDT) has been added along with lines marking the events of 9/11.

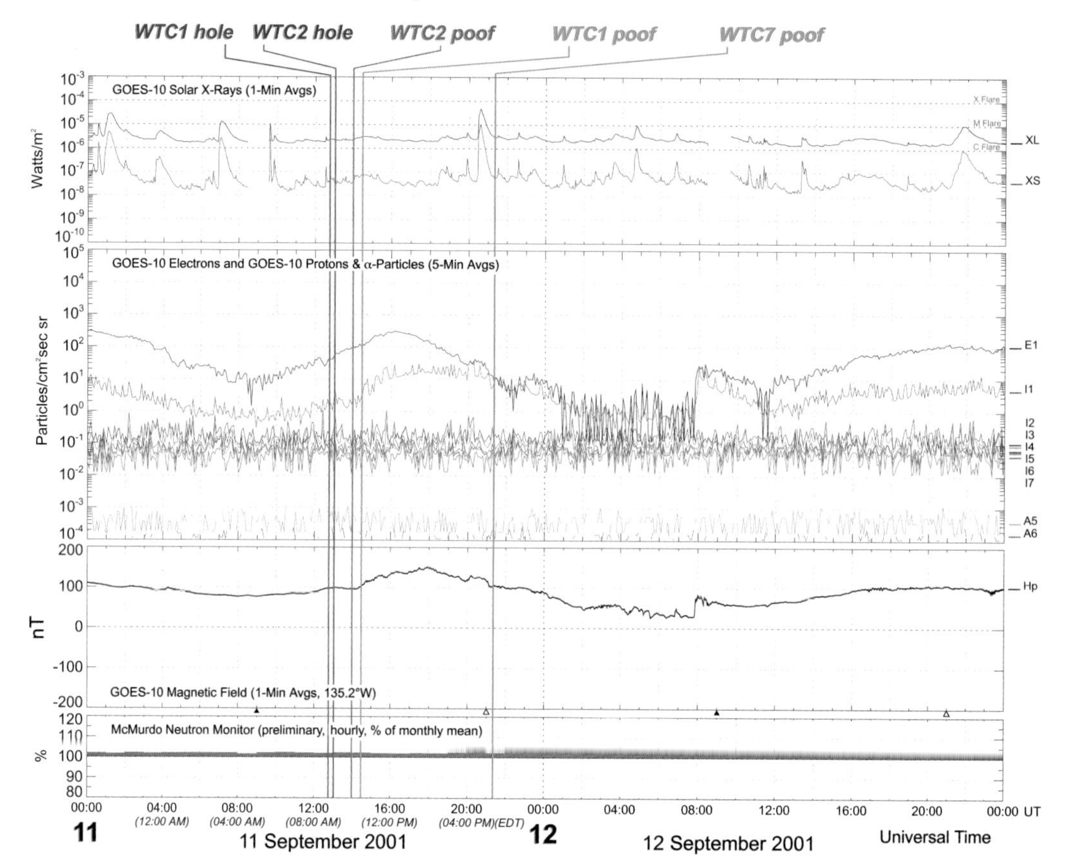

Figure 436. Space weather, 9/11/01-9/12/01. The magnetic field, as measured from the GOES-10 satellite, is plotted in lower third of the figure in nano-Tesla units (nT). The vertical lines indicate the time of each event on 9/11, as identified at the top.
http://goes.ngdc.noaa.gov/data/plots/2001/GOES-200109.pdf

D. Field Effects?

The fumes we have seen previously may not be indications of the presence of a magnetic field. But they *are* an indication of the presence of *some sort* of field effect. Evidence for such a presence comes from the general wind direction on 9/11/01. The wind that day was from the north-northwest and averaging 9 mph. Yet some images show fumes vigorously flowing directly east or even east-northeast from the World Trade Center Towers.

It seems rather extraordinary that the fumes—as in what I've called "lathering"—emerged from just one side of the buildings, and also fairly uniformly from the entire face on that side. This phenomenon can be seen in Figure 437, a photograph taken on the afternoon of 9/11. A larger image is in Figure 440 on page 429. The image shows the western face of WTC7. Midway down the building, there is a blackened region over two floors. Above this region, the fumes emerge from the building and travel at about a 45-degree angle upward. Below this region, the fumes appear to emerge at about a 45-degree angle downward. Around the darkened region, the fumes seem to be emerging radially.

Figure 437. (9/11/01) WTC7 lathering up. Note the upward and downward movement of the fumes.
http://www.jnani.org/mrking/writings/911/king911_files/image001.jpg

Figure 438. (9/11/01) WTC7 lathering up.[26]
http://www.magnumphotos.com/CoreXDoc/MAG/Media/TR3/F/W/L/Y/NYC14148.jpg

In Figure 438, these fumes from WTC7 emerge in a reverse pattern. Below the darkened region, the fumes seem to angle upwards. Above the blackened zone, the fumes appear to emerge in a downward direction. At the level of the darkened area, the fumes emerge straight out of the building.

The fumes poured out of WTC7 for approximately 7 hours (10:29 AM to 5:20 PM). For most of that time, they poured out of only the south face of the building, but from that face they poured both densely and uniformly, from the entire face, top to bottom. About 90 minutes before the building stopped standing, the fumes poured out of the entire *east* face as well, once again uniformly. The fuming eastern face is shown in Figure 439.

Figure 439 is a frame from a video shot from near the corner of West Broadway and Barclay Street between 3:53 p.m. and 4:02 p.m., showing the eastern side of the north face of WTC7. It appears by the discolored regions that fumes emerged also from the north face at some point, but it doesn't look as if they did so for very long.

Figure 439. Fumes from the east face of WTC7, 3:53 p.m. - 4:02 p.m.[27]
Page 227 (pdf page 271 of 404), http://wtc.nist.gov/media/NIST_NCSTAR_1-9_Vol1_for_public_comment.pdf, Figure 5–141.

It is remarkable that in Figure 439 no fumes are pouring out of the north face on floors 12 and 13, even though a glow can be seen inside the floors where the windows are missing. It is as though the fumes are waiting in a traffic jam to get out via the east face instead of flowing freely out of the north face. If the building is suffering a raging fire, and since we know that fire needs oxygen, this fire should have been fed by the open windows. We would expect to see fire lapping out of the broken windows on the north face, the path of least resistance. The breeze that day was light, coming from the northwest. *Thus, the force and direction of the fumes emerging from the building do not appear related to the 9 mph breeze from the north-northwest.*

Figure 440. (9/11/01) WTC7 Lathering up. Note the Upward and Downward Movement of the Fumes identified in Figure 437.
http://www.jnani.org/mrking/writings/911/king911_files/image001.jpg

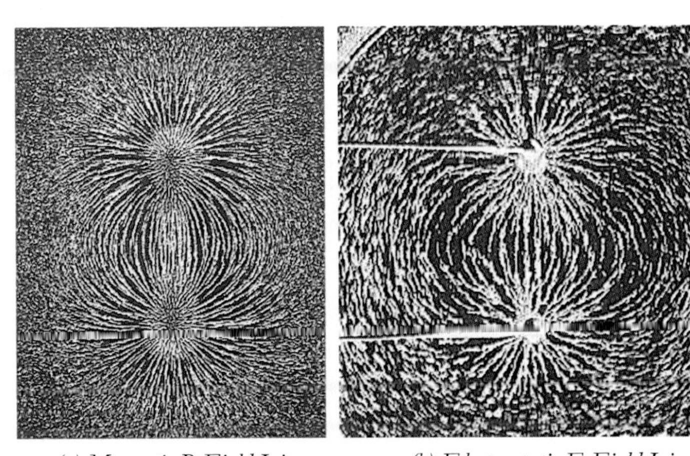

(a) Magnetic B Field Lines *(b) Electrostatic E Field Lines*

Figure 441. Field lines from magnetic and electrostatic fields.
(a)http://home.netcom.com/~sbyers11/MagPoles2.gif, (b)http://home.netcom.com/~sbyers11/posneg4.jpg, http://home.netcom.com/~sbyers11/images.htm#E_FieldLinesImage

What may one conclude from fumes that seem to defy wind currents by emerging from the face of a building where they more or less fly *into* the current? What may one conclude from fumes moving upward and downward? These phenomena may be indicators of the presence of strong magnetic fields, electrostatic fields, or yet some other type of field. It may be that the fumes are responding to the different polarities—north and south—of those powerful fields, of whatever kind, that were in play on 9/11.

---

[1] Data source: *http://magnet.gi.alaska.edu/table_index/2001_table.html*

[2] Data source: *http://magnet.gi.alaska.edu/table_index/2001_table.html*

[3] 8:46:26±1, 09:02:54 AM±2, From seismic data, Lamont-Doherty Earth Observatory: *http://www.ldeo.columbia.edu/LCSN/Eq/20010911_WTC/fact_sheet.htm*

[4] 08:46:30 by NIST, Page 37 (pdf page 87 of 298), *http://wtc.nist.gov/NISTNCSTAR1CollapseofTowers.pdf*

[5] 8:46:40 by The National Commission on Terrorist Attacks Upon the United States (The 9/11 Commission), *http://www.9-11commission.gov/report/911Report_Ch1.htm*

[6] 09:02:59 by NIST, Page 38 (pdf page 88 of 298), *http://wtc.nist.gov/NISTNCSTAR1CollapseofTowers.pdf*

[7] 9:03:11 AM *by The National Commission on Terrorist Attacks Upon the United States* (The 9/11 Commission), *http://www.9-11commission.gov/report/911Report_Ch1.htm*

[8] 9:37:46 AM *by The National Commission on Terrorist Attacks Upon the United States* (The 9/11 Commission), *http://www.9-11commission.gov/report/911Report_Ch1.htm*

[9] Page 241 (pdf page 285 of 404), *http://wtc.nist.gov/media/NIST_NCSTAR_1-9_Vol1_for_public_comment.pdf, Figure 5–157.*

[10] "Hit" refers to being affected by whatever caused the damage.

[11] Data source: *http://magnet.gi.alaska.edu/table_index/2001_table.html*

[12] *http://www.ngdc.noaa.gov/geomag/WMM/soft.shtmls*

[13] Data source: *http://magnet.gi.alaska.edu/table_index/2001_table.html*

[14] Data source: *http://magnet.gi.alaska.edu/table_index/2001_table.html*

[15] Data source: *http://magnet.gi.alaska.edu/table_index/2001_table.html*

[16] Data source: *http://magnet.gi.alaska.edu/table_index/2001_table.html*

[17] *http://en.wikipedia.org/wiki/North_Magnetic_Pole*

[18] *http://en.wikipedia.org/wiki/North_Magnetic_Pole*

[19] *http://www.realestate3d.com/gps/latlong.htm*

[20] *http://magnet.gi.alaska.edu/*

[21] Britannica Student Encyclopedia, *http://student.britannica.com/comptons/art-53251/The-electromagnetism-of-a-current-carrying-solenoid-the-ferromagnetism-of*

[22] Data source: *http://english.wunderground.com/history/airport/KJFK/2001/9/11/DailyHistory.html?MR=1*

[23] Data source: *http://magnet.gi.alaska.edu/table_index/2001_table.html*

[24] News from Fox 5, Washington D.C., 8:33:06 am - 8:34:08 am (September 11, 2001), *http://www.archive.org/details/fox5200109110831-0912*, recorded by the Television Archive, a non-profit archive, (01:53 - 02:55 in video segment) *http://www.archive.org/details/sept_11_tv_archive*

[25] *http://goes.ngdc.noaa.gov/data/plots/2001/GOES-200109.pdf*

[26] *http://www.magnumphotos.com/Archive/C.aspx?VP=Mod_ViewBox.ViewBoxZoom_VPage&VBID=2K1HZOQIW7CLO&IT=ImageZoom01&PN=985&STM=T&DTTM=Image&SP=Search&IID=2K7O3RBZ4JP6&SAKL=T&SGBT=T&DT=Image*

[27] The intensities have been adjusted, column and floor numbers have been added, as stated in the NIST report.

20.
TESLA-HURRICANE-MAGNETOMETER CORRELATION

I boarded the Kings' ship; now in the beak,
Now in the waist, the deck, in every cabin,
I flamed amazement; sometime I'd divide,
And burn in many places; on the topmast
The yards and bowsprit, would I flame distinctly,
Then meet and join.
–William Shakespeare –*The Tempest* (Act I, Scene 2)

A. Introduction

In Chapter 17, we discussed the Tesla-Hutchison Effect, in Chapter 18, we discussed Hurricane Erin, and in Chapter 19, we discussed the magnetometer data. In this chapter, we will discuss all of these, together, and their relationship to other known phenomena.

Figure 442. Hurricane Erin is in the distance, overlooking New York City on 9/11.
http://hereisnewyork.org//jpegs/photos/5748.jpg

The purpose of this chapter is to observe known and documented phenomena that appear *similar* to what we have observed, not necessarily identical. This is to provide insight into the reality of these phenomena that many may not realize actually exist.

Figure 443 shows the track of Hurricane Erin, where the bright-green zones along Erin's track correspond with intervals of active fluctuations in the magnetometer readings. Curiously, these correspond with the changes in course of Hurricane Erin. The blue concentric circles in Figure 443 show the approximate area of *cloud cover* by Hurricane Erin on the morning of September 11, 2001. The lighter sets of concentric circles show the approximate locations of Erin the day before and the day after September 11. Hurricane Erin spread out on September 11, and then tightened back up on September 12, as she headed back out. Satellite images for these three days are shown in Figure 445.

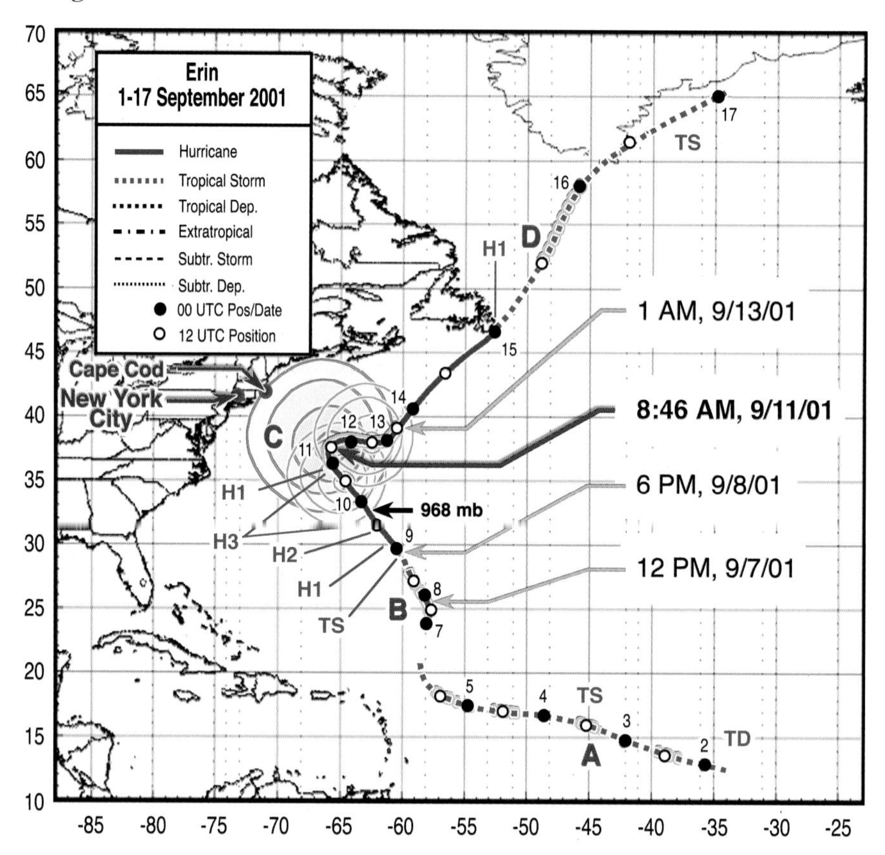

Figure 443. Best track of Hurricane Erin, September 1-17, 2001.
Source of map: http://www.nhc.noaa.gov/prelims/2001erin1.gif

Hurricanes rarely travel on courses perpendicular to the United States east coast. But Erin did. Not only that, but she sped up in this direction, as if "getting in the groove." The bright green zones on the track indicate approximately when the magnetometer fluctuations were more active than usual. The early green zones (September 2-5) have much less dramatic fluctuations than the later ones. Hurricane Erin re-formed, then sped up along a fairly straight line from time B to time C, in Figure 443. On September 9, Erin went in 24 hours from a tropical storm to a category 3 hurricane. On the night of September 10, she slowed in forward movement. At 8:46 AM on September 11, when Erin began to make the hard right turn, the magnetometer anomalies appeared, as if the storm were a runaway train spitting sparks on the rails as she tried to take a turn too fast.

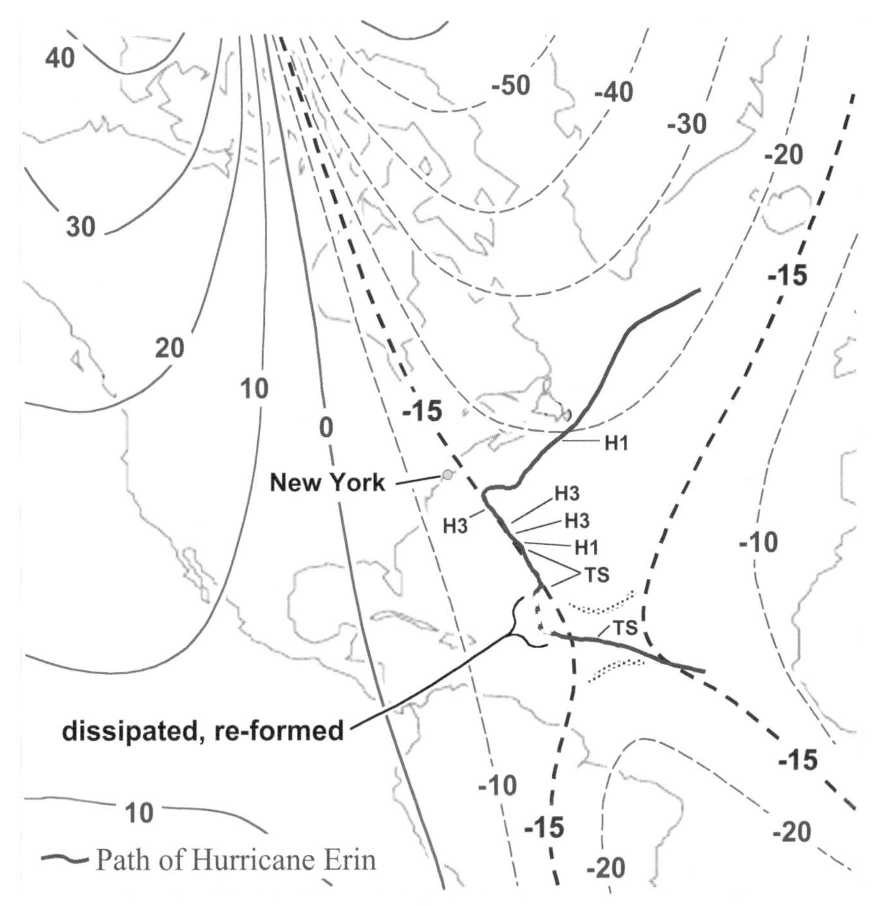

Figure 444. Magnetic declination and the path of Hurricane Erin.
Adapted from: http://upload.wikimedia.org/wikipedia/commons/6/68/IGRF_2000_magnetic_declination.gif

It appears that the turns in the hurricane's course are closely related to the fluctuations in the magnetometer readings. We need more data before saying one causes the other, or that both are the result of something we haven't considered, or if it is all just a coincidence. ☺ We do know, as Figure 444 shows, that "something" dramatic happened to get Erin off from what looked like a track locked on to the minus-15 declination line, and with such a dramatic turn.

Curiously, Erin became larger on 9/11, as if doing battle with the cold front approaching from the midwest and getting torn up by it. Then, the hurricane tightened back up as it turned around and began heading east, as shown in Figures 445a to 445c. The yellow circles are of equal size on each of the three photos, illustrating that Erin, filling the entire circle on 9/11, covered the greatest area on 9/11 itself.

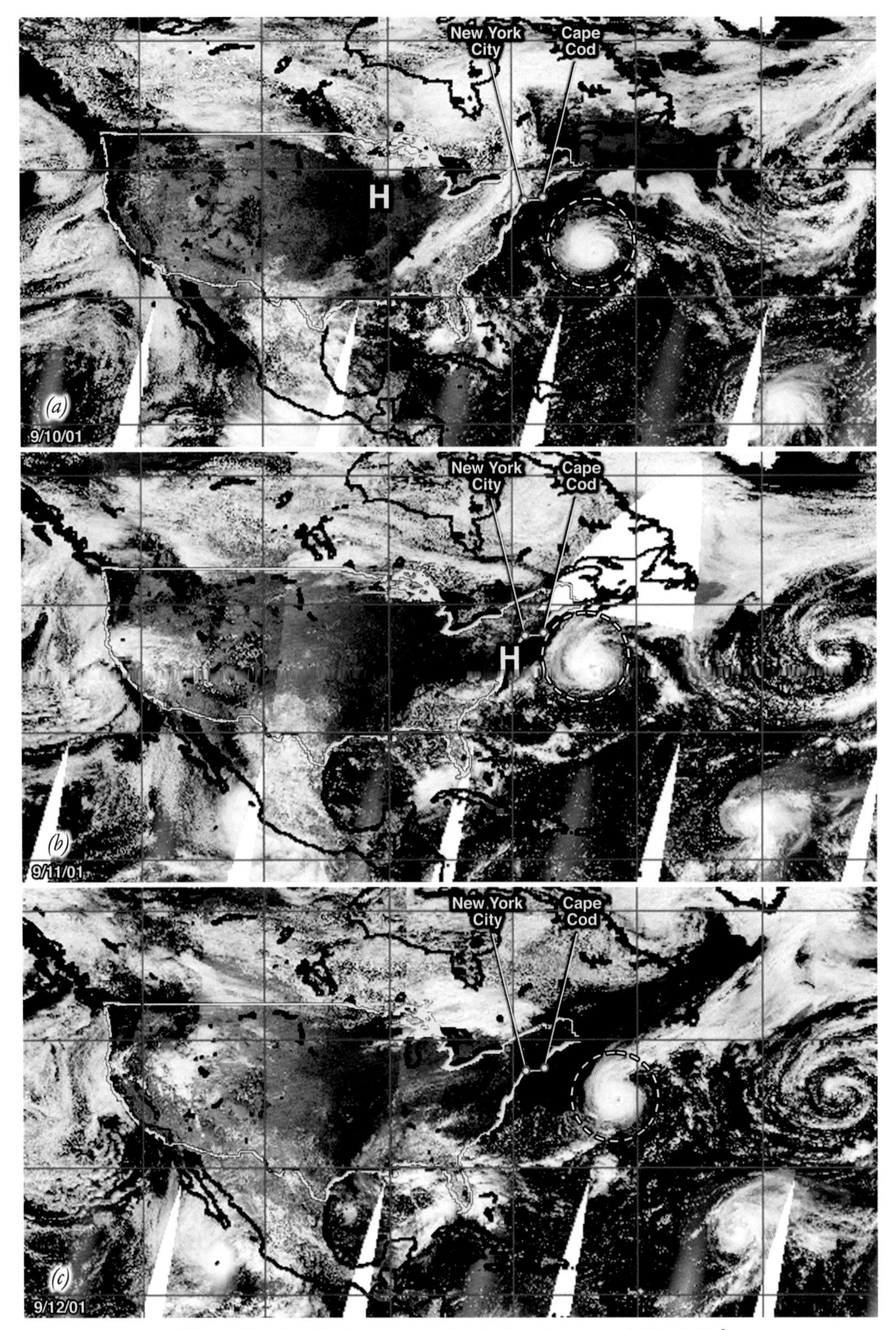

Figure 445. Satellite images from September 10, 11, and 12, 2001.[2]

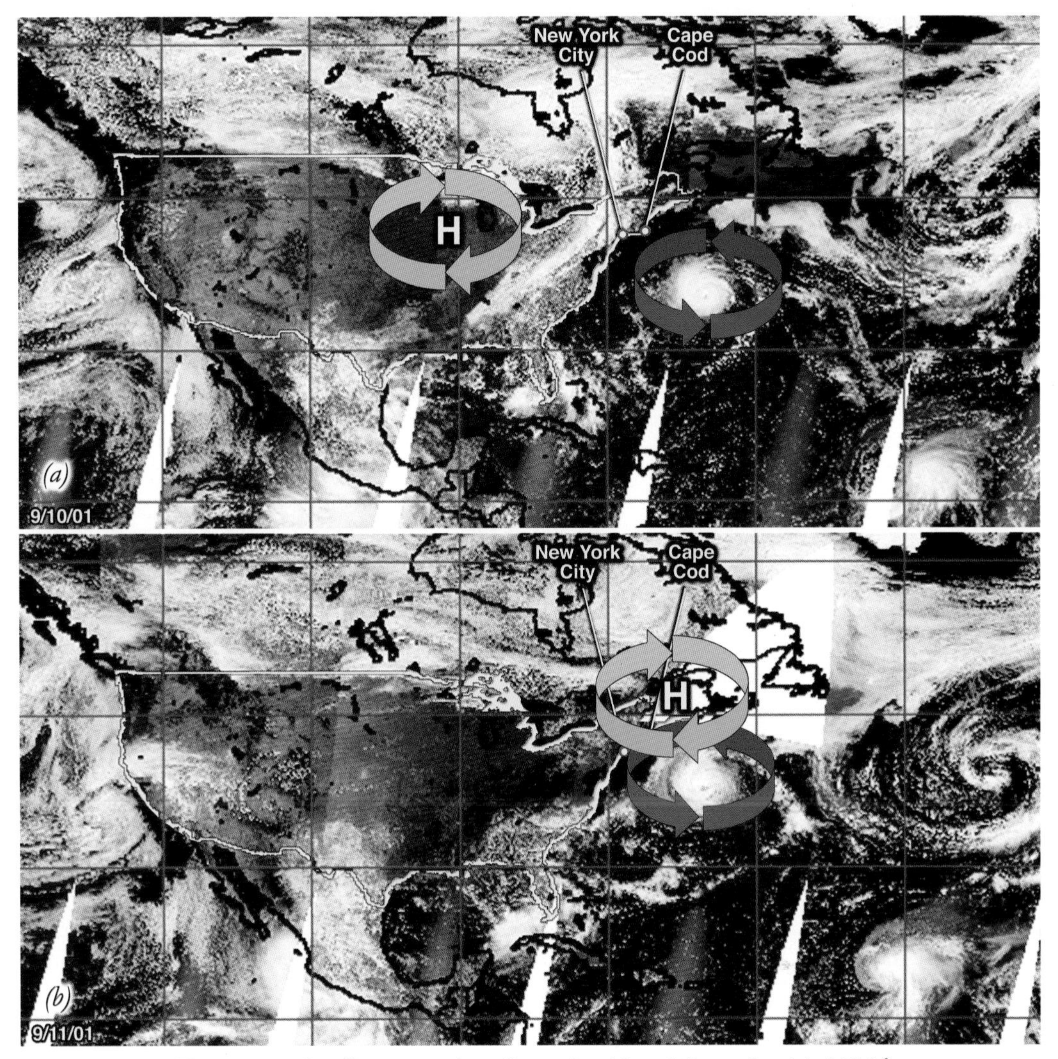

Figure 446. Satellite images from September 10 and September 11, 2001.[3]

Now, let us consider the high-pressure system that arrived along the east coast the morning of 9/11/01, as shown in the FOX5 weather map in Figures 414 and 415 (pages 400 and 401, respectively). The approximate location of the high-pressure system relative to Hurricane Erin is shown in Figure 446. High-pressure systems rotate clockwise in the northern hemisphere and are referred to as "anticyclonic" because they rotate in the opposite direction of cyclones or hurricanes. [4]

From Wikipedia: **Anticyclone**

> In meteorology, an **anticyclone** (that is, opposite to a cyclone) is a weather phenomenon in which there is a descending movement of the air, and with surface systems, higher than average atmospheric pressure over the part of the planet's surface. Effects of surface-based anticyclones include clearing skies as well as cooler, drier air. Fog can also form overnight within a region of higher pressure. Mid-tropospheric systems, such as the subtropical ridge, deflect tropical cyclones around their periphery and inhibit free convection near their center, building up surface-based haze under their base. Anticyclones aloft can form

within warm core lows, such as tropical cyclones, due to descending cool air from the backside of upper troughs, such as polar highs, or from large scale sinking, such as the subtropical ridge. *Anticyclonic flow spirals in a clockwise direction in the Northern Hemisphere* and counter-clockwise in the Southern Hemisphere.[5]

Now let us consider various phenomena associated with rotating weather systems.

B. Phenomena Caused by Rotating weather systems

1. Fusion of Dissimilar Materials

Now let us look at a phenomenon seen after Hurricane Andrew in 1992. Figure 447a shows a piece of lumber, commonly referred to as a two-by-four, to have penetrated through a palm tree.

Figure 447. Hurricane Andrew 1992.
(a)http://www.nasa.gov/images/content/116676main_plank_palm_lg.jpg,(b)http://www.nhc.noaa.gov/gifs/1992andrew6.gif, http://www.nhc.noaa.gov/1992andrew.html

Similarly, Figure 447b shows a piece of plywood to have penetrated another palm tree. This phenomenon may be most familiar to us with the description of finding "straw through trees" after a hurricane or a tornado. It is generally assumed through popular opinion that either the straw was flying so fast that it pierced the tree or that the wind bent the tree over, opening a small fissure and the straw was caught in that fissure.[6] But with closer examination, that does not appear to be the case. It certainly does not explain the plywood and lumber through the trees in Figure 447. The National Oceanic and Atmospheric Administration (NOAA) provides answers to Frequently Asked Questions (FAQ) about tornadoes and hurricanes. In response to a question about "some weird things," including straw into trees, it is acknowledged that there are *weird* phenomena that they do not fully understand.

> **How do tornadoes do some *weird things*, like drive straw into trees, strip road pavement and drive splinters into bricks?** [emphasis added]
>
> The list of bizarre things attributed to tornadoes is almost endless. Much of it is folklore; but there are some weird scenes in tornado damage....Intense winds can bend a tree or other objects, creating cracks in which debris (e.g., hay straw) becomes lodged before the tree straightens and the crack tightens shut again.

All bizarre damage effects have a physical cause inside the roiling maelstrom of tornado winds. *We don't fully understand what some of those causes are yet, however; because much of it is almost impossible to simulate in a lab.*[7]

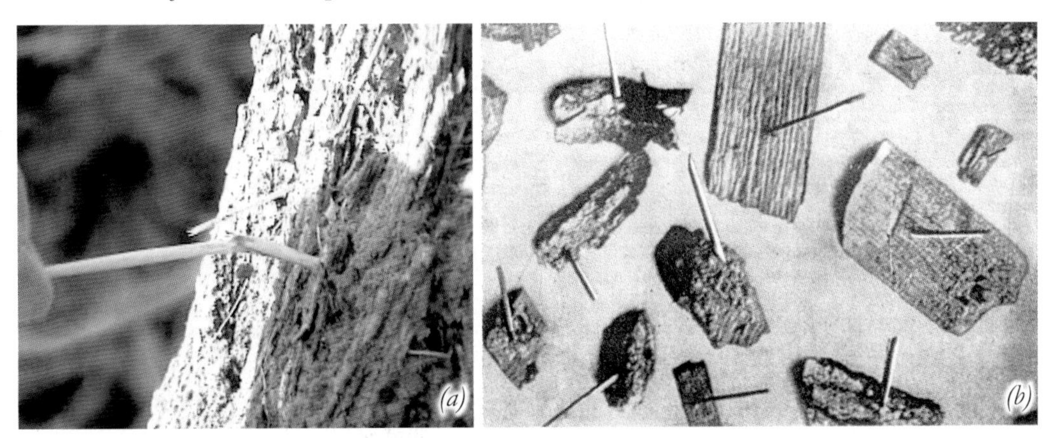

Figure 448. (a) Straw into tree, (b) V.I. Merkulov's straw-through-wood collection.[8]
(a)http://www.hydrolance.net/Common/Straw-BlowThroughTree.jpg, (b)http://www.math.nsc.ru/directions/Figure5.jpg

It is interesting that these appear to be the effects that have been produced by John Hutchison, such as the wood embedded in aluminum, shown in Figure 376 on page 361, as a result of the Hutchison Effect. The so-called "meteorite" found in the debris of the WTC contained a stack of paper embedded in steel, concrete, and other material. (See Figure 378b page 363.)

V.I. Merkulov, a Russian researcher, collected and studied samples of the straw-into-trees phenomenon shown in Figure 448b. Merkulov developed a mathematical model to explain this and the other physical anomalies associated with weather events, including straw in trees.[9]

2. Levitation

After a tornado, heavy objects such as vehicles may be found balancing precariously on top of trees or telephone poles. After a cluster of tornadoes swept through the midwest in May 2008, a survivor was quoted as saying,

> "I swear I could see cars floating," said Herman Hernandez, 68. "And there was a roar, louder and louder."[10,11]

It is generally assumed through popular opinion that a funnel cloud sucked up a vehicle and then dropped it in a particular location. Similarly, after a hurricane, it is generally assumed that the wind blew it to where it was found. But consider the images in Figure 449. How did the car in Figure 449a end up sitting on the wooden fence? The car itself shows no obvious signs of having tumbled there. The fact that the residential fence remained vertical implies there was no significant lateral load. In addition, the lower part of the doors and the undercarriage of the car look more like a misplaced jack location was used to lift the car up than an impact from having been dropped onto a narrow fence. That is, there are strong indications that the car did not tumble into the fence or slam into the fence, but was gently set down on the fence, clearing the nearby roof.

Figure 449. (15-25 October 2005) Cars relocated by Hurricane Wilma.[12]
(a)http://img503.imageshack.us/img503/7780/dsc07405fu.jpg, (b)http://www.high-techproductions.com/Wilma_54b.JPG

None of the three silver cars in Figure 449b appear to have tumbled, either. The one on the far left appears totally unharmed while the one on the right appears to have a broken back window but no other damage. How could the first and third cars be virtually undamaged if the middle car had blown in from elsewhere? If that middle car had been parked there, how could it have been flipped up and rotated while the other two cars remained virtually undamaged? It is also interesting why the front end of the flipped car is not as far forward as the other two. And where did that wood structure come from? If it was blown through the parking lot from elsewhere, why didn't it damage the parking lot and leave scattered debris? So it seems most likely that it was part of the roof structure over the cars and was lifted up, not blown. Had it been blown, we would expect to see a lot of other debris blown around as well, especially under the car. An open trash can on a windy day scatters trash all over the neighborhood; it does not simply relocate it to another place.

The Russian physicist, V.I. Merkulov, researching physical anomalies associated with tornadoes and hurricanes describes a *vacuum domain* that can carry an electrical charge and cause things to flip over.

> A vacuum domain in a gravitational field of the Earth is subject to gravitational polarization, that creates a strong localized change in the gravitational field, which is sufficient to turn over dishes or move furniture. *In combination with this capability to pass through a wall or a closed window and to carry an electrical charge, a vacuum domain can fully explain all phenomena ascribed to poltergeist.*[13]

In addition, Albert Budden, in an article in Nexus Magazine, came to the same conclusion, using the work of John Hutchison to demonstrate these phenomena.

> **THE POLTERGEIST MACHINE**
>
> Poltergeist effects may be as much the result of electromagnetic anomalies as the workings of mischievous discarnate spirits, as inventor John Hutchison has been able to demonstrate in his laboratory....
>
> The Hutchison device proves that poltergeist-type effects can be created in the laboratory. Indeed, many paranormal displays could be the result of electromagnetic anomalies.[14]

V.I. Merkulov recalled this event where coins were carried several kilometers before being dropped.

> On June 17, 1940 in Meschery village of [the] Gorky region, Russia [a] tornado poured out about a thousand XIV century silver coins. The coins were falling from the cloud, but not from the funnel itself. The treasure was transported at several kilometers and was poured out at a compact area [28][15, 16]

Similarly, residents of Brooklyn, New York, reported paper from the World Trade Center falling in their neighborhood.

Brooklyn resident Jerry Gross

> ...By the way on that day my neighborhood was littered with ashes from the WTC. Is that normal for papers fly 7 miles away?[17]

3. Feel and Appearance of a Funnel Cloud

David Handschuh, a photographer standing near the base of WTC2, describes his experience when WTC2 disintegrated.

Photographer David Handschuh

> ... and as I hit the corner of Liberty Street, I was almost being picked up by a tornado ... it was like being picked up... the black cloud had substance... it was like night – but it had a solid feel to it – like gravel... hot gravel and [it] just picked me up and tossed me about a block. Just... one second I was running and the next second I was airborne...[18]

It is interesting that Firefighter Angel Rivera described what he experienced on the morning of 9/11/01 as "*...like being inside a storm or volcano or something, something horrendous.*"[19] He was in WTC3 during the destruction of WTC2 and perhaps WTC1 as well.

Firefighter Angel Rivera, survivor in WTC 3

> When the second tower came down, we had no idea what was going on. We thought another plane, another bomb, another as a [sic] second device.
>
> We thought, this is it, we are dead.
>
> When that happened, as I told you before, everything was black. It was like being inside a storm or volcano or something, something horrendous. [emphasis added] We said we're going to die, we're going to die, God help us. I thought that was it, after that it was no more, no more talking.
>
> That was it. We were all quiet and said where are we, what happened.
>
> The smoke cleared up, the ashes or the cement or whatever. There was a little opening.[20]

Angel Rivera does not describe his experience as a banging from large sections of WTC2 crashing onto WTC3, where he was. He describes being *"inside a storm or volcano, something horrendous."* We might assume Angel Rivera's description that "everything was black,"[21] resulted from the volume of dust that went upward and blocked out all sunlight. But why did so much of it go up?

Figure 450. (9/12/01) (A close-up image is in Figure 101b, page 108.)
http://archive.spaceimaging.com/ikonos/2/kpms/2001/09//browse.108668.crss_sat.0.0.jpg

The plume of dust shown in Figure 450 does not behave as dust wafting away from a "collapsed" building in a light breeze, but better resembles a funnel cloud from a torsional weather event, such as a tornado or a large dust devil. So let us look at a funnel cloud.

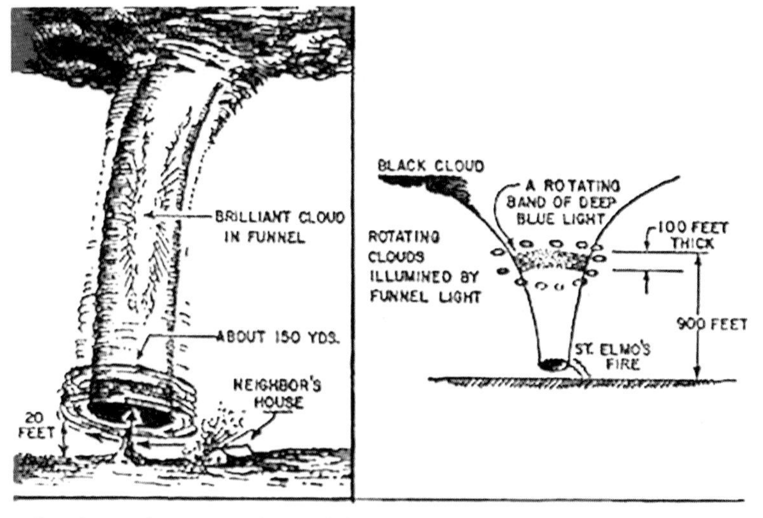

Luminous phenomena observed during a Blackwell, Oklahoma, tornado in 1955.

Figure 451. St. Elmo's Fire inside a tornado.
http://www.flatrock.org.nz/topics/environment/assets/tornado.jpg, [22]

The 25 May 1955 tornado at Blackwell, Oklahoma, was a particularly strong tornado. Lee Hunter saw the light-column effect vividly.

> "The funnel from the cloud to the ground was lit up. It was a steady, deep blue light—very bright. *It had an* orange color fire *in the centre from the cloud to the ground.* As it came along my field, it took a swath about 100 yards wide. As it swung from left to right, it looked like a giant neon tube in the air or a Bagman at a railroad crossing. As it swung along the ground level, the orange fire or electricity would gush out from the bottom of the funnel, and the updraft would take it up in the air causing a terrific light—and it was gone!"
>
> (*Journal of Meteorology*, 14:284, 1957)
>
> Source: Handbook of Unusual Phenomena: Eyewitness Accounts of Nature's Greatest Mysteries by William R Corliss[23]

The Russian physicist, V.I. Merkulov, presents that tornadoes are an electrical system. And, consistent with this, it appears that there may be a grounding effect.

> The interior hollow of a funnel has a clearly defined air walls, with lightning flashing between them. [sic] or water surface, then the action of current sharply displays itself. At the same time, when a funnel doesn't touch the ground, there's no vertical flow. In 1951 in Texas a funnel passed over an observer at 6 meter height, the interior having diameter about 130 m with walls of 3 meter width. Inside the hollow there was a brilliant cloud. There was no vacuum inside, because it was easy to breath. The walls were rotating with a very high speed, ant [sic] the rotation might be seen up to the top of the column. A bit later the funnel touched the neighbor's house and immediately took it off [20][15].
>
> *This description is similar to many others [21][15], [22][15], [23][15] and require the explanation of the fact that rotation of the air necessarily leads to decrease of pressure. Why, being 6m above the ground, the funnel end causes neither damage nor intense air motion, while, upon touching the ground, destroys and moves off a house?*
>
> Direct measurements show that there is a low pressure area inside the funnel (951 mb, Topica June 8, 1966). Such pressure decrease may be resulted at air rotation speed of about 100 m/sec [224 mph] and should make the breath difficult. Why [does] the funnel uncovers the river bed sucking out it's [sic] water, while at the same time the observers even do not notice a wind when the funnel passes above them?[24]...
>
> *The funnel, when it's not touching the ground, emits buzzing or hissing noise.* Faye [Fa][15] describes several cases when tornado was accompanied by ball lightning. Sometime short and wide sheet lightning surround a funnel. *Sometime all the surface of a funnel shines a* strange yellow glow. Sometime observers describe a bluish ball-like formations like ball lightning, but much larger, visible in a cloud. *Sometime a slowly moving fire columns are seen.* [VoM, Vo60, Fr][15]?[25]

It is generally assumed that the center of a funnel cloud contains a vacuum that sucks the material up, but the evidence does not support this. As a child, I saw the aftermath of the tornado that swept through Topeka, Kansas, June 8, 1966. It is a sight that I have never forgotten. An apartment building appeared to have been sliced in two, with one half of the building missing. You could see someone's bedroom, with the bed still made, some things on the dresser, and clothes hanging in the closet. This is what came to mind when I saw what had happened to WTC4, shown in Figures 179 and 185 on pages 188 and 192, respectively.

The strange and powerful phenomena associated with tornadoes remains unexplained by our traditional education. Curiously, the statements made by David Handschuh (page 304), describe the sound when the South Tower got its hole and then, later, the sound when the South Tower "*disintegrated.*"

> *And then, this noise filled the air that sounded like a high-pressure gas line had been ruptured.* And it seemed to come from all over, not from any one direction. *And everybody's looking around like "what's that?" And then all of a sudden the second tower explodes [as it gets its hole].*[26]

Shortly after that, David Handschuh described *"another loud noise that echoed the first noise"* as the building was *"just disintegrating."*

> *I was just walking by myself and heard …another loud noise that kind of* echoed the first noise… *Looked up and [thought] OK, it's another aircraft… and the whole building was just disintegrating…*

Although we may not completely understand how a tornado works, we do know it is real and can destroy things with enormous amounts of energy–energy that seems to come out of nowhere.

4. First Responders describing a tornado or hurricane

Firefighter Derek Brogan

> *We just had the lead guy in front of us yelled back, there's* no more stairs, *they're all gone.*
>
> *So in the darkness we believed everything we heard.* I was sharing my mask by putting it on people's faces. Gerard Gorman was sharing his mask by purging it. So maybe with that loud hissing noise going on, he didn't hear that we were getting off at the fourth floor.[27]

Firefighter Paul Quinn,

> I stayed there until the–it felt like a damn **tornado**. It was black for a long time.[28]

EMT Kevin Barrett

> I felt like I was in a movie at first, when you see everything, you know, typical movie. You would see a movie and everything is like *a storm and a* tornado and everything just turns gray and pitch black and I'm in the middle of this. It was like I thought it was a dream and then it was reality.[29]

Firefighter Adrienne Walsh

> As I got to the back of the rig, I don't know why, but I looked to my right, and *I saw a black cloud the size of the biggest skyscraper I've ever seen actually coming, hurdling at us like a* tornado, and I just yelled, "Run," and I took off.[30]

Firefighter Edward Kennedy

> So I was trying to hold my breath. I was trying to get the stuff out of my mouth. I opened my eyes twice. It was pitch black and it sounded like a hurricane coming through. The glass on the car was breaking and it was just, you know, as things settled, it seemed to me it was like 15 minutes…[31]

Lieutenant Richard Smiouskas

Everybody started running north, and this huge volume like ten stories high billowing, pushing black smoke *and like a glitter.* I guess it was glass that was *glitter that was in the cloud* of smoke. I saw everything flying around. I looked back, and there was this thing flying, coming at me.

I started running north. I got to the corner, and it was engulfing all the buildings around, maybe 50 feet away from me. I tried a couple of doors, and then I went into a bodega on the corner and shut the door. There were maybe ten people in that bodega. I held the door shut and it just kind of blew by like a hurricane, all this debris and paper, thousands of paper in it. Then it got pitch-black again. It was midnight again. [32]

5. St. Elmo's Fire

Another phenomenon associated with storms is St. Elmo's Fire[33], which is shown in the drawing describing what Lee Hunter saw in the Blackwell, Oklahoma funnel cloud in 1955. St. Elmo's fire is usually described as a blue glow, but many have described it as a fire, although, similar to the "weird fires" seen on 9/11/01, it has not been described as burning things.

> *The funnel, when it's not touching the ground, emits buzzing or hissing noise.* Faye [Fa][15] describes several cases when tornado was accompanied by ball lightning. Sometime short and wide sheet lightning surround a funnel. *Sometime all the surface of a funnel shines a* strange yellow glow. Sometime observers describe a bluish ball-like formations like ball lightning, but much larger, visible in a cloud. *Sometime a slowly moving fire columns are seen.* [VoM, Vo60, Fr].[15,34]

And then there is William Beaty's article in *Scientific American*, for September 22, 1997, asking, "What causes the strange glow known as St. Elmo's Fire? Is this phenomenon related to ball lightning?"

> St. Elmo's Fire is a type of continuous electric spark called a "glow discharge." You've seen it many times before, since it is almost exactly the same as the glows found inside fluorescent tubes, mercury vapor streetlights, old orange-display calculators and in "eye of the storm" plasma globes. When it occurs naturally, we call it St. Elmo's Fire, but when it occurs inside a glass tube, we call it a neon sign.
>
> St. Elmo's Fire and normal sparks both can appear when high electrical voltage affects a gas. St. Elmo's fire is seen during thunderstorms when the ground below the storm is electrically charged, and there is high voltage in the air between the cloud and the ground. The voltage tears apart the air molecules and the gas begins to glow. It takes about 30,000 volts per centimeter of space to start a St. Elmo's fire (although sharp points can trigger it at somewhat lower voltage levels.)
>
> St. Elmo's Fire is plasma. A normal gas is composed of molecules. The molecules are composed of atoms, which in turn are composed of electrons and clusters of proton particles. If the electric force applied to each bit of gas is

greater than a certain level of voltage, it causes the electrons and protons of the gas molecules to be pulled away from each other. High voltage transforms the gas into a glowing mixture of separate proton clusters and electrons. *We call this mixture of particles by the name "plasma," and it is conductive. It also fluoresces with light.*

The color of the glow depends on the type of gas involved. If we lived in an atmosphere of neon gas, then St. Elmo's fire would be red/orange, and lightning would be white with orange edges. Our atmosphere is nitrogen and oxygen, and this mixture glows blue/violet when exposed to high voltage fields. If a neon sign tube was [sic] filled with nitrogen/oxygen instead of neon, it would light up blue/violet rather than red/orange.

Is this phenomenon related to ball lightning? No one knows, because no one knows what ball lightning is, and it might not be a spark at all. St. Elmo's fire is sometimes mistaken for ball lightning. Among other differences, ball lightning can drift around like a soap bubble, while St. Elmo's Fire always remains attached to an object.[35]

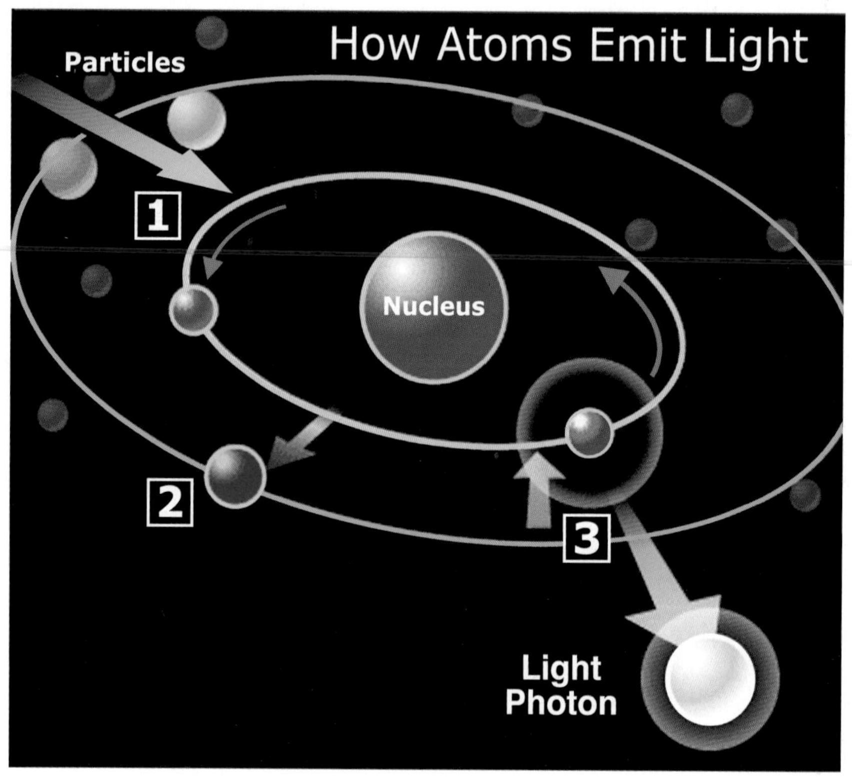

Figure 452. How atoms emit light. [redrawn from source]. [36]

1. A collision with a moving particle excites the atom.
2. This causes an electron to jump to a higher energy level.
3. The electron falls back to its original energy level, releasing the extra energy in the form of a light photon.[36]

A fluorescent lamp emits a bright glow without getting extremely hot. For this reason, fluorescent lights are more efficient than incandescent lights.

C. Plasma

From *How Stuff Works:*

Plasma, in physics, a highly ionized gas, *usually* at high temperatures, that conducts electricity and is affected by a magnetic field. Plasma is sometimes called the fourth state of matter. Plasma is the most common form of matter in the universe; the sun and all other stars, as well as some types of interstellar matter, consist of plasmas. Plasmas occur in lightning bolts and in the regions of the earth's upper atmosphere called the ionosphere and the Van Allen radiation belts. Man-made plasmas occur in electric discharge lamps and in electric arcs used for welding.

A plasma is composed of equal numbers of positively charged ions (atoms that have lost one or more of their electrons) and free electrons, which have a negative electrical charge. The ions and electrons have a strong tendency to recombine and form electrically neutral atoms of ordinary gas.

Figure 453. Glowing plasma.
http://www.stdi.ca/amazon/plasma-globes/Plasma-Globe-Color-Comparison.jpg[37]

Two factors can prevent a plasma from becoming an ordinary gas. One is a large supply of energy, *usually* in the form of heat, which can keep the ions and electrons moving too fast to recombine easily. The other is an extremely low density in the plasma. If the density is low enough, the space between the particles is so great that there is little chance for them to recombine.[38]

As discussed in Chapter 17, John Hutchison said that he uses magnetic fields to direct and guide the interfering energy to a particular zone.

John Hutchison

I use a different combination of radio waves, along with threshold high voltage, and electrostatic operators. And I only use weak magnetic fields to steer the electrostatic [field] around other... and to conform to certain patterns. Now mind you, this is a fairly accidental discovery I made. And I had a real problem trying to replicate it in the early years. And I didn't think it was important, but it leaked out into the scientific community and they started coming down with other scientists, and they started asking for demonstrations.[39]

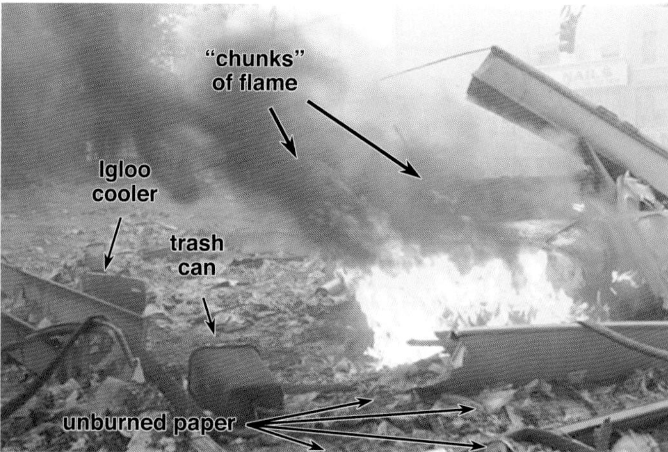

(a) Plasma *(b) (9/11/01) Location D of map in Figure 303 (page 290.)*

Figure 454. (a) Plasma globe.[40] (b) Near Church and Liberty Streets.[40]
Note the "chunks" of flame seen at the ends of both.

(a)http://uw.physics.wisc.edu/~wonders/PlasmaB.jpg, (b)http://images30.fotki.com/v478/photos/1/115/34570/DSC_0194-vi.jpg,

Now, let us review what Firefighter Armando Reno observed:

Firefighter Armando Reno

> I was working by the south bridge. There were numerous car fires there. I was located by the south bridge and the chauffeur from 1 Engine was with me. There were two lengths of a 2-1/2 inch line stretched off the hydrant there on the south side of Liberty Street. *We were putting out the car fires, or attempting to, and there was no*—the water had no effect on the car fires *at the time.* I started thinking about getting the foam off the rig, and I also noticed there were numerous bodies by Cedar Street, and I was thinking of getting the EMS equipment off the rig, putting gloves on and starting to get the bodies, putting them in bags. Well, body pieces. [41]

1. What is St. Elmo's Fire?

If you were to look outside your home during a thunderstorm and see a tall streetlamp glowing with blue flames, you might be tempted to call the fire department. Then you might notice that the streetlamp *is on fire but isn't actually burning—and the* water from the fire hose isn't putting out the flames. At this point, you might be about ready to call a priest, but that, like the call to the fire department, would be unnecessary. The phenomenon you're witnessing is actually **St. Elmo's Fire**. (Which has nothing to do with a 1980s coming-of-age film starring a young Emilio Estevez.)

St. Elmo's Fire is a weather phenomenon involving a gap in electrical charge. It's like lightning, but not quite. And while it has been mistaken for ball lightning, it's not that, either -- *and it's definitely not fire.* …..

As with all electrical phenomena, *St. Elmo's Fire is about electrons.* So, what is St. Elmo's Fire if it's not a form of lightning?[42]

2. Causes of St. Elmo's Fire: The Fire That's Not a Fire

Like lightning, St. Elmo's Fire is plasma, *or ionized air that emits a glow.* But while

> lightening is the movement of electricity from a charged cloud to the ground, St. Elmo's Fire is simply sparking, something like a shot of electrons into the air. It's a **corona discharge**, and it occurs when there is a significant imbalance in electrical charge, causing molecules to tear apart, *sometimes resulting in a slight* hissing *sound.*
>
> [...]
>
> Once the air is conducive to the movement of electrons, those electrons continue to increase the distance between their positively charged counterpart, protons. *This is* ionization*, and plasma is simply ionized air. The phenomenon that causes St. Elmo's Fire is a dramatic difference in charge between the air and a charged object...*[43]

In Chapter 11 (Toasted Cars) and Chapter 13 (Weird Fires), we saw that the *weird fires* away from the WTC affected vehicles without appearing to ignite paper or trees or people, or anything else around. For example, the cars burn but not the street signs in Figures 266 and 304 (pages 258 and 291). Also, the trees don't appear to have burned in Figures 265 and 243 (pages 257 and 235). The cars, on rubber tires, were not grounded, so if the air was highly ionized, the cars may have built up more of a charge than the grounded objects. This might also explain the abrupt boundaries of affected zones on vehicles such as the police car in Figure 214, page 214.

3. Plasma and transmission of energy

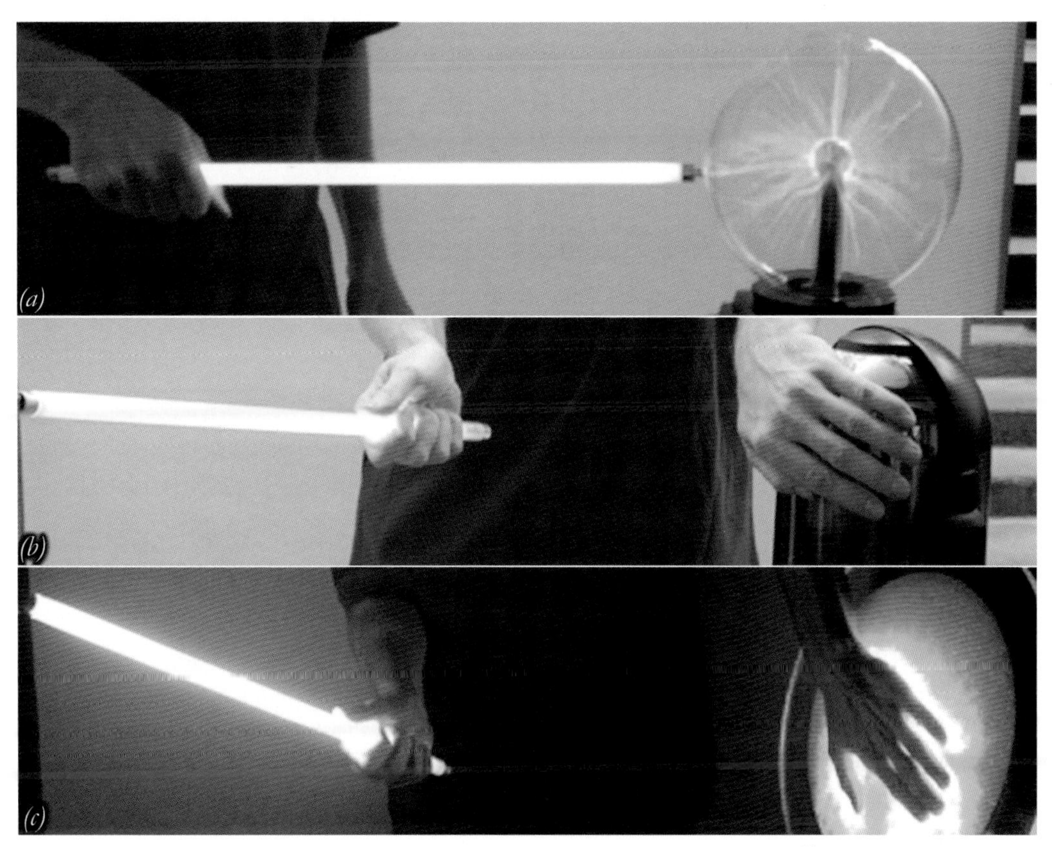

Figure 455. Physics demonstration of energy transmission.[44]
(a)http://uw.physics.wisc.edu/~wonders/PlasmaB1.jpg, (b)http://uw.physics.wisc.edu/~wonders/PlasmaA9.jpg, (c)http://uw.physics.wisc.edu/~wonders/Borg6.jpg,

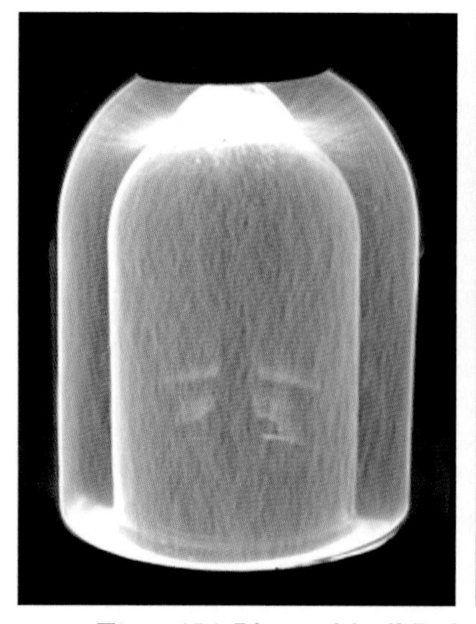
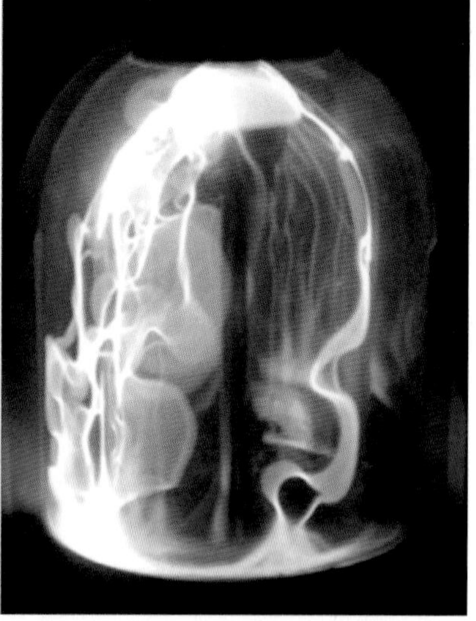

Figure 456. Plasma globes.[45] *Perhaps this is related to "furry lather."*
(a)http://uw.physics.wisc.edu/~wonders/PlasmaA3.jpg, (b)http://uw.physics.wisc.edu/~wonders/PlasmaA.jpg

D. Transmission of Energy

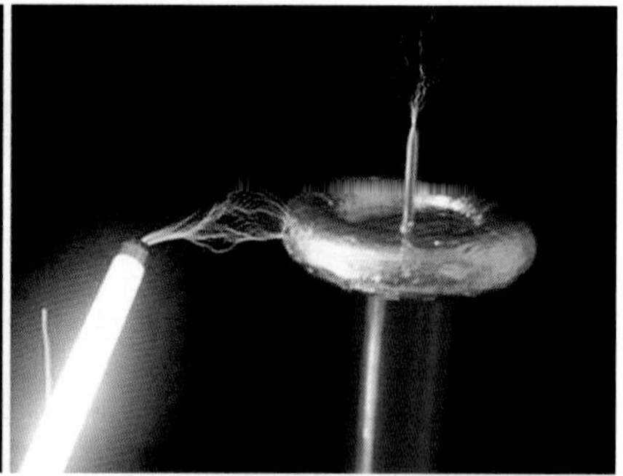

Figure 457. Transmission of electricity through the air.
(a)http://www.badbros.net/images/tesla_coil/tesla_coil/1f_1.jpg, (b)http://www.badbros.net/images/tesla_coil/tesla_coil/6c_1%20copy.jpg [46]

Figure 458. (a) Fluorescent tube with a Tesla coil. (b) Wardenclyffe Tower.
(a)http://uw.physics.wisc.edu/~wonders/Tesla6.jpg [47]*,*
(b)http://thinkorthwim.com/wp-content/uploads/2007/01/tesla-wardenclyffe-tower-shot.jpg

E. Lift or Disruption

1. Physicists have 'solved' mystery of levitation

By Roger Highfield, Science Editor, (08/08/2007)[48]

Figure 459. A rotating magnet in a magnetic field is antigravitic.
http://i.telegraph.co.uk/telegraph/multimedia/archive/00642/news-graphics-2007-_642176a.jpg

Figure 460. Compare this with the flipped car in front of the SUV in Figure 246 (page 237).
http://members.iinet.net.au/~pterren/CancrushingRedBull4labels.jpg[49]

Electricity, magnetism, and gravity are all interrelated. Moving a conductor through a magnetic field can produce electricity (e.g. a hydroelectric plant). Electricity can also produce a magnetic field (wrap wire around a nail and apply electrical current to the wire). A spinning magnet is shown to defy gravity as it floats above a magnetic field, shown in Figure 459. Aluminum Red Bull cans are crushed by magnetic induction, shown in Figure 460.

It would appear that if they can stop a tornado, they can start one.

2. Taking the Twist Out of a Twister

By Leonard David, Senior Space Writer, March 3, 2000, 03:56 p.m. ET

ALBUQUERQUE, NEW MEXICO—A blast of microwave energy beamed down from a space satellite could be used to tame the destructive nature of a tornado, a scientist said this week. Such weather-watching duty might be assigned to future solar power stations that would circle Earth.

Each year, about 1,200 tornadoes are reported in the United States, according to the American Meteorological Society. An average of 55 people die annually as a result of twisters, and billions of dollars worth of property are destroyed or damaged.

Their extremely high winds propel debris, destroy homes, collapse buildings and overturn vehicles. There is growing evidence that global warming may spawn increasing amounts of nasty weather, including tornadoes, at an even greater intensity in years to come.

But the tornado-nuking concept advocated here this week flies in the face of being at the mercy of Mother Natures [sic] fury. Called a Thunderstorm Solar Power Satellite, the concept was presented at the Space 2000 Conference and Exposition on Engineering, Construction, Operations and Business in Space, sponsored by the American Society of Civil Engineers.

The proposal calls for beaming microwave energy into the cold, rainy downdraft of a thunderstorm where a tornado could originate. That pulse of power would disrupt the convective flow needed to concentrate energy that forms a tornado, said Bernard Eastlund, president of Eastlund Scientific Enterprises Corp, based in San Diego, California.

He has teamed up with Lyle Jenkins, a 37-year NASA veteran who now heads his own firm, Jenkins Enterprises in Houston, Texas.

The two researchers envision surgical strikes of microwave energy that could modify the temperature and fine structure of storm systems.

"We call it taming the tornado," Jenkins said. "With just a little burst of microwave energy, we think we see a way to negate the trigger point in tornado creation. We want to heat the cold rain. By tailoring the beam, it can absorb the rain that is part of the tornado-making process."

Eastlund has looked at data provided by the Advanced Regional Prediction System at the Center for Analysis and Prediction of Storms Center at the University of Oklahoma.

These numerical simulation data were used to study the formation of conditions suitable for "tornado-genesis." And he and Jenkins used them to see the effects of zapping an incipient storm with electromagnetic radiation beamed from a proposed Thunderstorm Solar Power Satellite.

Another aspect of their proposal could address the need for orbiting Doppler radar that could see tornado conditions forming. These data would be fed into a tornado-stopping satellite, perhaps positioned in geosynchronous orbit above the areas most affected by severe weather. By using a specially-tuned microwave pulse, rather than laser or infrared beams, that energy can be targeted within a storms [sic] interior, not through it or reflected away.

"You cant [sic] wave your hands about this idea," Eastlund said. "You've got to use real numerical modeling. My research shows that by heating the falling rain, we can turn off the downdraft that drives a tornado." More research is needed, he said, to further determine just how much energy would yield a knockout punch to a tornado on the brew.

But is it nice to fool with Mother Nature? "This is a new science we're talking about of weather modification...a new paradigm which seeks to mitigate these violent weather systems," Eastlund said.

"If it does prove possible to prevent tornadoes," Eastlund continued, "then systems could be envisioned in which severe storm phenomena such as hurricanes and typhoons are also modified in some beneficial fashion, and weather modification could be routine in the 21st century."[50]

F. Conclusions

We're talking about the fact that most people see what they expect to see, what they want to see, what they've been told to see, what conventional wisdom tells them to see, not what is right in front of them in its pristine condition.–Vincent Bugliosi

Empirical evidence is the truth that theory must mimic.[51] I have repeated this statement several times in this book because its importance cannot be over emphasized. In today's culture of over simplification and standardized multiple-choice testing, many have an impulse to *name* a known technology (e.g. thermite, TNT, RDX, nukes, progressive *collapse*, HAARP, scalar weapons, torsion physics, Nazi Bell, etc.) instead of *looking* at the evidence that the use of one technology or another has left behind. It is like someone trying to force a square peg into a round hole because no one they know has ever seen a square peg, much less given it a name, so the possibility of a square peg's existence is not considered—even as they struggle to force it into the round hole.

Some people feel they are being more *scientific* when they use the name of a *known technology* to describe *unknown* phenomena, but the opposite is true. Such an approach omits evidence that does not *fit* any known technology. For some people, the term "HAARP" or the term "scalar weapons" or the term "Nazi Bell" is used as a catch-all weapon that can be blamed for whatever evidence needs to be explained, like the ultimate "boogieman," and without their even knowing what these weapons can do. Furthermore, if the full capabilities are classified information, they would not be publicly known. And a weapon that could produce all of the effects we saw on 9/11 would certainly not be in the public domain, no matter *whose* weapon it was. For these reasons, I have tried to focus on the phenomena, not on a trendy name of a particular technology. The evidence must come before the theory. It is understanding what the technology can do that matters, not the name of it. For these reasons, I have resisted the impulse to *name* a known technology and instead have focused on the physical evidence.[52] There will likely be those who will not be as successful in resisting the impulse to put a *name* of a known technology on the producer of this evidence. This *naming*, however, will only serve to pull a veil of mystery over it.

Any sufficiently advanced technology is indistinguishable from magic.—Arthur C. Clarke

Once the mechanism of something is understood, it is no longer seen as magic. Each of the phenomena that resulted from whatever took place on 9/11 can be understood as something many people have seen before, although in a different and usually far, far less massive context. But one thing is perfectly clear: 9/11 was the demonstration of a technology of horrendous magnitude. Such a technology, however monstrously destructive we've seen it to be, can also be used for constructive and benevolent purposes, such as, for just one obvious example, providing clean energy. And, like "magic," understanding it will only empower us.

In Chapter 17, we saw that all of the phenomena seen at the WTC on 9/11 could be produced by the Hutchison Effect. Although they are not on the same scale, they appear to be the same phenomena or to involve the same type of physics. The Hutchison Effect is produced by interfering radio-frequency energy within a static field—for example, by interfering microwave energy within a static field. The static field used by John Hutchison is produced by a Tesla coil or by a Van de Graaf generator. In Chapter 18, it was shown that Hurricane Erin was centered just outside of New York City on the morning of 9/11/01 with her outer bands reaching over Cape Cod as well as reaching the end of Long Island. On 9/11, thunder, which is evidence of an electrical field, was reported at JFK, LaGuardia, and Newark Airports, the three major airports around New York City. In Chapter 19, it was shown that magnetometer readings in Alaska exhibited excursions of the earth's magnetic field that were coincident with the time of each event at the WTC on 9/11. And we know that rotating electrical fields can produce a magnetic field. And an interference of these fields can cause molecular dissociation. And, referring to the article quoted on page 450, if it is possible to stop a vortex, then it is certainly possible to start one. And levitation as well as fusion of dissimilar materials and destruction of materials can result from an energy vortex.

What reason is there that a technology based on the same type of physics as the Hutchison Effect, and one that produces the same phenomena as the Hutchison effect, cannot be super-sized?

---

[1] *http://www.nhc.noaa.gov/2001erin.html*

[2] *(a)http://ladsweb.nascom.nasa.gov/allData/5/MOBRGB_C00/2001/253/MOBRGB_C00.A2001253.005.2006303143612.jpg, (b)http://ladsweb.nascom.nasa.gov/allData/5/MOBRGB_C00/2001/254/MOBRGB_C00.A2001254.005.2006303153504.jpg, (c)http://ladsweb.nascom.nasa.gov/allData/5/MOBRGB_C00/2001/255/MOBRGB_C00.A2001255.005.2006303162726.jpg*

[3] *(a)http://ladsweb.nascom.nasa.gov/allData/5/MOBRGB_C00/2001/253/MOBRGB_C00.A2001253.005.2006303143612.jpg, (b)http://ladsweb.nascom.nasa.gov/allData/5/MOBRGB_C00/2001/254/MOBRGB_C00.A2001254.005.2006303153504.jpg*

[4] *Anticyclone, http://en.wikipedia.org/wiki/Anticyclone*

[5] *Anticyclone, http://en.wikipedia.org/wiki/Anticyclone*

[6] Roger Edwards, Frequently Asked Questions about Tornados, *http://spc.noaa.gov/faq/tornado/*

[7] Roger Edwards, Frequently Asked Questions about Tornados, *http://spc.noaa.gov/faq/tornado/*

[8] *http://www.math.nsc.ru/directions/tornado-eng.htm*

[9] V.I.Merkulov, "*Electrogravidynamical model of UFO, tornado and tropical hurricane,*" Institute of Theoretical and Applied Physics of Siberian Branch of Russian Academy of Sciences, (1998?) Source: *http://www.math.nsc.ru/directions/tornado-eng.htm*, (archived: *http://drjudywood.com/articles/erin/erin4.html*)

[10] ROXANA HEGEMAN, "Survivors Dig Out After Tornadoes," 05/12/2008, *http://news.aol.com/story/_a/survivors-dig-out-after-tornadoes/20080510221609990001*

[11] Tornado, May 10, 2008, *http://www.chicagotribune.com/news/nationworld/nation/la-na-weather12-2008may12,0,1537389.story*

[12] *http://www.nhc.noaa.gov/2005atlan.shtml, http://www.nhc.noaa.gov/pdf/TCR-AL252005_Wilma.pdf*

[13] V.I.Merkulov, "*Electrogravidynamical model of UFO, tornado and tropical hurricane,*" Institute of Theoretical and Applied Physics of Siberian Branch of Russian Academy of Sciences, (1998?) Source: *http://www.math.nsc.ru/directions/tornado-eng.htm*, (archived: *http://drjudywood.com/articles/erin/erin4.html*)

[14] Albert Budden, B.Ed., *"THE POLTERGEIST MACHINE"* (1996) vol. 4, n. 1, *http://www.nexusmagazine.com/articles/polter.html*, 17 Brook Road South, Brentford, Middlesex TW8 ONN, United Kingdom, Telephone +44 0181 560 9497, and *http://www.nexusmagazine.com/index.php?searchword=Hutchison&ordering=&searchphrase=all&Itemid=44&option=com_search*

[15] [20]. Hall R. S., Inside a Texas tornado. Watherwise, 1951.-v. 4, No 3, pp. 54-57, 65,

[21]. Flora S. D., Tornadoes of the United States. Oklahoma, 1953, 194 pp.,

[22]. Justice A. A., Seeing the inside of a tornado. Monthly Weather Rev.1930.- v. 58, pp.57-58,

[23]. Hoecker W. H., Jr., Three-dimensional pressure pattern of the Dallas tornado and some resultant implications. Monthly Weather Rev.

[28]. Hurd W. E., Some phases of waterspout.1950, v. 3, N 4, pp.75-78

[Fa].Faye H., Nouvelle etude sure les tempetes, cyclones, trombes ou tornado. Paris, 1897, 142 p.

Jones H. L., The tornado pulse generator. Weatherwise, 1965, v. 18, N 2, pp.78-79, 85

[VoM]. Vonnegut B. and Meyer J. R., Luminous phenomena accompaning tornadoes Weatherwise, v. 19, N2, 1966, pp. 66-68

[16] V.I.Merkulov, "*Electrogravidynamical model of UFO, tornado and tropical hurricane,*" Institute of Theoretical and Applied Physics of Siberian Branch of Russian Academy of Sciences, (1998?) Source: *http://www.math.nsc.ru/directions/tornado-eng.htm*, (archived: *http://drjudywood.com/articles/erin/erin4.html*)

[17] Personal communication from Jerry Gross, resident of Brooklyn, New York.

[18] Charlie Rose Show, October 2001, *http://youtube.com/watch?v=ulE9OiZqQwg*, 03:32

[19] File No. 9110489, "Firefighter Angel Rivera," January 22, 2002, p. 9, *http://graphics8.nytimes.com/packages/pdf/nyregion/20050812_WTC_GRAPHIC/9110489.pdf*

[20] File No. 9110489, "Firefighter Angel Rivera," January 22, 2002, p. 9, *http://graphics8.nytimes.com/packages/pdf/nyregion/20050812_WTC_GRAPHIC/9110489.pdf*

[21] File No. 9110489, "Firefighter Angel Rivera," January 22, 2002, p. 9, *http://graphics8.nytimes.com/packages/pdf/nyregion/20050812_WTC_GRAPHIC/9110489.pdf*

[22] *http://www.flatrock.org.nz/topics/environment/wipeout.htm*

[23] *http://www.flatrock.org.nz/topics/environment/wipeout.htm*

[24] V.I.Merkulov, "*Electrogravidynamical model of UFO, tornado and tropical hurricane,*" Institute of Theoretical and Applied Physics of Siberian Branch of Russian Academy of Sciences, (1998?) Source: *http://www.math.nsc.ru/directions/tornado-eng.htm*, (archived: *http://drjudywood.com/articles/erin/erin4.html*)

[25] V.I.Merkulov, "*Electrogravidynamical model of UFO, tornado and tropical hurricane,*" Institute of Theoretical and Applied Physics of Siberian Branch of Russian Academy of Sciences, (1998?) Source: *http://www.math.nsc.ru/directions/tornado-eng.htm*, (archived: *http://drjudywood.com/articles/erin/erin4.html*)

[26] *http://digitaljournalist.org/issue0110/video/dh03.mov*, (video 3, 0:0 3:38)

[27] File No. 9110414, "Firefighter Derek Brogan," December 28, 2001, p. 17, *http://graphics8.nytimes.com/packages/pdf/nyregion/20050812_WTC_GRAPHIC/9110414.pdf*

[28] File No. 9110422, "Firefighter Paul Quinn," January 8, 2002, p. 5, *http://graphics8.nytimes.com/packages/pdf/nyregion/20050812_WTC_GRAPHIC/9110422.pdf*

[29] File No. 9110464, "EMT Kevin Barrett," January 17, 2002, pp. 8-9, *http://graphics8.nytimes.com/packages/pdf/nyregion/20050812_WTC_GRAPHIC/9110464.pdf*

[30] File No. 9110353, "Firefighter Adrienne Walsh," December 17, 2001, pp. 4-5, *http://graphics8.nytimes.com/packages/pdf/nyregion/20050812_WTC_GRAPHIC/9110353.pdf*

[31] File No. 9110502, "Firefighter Edward Kennedy," January 17, 2001, p. 8, *http://graphics8.nytimes.com/packages/pdf/nyregion/20050812_WTC_GRAPHIC/9110502.pdf*

[32] File No. 9110210, "Lieutenant Richard Smiouskas," November 27, 2001, p. 12, *http://graphics8.nytimes.com/packages/pdf/nyregion/20050812_WTC_GRAPHIC/9110210.pdf*

[33] *http://www.cartage.org.lb/en/themes/sciences/physics/Electromagnetism/Electrostatics/ElectroSpectrum/StElmosFire/StElmosFire.htm, http://www.connectonline.com*

[34] V.I.Merkulov, "*Electrogravidynamical model of UFO, tornado and tropical hurricane*," Institute of Theoretical and Applied Physics of Siberian Branch of Russian Academy of Sciences, (1998?) Source: *http://www.math.nsc.ru/directions/tornado-eng.htm*, (archived: *http://drjudywood.com/articles/erin/erin4.html*)

[35] William Beaty, "What causes the strange glow known as St. Elmo's Fire? Is this phenomenon related to ball lightning?" *Scientific American*, September 22, 1997, *http://www.scientificamerican.com/article.cfm?id=quotwhat-causes-the-stran*,

[36] *http://static.howstuffworks.com/gif/fluorescent-lamp-atom.gif, http://home.howstuffworks.com/fluorescent-lamp.htm*

[37] *http://www.stdi.ca/amazon/plasma-globes/plasma-globe-comparison-video.htm*

[38] *http://science.howstuffworks.com/plasma-info.htm*

[39] John Hutchison gave this information on *The Ralph Winterrowd Show*, April 4, 2010.

[40] (a) Color adjusted, (b) (brightness and contrast adjusted)

[41] File No. 9110448, "Firefighter Armando Reno," January 13, 2002, p. 3, *http://graphics8.nytimes.com/packages/pdf/nyregion/20050812_WTC_GRAPHIC/9110448.pdf*, emphasis added.

[42] Layton, Julia, "What is St. Elmo's Fire?." 24 June 2008. *HowStuffWorks.com. http://www.howstuffworks.com/st-elmo-fire.htm.*

[43] Julia Layton, *http://science.howstuffworks.com/st-elmo-fire1.htm*

[44] *http://uw.physics.wisc.edu/~wonders/DemoPB.html,*

[45] *http://uw.physics.wisc.edu/~wonders/DemoPB.html*

[46] *http://www.badbros.net/teslacoil.html*

[47] *http://uw.physics.wisc.edu/~wonders/DemoTesla.html*

[48] *http://www.telegraph.co.uk/core/Content/displayPrintable.jhtml;jsessionid=P2NBMUYQ2V1AJQFIQMFCFFWAVCBQYIV0?xml=/news/2007/08/06/nlevitate106.xml&site=5&page=0*

[49] *http://members.iinet.net.au/~pterren/*

[50] Leonard David, Taking the Twist Out of a Twister, 3 March 2000, *http://www.space.com/scienceastronomy/planetearth/tornado_taming_000303.html*

[51] A powerful statement by someone who has taught me well.

[52] The technology used on 9/11/01 can however be identified as involving energy, that was directed, and was used as a weapon. In other words, a **D**irected **E**nergy **W**eapon, or **DEW**. The category of **DEW** includes *hot* DEW (**hDEW**) and *cold* DEW (**cDEW**) as well as **D**irected **F**ree-**E**nergy **W**eapons (**DFEW**). The evidence shows that some type of *cold* DEW was used. The evidence contradicts the use of *hot* DEW on 9/11/01 (e.g. lasers and masers) because there is no indication of high heat (see Chapter 13). The data also shows that the technology used on 9/11/01 is capable of producing *free-energy* for the world. That is, the technology that was demonstrated on 9/11 was at that time used for destructive purposes, but this technology can also be used to provide free energy to the world. The use of this technology on 9/11 demonstrated to the world that *free-energy* technology exists. It's up to us to determine what uses are made of it.

21.
Rolled-Up Carpets

One of the saddest lessons of history is this: If we've been bamboozled[1] long enough, we tend to reject any evidence of the bamboozle. We're no longer interested in finding out the truth. The bamboozle has captured us. It is simply too painful to acknowledge–even to ourselves–that we've been so credulous.
—Carl Sagan, *The Fine Art of Baloney Detection*

The problem is never how to get new, innovative thoughts into your mind, but how to get old ones out.
—Dee Hock, Founder, Visa

A. Introduction

I use the term "wheatchex" to refer to the prefabricated sets of three outer columns on WTC1 and WTC2, each three stories tall, connected with spandrel plates as shown in Figure 461. The NIST NCSTAR 1 summary document stated that the outer columns bowed and buckled. However, that behavior is not consistent with the evidence.

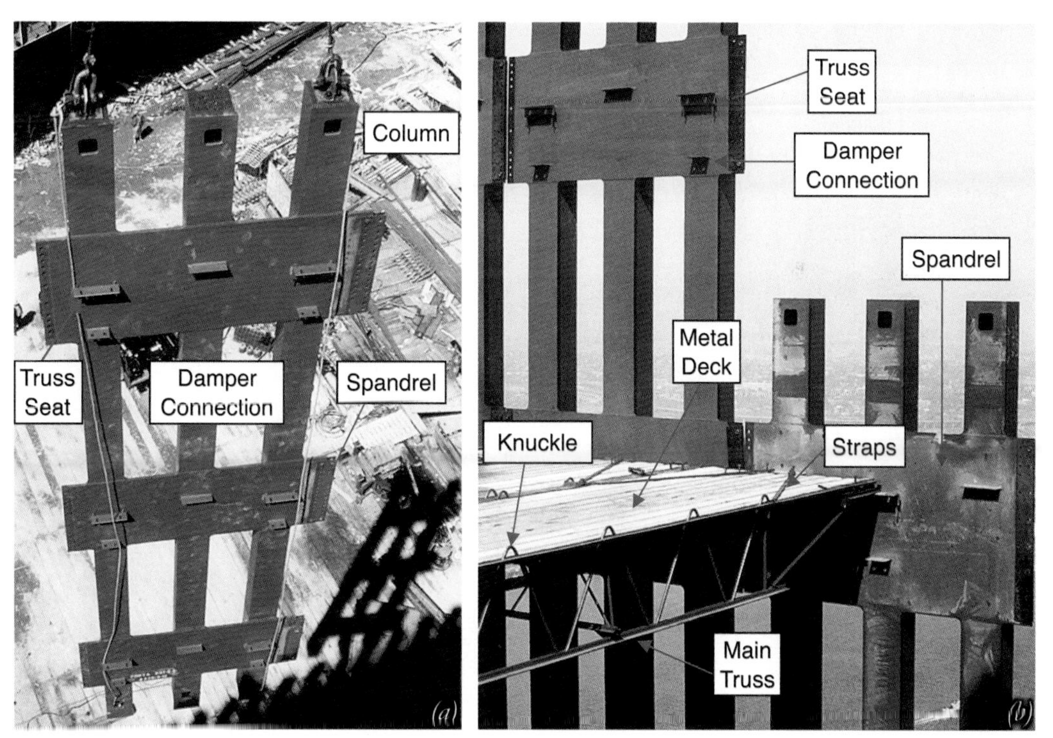

Figure 461. Characteristic perimeter column panel consisting of three full columns connected by three spandrels. Perimeter column/spandrel assembly and floor structure.
Figure 1-4, NISTNCSTAR1Draft.pdf, p. 8 (pdf 62 of 292), http://wtc.nist.gov/pubs/NISTNCSTAR1Draft.pdf

[1] Bamboozle - 'trick' or 'hoodwink'[Webster], Hoodwink - 'to deceive by false appearance'[Webster]

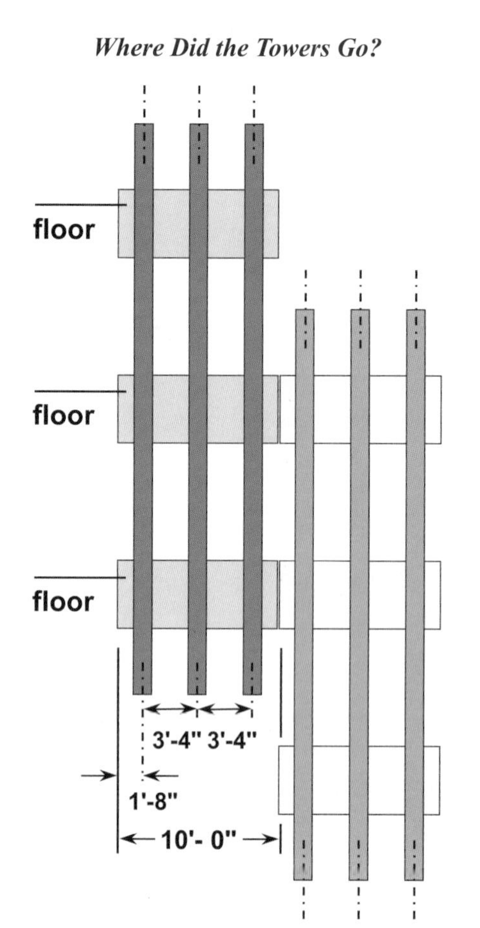

Figure 462. Diagram of a pair of "wheatchex."[1]

Figure 463. World Trade Center construction with core and structural perimeter of tower.
(a)http://data.GreatBuildings.com/gbc/images/cid_wtc_mya_WTC_const.4.jpg, (b)http://www.hybrideb.com/images/newyork/core_2.jpg

Figure 464 shows an example of vertical loading on these wheatchex.

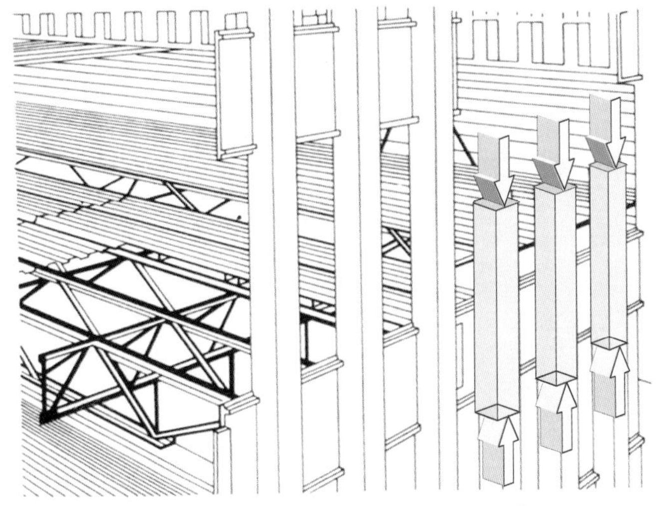

Figure 464. Vertical loading on perimeter columns.
http://www.studyof911.com/gallery/albums/userpics/10002/assembly.jpg

B. Expected Results from a Gravity Collapse

If the WTC had been destroyed by a gravity collapse, what would engineers expect to see?

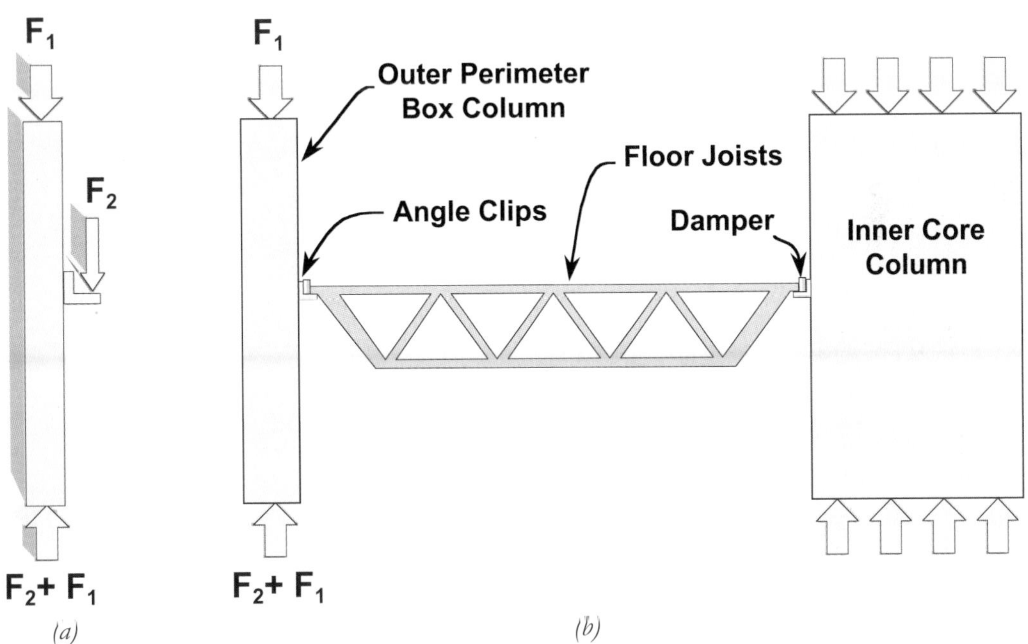

Figure 465. Cross-section diagram of floor truss connections.

Either the truss supports (the angle clips shown in Figure 465) hold or they don't. If they don't hold and there is pancaking of the floors/floor joists, there will not be column failure, because if the floors pancake down, the columns will no longer be carrying a significant load. If the truss supports do hold and there is no pancaking, and if the columns are overloaded and/or tremendously weakened, there may be column buckling. Consider the case of buckling, a process exemplified by the can in Figure 466.

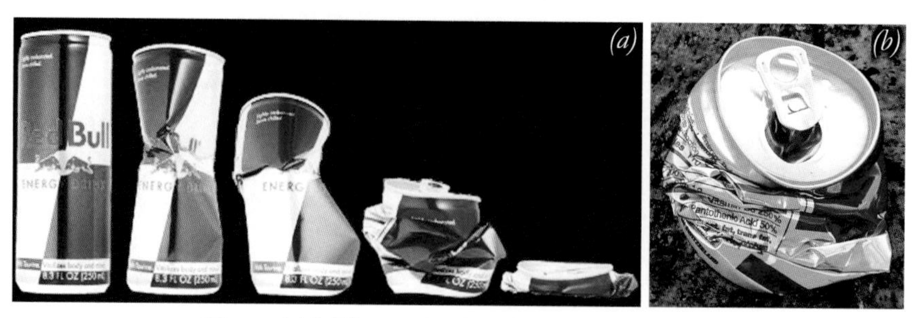

Figure 466. The can kinks. The kinks are sharp.
(a)http://www.widgetslab.com/wp-content/uploads/2007/09/redbull_mac_widget.jpg,
(b)http://www.worldbehind.com/Images/Backgrounds/CrushedCan1024x768.jpg

Note that the can kinks, and that the kinks are sharp, the result being a "smushed" can. Now let's consider what causes buckling, beginning with the formula for critical buckling:

Critical load formula: $P_{cr} = \dfrac{EI\pi^2}{(KL)^2}$, *where*

P_{cr} = *critical or maximum axial load on the column just before it begins to buckle. This load must not cause the stress in the column to exceed the proportional limit, and*
E = *modulus of elasticity for the material*
I = *least moment of inertia for the column's cross-sectional area,*
L = *unsupported length of the column, whose ends are pinned, and*
K = *effective-length factor.*

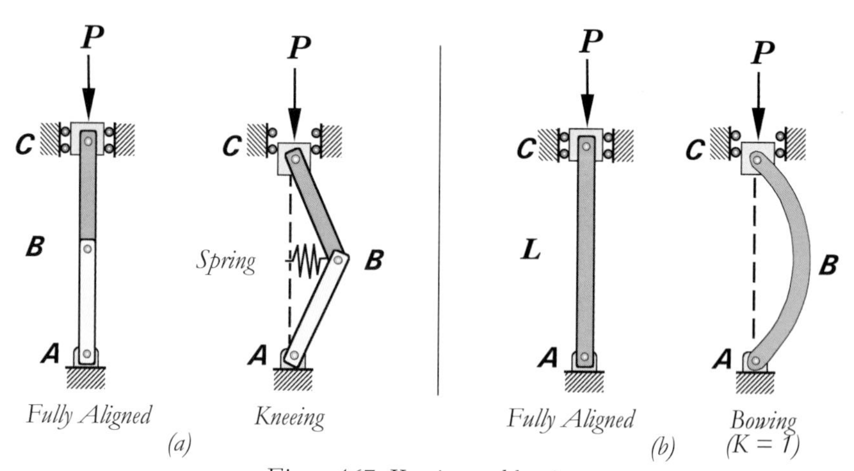

Figure 467. Kneeing and bowing.

What the math reveals is that if there is more force than the column can carry (through overload and/or weakening from high temperature), it will bow outward or inward, depending on the minor lateral loading. Figure 467a is a simplified model of a buckling column where the spring represents the lateral stiffness of the structure shown in figure 467b. When the load, P, is too great, or when the spring is not stiff enough, the column will buckle.

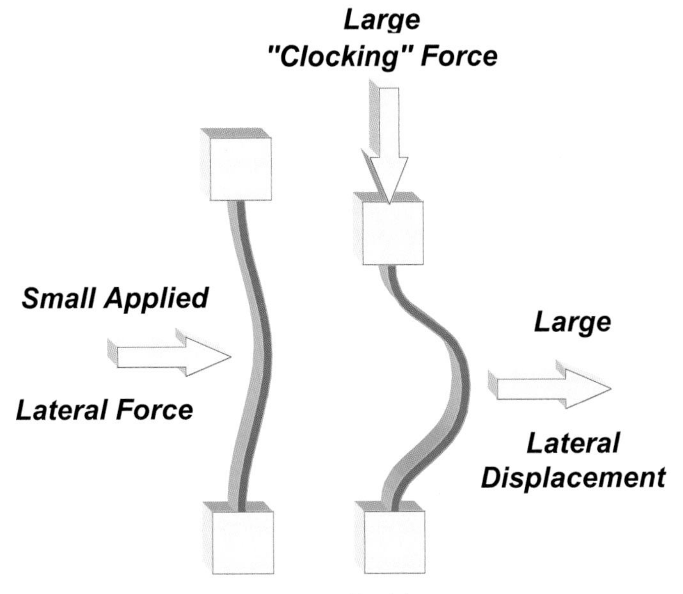

Figure 468. Buckling.

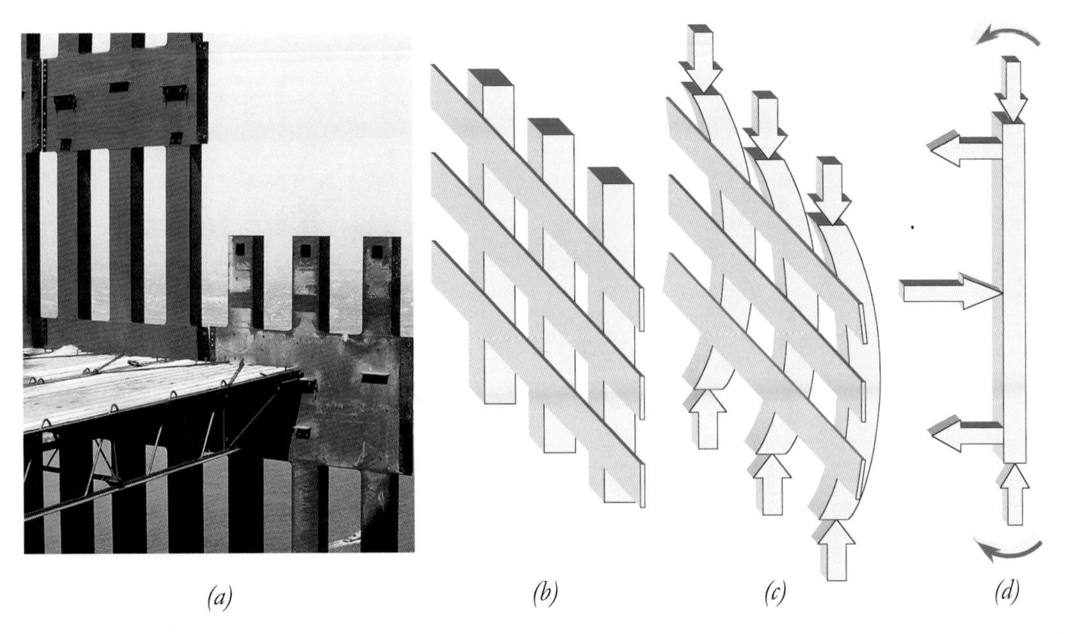

Figure 469. (a) Floor framing, (b) before, (c) buckling, (d) loading
(a)http://studyof911.com/gallery/albums/userpics/10002/WTC-constructionimage017.jpg,

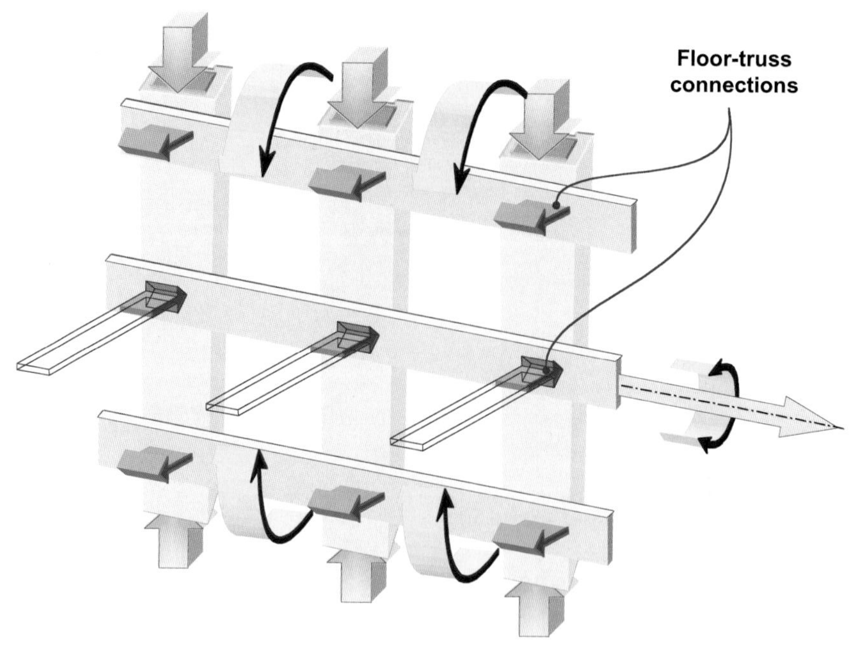

Figure 470. The outer columns are subjected to a vertical axial load. If overloaded, bending is expected to occur as shown.

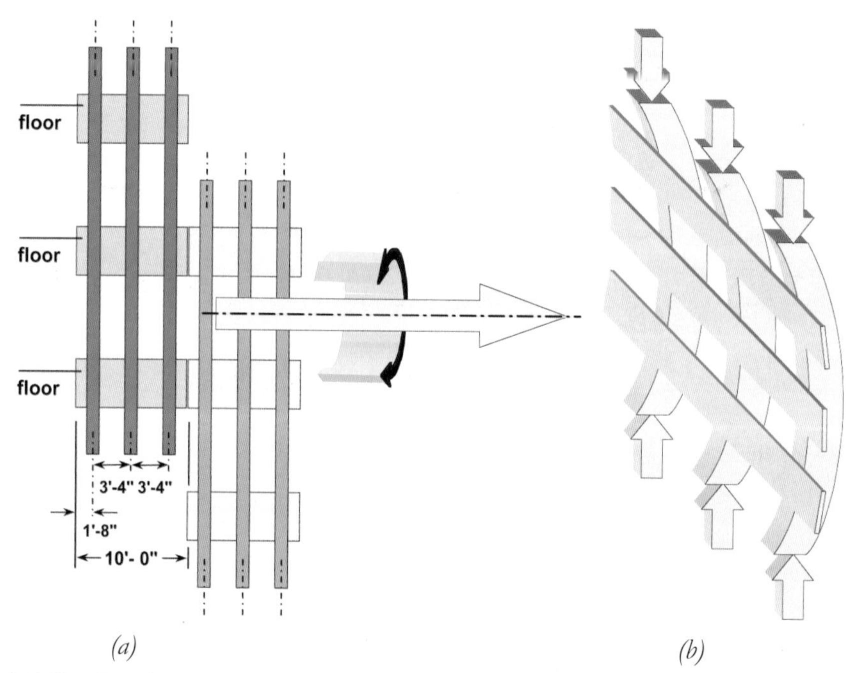

Figure 471. Bending about a horizontal axis (expected in a "collapse" from overload and/or weakening).

But this kind of bending or bowing is not at all what we see in the remains of the WTC. Consider the picture in Figure 472:

C. What We Observe

Figure 472. WTC beams are not consistent with a gravity collapse or a conventional explosion.
(NIST[2]) *http://wtc.nist.gov/images/steel11_hires.jpg*

Contrary to the preceding analysis about buckling, we encounter this extraordinary phenomenon. It is explained below. I call this the "rolled-up-carpet" phenomenon because the rolled-up assemblies of columns, with spandrel belts wrapped around them, resemble rolled-up carpets, emphasizing that their rolled-up appearance is different than a buckled appearance.

Figure 473. WTC beams. This deformation is not consistent with a gravity collapse or conventional explosion. (NIST[2])
http://wtc.nist.gov/images/WTC-003_hires.jpg,

The cluster of beams in Figures 472 to 475, wrapped in their own spandrel plates, look more like rolled-up carpets than perimeter columns of the WTC (wheatchex).

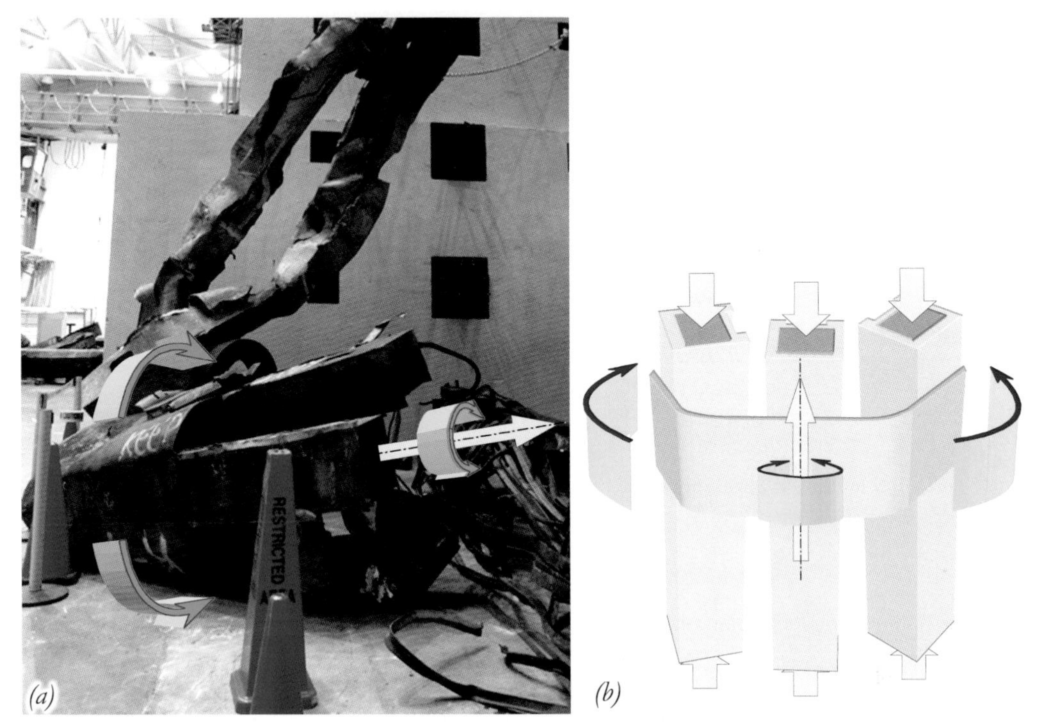

Figure 474. (a) These columns are bent around the wrong axis.[3] *(NIST*[2]*) (b) In a gravity collapse, the outer columns are essentially subjected to no loading in this direction.*

(a)Adapted from: http://wtc.nist.gov/images/steel11_hires.jpg

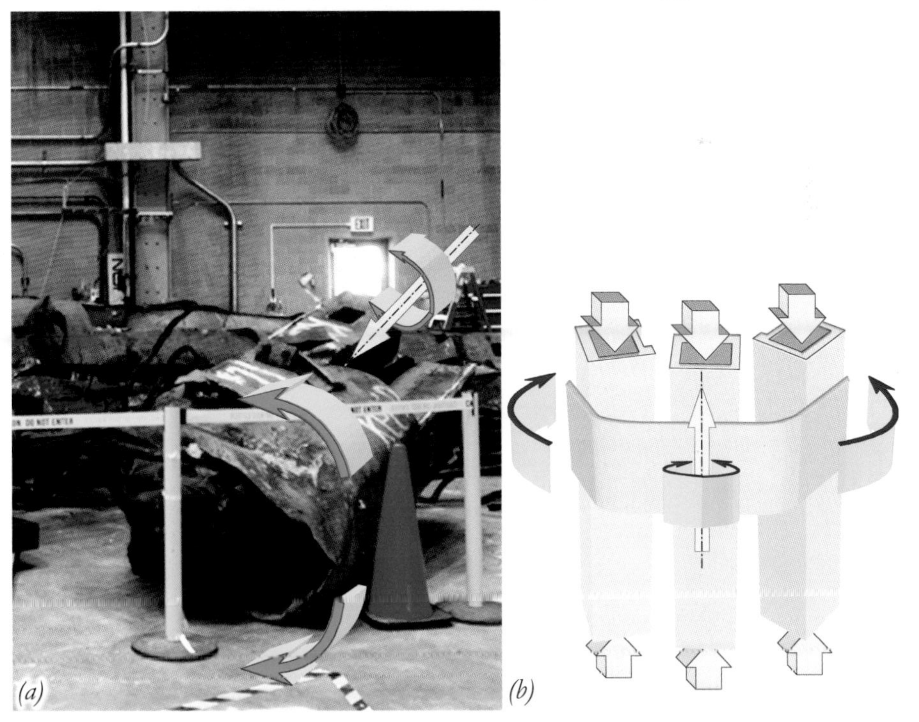

Figure 475. (a) Spandrel belts wrap around columns. (b) In a gravity collapse, the outer columns are essentially subjected to no loading in this direction. (NIST[2]*)*

(a)Adapted from: http://wtc.nist.gov/images/WTC-003_hires.jpg

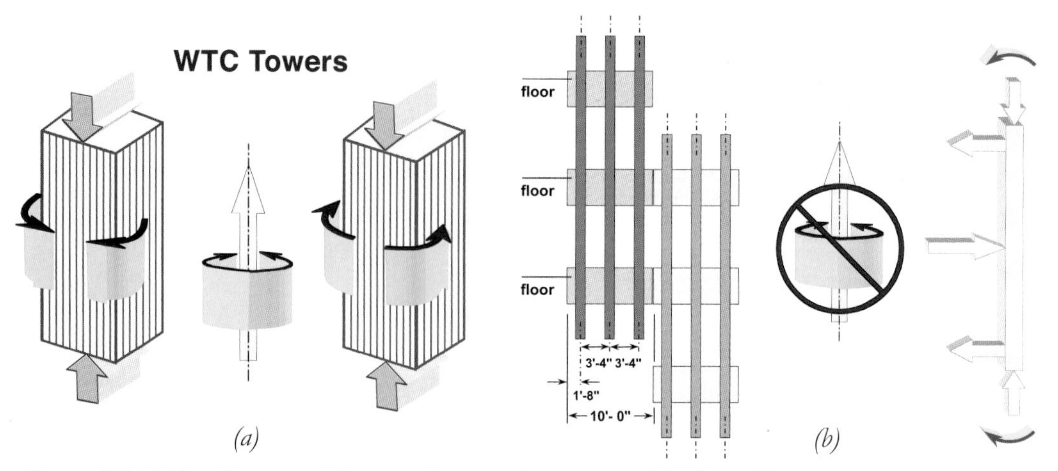

Figure 476. (a) Bending around the vertical axis makes no sense. (b) The building is not loaded around the vertical axis.

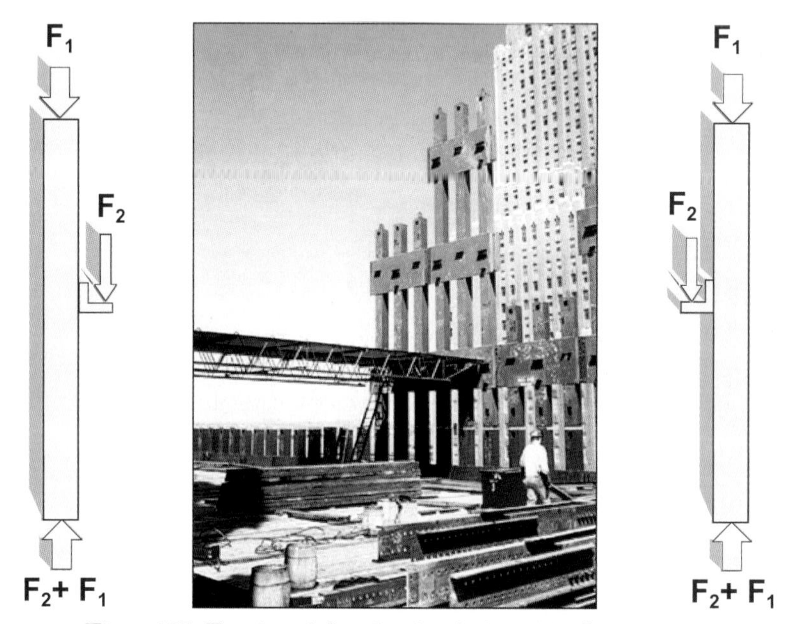

Figure 477. Erection of floor framing during original construction.
Fema Photo[4]

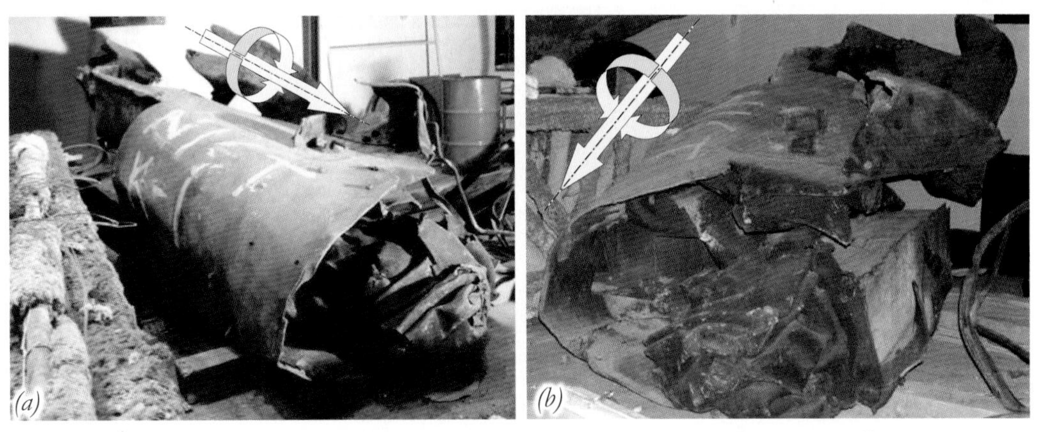

Figure 478. (a) WTC beams wrapped like a burrito (b) or Pepperoni roll.
Adapted from: (a)F-A-2a: Piece K-1. NISTNCSTAR1-3C Appxs.pdf, Attachment A, WJE No. 2003.0323.0, Page A-497, NISTNCSTAR 1-3C Appxs.pdf, File page (211 of 258), http://wtc.nist.gov/WTCfinal1-3.zip, (b)F-A-2b: Piece K-1. NISTNCSTAR1-3C Appxs.pdf, Attachment A, WJE No. 2003.0323.0, Page A-497, NISTNCSTAR 1-3C Appxs.pdf, File page (211 of 258), http://wtc.nist.gov/WTCfinal1-3.zip

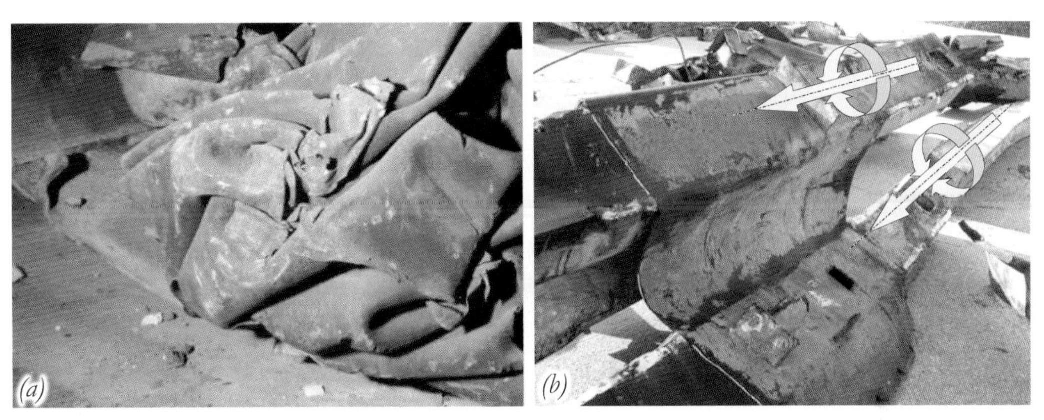

Figure 479. (a) WTC beams that look like poured cake batter. (b) Straight WTC outer columns look like the spandrel belts were draped over them like papier-mâché, then solidified and spray-painted fire-engine red.
Adapted from: (a)F-A-2c:Column 210. NISTNCSTAR1-3C Appxs.pdf, Attachment A, WJE No. 2003.0323.0, Page A-497, NISTNCSTAR 1-3C Appxs.pdf, File page (211 of 258), http://wtc.nist.gov/WTCfinal1-3.zip, (b)Attachment A, WJE No. 2003.0323.0, Page B-520, NISTNCSTAR 1-3C Appxs.pdf, File page, (234 of 258), http://wtc.nist.gov/WTCfinal1-3.zip

Figure 480a shows an outer column that is not buckled from an axial load. It looks like a curled ribbon on a gift-wrapped package. Decorative ribbon is curled by scraping one side with a sharp object, such as the blade from a pair of scissors. This deforms one side. If one side shrinks along its length, the ribbon curls. But there are other ways to alter the properties of a material on just one side, such as by means of heat[56], chemicals, energy fields, or loading (e.g., as in a bi-metal thermostat[709]). Column buckling occurs at a particular point, like a hinge, and is not uniformly distributed over the length, as demonstrated by the Red Bull can in Figure 466. The column shown in Figure 480a, on the other hand, is smoothly curled along approximately half its length like a curled ribbon.

Figure 480. (9/05) Curled columns.
(a) This outer column looks like it became a limp lasagna noodle. (b) Straight columns, no buckling, but curled spandrel belts.
Adapted from: (a)NISTNCSTAR 1-3C chaps.pdf, Figure 4-3, p. 203 (pdf 253 of 336), http://wtc.nist.gov/WTCfinal1-3.zip, (b)http://graphics.jsonline.com/graphics/news/img/sep02/snapabig090702.jpg, http://www.arrse.co.uk/cpgn2/Forums/viewtopic/t=92059/postdays=0/pos

D. Straight Columns – Not Buckled

Figure 480b shows outer columns that are not buckled from an axial load, but that do have spandrel belts folded about a vertical axis indicating a rotational load or torque having been applied, or a uniform field effect that altered the material properties on one surface more than the other. There is no gravitationally applied mechanical load that would cause this configuration.

(a) (9/13/01) *(b) (9/11/01)*

Figure 481. Wheatchex stabbed into West Street[10], viewed (a) from the north, (b) from the south.
(a)http://stopviolence.com/images/9-11/groundzero-cut.jpg, (b)http://www.photolibrary.fema.gov/photodata/original/3887.jpg

Figure 481a shows the wheatchex that stabbed into West Street. The photo gives a view down the straight columns from the lower portion of WTC1 that lay down across West Street. Figure 481b shows the same columns from the south. These columns that stabbed into West Street were about 200 feet away from WTC2 and were not buckled.

Figure 482. These WTC beams (wheatchex) stabbed into Church Street without buckling and are almost 400 feet away from WTC2.
http://www.sharpprintinginc.com/911_math/mech_room_perimeter.jpg

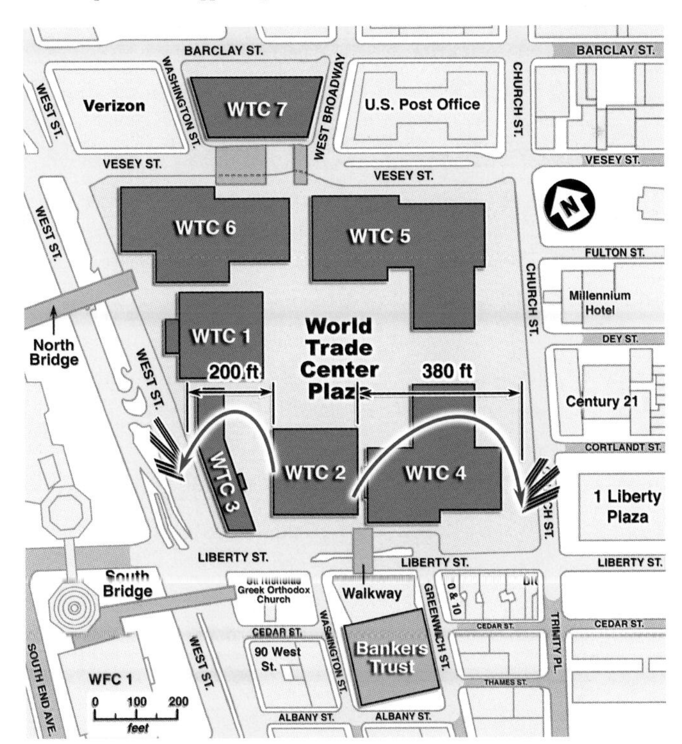

Figure 483. The wheatchex on Church Street flew approximately 400 feet.
(Map redrawn from NIST Report.[11])

Figure 482 shows that the wheatchex that stabbed into Church Street were straight, while the only remaining portion of WTC4's main body is slumped over, as if wilted. A long wooden platform can be seen under the two large cranes in Figure 484. This is used to distribute the weight of the cranes and avoid damaging the delicate bathtub wall.

Figure 484. (9/22/01) These last standing columns are not buckled. Note the hole on the far right.
Adapted from: http://bocadigital.smugmug.com/photos/10698621-D.jpg

The wooden platform itself is outside of the bathtub's western wall, and the long booms of the cranes are cantilevered over the delicate wall to avoid damaging it. Even the small yellow grappler (just above the time stamp in the lower-right of the photo) is working from outside of the bathtub and reaching over the wall. The holes indicated in the photo are just inside the bathtub wall. The hole at the far right is much larger than the full length of the extended grappler.

As described in chapter 17, the Tesla-Hutchison Effect, changing the molecular structure in just one area of a material, can produce anomalous bending of the kind shown in Figure 485.

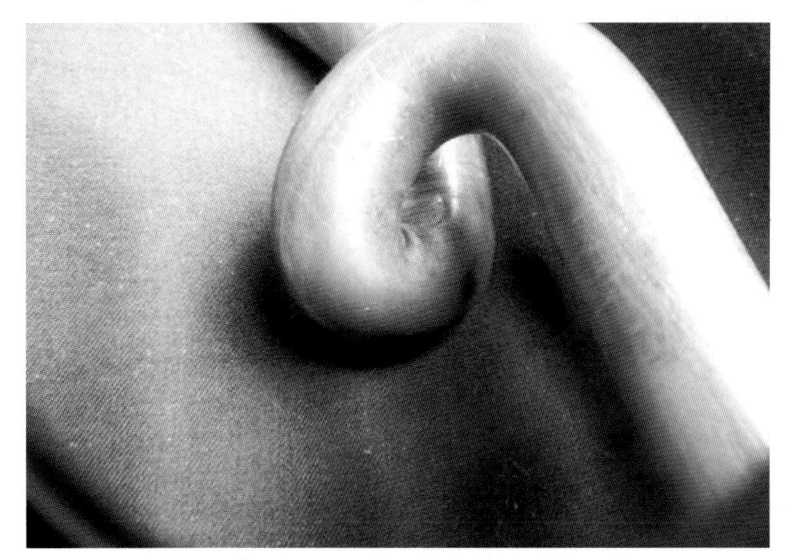

Figure 485. Solid molybdenum Tesla-Hutchison-Effect bar. (2.5- 3-inch diameter)
Photo courtesy of John Hutchison[12]

E. I-Beams Bent Around Wrong Axis

Figure 486. I-beams deformed in the wrong direction. This deformation is not consistent with overload.
(NIST[7]*) http://wtc.nist.gov/images/DSCN0941_hires.jpg*

In this context, it is interesting to note that there are several pictures of "I-Beams" that are bent around the wrong axis, as shown in Figures 486 and 487, which is inconsistent with the Twin Towers having been destroyed by a gravitational collapse and pancaking floors.

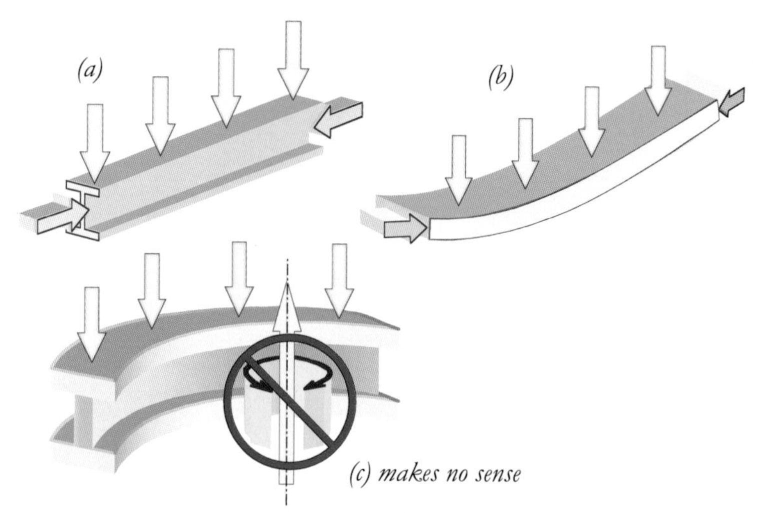

Figure 487. (a) Typical loading of an I-beam, (b) expected deformation from this loading, (c) actual deformation *seen in the beams found at GZ, as seen in Figure 486. The vertical loading should not cause the I-beams to bend around a vertical axis.*

F. Over-Curled Beams

Figure 488. (2002) (a) Massive core column with anomalous smooth bend. (b) WTC core column curled, not buckled. A gravity-driven "collapse" could not do this. The beam above has smooth curves, without kinks.

(a)http://www.amny.com/media/photo/2006-08/24912250.jpg,[13] *(b)http://www.studyof911.com/gallery/albums/userpics/10002/core1.jpg*

Buckled beams characteristic of a gravity-driven collapse were virtually non-existent at the WTC site. Some of the beams are bent more than 180°. This extent and type of distortion in the steel cannot be the result of a gravity-driven collapse.

G. Bankers Trust Building Evidence

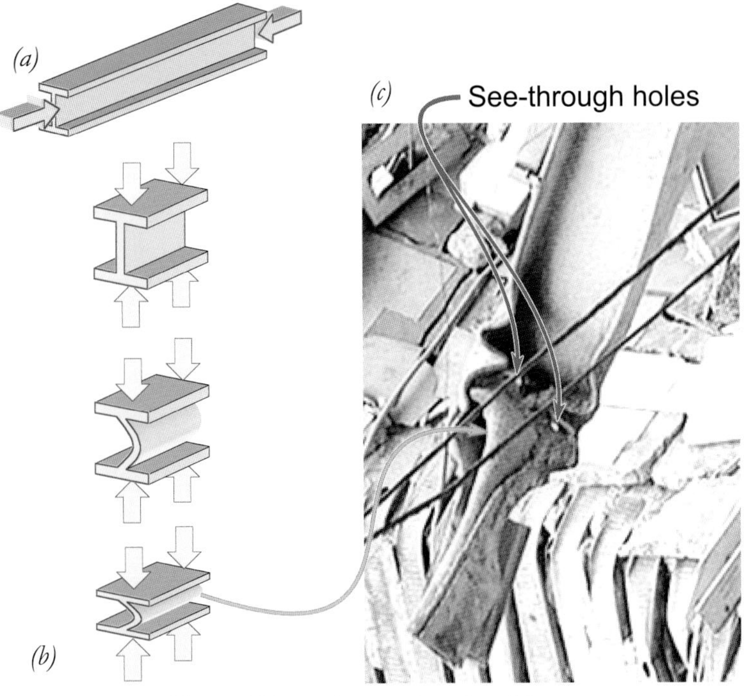

Figure 489. A close-up view of an I-beam
Source: *http://www.fema.gov/pdf/library/fema403_ch6.pdf*

In Figure 489, we see a shriveled up I-beam inconsistent with vertical loading or with being hit by a falling object. The Bankers Trust building was directly across the street from WTC2 and was reported to have had no fires.[14] But fire damage could not explain this deformation. The shriveled up beam in Figure 489c could not be the result of axial loading, as the diagram in Figure 490 illustrates:

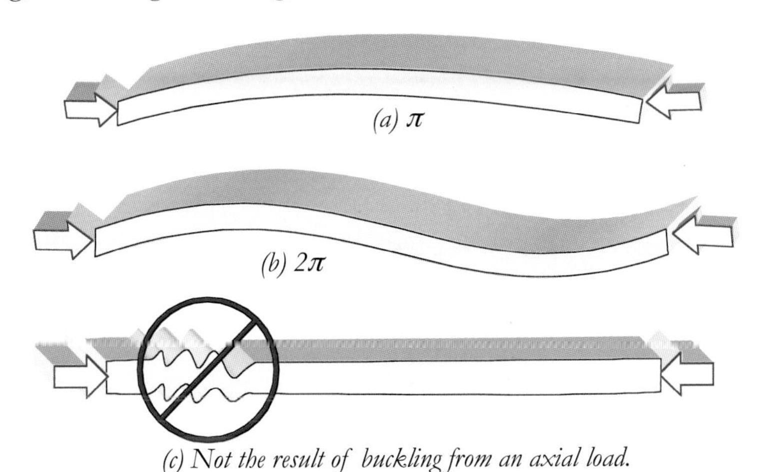

Figure 490. In buckling, a beam deforms into (a) a half sine wave, π, or (b) a full sine wave, or 2π. The random deformation in (c) is not associated with buckling.

H. Examples of Buckling from Other Events

Figure 491. Buckled column.
http://sophies.maze.free.fr/lab2/pct/baud4.jpg

Existing 9/11 photography strongly indicates that the Trade Center Towers were demolished but did not "collapse." Or, more exactly, the existing video and still photos prove beyond doubt that the towers did not collapse as a result of mechanical load failure caused by damage from airplanes or fires:

1) When a building collapses as a result of load failure, its combined dead and live loads (typically gravity, earthquake, and/or wind loads) overcome the load-bearing capacity of its structural members. When this happens, each failed structural member shows a distinct pattern of failure, such as the buckled column seen in Figure 491.

Figure 492. World Trade Center construction of core and structural perimeter.
http://data.GreatBuildings.com/gbc/images/cid_wtc_mya_WTC_const.3.jpg[15]

2) In the official WTC explanation promoted by the media and federal agencies such as the NIST, we are asked to believe that plane crashes and fuel fires weakened local core and perimeter columns (both visible in the large construction photo in Figure 492) to the point where gravity overcame them. After that, in an as-yet unexplained chain reaction, the rest of both the core and perimeter members also failed, nearly simultaneously.

Figure 493. Flying beams, not Buckled, but essentially dissolving.
(a)http://bocadigital.smugmug.com/photos/10746211-D.jpg,
(b)figure 3-6, page 3-6, www.fema.gov_pdf_library_fema40, (Richard Drew photo)

3) The existing 9/11 photography shows clearly that this kind of "collapse due to weakening or overload" did *not* occur. We know this because the columns that can be seen flying out of the buildings before and during the terminations, and that can later be seen lying or standing in the debris, are clearly *undeformed.* That is, they have not experienced any kind of vertical load failure. A few familiar examples can be seen in Figure 493, while a column that actually *is* buckled (from the 1964 Alaska earthquake) can be seen in Figure 494.[16]

4) If there is no evidence of structural load failure, and if on the other hand there is considerable evidence of structural members that have *not* experienced load failure (as has been seen), then the buildings cannot have collapsed as a result of load failure.

Figure 494. A buckled column, from the 1964 Alaska earthquake. This image is useful not only for illustrating buckling in general, but also the effects of support restraint.
http://urban.arch.virginia.edu/~km6e/tti/tti-summary/full/buckle-col.jpeg[17]

I. Conclusions

If the WTC "collapsed" from a vertically-induced mechanical overload, columns would exhibit typical buckling failure. However, we find an absence of buckled columns in the "rubble pile." Additionally, for that matter, there is the lack of a significant rubble pile commensurate with the sheer *amount* of material that comprised each building.

1. No matter what hypothetical fire event is posited, it cannot explain the "rolled-up carpets."
2. Similarly, a gravity collapse with or without heat will not cause this type of deformation during failure.
3. Bombs will not cause this type of deformation during failure.
4. A nuclear explosive will not cause this type of deformation during failure.
5. Cutting torches will not cause this type of deformation during failure.
6. Thermite, thermate, super thermate, spray-on thermite, or nano-thermite, cannot cause this type of deformation during failure.

Considering the very nature of the case and all of the evidence presented in this book so far, it is impossible not to conclude that the destruction of the World Trade Center towers came about through a form of directed energy technology. That is, the destruction was caused by energy that is directed and used as a weapon, as in the phrase Directed Energy Weaponry, or DEW [18].

---

[1] Drawn from dimensions given in the NIST NCSTAR 1 report., *http://911research.wtc7.net/wtc/arch/docs/spandrel.jpg*

[2] *http://wtc.nist.gov/media/gallery.htm*

[3] Cropped and lightened photo.

[4] Figure 2-8, World Trade Center Building Performance Study: Data Collection, Preliminary Observations, and Recommendations, Federal Emergency Management Agency, FEMA 403/May 2002, Chapter 2, File page 2-9 (pdf 9 of 40), *http://www.fema.gov/pdf/library/fema403_ch2.pdf*

[5] Wood, J. D., "Determination of Thermal Strains in the Neighborhood of a Bimaterial Interface," Ph.D. Dissertation, Virginia Polytechnic Institute and State University, Blacksburg, Virginia (1992).

[6] Post, D., J. D. Wood, B. Han, V. J. Parks and F. P. Gerstle, Jr., "Thermal Stresses in a Bimaterial Joint: An Experimental Analysis," ASME *J. Applied Mechanics,* Vol. 61, No. 1 (1994).

[7] S. Timoshenko, "Analysis of Bi-Metal Thermostats," J. Opt. Soc. Am. 11, 233-233 (1925)

[8] E. Suhir, : "Predictive Analytical Thermal Stress Modeling in Electronics and Photonics," Applied Mechanics Reviews, Transactions of the ASME, Vol. 62, March 2009, pp.(preprint), *http://www.isas.tuwien.ac.at/upload/downl/amr_020709.proofms1_copy1.pdf*

[9] Post, D., and Wood, J. D., 1989, "Determination of Thermal Strains by Moire Interferometry," Exp. Mech., 29(3), pp. 318–322.

[10] Photo is very similar to another by NIST: *http://www.photolibrary.fema.gov/photodata/original/3887.jpg*

[11] Map redrawn from p. 3 (pdf p. 53 of 298), *http://wtc.nist.gov/NISTNCSTAR1CollapseofTowers.pdf*

[12] *http://bp0.blogger.com/_SkyVfBO47ts/R1eSaNgnw9I/AAAAAAAAHHo/vSu4LMFqr6g/s1600/y1pl90TVU-OcIwN4ecOXONeyFZgomAw9RZx2PHqelwTKFLrC0ynOHnyr5TI2t5vgeTY.jpg, http://hutchisoneffect2008.blogspot.com/2007/11/starting-few-metals-more-to-come.html*

[13] *http://www.amny.com/news/local/groundzero/am-wtcrelics-pg2006,0,17616.photogallery?index=1*

[14] *http://www.fema.gov/pdf/library/fema403 ch6.pdf*

[15] *http://www.GreatBuildings.com/cgi-bin/gbi.cgi/World_Trade_Center_Images.html/cid_wtc_mya_WTC_const.3.gbi*

[16] The Earthquake Engineering Online Archive
Karl V. Steinbrugge Collection: S2166
* Title: Westward Hotel, buckled steel column
* Creator(s): Steinbrugge, Karl V.
* Date: 1964-04-05
* Location: NORTH AMERICA/United States/Alaska/Anchorage
* Earthquake: Anchorage, Alaska earthquake, Mar. 27, 1964 Magnitude: 9.2
* Structure: Anchorage Westward Hotel
* Description: Buckled steel column at room 241, east wall. Room 241 is the northeast corner room. (Anchorage, Alaska)

[17] (original image from the EqIIS collection, the image shown has been cropped and enhanced) *http://urban.arch.virginia.edu/~km6e/tti/tti-summary/part-3.html*

[18] I use the most general description of this by identifying it as a Directed Energy Weapon (DEW), which is Energy that is Directed, and is used as a Weapon.

22.
Conclusions and Summary

See the world as a child sees it, with wonder and with hope.[1]

A. Introduction

When I was seven years old, my brother took me outside to watch a total eclipse of the moon through a telescope my mother had made. I was quite impressed indeed. It seemed that the folks who lived on the moon were much more advanced than we were, because they were able to organize turning off all their street lights at nearly the same time, in sequence. Folks living on the moon was something entirely possible—until I was taught to believe that it wasn't. At school the next day I was ridiculed for suggesting that folks can live on the moon. And yet just a few years later we were told about two people (Neil Armstrong and "Buzz" Aldrin) walking around on the moon's surface, very much alive indeed.[2] Ironically, our "educational system" teaches us to limit our thinking as much as (or more than) it teaches us to open and expand it. I know that there are plenty of things that we've simply *got* to learn—but why it is that intellectual circumscription and imagination-denial have to go along with that learning is not only anathema to me but socially and culturally most, most counterproductive. Let us return to seeing the world as a child sees it, with wonder and with hope.

Figure 495. WTC2 Disappearing top.
http://sf.indymedia.org/uploads/const_in_foreground.jpg

The image in Figure 495, for example, when I first saw it, made me think of water turning into steam, a bit the way the moon's eclipse made me think of street lights going out. Always having been a student who wanted to go out of bounds, I let myself keep on looking at WTC2 through the metaphor of boiling water.

Figure 496. WTC2 Tipping top.
http://www.waarheid911.nl/wtc2collapse.jpg

It looked to me as if the entire building was boiling away and becoming another substance. And it turned out, in the main, that I was right.

Figure 497. The building material appears to transform into a kind of foam.
Figure 6-26, NCSTAR1-6, page 183 (pdf page 265 of 470), http://wtc.nist.gov/NISTNCSTAR1-6.pdf

In boiling water, the water's hydrogen and oxygen molecules repel one another, and, in the case of WTC2, it looked as if the building's own molecules were suddenly repelling each other. What I saw, and what we all now see, in Figure 495, is *not* rising smoke or trailing cloud, but, instead, what we see is an *energized propulsion.* We are not looking at signs of a force that caused the building to slam down into the ground at a rate faster than "free fall." We are looking at signs of a force, instead, that turned the building to dust before it even had a chance to *get* to the ground. The official explanation—along with all theories of controlled demolition, thermite, and so on—would have us believe that we are looking at a gravity-driven collapse. And yet, hard as it may be to believe, what we are seeing in the Figure 497 photograph is not a building in collapse, but a building being disintegrated into parts so small that it will, in practical terms, disappear. As for it not being a "collapse" we're looking at, consider the simple fact that some of the "dust" ejecta is actually being propelled *upward.*

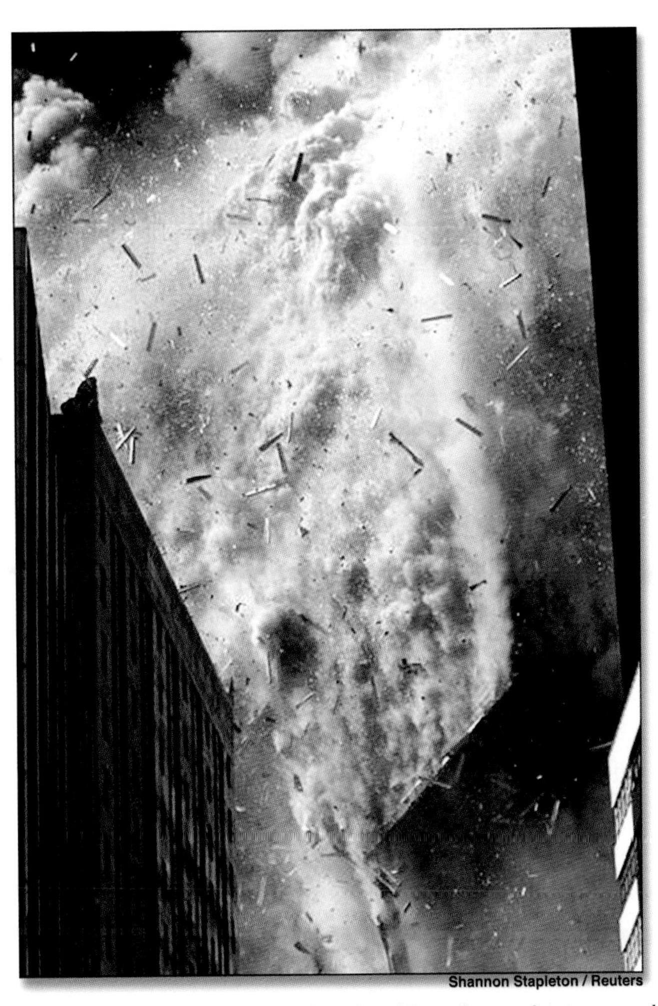

Figure 498. Material leaves the wheatchex like exhaust of a jet propulsion.
Photo by Shannon Stapleton, Reuters[3]

B. WTC Evidence that Must be Explained[4]

In addition to explaining why ejecta is being propelled *upward* in what is officially said to be a downward "collapse," any model of the WTC's destruction, if that model is to be taken seriously, must seek to explain not some but all of the following facts, although these "facts" may also be thought of as occurrences, questions, things, and anomalies. The alert reader may notice that not even this highly detailed book itself has been able to cover all of them:

1. FACT: Although Hurricane Erin was located just off Long Island throughout the day of 9/11/01, both the approach in days before and the presence of the storm on that day went almost totally unreported. Hurricane Erin was omitted on the morning weather map, even though that portion of the Atlantic Ocean where she stood was covered by the map. Astronauts gazing down said they could see the drifting plume from the destruction of WTC2 and WTC1 but made no mention of the highly visible Erin. WHY?

2. FACT: Approximately 1,400 motor vehicles were toasted in strange ways during the destruction of the Twin Towers. WHY AND HOW?

3. FACT: During destruction, there appeared alongside the buildings curious cork-screw trails, called in this book *Sillystrings*. WHY?

4. FACT: During the demise of each tower, large enough volumes of dust made of nano-sized particles went up, enough to block out 100% of sunlight in some areas. This nano-sized particulate dust in volume enough to achieve sun-light-blocking density constituted the remains of the greatest part of the destroyed buildings' material substance. WHAT CAUSED THIS DUST TO FORM?

5. FACT: During the destruction, there was an absence of high heat. Witnesses reported that the initial dust cloud felt cooler than ambient temperatures. Additionally, there was scant evidence of burned bodies, although in one case a man was described as "crisped" even while his jacket remained uncrisped, indicating an "inside-out" combustion not possible with conventional fire. WHAT CAUSED THESE PHENOMENA?

6. FACT: Evidence that the WTC dust continued to break down and become finer and finer long after 9/11 itself came through the observable presence of *Fuzzballs*. WHAT CAUSES THIS PHENOMENON?

7. FACT: First responders on 9/11 testified as to toasted cars, spontaneous "fires" (including the flaming heavy coat of a running medic, who survived), the instant

disappearance of people, a plane turning into a fireball in mid-air, electrical power cut off moments before WTC 2 destruction, and the sound of explosions. WHAT CAUSED THESE PHENOMENA?

8. FACT: For more than seven years, regions in the ground under where the main body of WTC4 stood have continued to fume. WHY?

9. FACT: Hazy clouds, called *Fuzzyblobs* in this book, appeared in the vicinity of material undergoing destruction. WHY?

10. FACT: Magnetometer readings from six stations in Alaska recorded abrupt shifts in the Earth's magnetic field as each of the major destructive events unfolded at the WTC on 9/11. WHY?

11. FACT: Many cars in the neighborhood of the WTC complex were flipped upside down. They couldn't have been flipped by hurricane-force winds, since they stood adjacent to trees with full foliage, not stripped by high wind. WHY?

12. FACT: More damage was done to the bathtub by earth-moving equipment during the clean-up process than from the destruction of more than a million tons of buildings above it. WHY?

13. FACT: Most of the destroyed towers underwent mid-air pulverization and were turned to dust before they hit the ground. WHAT FORCE CAUSED THIS "DUSTIFICATION"?

14. FACT: Near-instant rusting of affected steel provided evidence of molecular dissociation and transmutation. WHY?

15. FACT: Of the estimated 3,000 toilets in WTC1 and WTC2, not one survived, nor was any recognizable portion of one whatsoever found. WHY?

16. FACT: Only one piece of office equipment in the entire WTC complex, a filing cabinet with folder dividers, survived. WHY?

17. FACT: Only the north wing of WTC4 was left standing, neatly sliced from the main body, which virtually disappeared. FACT

18. FACT: Rail lines, tunnels and most of the rail cars at levels under the WTC complex had only light damage, if any. WHY?

19. FACT: Cylindrical holes were cut into the vertical faces of buildings 4, 5 and 6. They were cut also into Liberty Street in front of Bankers Trust and into Vesey Street in front of WTC6. In addition, a cylindrical arc was cut into the façade of Bankers Trust. WHY AND HOW?

20. FACT: Scott-Paks—portable air-tanks for firemen—frequently exploded for no visible reason. Entire fire trucks themselves that were parked near the WTC exploded. WHY? HOW?

21. FACT: Sheets of plain office paper were omnipresent throughout lower Manhattan after each tower's destruction. This paper, however, remained unburned, even though it was often immediately adjacent to flaming cars or to steel beams glowing red, yellow, and even white. WHY?

22. FACT: Some steel beams and pieces of glass at and near GZ had what this book calls a *Swiss-Cheese* appearance. WHY?

23. FACT: Steel columns from the towers were curled around vertical axes like *rolled-up carpets*. Steel columns of this kind, however, when they buckle from being overloaded, would be bent around the horizontal, not the vertical, axis. WHY?

24. FACT: The "collapse" of the towers took place with remarkably little damage to neighboring buildings. The only seriously damaged or entirely destroyed buildings, in fact, were those with the WTC prefix, only those, that is, that were a part of the WTC complex. WHY?

25. FACT: The destruction of WTC7 in late afternoon on 9/11 was whisper quiet. The seismic signal during its disappearance was not significantly greater than background noise. WHY?

26. FACT: The facades of WFC1 and WFC2 showed no apparent structural damage from the destruction of WTC1 and WTC2. However, the decorative marble façade around the entry to the buildings was completely missing, entirely gone. WHY? FROM WHAT FORCE?

27. FACT: In the dirt pile, *the Fuming* was unusual for its quality of immediately decreasing when watered, contrary to fumes caused by fire or heat, where an initial steam-up is the response to watering. WHY?

28. FACT: The majority of the towers (WTC1, WTC2, WTC3, WTC7) did not remain as rigid bodies as they "fell." WHY NOT?

29. FACT: The method of destruction in the case of each tower *minimized* damage to the bathtub and adjacent buildings, whereas terrorists would have been expected to maximize damage, including that of infrastructure.

30. FACT: The protective bathtub was not significantly damaged by the destruction of the Twin Towers.

31. FACT: The seismic impact was minimal during the destructions of WTC1, WTC2 and WTC7 and far too small to correspond with a conventional "collapse" as based on a comparison with the Kingdome controlled demolition.

32. FACT: The Twin Towers were destroyed from the top down, not from the bottom up.

33. FACT: The Twin Towers were destroyed in a shorter time than can be explained by physics as a "collapse" even at free-fall speed.

34. FACT: The upper 80 percent, approximately, of each tower was turned into fine dust and did not crash to the earth.

35. FACT: The upper 90 percent, approximately, of the inside of WTC7 was turned into fine dust and did not crash to the earth.

36. FACT: The WTC underground mall survived well, witnessed by Warner Brothers' Road Runner and friends.

37. FACT: The WTC1 and WTC2 rubble pile was far too small to account for the total mass of the buildings.

38. FACT: The WTC7 rubble pile was too small to account for the total mass of the building, and much of it consisted of mud.

39. FACT: Truckloads of dirt were hauled both into and out of the WTC site, a pattern that continues to this day.

40. FACT: What this book calls *lather*, thick clouds of dust and fumes, emanated from some faces of buildings before destruction, as if large volumes of the buildings' mass was dissolving into the air. Lather poured from WTC7 for several hours before its destruction. WHY?

41. FACT: What this book calls *weird fire* appeared frequently on 9/11. This "fire" flamed but gave no evidence of providing heat, not even enough to burn nearby sheets of paper. WHY?

42. FACT: Glass windows on nearby buildings received circular and other odd-shaped holes without the entire panes breaking. WHY?

43. FACT: Changes and alterations in materials on 9/11 were similar or even identical in a great many ways to the changes and alterations in materials caused by The Hutchison Effect. The Hutchison Effect is known to result in material-altering phenomena of the kinds we have listed here.

C. Summary

To determine what happened on 9/11, not just some but *all* available evidence must be considered. We cannot pick and choose which observable facts we may want to explain and then ignore the others. Any explanation must be integrated into a Unified Theory.

Before sundown on 9/11/01, seven buildings lay in ruins, some of which quite simply looked bombed, and one of those buildings in ruins, a 47-story tower, had not even, in its disappearance, produced a seismic impact that was significantly identifiable beyond normal background noise at several of the nearby seismic-recording stations. Audio/video recordings during the "collapse" of that building, WTC7, were reported to have recorded low-level street conversations but not the overwhelming sounds one would expect if 240,000 tons of material were in the process of slamming to the ground. As for the tallest two towers, they seemed simply to have vanished into vast clouds of dust.

Impossible as it may seem, most of the steel from those towers also vanished, literally turned into dust. From Day One of clearing ground zero, around-the-clock photos of the demolition were taken. Did *you* see any photos of the 3000 truckloads of steel being driven away? Neither did I.

A few surviving massive steel members were completely convoluted. Can anyone still think these incredible, impossibly-contorted shapes were the result of a jet-fueled, relatively low-temperature, fire?

One exceptionally astute scientist/experimentalist, John Hutchison, produces field effects that contort metal in manners comparable to what we see in official photos of those grossly-twisted structural members of heavy steel.

In my attempt to learn *what* happened on 9/11, I have examined Official Government photographs, photographs from news organizations, professional photographers, and private citizens, as well as other independent information from many entirely trustworthy and credible sources. *All* observable facts that are available *must* be considered carefully, seriously, and objectively if we are to establish *what* it was that happened on 9/11/2001.

None of the facts, events, anomalies, or phenomena that we have listed, discussed, and analyzed in this book can be explained by airliner crashes, jet fuel fires, or any scheme of controlled demolition. A comparison of the evidence we have collected with the evidence of results produced by Low Energy Nuclear Reactions (LENR), hurricanes and tornadoes, and the well-established Hutchison Effect has proven highly suggestive, pointing to a way of understanding the source of otherwise-inexplicable events and phenomena. It points to a way of helping us to find out *what happened* on 9/11.

This book is not about *naming* a known technology. This book is about observing the evidence of *what happened* and *understanding what happened.* As explained at the end of Chapter 20 (pages 451 to 452), there will likely be those who will not be able to resist the impulse to put a *name* of a known technology on what produced this evidence. To do so blindly assumes such technology is in the public domain. This *naming* introduces assumptions that distract away from understanding the actual evidence and will only serve to pull a veil of mystery over it.

My concern throughout this book has been solely and only with that single and scientific question of *what happened.* I have presented the evidence available to me and I have explored and analyzed it to the best of my ability. I trust that readers will follow me in the same spirit of objective scientific inquiry that is the spirit that has guided me in the writing of the book.

D. Final thoughts

We stand at the dawn of an entirely new age. By all the evidence, man has in his hands a method of disrupting the molecular basis for matter and the ability, very possibly, to split the earth in half on a moment's notice. The technology demonstrated on 9/11 can, indeed, split the earth in half,[5] or it can be used to allow *all* people to live fruitful, constructive, and non-polluting lives through their use of free energy.

He who controls the energy, controls the people. Control of energy, depending on what that energy is, can either destroy or sustain the planet.

We have a choice. And the choice is real. We can live happily and fruitfully and productively, or we can destroy the planet and die, every last one of us, along with every living being on this planet.

These are the reasons I have spent so much time studying the evidence of what happened on 9/11. 9/11 was a demonstration to the world of a new technology known as free energy. This is a force that can be used for demonic, ruinous, Earth-destructive purposes. Or it can be used for the good of us, of the Earth, and of our civilizations. We have a choice. And that choice is ours.

But in order to exercise that choice, in order to be *equipped* to exercise it, we must keep our eyes open wide and our minds, both our scientific and our social-humanist minds, unvaryingly on high alert. After all, two of the very tallest buildings in the world went missing on that awful, grievous day in September. And yet no one continued to ask, "Where did the towers *go*?"

Now we know the answer to that extraordinary question—the question that *should* have been asked by every one of us long before now.

---

[1] From a poster my roommate had in our dormitory room. Author unknown.

[2] The *belief* that man walked on the moon became acceptable on July 20, 1969.

[3] *http://upload.wikimedia.org/wikipedia/en/c/c5/DustifiedWTC2.jpg, (now removed), http://hereisnewyork.org//jpegs/photos/1539.jpg*

[4] *http://www.drjudywood.com/wtc/*

[5] The amount of energy the earth receives from the sun each day is enough to split the earth in half or enough to support all life on it.

Acknowledgements

At times our own light goes out and is rekindled by a spark from another person. Each of us has cause to think with deep gratitude of those who have lighted the flame within us. —Albert Schweitzer

Life is the first gift, love is the second, and understanding is the third. —Marge Piercy

A. Angels Who Have Been a Part of this Book.

Danny Suhr
Alayne Gentul
Melissa Doi
Bill Biggart*
Jonathan Briley
Steven Michael Hagis

Eugene F. Mallove
Michael Zebuhr
Emily J. Hilscher
G. V. Loganathan
Kevin P. Granata
Liviu Librescue

René Davila
Priscilla Wood
Jeffrey Nico
Keith Mothersson
Gerard Holmgren
Ambrose I. Lane

**I am deeply indebted to Bill Biggart's friend Chip East and wife/widow Wendy Biggart for making his photos available to the world. Bill spoke through them and someone was listening. Thank you.*

B. Angels among us.

Those to whom I literally owe my life. Thank you,

Kelly K.
Paul W.
Mark E.
Robert

Jeffrey C.
Eric C.
Chris J.

Terry T.
Kathy W.
Carol W.

Now you know the reason.

C. Other Angels among us.

Morgan
Eric
Frank
Andrew
Russ
Jerry
Alex
Ellen
Rose

Connie
Nathan
John
James
Diane
Jeff
David
Dan
Steven W.

Joey
Jesse
Pete
Clark
RT
ST
Herb
Yogi Bear
Wile E. Coyote

I wrote the following poem while writing this book. As discussed on page 295, it is true that "first impressions stick,"[1] so they will tend to bias our observations. This poem helped me to focus past the various interpretations we have been given, and to see and communicate what was really there.

Magic of the Heart

Look from your heart and you will see the truth.
Feel with your heart and you will know which way to go.

Speak from your heart and you will speak the truth.
Listen with your heart and you will understand.

Live from your heart and you will live in peace.
Love with your heart and you will love forever.

[1] Max Weisbuch, *Why do first impressions stick with us so much?,* September 9, 2009, *http://tuftsjournal.tufts.edu/2009/09_1/professor/01/:* "Psychologists have known for some time that expectations influence visual perception. For example, whether you see a '13' or a 'B' in response to seeing the set of characters '13' depends on your expectations. Still, even for those of us who are psychologists, it can be very difficult to disbelieve something we see with our own two eyes. Hence, rejecting a first impression can require a very difficult act–to literally 'unsee' what seems like ongoing evidence in support of that first impression.... Beyond behavior interpretation, memory is surprisingly fragile and subject to interpretation. Even those memories that seem most real ("flashbulb memories") are vulnerable to bias. For example, well-intentioned people can honestly but falsely remember being sexually abused as a child. It is hardly surprising that more mundane memories of others' behavior can be biased."

Index-Glossary and Supplemental Information

An unbiased and truly scientific study of observable phenomena requires a new vocabulary. As discussed in this book, language influences our observations. We tend to see what we expect to see. The use of a known term to refer to an unknown phenomenon can bias our observations. Therefore, adopted terms are used here to represent a group of characteristics associated with an unknown phenomenon. These unique terms provide for a more scientific study, shielding us from our expectations in the way double-blind studies are conducted to test a new drug's effectiveness. For a study of visual characteristics, and awkward label (such as "characteristic #349658-78756-3A") would in fact bias the observation because it is awkward to use. For these reasons, I have used adopted terms that are easy to remember and ten to describe what is seen.

The table below lists many of the terms I have used to describe various phenomena. A brief description is given as well as the page numbers in this book where the term is discussed. Throughout, I have used adopted words that have familiar meanings but no pre-known scientific relationship to the phenomena I am using them to refer to. For example, deformed steel columns are unlikely to be confused (as to *their scientific origins*) with Mexican food such as burritos or tortilla chips.

Never before have we seen *nearly* an entire building turn into dust, in mid flight, before hitting the ground. A new phenomenon requires a new word to identify it. That new word is *dustification.*

A. Loaded Terms

Fire: Implies a known cause and source.
Smoke: Implies a known cause.
Heat: Implies knowledge of temperature, separate from sensation or appearance. (cool acid can burn skin and feels like heat)
Burn, burned, burnt:
Implies high temperature as cause (cool acid can burn skin and feels like heat)
Hot: Implies knowledge of temperature
Glowing: Implies knowledge of temperature (unconsciously assumed)
Exploded: Implies the use of explosives.
Vaporize: Implies a high temperature process, which contradicts the evidence.

B. Index and Glossary of Terms

term	page	Description (not necessarily an exact definition)	image
Alkaseltzer® Shaving cream	340	Rapid degradation of material into dust, resembling an Alkaseltzer tablet's effect. The material separates into dust so quickly that it appears to be propelling the object from which it is emerging.	
Bubbler	1, 15	"Bubbler" is a midwestern term for *drinking fountain*. The original bubbler was developed in 1888 by the Kohler Company, in Kohler, Wisconsin.[1] "The original bubbler shot water one inch straight into the air, creating a bubbling texture, and the excess water ran back down over the sides of the nozzle..."[2] The image to the right resembles a *bubbler* more than it does a building *collapse*.	
Burritos	465	See rolled-up carpets.	
Cake batter	465	See rolled-up carpets.	
CANR	365, 366	Chemically Assisted Nuclear Reactions (See LENR and Cold Fusion)	
cDEW	454	See **D**irected **E**nergy **W**eapon (**c**old **D**irected **E**nergy Technology used as a **W**eapon)	
Cheetos®	265, 284	This term originated with the photo shown to the right. The orange object enlarged in the lower-left corner reminds me of a single *Cheeto* of the *Cheetos*® snack food made by Frito-Lay[3].	
Cold Fusion, CF	264, 357, 365, 374	Term coined for referring to Low Energy Nuclear Reaction (LENR) and Chemically Assisted Nuclear Reactions (CANR)	
DEW	454	See **D**irected **E**nergy **W**eapon (**D**irected **E**nergy Technology used as a **W**eapon.)	
Directed- Energy Weapon	289, 291, 292, 293, 346, 410, 475	The term, **D**irected **E**nergy **W**eapons, is a category of technology that involves **E**nergy that is **D**irected and is used as a **W**eapon.[4] DEW is a Directed Energy Technology used as a weapon. Microwaves and radio-frequency signals are a type of directed energy, but are not considered a DEW unless used as a weapon. The soviet term for this is, *energetics*[5], which includes conventional DEW as well as unconventional DEW.[6] These are not limited to line-of-sight, such as interfering field effects. Some types of DEW are the following: **cDEW** [7], cold **D**irected **E**nergy **W**eapons, (e.g. created by field effects) **hDEW** [7], hot **D**irected **E**nergy **W**eapons, (e.g. from lasers or microwaves) **DFEW** [7], **D**irected **F**ree-**E**nergy **W**eapons, (e.g. free-energy technology used as a weapon)	
DFEW	454	See **D**irected **E**nergy **W**eapon (**D**irected **F**ree-**E**nergy Technology used as a **W**eapon.)	
Directed- Free-Energy Technology	xxvii, 454, 485	The technology that was demonstrated on 9/11 was used for destructive purposes. However, this technology can also be used to provide free energy to the world. So the use of this technology on 9/11 demonstrated to the world that free-energy technology exists.	
Dirt, mudpie	288	Buildings do not collapse into piles of dirt, but large amounts of dirt regularly trucked in and trucked out from the site for months and years following 9/11.	
Disbelief	179	Where did the towers go? Survivors stand with postures of disbelief—as if questioning whether they are in The Twilight Zone.	
Donuts, bullet holes	139, 140, 304	Toroid shapes within a rapidly-expanding dust cloud. These appear as donut-shapes of denser dust around an apparent void ("bullet hole").	

Term	Pages	Definition
Dust rollout, energized clouds, attack clouds, rolling clouds	297, 300, 311, 313	The outward expansion of the dust cloud, e.g. resulting from the demise of a tower.
Dustification, dustify	131, 132, 133, 139, 178, 481	General catch-all term of turning the building into dust.
Elongation	328	This is the material's fracture strain, ε_f, and is expressed as a percent. Strain is the change in length, $L_f - L_o$, divided by the original length, L_o. $\varepsilon_f = \frac{L_f - L_o}{L_o} \times (100\%)$
Eutectic mixture	248, 369, 370, 370	A eutectic mixture contains a combination of materials in the right proportion to produce the minimum melting temperature.[8] Also, the entire eutectic mixture also melts at that minimum temperature. Solder made of 63% tin and 37% lead will melt at 183°C, where tin and lead have individual melting temperatures of 294°C and 327°C, respectively.[9]
EVO	357, 357, 438	Exotic Vacuum Object, Vacuum Object
Flipped cars	220, 221, 237, 242, 377	Vehicles oriented with the wheels away from the pavement instead of toward the pavement. That is, the orientation of the vehicles has been altered from its normal and operational one. Example: upside down (inverted), on a side, or on top of something not normally driven over.
Free Energy	454, 485	Free Energy is an unconventional energy source that is not metered and sold for profit.
Fumes, fuming	228	Fine particulate suspended in the air, emanating from or looming over a building or its remains. Fuming is of unknown cause and mechanism. The word "smoke" implies a known cause, a known mechanism, and is associated with heat or fire. Fuming shares none of these characteristics. Therefore the word "smoke" is misleading.
Fuming dirt	287, 288,	Fumes emanating from wet and often saturated dirt. (see fumes)
Furry lather	260, 304, 448	A thinner and wispier version of lather, seen shortly before a building "went poof." (To be addressed in next book.)
Furry rust	256	Extreme rust that is fairly uniform in color and texture. The uniform rust appears to have a the texture of a thick velvet or even fur, appearing to have a furry texture. A clean iron skillet soaking in water for a long time would develop this kind of uniform rust, but it is not expected of structural steel (e.g. A-36) in air.
Fuzzballs	321	Fine dust began to rise from the ground soon after coarse dust had landed. It was first noticed emerging from around the feet of survivors, appearing as fuzzy regions or *fuzzballs*. Later, this material emerged on its own. Dust this fine could not have settled out of the air in such a short time, leading to the conclusion that it was the result of an ongoing process of the dissociation of material

Term	Pages	Definition
Fuzzyblobs	481	Zones of haze with no immediately-recognized origin. In the photo to the right, the fuzzyblobs appear to adversely affect the firefighters. (This topic, again, will be addressed in my next book.)
Got its hole	3, 416	Referring to the moment when something left an airplane-shaped silhouette of passage in a building (created a hole in)
HE, Hutchison Effect	264, 349, 350, 353, 356, 357, 438	The Hutchison Effect (HE)
hDEW	454	See **D**irected **E**nergy **W**eapon (**h**ot **D**irected **E**nergy Technology used as a **W**eapon)
Jellified, jellied jellification, jelly	359, 387	Softening of a material to where it cannot hold its shape, approaching the consistency of jelly or jello.
KEW	73, 174	See **K**inetic **E**nergy **W**eapon
Kinetic-Energy Weapon	73, 174	The term, **K**inetic **E**nergy **W**eapons, is a category of technology that involves Kinetic Energy that is used as a Weapon. Physical objects, such as missiles, bombs, and bullets physically contact something. If they are also designed to explode they will send chunks of material flying.
Kitchen sink	126	The speculation that multiple methods of destruction were employed.
ksi	328	A unit of pressure or a unit of stress, which is kips per square inch, or 1,000 pounds per square inch, or 1,000 psi.
Lasagna	466	See rolled-up carpets.
Lather	339, 344	The material that resembles the material emerging from an Alkaseltzer® tablet, but is emerging from a building. It is a thick cloud that can block out all light.
LENR	357, 364, 373 - 375	Low Energy Nuclear Reaction (see Cold Fusion)
Magnetic precipitate	366, 371	A solid has formed (in tiny particles) in the electrolytic cell containing the reaction and that this solid is magnetic.
Mechanical floors	28, 135, 187, 190, 304	Each of the WTC Towers were like three separate buildings joined together by two apparent "seams" that can be seen at one third and two thirds of the height by their different coloration. These were *mechanical floors* that were the following: (7-8, 41-43, 75-76, 109-109)
Micrometer, micro-		Parts per million, or 1×10^{-6} μ
Molecular dissociation	247, 252, 255, 331	Molecules separate or even repel each other.
Nanodust	124, 315, 480	Very fine dust. Some dust is less than one micron, approaching the nano scale. It stays aloft and even floats upward.
Nanohaze	124, 315, 480	Looming dissociated material (see nanodust)

Nanometer, nano-		One thousandth of a part per million, or 1×10^{-9}	nm
NIST		National Institute of Standards and Technology	www.nist.gov
NOAA		National Oceanic and Atmospheric Administration (provided aerial image of the aftermath).	www.noaa.gov
Pepperoni roll	465	See rolled-up carpets	
Plasma		A plasma (often called the fourth state of matter) contains ions and free electrons. Ions are atoms that have gained or lost at least one electron, giving it positive or negative charge. Because of their charged particles, plasmas can be magnetic.	
Poof, poofing, poofed, went poof	5, 23, 132, 167, 168,	Dustification of the WTC in a very short amount of time	
Psyop, or psyops	271	Psychological operation or psychological manipulation.	
Rain	259, 275	See Weather.	
Rolled-up carpets, burritos, pepperoni roll, lasagna, tacos	465, 465	Wheatchex rolled-up, that resemble a rolled-up carpet. Other odd-shaped steel structural elements are referred to by food items they resemble.	
Rustification, tortilla chips	247, 248, 251, 255	Steel that has quickly eroded and rusted or disintegrated.	
Scott-Pak	110	Scott Paks[10] are a brand name of a Self-Contained Breathing Apparatus (SCBA[11]) worn by rescue workers, firefighters, and others to provide portable, breathable air in a hostile environment. They also contain a PASS alarm, such as the Scott Pak-Alert [12] which alerts rescuers when a firefighter is not moving (or breathing). These whistling alarms could be heard on 9/11 following the destruction of the buildings.	
Sillystring	480	Light-colored trails often seen in a corkscrew-like pattern. An example of a "sillystring" is shown in the photo to the right, but the subject will be covered in my next book. "Sprinkles" can also be seen in this image.	
Smoke	not smoke: 344	"Smoke" is a term typically refering to the output from hydrocarbon fires. For fumes that have unknown sources, or that may not even involve or be produced by oxidization, the term "smoke" is inaccurate and misrepresentative, connoting a known process whose cause is in fact not known.	Burning firewood, campfire
Smoke rings	343	segmented puffs of dark fumes reminiscent of a smoker blowing cigarette smoke in segmented puffs, i.e. blowing *smoke rings*. This is unnatural for "regular fires."	

Smushed	458	A combination of smashed and crushed.	
Snowball	306	Round *snowball*-shaped dust cloud (WTC2)	
Sprinkles	157, 163, 443	An example of a "sprinkles" is shown in the photo to the right, but the subject will be discussed more thoroughly in my next book.	
Squibs	111, 121, 114	See squirts.	
Squirts, mistaken as "squibs"	111, 114, 121,	Visible streams of material that appear to squirt out of the building just ahead of the destruction wave.	
Stress	328	Force divided by area. Example: pounds per square inch (psi) See Yield Stress and Ultimate Stress.	$Stress = \frac{Force}{Area}$
Swamp	230, 232, 235, 236, 258, 277	West Broadway looked like a swamp after fire hoses had saturated the paper and debris covering the entire street.	
Tacos	405	See rolled-up carpets	
TEW		See **T**hermal **E**nergy **W**eapon	
Thermal-Energy Weapon	100, 102, 123, 153, 262	The term, **T**hermal **E**nergy **W**eapons, may be viewed as a category of technology that involves Thermal Energy (heat) that is used as a Weapon to *cook* or *burn* its target. Example: thermite, thermate, nano-thermite, welding materials, etc.	
Toasted, toasted cars, toasted street, toasted lot,	213	Generic term for the odd destruction seen on vehicles, but not limited to vehicles. A "toasted" vehicle means that "that vehicle is history; it's toast." It's a general term for an unknown process and, an important fact, does not necessarily imply heat.	
Ton, metric ton		A metric ton is a unit of mass equal to 1,000 kg, or 2,204.62262 lb.[13] A *short ton* is 2,000 lb	t
Tortilla chips		See rustification	
Transmutation	253, 349, 357, 364, 366, 481	Transmutation is the conversion of one chemical element into another.[14]	
Two-toned	316	The phenomenon of bi-colored dust-filled clouds	
Ultimate Strength	328	The maximum stress carried before failure.	
Vacuum domain	438	Also see EVOs	
Venting	143, 177, 203, 285, 323	An Alkaseltzer-like phenomenon, emerging from a stationary object or shooting up from an opening (see Alkaseltzer or shaving cream).	

Weather during cleanup	259, 275	Rain dates and amount in inches, 9/11 through October.[15] 9/14/01 1.21 in.; 9/15/01 0.69 in.; 9/20/01 0.71 in.; 9/21/01 0.48 in. 9/23/01 0.01 in.; 9/25/01 0.42 in.; 9/26/01 0.03 in.; 9/30/01 0.22 in. 10/1/01 0.23 in.; 10/6/01 0.12 in.; 10/15/01 0.35 in.; 10/17/01 0.10 in.
Wheatchex	203, 204, 455, 456, 466	Sets of three outer columns on the WTC towers, three stories tall, and connected by spandrel belts. These sections look like Wheatchex®[16] cereal (or Shreddies®[17] in the UK). I first heard this term used by Chip East[18],[19]
Where did the towers go?	179	See Disbelief
Yield Strength	328	The stress at which permanent deformation begins. If stresses greater than this are applied, then removed, the material will not return to its original dimensions.

Table 25. Glossary

---

[1] *http://csumc.wisc.edu/wep/map.htm, http://www.us.kohler.com/onlinecatalog/detail.jsp?item=428202, http://en.wikipedia.org/wiki/Bubbler*

[2] *http://csumc.wisc.edu/wep/map, http://en.wikipedia.org/wiki/Bubbler*

[3] *http://www.fritolay.com/cheetos.html*

[4] *http://drjudywood.com/articles/DEW/StarWarsBeam7.html#DEW*

[5] *http://www.freedomdomain.com/weathercontrol/scalarweapon01.html*

[6] "The ordinary Soviet name for this type of weapons science [DEW] is energetics. In the west that term is believed to be associated with conventional directed energy weapons (DEWs) such as particle beam weapons, lasers, radio-frequency (RF) directed energy devices, etc. The Soviets do not limit the term in this way.

[...]

However, it is possible to focus the potential for the effects of a weapon through spacetime [sic] itself, in a manner so that mass and energy do not 'travel through space' from the transmitter to the target at all. Instead, ripples and patterns in the fabric of spacetime [sic] itself are manipulated to meet and interfere in and at the local spacetime [sic] of some distant target. There interference of these ripple patterns creates the desired energetic effect (hence the term energetics) directly in and through the target itself, emerging from the very spacetime [sic] (vacuum) in which the target is imbedded at its distant location. As used by the Soviets, energetics refers to these eerie new superweapons, as well as to the more mundane DEWs known to the west." *http://www.freedomdomain.com/weathercontrol/scalarweapon01.html*

[7] **cDEW, hDEW, and DFEW,** *are terms coined by Andrew Johnson.*

[8] *http://www.thefreedictionary.com/eutectic,*

[9] *http://www.indium.com/products/alloy_sorted_by_temperature.pdf*

[10] *http://www.scottint.com/Americas/en/Products/SuppliedAir/SuppliedAir.aspx*

[11] *http://www.scottint.com/AMERICAS/en/Resources/News/pressrelease/press09207_futureofscba.aspx*

[12] *http://www.hellotrade.com/industrial-protection-services/scott-pak-alert.html*

[13] *http://en.wikipedia.org/wiki/Metric_ton*

[14] Transmutation, *http://www.britannica.com/EBchecked/topic/602997/transmutation*

[15] *http://www.almanac.com/weather/history/NY/New%20York/2001-09-01*

[16] Wheatchex® is a breakfast cereal introduced in 1937 and currently manufactured by General Mills. (*http://www.chex.com/products/products.aspx*) It was originally owned by Ralston Purina, and the name reflects the "checkerboard square" logo of Ralston Purina. Ralston Purina started *Ralcorp* in 1994 and was sold to General Mills in 1997. (*http://en.wikipedia.org/wiki/Chex*)

[17] Shreddies is a breakfast cereal sold in the United Kingdom, Ireland, Canada and New Zealand, produced by Post Cereals and General Mills. Shreddies® was introduced in the UK in 1955 under Nabisco, until it was sold to Post Cereals in 1993. It was sold to *Ralcorp* in 2008. Shreddies® *look like* Wheatchex®.

[18] Chip East's reference to "wheatchex" is in video segment 8. *http://digitaljournalist.org/issue0111/video/ce8.mov*

[19] *http://digitaljournalist.org/issue0111/biggart_intro.htm,*

C. Magnitude Abbreviations

T	= tera (one trillion)	$= 10^{12}$	1,000,000,000,000
G	= giga (one billion)	$= 10^{9}$	1,000,000,000
M	= mega (one million)	$= 10^{6}$	1,000,000
k	= kilo (one thousand)	$= 10^{3}$	1,000
1	= one	$= 10^{0}$	1
d	= deci (one tenth)	$= 10^{-1}$	0.1
c	= centi (one hundredth)	$= 10^{-2}$	0.01
m	= milli (one thousandth)	$= 10^{-3}$	0.001
μ	= micro (one millionth)	$= 10^{-6}$	0.000 001
n	= nano (one billionth)	$= 10^{-9}$	0.000 000 001
p	= pico (one trillionth)	$= 10^{-12}$	0.000 000 000 001

Table 26. Magnitude Abbreviations.

D. Melting and Boiling Temperatures for Selected Elements

element	Symbol	Atomic number	melting point °C[1]	boiling point °C[2]	boiling point °F[2]
Carbon (graphite)	C	6	3,675 (sublimates)	4,027	7,280
Carbon (diamond)	C	6	3,550	4,827	8,720
Tungsten	W	74	3,422	5,555	10,031
Molybdenum	Mo	42	2,617	4,639	8,382
Vanadium	V	23	1,910	3,407	6,165
Chromium	Cr	24	1,857	2,671	4,840
Titanium	Ti	22	1,660	3,207	5,949
Iron	Fe	26	1,535	2,861	5,182
Nickel	Ni	28	1,453	2,913	5,275
Silicon	Si	14	1,410	3,265	5,909
Manganese	Mn	25	1,246.35	2,061	3,742
Copper	Cu	29	1,084.6	2,562	4,643
Gold	Au	79	1,064.18	2,856	5,173
Silver	Ag	47	961.78	2,162	3,924
Calcium	Ca	20	839.	1,484	2,703
Strontium	Sr	38	777	1,382	2,520
Barium	Ba	56	727	1,897	3,447
Aluminum	Al	13	660.25	2,519	4,566
Magnesium	Mg	12	650	1,090	1,994
Zinc	Zn	30	419.73	907	1,665
Lead	Pb	82	327.46	1,749	3,180
Tin	Sn	50	231.93	2602	4,716
Sulfur	S	16	115.36	444.6	832.3
Sodium	Na	11	98.	883	1,621
Potassium	K	19	63.35	759	1,398
Phosphorus (white)	P	15	44.1	280	536
Rubidium	Rb	37	39.31	688	1270
Bromine	Br	35	-7.2	58.8	137.8
Chlorine	Cl	17	-100.84	-34.04	-29.27
Nitrogen	N	7	-209.86	-195.79	-320.42
Fluorine	F	9	-219.52	-188.12	-306.62
Oxygen	O	8	-222.65	-182.95	-297.31
Neon (orange glow)	Ne	10	-248.447	-246.08	-410.94
Helium	He	2	-272.20	-268.93	-452.07

Table 27. Melting and boiling points for various elements.

E. Tritium Values

Tritium Levels:	Bq/L	nCi/L
Agency Tritium limits		
Canada (1978)drinking water[3]	40,000.	1,080.
World Health Organization (WHO)[4]	10,000.	270.
Canada (1980)drinking water[5],[6]	7,000.	190.
US EPA (1999).[7],[8]	740.	20
European Union (1998)[9],[10]	100.	2.7
US State of Colorado, surface water[11],	18.	0.50
US State of California, surface water[12],	15.	0.40
A. Atmospheric Atomic Bomb Testing		
1963 - late 1963,[13,14]	51.80	1.400
1963 - late 1966,[13,14]	7.40	0.200
1963 - late 1970,[13,14]	3.70	0.100
1963 - late 1975,[13,14]	1.85	0.050
1963 - late 1990,[13,14]	1.00	0.027
B. Great Lakes Tritium data		
Lake Superior[15]	2.	0.05
Lake Michigan[15]	3.	0.08
Lake Huron[15]	7.	0.19
Lake Erie[15]	5.5	0.15
Lake Ontario[15]	7.1	0.19
The Great Lakes ca. 1960 from *atmospheric* atomic bombs.[16]	20 - 30	0.5 - 0.8
C. WTC Tritium data[17],[18]		
WTC storm sewer (9/14/01) (day 4)	6.07±2.74	0.164±.074
WTC 6, basement B5 (9/21/01) (day 11)	130.6±6.3	3.53±0.17
WTC 6, basement B5 (9/21/01) (day 11)	104.7±5.6	2.83±0.15
Background for WTC	4.1-4.8	0.11-0.13
D. Tritium levels in LENR Cell[19] ***("cold fusion")***		
Day 1	1,730.	47.
Day 2	2,480	67.
Day 16	5,700.	154.
Day 24	6,500.	176.
Day 44	4,330.	117.
E. Tritium levels reported resulting from Nuclear-Plant Accidents ***(hot fusion)***		
In 1979, tritium *groundwater* concentration reached 2.15 MBq/L following a *release* of 666 TBq at Pickering.[20]	2,150,000	58,110
In May 1994, Ontario Hydro found a tritium *groundwater* concentration of 0.7 MBq/L from a *leak* at Pickering.[20]	700,000	19,000
In September 1983, a *leak* of 222 TBq of tritium at the Douglas Point reactor on Lake Huron caused Port Elgin, *drinking water* levels to reach 1,600 Bq/L for 2-days.[20]	1,600	43
In June 1991, a *leak* from Chalk River Nuclear Laboratories into the Ottawa River, tritium in *drinking water* at Petawawa was ~400 Bq/L. The Ottawa (200 km downstream) tritium level was ~150 Bq/L.[20]	400 150	11 4
In August 1992, a *tube break* at Pickering 1 caused the release of 2,300 TBq of tritium into Lake Ontario. A nearby *drinking water plant* was shut down and elevated levels of tritium (up to 195 Bq/L) were found in Toronto *drinking water.*[20]	195	5.27

Table 28. Tritium values resulting from various causes.

[1] *http://en.wikipedia.org/wiki/List_of_elements_by_melting_point*
[2] *http://en.wikipedia.org/wiki/List_of_elements_by_boiling_point*

[3] Ian Fairlie, *Tritium Hazard Report: Pollution and Radiation Risk from Canadian Nuclear Facilities, June 2007*, p. 18, *http://www.greenpeace.org/raw/content/canada/en/documents-and-links/publications/tritium-hazard-report-pollu.pdf*

[4] World Health Organization: 10,000 Bq/L., *http://en.wikipedia.org/wiki/Tritium*

[5] Ian Fairlie, *Tritium Hazard Report: Pollution and Radiation Risk from Canadian Nuclear Facilities*, June 2007, p. 18, *http://www.greenpeace.org/raw/content/canada/en/documents-and-links/publications/tritium-hazard-report-pollu.pdf*

[6] Canada: 7,000 Becquerel per liter (Bq/L), *http://en.wikipedia.org/wiki/Tritium*

[7] United States: 740 Bq/L or 20,000 picocurie per liter (pCi/L) (Safe Drinking Water Act), *http://en.wikipedia.org/wiki/Tritium*

[8] Ian Fairlie, *Tritium Hazard Report: Pollution and Radiation Risk from Canadian Nuclear Facilities*, June 2007, US EPA limit of 20nCi/L ([7]reported as 20,000 picocuries per litre), p. 18, *http://www.greenpeace.org/raw/content/canada/en/documents-and-links/publications/tritium-hazard-report-pollu.pdf*

[9] European Union: 'investigative' limit of 100 Bq/L., *http://en.wikipedia.org/wiki/Tritium*

[10] Ian Fairlie, *Tritium Hazard Report: Pollution and Radiation Risk from Canadian Nuclear Facilities*, June 2007, European Commission (1998), p. 18, *http://www.greenpeace.org/raw/content/canada/en/documents-and-links/publications/tritium-hazard-report-pollu.pdf*

[11] *Ibid*, [8](reported as 500 picocuries per litre.), [9]For example, the US Department of Energy has specified the Colorado state action level for tritium in surface water in its clean-up program at the Rocky Flats plutonium plant in Colorado), p. 18, *http://www.greenpeace.org/raw/content/canada/en/documents-and-links/publications/tritium-hazard-report-pollu.pdf*

[12] Ian Fairlie, *Tritium Hazard Report: Pollution and Radiation Risk from Canadian Nuclear Facilities*, June 2007, ([10]reported as 400 picocuries per liter), p. 18, *http://www.greenpeace.org/raw/content/canada/en/documents-and-links/publications/tritium-hazard-report-pollu.pdf*

[13] Cordy, G.E., Gellenbeck, D.J., Gebler, J.B., Anning, D.W., Coes, A.L., Edmonds, R.J., Rees, J.A.H., and Sanger, H.W., *2000, Water Quality in the Central Arizona Basins, Arizona, 1995–98: U.S. Geological Survey Circular 1213, 38 p., on-line at http://pubs.water.usgs.gov/circ1213/*, Tritium in precipitation from 1950 to 1998, *http://pubs.usgs.gov/circ/circ1213/images/tab_fig18a.gif, http://pubs.usgs.gov/circ/circ1213/major_findings2.htm*

[14] Toxic Substances Hydrology Program, *http://toxics.usgs.gov/definitions/tritium.html*

[15] Ian Fairlie, *Tritium Hazard Report: Pollution and Radiation Risk from Canadian Nuclear Facilities, June 2007*, Table 5.1, [Sources: King et al (1998, 1999)], p. 20, ([12]It appears that there are no published *average* levels more recent than 1998.), *http://www.greenpeace.org/raw/content/canada/en/documents-and-links/publications/tritium-hazard-report-pollu.pdf*

[16] Ian Fairlie, *Tritium Hazard Report: Pollution and Radiation Risk from Canadian Nuclear Facilities, June 2007, (p. 20 0f 92)* [11]Tritium levels in the Great Lakes increased to about 20–30 Bq/L in the late 1950s and early 1960s as a result of atmospheric atomic bomb testing. This has now mostly decayed to about 1 Bq/L and constitutes a part—perhaps as much as a half—of background tritium levels., *http://www.greenpeace.org/raw/content/canada/en/documents-and-links/publications/tritium-hazard-report-pollu.pdf*

[17] Thomas M. Semkow, Ronald S. Hafner, Pravin P. Parekh, Gordon J. Wozniak, Douglas K. Haines, Liaquat Husain, Robert L. Rabun and Philip G. Williams, "Study of Traces of Tritium at the World Trade Center," Proceedings of the Symposium on Radioanalytical Methods at the Frontier of Interdisciplinary Science: Trends and Recent Achievements. 223rd American Chemical Society National Meeting, Orlando, FL, April 7-11, 2002, October 1, 2002 Preprint NYS DOH 02-116, *http://www.osti.gov/energycitations/servlets/purl/15002340-YM5IJp/native/15002340.PDF*

[18] Pravin P. Parekh, Thomas M. Semkow, Liaquat Husain, Douglas K. Haines, Gordon J. Wozniak, Philip G. Williams, Ronald S. Hafner, and Robert L. Rabun, "Tritium in the World Trade Center September llth, 2001 Terrorist Attack, It s Possible Sources and Fate," Proceedings of the Symposium on Radioanalytical Methods at the Frontier of Interdisciplinary Science: Trends and Recent Achievements. 223rd American Chemical Society National Meeting, Orlando, FL, April 7-11, 2002, May 3, 2002 Preprint NYS DOH 02-116, *http://www.llnl.gov/tid/lof/documents/pdf/240430.pdf*

[19] Tritium production reported in electrochemical cells (Cell 73) by Drs. Edmund Storms and Carol Talcott of the Los Alamos National Laboratory. (Courtesy of Drs. Edmund Storms and Carol Talcott) in Eugene F. Mallove, *Fire from Ice: Searching for the Truth Behind the Cold Fusion Furor,* 1991, John Wiley & Sons (p. 227).

[20] Ian Fairlie, *Tritium Hazard Report: Pollution and Radiation Risk from Canadian Nuclear Facilities*, June 2007, (p. 21 0f 92) *http://www.greenpeace.org/raw/content/canada/en/documents-and-links/publications/tritium-hazard-report-po*

Maps and Building Information

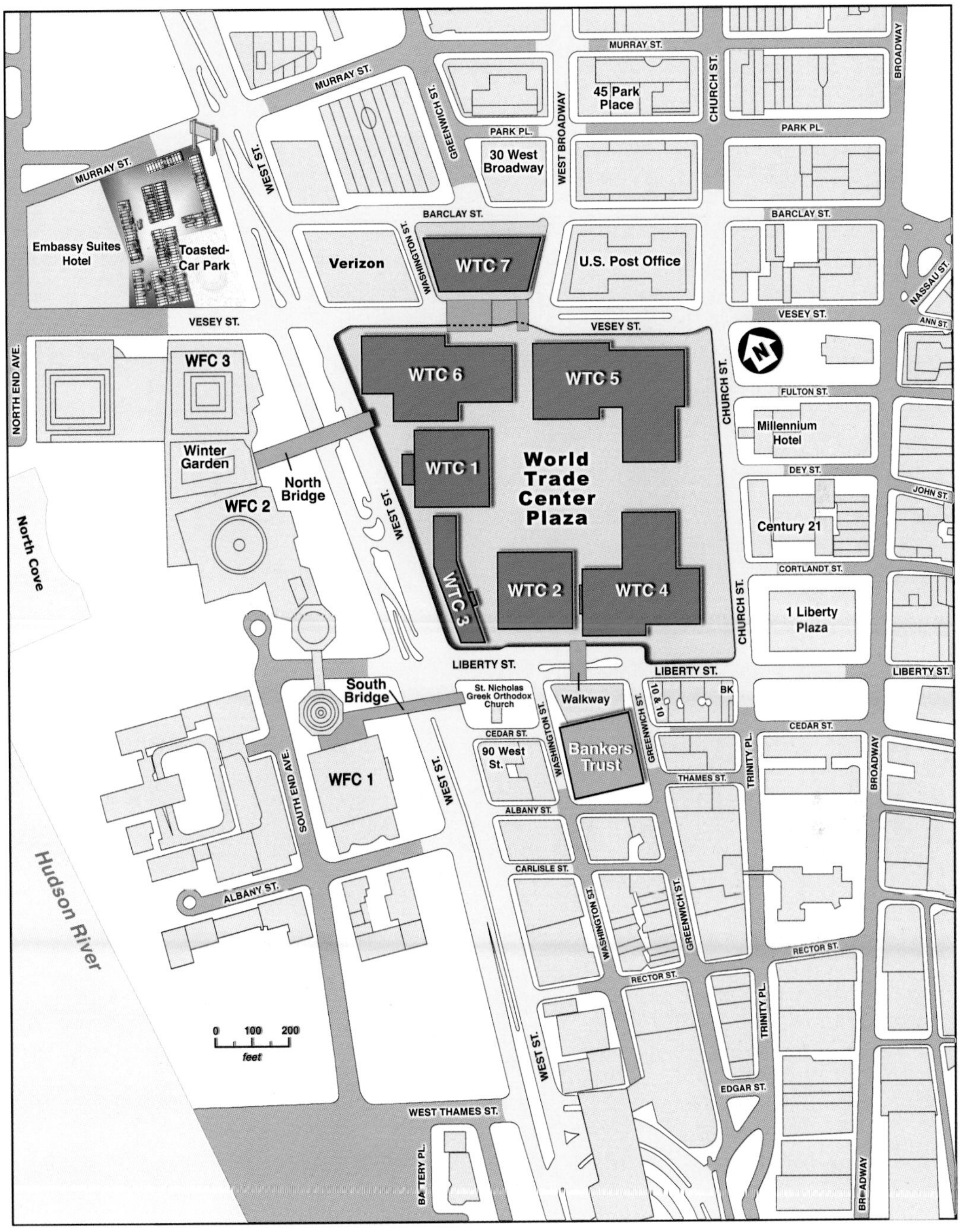

Figure 499. Map of Lower Manhattan.
(Map redrawn from the NIST Report.)

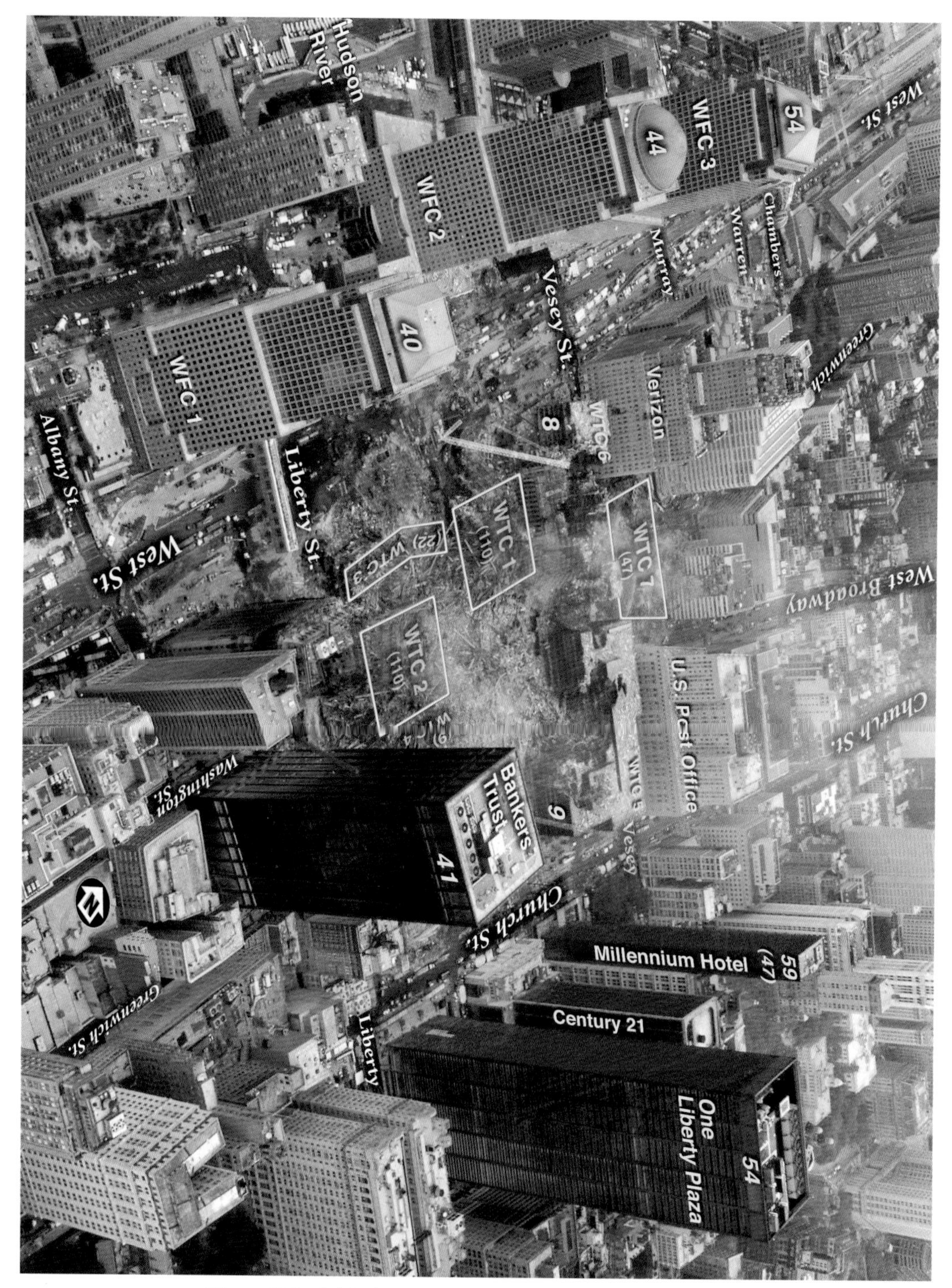

Figure 500. (9/27/01) What remained of the WTC complex.
Figure 500. The numbers added to near the top of remaining buildings shows the number of stories tall that building is. The corresponding number of stories of former buildings is shown in their footprint. The Millennium Hotel has 59 stories, but is approximately the height of a 47 story building.
(Brightness and Contrast adjusted)
http://bocadigital.smugmug.com/photos/10948720_ejP7a-O.jpg